Praise for Biodiversity and Climate Change: Transforming the Biosphere

"The evidence assembled in Biodiversity and Climate Change leaps out at us like a scene from a 3-D movie. It's warning us—compelling us—to act. Nowhere is the science clearer or the facts more compelling than in this meticulously researched volume. When conservationists and scientists of Lovejoy's and Hannah's caliber warn that if we don't press forward faster, we'll see greater species extinction, more ocean acidification, more biodiversity loss, more strains on agriculture and fishing—let me tell you: we need to listen. This book isn't just a call to heed the science; it's a call to citizens everywhere to live up to their responsibilities and protect this fragile planet we share."
—John Kerry, United States Secretary of State, 2013–2017

"Mankind's heedless extraction and pollution of our planet's resources is tearing apart the web of natural systems that has sustained our species throughout the long course of human development. Tom Lovejoy and Lee Hannah have assembled a book that chronicles these emerging ecological and climatic disasters yet gives hope that we can still help Earth's systems heal and blunt the suffering of coming generations."
—Sheldon Whitehouse, United States Senator for Rhode Island

"Biodiversity and Climate Change: Transforming the Biosphere serves as a comprehensive account of this greatest of threats to humanity's future. It will serve both as a textbook and a call to action."
—From the Foreword by Edward O. Wilson

"An authoritative analysis of the increasing speed and scale of climate-change impacts on our biodiversity, together with an illuminating set of specific ways to use our biodiversity to address climate change. A powerful coupling."
—Christiana Figueres, Executive Secretary of the United Nations Framework Convention on Climate Change, 2010–2016

"In Biodiversity and Climate Change, the renowned Tom Lovejoy and Lee Hannah blend leading voices to form a clear case for climate action, highlighting a powerful though underutilized natural defense: ecosystem restoration."
—Henry M. Paulson Jr., Chairman of Paulson Institute, and United States Secretary of the Treasury, 2006–2009

"Biodiversity and Climate Change lays out the latest science on the central challenges of our time. It is clear, comprehensive, and utterly compelling—an essential addition to the literature."
—Elizabeth Kolbert, author of The Sixth Extinction: An Unnatural History

"We've all heard the clanging alarms about climate change. This book is an extraordinary scientific portrait, expert, multifaceted, and up-to-date, of how those changes are affecting biological diversity—upon which we humans depend, of which we are part. Don't send to know for whom the bell tolls. It tolls for us."

—David Quammen, author of *The Tangled Tree: A Radical New History of Life*

"Lovejoy and Hannah generate a compelling story of the species extinctions that will accompany ongoing, rapid changes in Earth's climate, coupled with the unrelenting pressure of human population growth."

—William H. Schlesinger, President Emeritus of the Cary Institute of Ecosystem Studies

"As the global community prepares a post-2020 Global Deal for Nature, this book reminds us of the importance of ecosystems and nature-based solutions to advance the Paris Agreement, the Sustainable Development Goals, and the 2050 Biodiversity Vision."

—Dr. Cristiana Paşca Palmer, United Nations Assistant Secretary-General and Executive Secretary of the Convention on Biological Diversity

"Come explore this comprehensive and insightful synthesis that underscores the intimately interconnected nature of biodiversity and climate change—its past, present, and future."

—Jane Lubchenco, Oregon State University

"Escape to Mars is not a realistic option for any species. Lovejoy and Hannah's assessments will help us chart a feasible pathway to preservation of our remarkable world."

—James Hansen, Columbia University Earth Institute

Biodiversity and
Climate Change

Biodiversity

EDITED BY
THOMAS E. LOVEJOY
& LEE HANNAH

and

Climate

Change

Transforming the Biosphere

FOREWORD BY EDWARD O. WILSON

Yale UNIVERSITY PRESS
NEW HAVEN & LONDON

Published with assistance from Gordon and Betty Moore, Conservation International, and with assistance from the foundation established in memory of Calvin Chapin of the Class of 1788, Yale College.

Yale University Press books may be purchased in quantity for educational, business, or promotional use. For information, please e-mail sales.press@yale.edu (U.S. office) or sales@yaleup.co.uk (U.K. office).

Set in Joanna type by Newgen.
Printed in the United States of America.

Library of Congress Control Number: 2018954278
ISBN 978-0-300-20611-1 (paperback : alk. paper)

A catalogue record for this book is available from the British Library.

This paper meets the requirements of ANSI/NISO Z39.48-1992 (Permanence of Paper).

10 9 8 7 6 5 4 3 2 1

For Betsy, Kata, Annie, Tia, and Jay—
and the generations to come

Contents

.

Foreword

EDWARD O. WILSON

Biodiversity and Climate Change: Transforming the Biosphere is the definitive statement composed by a large percentage of the leading scientific experts on the most urgent global problem short of the destruction of Earth by a planet-sized asteroid. A remark cited in *Biodiversity and Climate Change* states the issue cogently: if *nothing is done about climate change, you can forget about biodiversity.* Let me expand on that fundamental truth with different words: you can forget about millions of species, thus about the living part of nature, about the stabilizing power of ecosystems, and not least about an autonomously secure future for the human species.

The world has already lost irrevocably a substantial portion of the natural biosphere. The consequences, now and increasingly into the future, are outlined in this book.

To assist the nontechnical readers of the current status, allow me to offer a primer of some of the essential facts and concepts.

First, what is biodiversity? The word, short for "biological diversity," means the commonality of all heritable (gene-based) variation, both natural and domestic—in other words, globally the biosphere. Biodiversity consists of a biological hierarchy. At the top are the ecosystems, such as forest patches, ponds, coral reefs, and the bacterial flora of your intestine. Next are the species that compose the ecosystems; and finally, at the base, are the genes that prescribe the traits that define the species that compose the ecosystems.

Species are the units on which conservation biology is most concretely based. Most biological research starts with known species and ends with them, including disciplines as different from each other as ecology and molecular biology. Species further have the greatest exactitude and the most relevant information for conservation science. In contrast, the boundaries of ecosystems, which are vitally important to understand, are nevertheless characteristically more subjective. "Ecoregions," which are based in part on geographic data, are even more subjective and more difficult to classify.

How many species still exist on Earth? A bit more than two million have been discovered, formally named, and given a Latinized name. (The common housefly is denoted, for example, as *Musca domestica*.) We recognize that understanding of this element of biodiversity, begun by the Swedish naturalist Carl Linnaeus in 1735, is still woefully incomplete except for plants and vertebrates (mammals, birds, reptiles, amphibians, fishes). The actual number of species on the planet, deduced statistically, is roughly 10 million.

Of the species known, how fast are species disappearing due to human activity? The estimates range between one hundred and one thousand times the rate that existed before humanity spread around the world. And it is accelerating.

What exactly are the causes of extinction? They are often summarized by the initial-

ism HIPPO, with the position of each letter representing an extinction factor and (usually) in the order of its importance: thus, *habitat* reduction, *invasion* by nonnative species, *pollution* of habitat, *population* increase of humanity, and *overhunting* by excessive fishing, or hunting, or removal for other purposes. Climate change is perhaps the most important dimension of pollution impact.

Finally, how well are conservation projects doing? In the case of land-dwelling vertebrates globally assessed at the time of this writing, approximately one-fifth of land-dwelling vertebrate species are classified as threatened to some degree, and within that only one-fifth of the one-fifth have been stabilized or improved in status.

The authors of the present work agree overall that climate change is the principal threat to the survival of biodiversity even in its early stages. *Biodiversity and Climate Change: Transforming the Biosphere* serves as a comprehensive account of this greatest of threats to humanity's future. It will serve both as a textbook and a call to action.

Preface

The field of climate change biology dates to the mid-1980s, when the first symposium was held on the subject at the National Zoo in Washington, DC. The World Wildlife Fund convened that meeting, and one result was the book *Global Warming and Biodiversity*, edited by Robert L. Peters and Thomas E. Lovejoy and published by Yale University Press in 1992. At that point, the contributors and editors were largely confined to examining how past biological responses to climate change might predict change to come. It was nonetheless clear, as one of us remarked to Mostafa Tolba at a 1987 gathering to help plan the Convention on Biological Diversity, that if nothing was done about climate change, one "could forget about biological diversity."

As the field grew and the imprint of climate change on the planet's biodiversity became more apparent, little more than a decade after that initial meeting it was already clear that a completely new synthesis of the field was in order. In 2001, we began several years of gathering the best authors and writing, a process that culminated in *Climate Change and Biodiversity*, which we coedited and which was published by Yale University Press in 2005. That volume highlighted many of the numerous ways climate change was affecting biodiversity and biological systems. But change was still happening, both in the real world and in our understanding of impacts, faster than it could be synthesized into print. Ocean acidification was recognized as a major global environmental problem by the time the book appeared in bookstores, but it was treated in part of only one chapter.

With the passage of another decade, a truly staggering body of evidence suggested that the planet was headed for serious biological trouble, so in 2014 we realized that a new overview of this centrally important topic was once again needed. Indeed, what we first thought of as a second edition very quickly morphed into an entirely new book. Today, as the world is pushed through the critical barrier of 1° centigrade of global warming, the fingerprints of climate change can be seen in biodiversity everywhere. Already some major ecosystems are experiencing abrupt change, and major portions of the planet's biology such as the Amazon are approaching tipping points.

The impact of climate change in unfortunate synergy with other forms of human-driven change means that humanity is, in fact, changing the biosphere. This picture is laid out well by this new team of leading scholars in the following pages, as are some of the possible forms of adaptation that might lessen the biological impact.

This volume opens with an introduction to the topic and the relevant climate changes. Our chapter authors then look at the world around us and the mounting body of evidence that climate change is already rewriting biological history. From hundreds of millions of tree deaths to massive mortality in coral reefs, the evidence of climate change is already all around us, and it is writ large.

Placing these stunning changes in context requires understanding the past, and that perspective is provided by an outstanding set of paleoecological researchers, examining both marine and terrestrial realms. Movement has been the hallmark of past biotic response to climate change, but genetic responses may manifest themselves when bottlenecks are tight, as they will be in the future thanks to human conversion of natural habitats to other uses.

We then move from the past to the future. Our learning from models of the future and theory are explored in a series of chapters exploring possible biotic consequences of change in a variety of systems. The evidence in these chapters suggests the importance of questioning the global target of limiting climate change to 2° centigrade. Many systems will be past tipping points or past acceptable limits of change at that point.

Finally, we and our contributors examine the importance of being more proactive in conservation and policy. Where in the world can protected areas be placed for climate change? How should the protected-area estate be redesigned to be resilient in the face of climate change? How can we manage systems through extraordinary change, and which policies are needed to facilitate that process? These chapters recognize the positive contribution that biodiversity can make; for instance, transforming the biosphere through ecosystem restoration at a planetary scale can reduce the amount of climate change that biodiversity and humanity must cope with in the future. The volume concludes by drawing all the pieces together at a planetary scale and asking how we can feed the still-increasing billions of people, maintain robust biodiversity, and still accommodate change in both.

This book presents a current assessment of knowledge about climate change and biodiversity. More will continue to be learned and understood on an ongoing basis about the interaction between human-driven climate change and the biological fabric and systems of the planet. It is our fervent hope that this synthesis will lead to a widespread understanding that our planet works as a linked biological and physical system, and therefore will lead to less climate change and more surviving biological diversity than would otherwise be the case.

Acknowledgments

We would like to extend our deepest thanks to all those who have contributed to the creation of this book. The participants in a special meeting called by the Science and Technical Advisory Panel of the Global Environment Facility provided inspiration, including Camille Parmesan, Joshua Tewksbury, Gustavo Fonseca, Miguel Araújo, Mark Bush, Rebecca Brock, Joanie Kleypas, Rebecca Shaw, and Tom Hammond. Elizabeth Hiroyasu went above and beyond, overseeing and coordinating much of the production of the book. Deepest thanks go to Jean Thomson Black and Michael Deneen for their support through the editing process.

We are indebted to Ed Wilson for his continuing inspiration and insights contained in the foreword to this book. We are grateful to the Moore Center for Science and the Climate Change Strategy Team at Conservation International for their insights and support. For hard work on coordinating figures and assisting with graphics, we thank Monica Pessino and the rest of the Ocean o' Graphics team at the University of California, Santa Barbara. Funding from Cary Brown and the Fiddlehead Fund has supported many aspects of this project. The National Science Foundation, the National Science and Engineering Research Council of Canada, and the Belmont Forum provided support that enabled major insights in freshwater biology and ecological theory presented here. Special thanks to Carmen Thorndike for all that she does. And finally, we thank a long list of colleagues, including those who contributed chapters, without which this endeavor would have been impossible.

PART I

Overview: What Is Climate Change Biology?

CHAPTER ONE

Changing the Biosphere

THOMAS E. LOVEJOY AND LEE HANNAH

People are changing the biosphere. Expanding agricultural production, growing cities, and increasing resource use and disposal are among the human activities that are having profound effects on our planet. Some of these effects are exerted and felt locally, whereas others have global reach. Some human impacts, such as pollution, have been effectively addressed at local or national scales, and others, such as depletion of the ozone layer, have been effectively addressed through global cooperation. But many of the effects people are having on the biosphere have not been fully addressed, and some of these have global consequences of unprecedented proportion.

Two global environmental problems—biodiversity loss and climate change—are the preeminent environmental issues of this millennium. Biodiversity loss—the extinction of millions of species—is a very real consequence of increasing human use of the planet. Climate change due to greenhouse gas emissions threatens to alter the composition of the atmosphere and the living conditions on the surface of the planet for both people and all other species.

The purpose of this book is to explore multiple dimensions of the interaction between climate change and biodiversity. Climate change biology is an emerging discipline that boasts an explosion of literature reflecting the importance and abundance of climate change effects on the natural world. The great interest in this topic, and the large number of research findings emerging about it, combine to make a synthesis of the whole field invaluable for researchers and students, as well as for professionals in land use and conservation who increasingly need to understand the field in broad overview.

ISBN 978-0-300-20611-1.

We present the fundamentals of climate change for biologists in Chapter 2. The headline is that human burning of fossil fuels for heat, transportation, and other purposes, combined with releases of other pollutants, is altering the thermal balance of the atmosphere. These pollutants, including carbon dioxide (CO_2), methane, and others, are collectively known as greenhouse gases, because they absorb and reradiate outgoing longwave radiation, effectively warming the Earth's atmosphere. This warming has effects on atmospheric circulation, resulting in a vast number of changes to climate, such as changes in precipitation, seasonality, storm intensity and frequency, and many other factors, in addition to a pronounced warming of the planet's surface. Increases in atmospheric concentrations of CO_2 and other greenhouse gases have been measured beyond question, and the warming of the atmosphere and its consequences are now being observed.

The next twenty chapters address the past, present, and future of biodiversity response to climate change. Plants and animals all have climatic conditions that limit their distribution, so this major and complex change in climate will result in massive shifts in the distribution of species and ecosystems across the planet. Many shifts in individual species are already being observed. Observations of changes in ecosystems and changes in interactions between species are more complicated and therefore more difficult to document, but they are now being recorded in many regions. Marine, terrestrial, and freshwater systems are all being affected, and the conservation consequences of these biological changes are of major significance.

We explore the conservation and policy responses to these challenges in Part V (Chapters 22–28). Understanding climate change biology can be a complex task for policy makers and resource managers, particularly when the international response to climate change and biodiversity loss is segregated into separate international conventions and separate national policy-response

streams. The need for integrated response to climate change and biodiversity has never been greater, because changes in the biological world have profound implications for regional and global climate. The way in which people relate to nature will be forever altered by climate change.

A new synthesis is needed—a management-relevant analysis of climate change and biodiversity that transcends categories and conventions—to arrive at solutions that provide a robust future for both people and ecosystems. We now offer four themes that set the stage for the chapters that follow.

PLANETARY BOUNDARIES

Planetary boundaries represent the "safe operating space" for the Earth system (Rockström et al. 2009). Climate change and biodiversity are two of the nine planetary boundaries defined to date and represent half of the planetary boundaries that are estimated to have already been exceeded (the others being deforestation and nitrogen deposition). The planetary boundary for biodiversity is being exceeded because we are losing genetic diversity and robust populations of various species, which means there are ongoing and expected extinctions (Urban et al. 2016; Steffen et al. 2015).

In many ways, all four of the planetary boundaries that have been exceeded to date are related to climate change and biodiversity. The third exceeded boundary (after climate and biodiversity) is deforestation, a major contributor to both climate change and biodiversity loss (Steffen et al. 2015). The final exceeded boundary is disruption of nutrient cycles, especially that of nitrogen. One of the leading causes of nitrogen release into the atmosphere is the burning of fossil fuels, and species loss is one of the major consequences of a disrupted nitrogen cycle. Nitrogen overuse in agriculture, especially in developed countries, is leading to nitrogen runoff impacts on freshwater systems and massive dead zones in the oceans. Biodiversity loss and climate change are thus strongly linked to deforestation and the disruption of nutrient cycling.

In this most global sense, biodiversity loss and climate change are intimately connected and affect all people on the planet. It is important that we find solutions to these two great environmental problems that are commensurate with the global scale of the problems and that integrated solutions are pursued, not solutions in isolation. Although local and regional solutions are critical for affected populations, true success in conserving biodiversity and combating climate change can be achieved only when the results are measurable at a global scale. Extinction rates need to be lowered so that they are commensurate with background rates, and atmospheric CO_2 concentrations need to be rapidly stabilized and probably gradually decreased thereafter.

We will see in the chapters that follow that many parts of the solution space include both climate and biodiversity benefits. For example, reducing deforestation maintains habitats and also reduces CO_2 releases into the atmosphere, while lowering greenhouse gas emissions mitigates climate change, which reduces the magnitude of species range shifts and thus the conflict between species needs for movement and human land uses, such as agriculture, that stand in the way of those movements.

Although the climate-biodiversity nexus is deep and affects all people on the planet, this is only the first and highest-order dimension of the need for integrated response. Integrated responses are needed that address the global dimension but also the relationship of people with nature in local situations, where ecosystems are an effective tool to meet adaptation needs.

PEOPLE AND NATURE ADAPTING TOGETHER

Adaptation is a response to climate change that seeks to maintain individual or societal goals. There are many dimensions to adaptation, including maintenance of living space in societies challenged by sea-level rise, incomes for individuals and families facing adverse impacts of climate change, and ecological functioning and services. These dimensions are all often addressed in isolation—for instance, through engineering responses such as building dams to respond to water shortage—but there are major advantages to addressing the needs of people and nature in a more integrated fashion.

Wherever they live, people have a relationship with nature. That relationship may be close and intimate, as in communities that grow or harvest their own food, or it may be remote and less obvious, as in cities or highly structured economies. Whatever the relationship, people and nature are closely tied together, which involves food (produced locally or grown elsewhere and then transported to where we consume it), living space, recreation, spiritual connectedness, climatic conditions, and clean air and water.

Climate change will alter the relationship between people and nature everywhere on the planet. People in rural settings may find that the food they can grow and where they can grow it are changed, resulting in different patterns of agricultural use and natural lands. People in cities may find that there are higher costs for importing fuel for heating or cooling, and that supply chains for goods that are consumed have greatly different environmental impacts.

Nature provides both tangible and intangible benefits that accrue to people near and far in complex ways. Nature may have benefits for people far away (e.g., sequestration of carbon for climate stabilization, biodiversity existence value) at the same time that it has benefits for people close at hand (e.g., fresh water, tourism revenue) (Kremen et al. 2000).

This makes understanding the effect of climate change on the relationship between people and nature complex and important.

The effects of climate change occur in an already-dynamic situation, in which humans and their activities are interconnected locally, nationally, regionally, and internationally. Maintaining positive outcomes for people and ecosystems across all of these spheres as the climate changes is a major goal for people and nature adapting together.

For example, communities living with coral bleaching may need to adjust to both reduced fisheries output and reduced opportunities for tourism income. Although the extent of coral bleaching suggests that completely maintaining current levels of fishery output and tourism income may be impossible, a new balance may be struck in which protecting reefs most able to survive bleaching can help in the recovery of neighboring reefs and fisheries following bleaching while still maintaining tourism revenues. Actions to create habitats for remaining coral reef fish species may be important in maintaining fisheries production. Life with climate change may look considerably different for a community that is dependent on a coral reef. People's relationship with nature may change, through changes in fisheries livelihoods and through reduced reef extent. But a new relationship may maintain some reef fishery, some reef tourism, and a coastal lifestyle with sound planning, whereas an unplanned response may result in total loss of the reef and its tangible and intangible benefits.

Similar scenarios are played out across other coastal environments, rangelands, forests, lakes, and rivers. For instance, areas denuded of mature trees due to climate-driven bark-beetle outbreaks may see timber revenues, recreational opportunities, and habitat quality decline. A fundamental restructuring of the relationship between people and nature may be required on multiple levels in such settings.

Understanding the relationship between people and nature and how it is affected by climate change can help policy makers, managers, and communities living with nature understand how the best elements of a relationship with nature can be maintained to the greatest extent possible and where new relationships with nature will be needed because of climate change.

ECOSYSTEM-BASED ADAPTATION

One aspect of people and nature adapting together is the use of ecosystems to help meet human adaptation goals, or ecosystem-based adaptation (Jones et al. 2012). This is a second example of the interaction of solutions to both climate change and biodiversity loss. It is the special case of ecosystems being a functional part of needed human responses to climate change, with benefits that are usually focused locally or regionally.

Examples of ecosystem-based adaptation include coastal protection provided by mangroves and coral reefs, freshwater provisioning provided by cloud forests, flood control provided by upland habitats, and drought resistance provided by healthy range and natural wetlands. The scope of ecosystem-based adaptation literally covers all ecosystems in all regions of the world, even though it is only one small part of people and nature adapting together.

Whereas the mechanisms of ecosystem-based adaptation are direct, the management actions needed to maintain or improve it may be complex (Vignola et al. 2009). For instance, offshore coral reefs provide protection from rising sea levels and increasing storm surges along many coastlines in the world. But coral bleaching also affects these reefs. Reef recovery after bleaching is critical to the continued provision of coastal protection services. After bleaching, reef recovery is a race between recolonizing corals and the algae that grow on the dead reef (McClanahan et al. 2001). Healthy populations of fish that eat the algae are required to help the corals win the race and outcompete algae in the battle to recolonize the dead reef. Large parrotfish are among the most effective consumers of algae, but they are also the targets of fishing. Helping protect enough parrotfish to let corals recolonize the reef maintains coastal protection and in the long run improves the overall fishery.

This example illustrates that sometimes there are trade-offs inherent in achieving ecosystem-based adaptation benefits. In this case, a short-term reduction in fishing is required to maintain both the long-term health of the overall fishery and coastal protection (McClanahan et al. 2001). Establishing mechanisms for benefits transfer across time and between beneficiaries can therefore be critical to the success of ecosystem-based adaptation.

Ecosystem-based adaptation approaches are now being combined with engineering approaches in "green-gray adaptation." This combination allows for the delivery of benefits in the near and long terms, improving utility through time and across generations (Bierbaum et al. 2013). An example might be the combination of upland forest protection with a downstream flood-retention dam. The forest provides flood amelioration benefits for small and medium-sized storms, whereas the retention dam provides added capacity to cope with large storms. In this setting, ecosystem-based adaptation can provide part of the adaptation solution while capturing many biodiversity benefits.

Although ecosystem-based adaptation is just one dimension of people and nature adapting together, it is a critically important part of that spectrum, with the most obvious benefits for people who live in close proximity to ecosystems. As ecosystem-based adaptation matures, the benefits of capitalizing on the nexus of biodiversity and climate change will become more apparent, and the extension of this principle to broader elements of people and nature adapting together will emerge.

RUNNING OUT OF TIME

A third example of mutual benefits from addressing biodiversity and climate change

together comes from the global race to protect the world's last remaining natural habitats. As global population increases, the footprint needed to feed the planet expands. As a result, the amount of natural habitat remaining in the world is progressively disappearing. The world's population roughly doubled between 1890 and 1950, doubled again by 1970 and again by 2010 (Gonzalo and Alfonseca 2016). The space required for people and their material needs is rapidly squeezing out space for natural habitats everywhere on the planet.

At the same time, the world is moving to protect natural habitats. The area of national parks and other protected areas on the planet has doubled approximately every 40 years, until now about 17 percent of the terrestrial surface of the planet is in some form of protection (Costelloe et al. 2016). Although this protection ensures the existence of remaining natural habitats, it also reduces the amount of natural habitat available for protection. So the amount of remaining natural habitat available for protection is a function of decreases due to habitat destruction to provide for agriculture and other human needs and increases in land that is already protected.

Climate change and biodiversity come together in this race to protect remaining natural habitat, because these habitats are critical for the survival of species at the same time that they are critical stores of carbon for climate stabilization. When forests or other ecosystems are destroyed for conversion to agriculture or cities, the carbon contained in the vegetation often enters the atmosphere during burning associated with land clearing. Burning converts the carbon in vegetation to CO_2 in the atmosphere, a process that contributes about one-third of the greenhouse gases entering the atmosphere each year. So by protecting natural habitats from destruction, we can both avoid the extinction of species living in those habitats and make a substantial contribution to constraining climate change.

The time remaining to capture these dual benefits is limited, however. Based on current rates of habitat loss to human uses and protection of habitat, the estimates are that by 2050 or shortly thereafter, remaining unprotected primary tropical habitat on the planet will be gone (Hannah et al. 2018). Including secondary, recovering habitats extends the amount of time remaining to protect habitat on the planet by 20 to 30 years, but by the end of this century the window of opportunity for new protection will be closed in both primary and secondary habitats.

After opportunities to protect large, land-sparing protected areas have closed, there will be opportunities to protect smaller parcels of natural or seminatural habitat in productive landscapes. These land-sharing approaches are an important component in preserving biodiversity and responding to climate change, as is land restoration. Restoration of ecosystems (Chapter 25) is an important way to sequester carbon and provide more habitat for biodiversity. But as human populations continue to grow and resource use intensifies, increasing pressure will be put on productive landscapes and degraded lands. The window for these types of win-win activities for biodiversity and climate change is closing as well unless we change our approach to planetary management.

Across conservation mechanisms, the scope for constructive action to have biodiversity conservation and climate change mitigation and adaptation work together will narrow as the century progresses. The greatest opportunity to identify management actions that help prevent extinctions and biodiversity loss while helping people adapt to climate change and reduce greenhouse gas emissions is therefore right now.

ACTION FOR AN INTEGRATED FUTURE

Staying within planetary boundaries, people and nature adapting together, ecosystem-based adaptation, and the limited time for action are all examples of the challenges and synergies operating at the nexus of climate change and biodiversity. Capitalizing on opportunities and responding to those challenges involve science, awareness, policy, and management action. Action on each of these fronts is needed to realize a world in which the relationship between people and nature remains positive.

Science creates the understanding of biodiversity-climate relationships needed to generate awareness, policy, and management action. The first sections of this book are devoted to the insights that science has already generated, from current observations to understanding of the past, and how those past and present insights can be applied as we manage change. The chapters of the later sections look at how we can understand and manage the future.

Many of our existing insights come from systems for which data are fortunately available. For instance, our understandings of climate change responses of birds and butterflies are particularly well understood because these taxa are particularly well studied, allowing for distributional responses to climate change to be tested. Despite a large number of biological responses to climate, however, these cases are not necessarily the most important ones, either for people or for biodiversity. There is an urgent need to generate information about particularly valuable systems and to generate policy and management action based on those findings (see Box 1.1). For example, how elephants respond to climate change is a complex mixture of habitat change due to climate change, changing human land uses, and an often-delicate balance in human-elephant relations. Illegal hunting for the ivory trade may be the major threat to elephants in the near term, but assuming that the threat of illegal trade can be addressed, in the long term it is habitat access and connectivity that are likely to determine the number of elephants that can be sustained on the planet. Scientific understanding of these and other

Box 1.1.   **Ecosystem Experiments Are Important**

Ecosystem experiments can be of major value in revealing potential climate change impacts and mechanisms. Without them we would be left in the defensive position of being able to understand effects only as they occur rather than anticipate them with preemptive policies as well as adaptation measures.

In many senses ecosystem experiments came of age with the Hubbard Brook experiment (Holmes and Likens 2016), although there had been important precursors, such as the Brookhaven Irradiated Forest and the Luquillo Forest of Puerto Rico (Odum 1970).

The first climate change experiment was initiated by John Harte at the Rocky Mountain Biological Laboratory, where an alpine meadow has been, and continues to be, artificially warmed (Harte and Shaw 1955), revealing changes in the plant community and soil biota. Since climate change will also include changes in precipitation, another experiment has involved simulating such changes in grasslands in northern California (Suttle, Thomsen, and Power 2007).

Other relevant ecosystem experiments have endeavored to understand the impacts that higher CO_2 levels might have on ecosystems. Will there or will there not be a "fertilization" effect, or if there is one, will it be short-lived because the ecosystem will encounter some other limitation (e.g., a key nutrient)? There have been a number of such experiments and some are ongoing. One Oak Ridge experiment showed a short-lived fertilization response (Norby et al. 2010), whereas others have not. The new free-air concentration enrichment (FACE) experiment in the central Amazon may reveal a similar nutrient constraint (in that case from phosphorous).

complex systems are needed to help guide stewardship of the planet through an era of unprecedented climate change.

Awareness is needed to complement scientific understanding, both about climate change itself and especially about how climate change and biodiversity interact. Beyond accepting the obvious—that climate change is real and is having real impacts today—the general public and policy makers alike need to be aware that the interface of biodiversity and climate change is complex and has a vitally important role to play in shaping people's relationship with nature. The latter sections of this book explore what is needed in policy, awareness, and management action.

Climate change biology is not just about the extinction of polar bears, or even just about extinction. Although the elevated risk of extinction due to climate change is a major international concern, there is much more at stake than just bringing along the full complement of nature's bounty with us as we move into the future. At stake is the relationship of all humans with nature, the ability to capitalize on ecosystems to help us adapt to climate change, and the ability to move quickly to realize the remaining scope for solutions before it is too late.

Policy is needed at the interface of these two great environmental concerns, to avoid having climate change biology and its consequences for people fall between the cracks. Since 1992, the world has had international conventions that deal with biodiversity and with climate change. These international conventions hold separate meetings and are often administered by separate agencies within individual countries. What has been lacking in the policy world has been an ability to work between and across the conventions. Mitigation and adaptation of climate change have massive implications for avoiding extinctions and conserving biodiversity, whereas maintaining biodiversity has major implications for managing the planet's atmosphere and helping people adjust their relationships with nature and use ecosystems to adapt to climate change.

Now that fundamental progress has been made on each issue in its own right, it is particularly critical that collaborative work at the interface of climate change and biodiversity occur. Just as in science at the interface, policy at this interface is beginning to emerge. For instance, the Green Climate Fund recognizes ecosystems as one of the four pillars of climate adaptation.

Managers, scientists, policy makers, and the public are not separate teams that lob questions or insights back and forth to one another. Rather, they need to form a single integrated team in which scientific understanding and awareness drive policy that supports sound management action. Because of this need for integration, it is critical that the climate change and biodiversity communities work together for mutual understanding and for the evolution of awareness, policy, and management that responds to the full scope of the implications climate change carries for ecosystems and biodiversity.

The role of climate change biology is to provide an understanding of the responses of the natural world to climate change. Insights from changes already being observed (Part II) put us on alert about possible disruptions to the balance in the relationship between people and nature. Understanding how species and ecosystems have responded to past climate change, both in deep time and in near time (Part III), provides a window onto how species and ecosystems have responded in a fully natural world, which we must now interpret, and landscapes increasingly dominated by human land uses. Modeling and experimentation to estimate future change (Part IV) can be combined with knowledge and models of changes in human production to help in-

form policy and management actions that maintain a healthy balance between people and nature as the climate changes (Part V). The combined power of these insights is to give humanity the opportunity to create a healthy relationship with nature of multiple scales—from planetary boundaries to ecosystem-based adaptation—that will endure as the climate changes.

REFERENCES

Bierbaum, Rosina, Joel B. Smith, Arthur Lee, Maria Blair, Lynne Carter, F. Stuart Chapin, Paul Fleming, Susan Ruffo, Missy Stults, Shannon McNeeley, Emily Wasley, and Laura Verduzco. 2012. "A comprehensive review of climate adaptation in the United States: More than before, but less than needed." *Mitigation and Adaptation Strategies for Global Change* 18 (3, October): 361–406.

Costelloe, Brendan, Ben Collen, E. J. Milner-Gulland, Ian D. Craigie, Louise McRae, Carlo Rondinini, and Emily Nicholson. 2016. "Global biodiversity indicators reflect the modeled impacts of protected area policy change." *Conservation Letters* 9 (1, January): 14–20.

Gonzalo, Julio A., and Manuel Alfonseca. 2016. "World population growth." *World Population: Past, Present & Future* 29.

Hannah, Lee, Patrick Roehrdanz, Guy Midgley, Jon Lovett, Pablo Marquet, Richard Corlett, Brian Enquist, and Wendy Foden. "Last call for tropical protected areas." In review.

Harte, John, and Rebecca Shaw. 1995. "Shifting dominance within a montane vegetation community: Results from a climate-warming experiment." *Science* 267: 876–880.

Holmes, Richard T., and Gene E. Likens. 2016. *Hubbard Brook: The Story of a Forest Ecosystem.* Yale University Press.

Jones, Holly P., David G. Hole, and Erika S. Zavaleta. 2012. "Harnessing nature to help people adapt to climate change." *Nature Climate Change* 2 (7): 504.

Kremen, Claire, John O. Niles, M. G. Dalton, Gretchen C. Daily, Paul R. Ehrlich, John P. Fay, David Grewal, and R. Philip Guillery. 2000. "Economic incentives for rain forest conservation across scales." *Science* 288 (5472): 1828–1832.

McClanahan, T., N. Muthiga, and S. Mangi. 2001. "Coral and algal changes after the 1998 coral bleaching: Interaction with reef management and herbivores on Kenyan reefs." *Coral Reefs* 19 (4): 380–391.

Norby, Richard J., Jeffrey M. Warren, Colleen M. Iversen, Belinda E. Medlyn, and Ross E. McMurtrie. 2010. "CO_2 enhancement of forest productivity constrained by limited nitrogen availability." *Proceedings of the National Academy of Sciences* 107: 19368–19373.

Odum, Howard T. 1970. "Summary: An emerging view of the ecological system at El Verde." *Tropical Rain Forest*: 1191–1281.

Rockström, Johan, Will Steffen, Kevin Noone, Åsa Persson, F. Stuart Chapin III, Eric Lambin, Timothy Lenton, et al. 2009. "Planetary boundaries: Exploring the safe operating space for humanity." *Ecology and Society* 14 (2): art. 32.

Steffen, Will, Katherine Richardson, Johan Rockström, Sarah E. Cornell, Ingo Fetzer, Elena M. Bennett, Reinette Biggs, et al. 2015. "Planetary boundaries: Guiding human development on a changing planet." *Science* 347 (6223): art. 1259855.

Suttle, K. B., Meredith E. Thomsen, and Mary E. Power. 2007. "Species interactions reverse grassland responses to changing climate." *Science* 315: 640–642.

Urban, M. C., G. Bocedi, A. P. Hendry, J. B. Mihoub, G. Pe'er, A. Singer, J. R. Bridle, L. G. Crozier, L. De Meester, W. Godsoe, and A. Gonzalez. 2016. "Improving the forecast for biodiversity under climate change." *Science* 353 (6304): p.aad8466.

Vignola, Raffaele, Bruno Locatelli, Celia Martinez, and Pablo Imbach. 2009. "Ecosystem-based adaptation to climate change: What role for policy-makers, society and scientists?" *Mitigation and Adaptation Strategies for Global Change* 14 (8): 691.

CHAPTER TWO

What Is Climate Change?

MICHAEL C. MACCRACKEN

Climate is the set of weather conditions prevailing over a region over a period of time. To be long enough to average across the natural, short-term variations of these conditions, the period over which the climate is defined has generally been three decades—roughly a human generation. By contrast, *weather* is the instantaneous state of the atmosphere; it is the statistical ensemble of the weather that is used to quantify the state of the climate. As some say, climate is what you expect and weather is what you get.

Weather conditions (and so climatic conditions, being the accumulated set of weather conditions) are largely determined by the dynamic interactions of the atmosphere, the oceans, the land surface, and the glaciers and ice sheets, which are often collectively referred to as the components of the climate system. For each location, climate generally represents the conditions with which a region is associated, such as southern California being warm and dry and the Pacific Northwest coast being cool and wet. However, scientifically, the climate of a region also includes the seasonal, annual, and longer variations and cycling (e.g., El Niño and La Niña oscillations) that result from the various internal intercouplings and linkages of the components of the climate system.

In addition to internal variability, the climate can also be affected by factors considered external to the climate system. External factors that are capable of causing changes in the system's energy fluxes and balances (generally referred to as radiative forcings) include changes in incoming solar radiation due to changes in solar output and/or time-varying changes in the distance to the Sun, aerosol injections into the stratosphere resulting from major and persistent volcanic eruptions, geological processes such as the shifting distribution of continents and mountain ranges, and changes in composition of the atmosphere due to processes ranging from weathering to sequestration of carbon in swamps or submergence of vegetation by sea-level rise. Human activities of various types are now, however, altering the Earth's energy balance in comparable ways. The largest warming influence is resulting from the increase in the CO_2 concentration due to combustion of fossil fuels (i.e., coal, petroleum, natural gas, oil shale-derived fuels) and deforestation and soil disturbance. In addition to CO_2, human activities have been increasing the atmospheric concentrations of methane (CH_4), nitrous oxide (N_2O), halocarbons (mainly chlorine containing compounds), and other gases that alter the Earth's infrared (heat-related) gases (collectively, greenhouse gases). Human activities have also been adding to the atmospheric loadings of dark aerosols like black carbon (e.g., from diesel engine exhaust and coal-fired power plants) that absorb additional sunlight and light-colored aerosols such as dust and sulfate (the latter created mainly from sulfur dioxide, SO_2, emissions by coal-fired power plants) that exert a cooling influence by reflecting additional sunlight back to space. Human activities have also caused changes in land surface characteristics such as deforestation, agriculture, and urbanization, which can affect surface roughness and the rate of evaporation from soils.

The responses of the climate system to these forcings not only are direct, but the direct responses also can trigger further interactions and feedbacks among climate system components that can amplify or moderate the direct response. Thus, working out the net effects of these forcings on temperature, precipitation, and the weather requires very careful quantitative analysis, especially in seeking to learn from what has happened in the past and projecting forward what could happen in the future.

Among the most important positive (or amplifying) feedbacks to consider in these analyses are water-vapor feedback (whereby warming leads to an increase in atmospheric water vapor that then exerts further warming) and snow- and ice-albedo (or reflectivity) feedbacks (whereby warming leads to melting of snow and ice that decreases surface reflectivity that, in turn, allows for additional absorption of solar radiation and so more warming). Countering these amplifying processes are ones that moderate how much change can occur. For example, the infrared radiation emitted upward from the surface is proportional to the fourth power of the surface temperature, although this negative feedback is moderated by a process referred to as the atmospheric greenhouse effect, whereby the atmosphere absorbs much of the upward radiation from the surface and lower atmosphere and then reradiates much of it (presently about 90 percent) back to the surface to induce further warming. Interacting together, these processes have created the range of climates across the Earth to which we have become accustomed and on which society has become dependent.

LESSONS FROM EARTH'S CLIMATIC HISTORY

Geological records, chemical and physical variations in ice cores, and many other types of evidence make it clear that the Earth's climate has varied greatly over its history. Indeed, the most important lessons from the paleoclimatic record are that significant climate change can occur, that these changes occur as a result of identifiable changes in forcing rather than as a result of random fluctuations, and that large changes generally tend to result from accumulated changes over many millennia.

An important example of the role of climate forcings in altering the climate is evident in the glacial-interglacial cycling that occurred over the past million years

(Berger 2001). Roughly 125,000 years ago, the global average temperature during the interglacial period before the present one was up to ~1°C–2°C higher than present, and sea level is estimated to have been 4–8 meters higher. This several-thousand-year warm interval was followed by transition into an increasingly glacial period that lasted ~100,000 years. At its coldest, the global average temperature was ~6°C below present and sea level dropped by ~120 meters, exposing large areas of continental shelves around the world, drying out the Mediterranean Sea, and creating a land bridge between North America and Asia. As the global average temperature rose to its present level about 6,000–8,000 years ago, ice sheets melted and sea level rose at an average rate of ~1 meter per century for 120 centuries. Large shifts in flora and fauna occurred throughout this period, demonstrating that ecosystems on land and sea are closely tied to the prevailing climate, albeit perhaps with a lag of centuries, and that such shifts can occur if climate change is extended over relatively long periods.

These glacial-interglacial changes in the climate were driven by cyclic changes in the Earth's orbit around the Sun, which redistributed incoming solar radiation by season and latitude in ways then amplified by a wide range of mostly positive feedback processes, including a natural carbon feedback that tended to amplify the orbitally induced warming. As a result of quite fortuitous timing of three important orbital periodicities (ellipticity of the orbit, tilt of the axis of rotation, and time of year of closest approach to the Sun), seasonal and latitudinal patterns of incoming solar radiation have been relatively stable over the past several thousand years, which is known as the Holocene epoch, and the astronomically determined seasonal and global patterns of solar radiation would be expected to vary little over the next several tens of thousands of years (Mysak 2008)—that is, absent human activities, the world might be a bit cooler,

but not experiencing extensive glaciation over this period.

During the Holocene, fluctuations in global average temperature have generally been less than ~0.5°C, driven primarily by extended periods of volcanic activity, small changes in orbital forcing and solar activity, and changes in land cover and agriculture. Although regional-scale fluctuations have been somewhat larger, the relatively favorable global climatic conditions that have existed well into the twentieth century have allowed time for the maturing of ecosystems and the development of agriculture, communities, and civilizations around the world.

HUMAN-INDUCED CLIMATE CHANGE

The relative steadiness of the climate is now being disrupted because of the changes in atmospheric composition caused mainly by the use of coal, petroleum, and natural gas to power industrialization and economic growth, and alteration of the landscape to allow for global agriculture. Figure 2.1 shows the changes in the global average surface air temperature from 1880 to 2016 compiled by the National Oceanographic and Atmospheric Administration. Detection-attribution analyses described in IPCC's Second Assessment Report made it possible to initially conclude that "there was a discernible human influence on the global climate" (Santer et al. 1995). Analyses of over a dozen measures of climate change (e.g., changes in atmospheric moisture; the height of the tropopause, which is the boundary between the troposphere and stratosphere; regional patterns and timing of land and ocean warming) and more since that initial assessment now support the much stronger conclusion that "it is *extremely likely* that human influence has been the dominant cause of the observed warming since the mid-20th century" (IPCC 2013, emphasis in original). With human

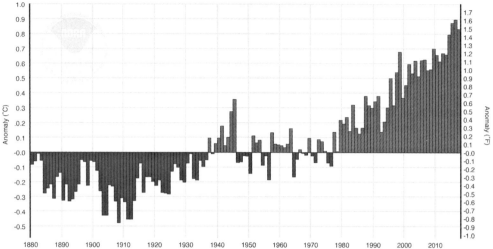

Figure 2.1. Time history (1880–2016) of the change in global average temperature over time based on the average of geographically weighted local departures of annual-average surface air temperature (°C) from the 20th century baseline; this compilation is maintained by the US National Oceanographic and Atmospheric Administration. The late 19th- and early 20th-century values were depressed by a higher-than-average number of significant volcanic eruptions; the mid-20th century was roughly level as greenhouse gas–induced warming was offset by an increased loading of sulfate aerosols and slowly declining solar irradiance; warming since the 1970s is mainly due to the rising concentrations of greenhouse gases. (Figure courtesy of NOAA.)

forcing now emerging as the dominant cause of climatic change, the period since roughly the mid-twentieth century is now being designated the Anthropocene epoch (Steffen et al. 2011).

According to the most recent data, the global average surface air temperature has increased nearly 1°C since the mid-nineteenth century (Plate 1), and suggestions of a break in the rate of warming during the first decade of the twenty-first century have proved mistaken (Santer et al. 2014). An increase of 1°C may seem small in contrast to diurnal and seasonal swings in local temperature on the order of 20°C in some regions, but the global warming to date (so the baseline around which diurnal and sea-

sonal temperatures swing) is now equal to about 15 percent of the 6°C warming from the Last Interglacial Maximum to the present and about the same fraction of the similar difference between the temperatures of the present and the very warm Cretaceous, when dinosaurs roamed the Earth.

Although international attention has focused on the change in the global average temperature, changes in climate involve much more than simply a change in the average temperature (for details, see IPCC 2013; Cramer and Yohe 2014; for the United States, see USGCRP 2017). Indeed, what people, plants, and wildlife experience is not the mathematical construct that makes up the three-decade average; it is changes in the intensity, occurrence, and patterns of the weather. As an example, in addition to recent changes in the average temperature, the likelihood of warm and cool summers over Northern Hemisphere land areas has been changing (Figure 2.2). Quite amazingly, in response to global warming of only ~0.5°C since the mid-twentieth century, the likelihood of the very warmest summers occurring has increased by more than three standard deviations from a likelihood of less than 0.1 percent during the period 1951–1980 to currently ~10 percent of the

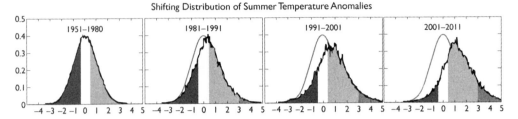

Figure 2.2. The frequency of occurrence of local departures of June-July-August (summertime) surface
air temperature (in standard deviations from the local
mean value) for land areas in the Northern Hemisphere
relative to the mean value for 1951–1980 (generally described as the seasonal climatological value). The distribution of frequencies of occurrence for the period 1951–
1980 closely matches a normal (i.e., bell-shaped curve)
distribution (solid line), which is used to subdivide the
set into cool (dark gray), typical (white), and warm
(light gray) occurrences, each subset having a probability of a third. The distribution of local summertime departures from the 1951–1980 normal has shifted to the
right as the three-decade climatology has moved closer
to the present. What were once cool summers now occur much less frequently, and what were once warm and
extremely warm summers now occur much more frequently. (Redrawn from Hansen et al. 2012.)

time—so an increase in incidence of such
extremes by over a factor of a hundred in
only about six decades (Hansen et al. 2012).

Other aspects of human-induced change
are also discernible. Broadly speaking,
warming in the Arctic has been at least
twice as large as the global average, causing
widespread impacts on ecosystems (Arctic
Climate Impact Assessment 2004). Also, the
dry subtropics have expanded poleward
and average precipitation has increased over
mid- and high-latitude land areas. The likelihood of intense rainfall has also increased,
with a greater fraction of precipitation coming in such events, causing increased runoff
and in turn causing higher river levels and
increased potential for flooding at those
times of year when stream flows are high.
With less precipitation occurring in small
rain events, periods of high evaporation of
soil moisture are now longer and warmer,
leading to faster and more frequent transitions to very dry and even drought conditions (USGCRP 2017). The occurrence
of hurricanes and typhoons is so low and

variable that trend statistics are not well established, but oceans are warmer, and there
appears to be a greater likelihood that the
storms that do develop will have higher
peak winds and greater precipitation.

Changes in mid-latitude weather patterns
illustrate how warming in high latitudes can
drive shifts in the climate (e.g., Francis and
Vavrus 2012; Francis and Skific 2015). The
variable weather in Northern Hemisphere
mid-latitudes is created by cold, dense Arctic air masses spreading southward and
interacting with warm, moist subtropical
air masses pushing poleward. Convection
and storms occur where moist air is forced
to rise over colder air. When the equator-
pole temperature gradient is strong, the
atmospheric circulation sets up to most efficiently move heat poleward and the Co-
riolis effect (i.e., the turning eastward of
meridional, or south to north, winds resulting from the rotation of the Earth) leads to
there being a stronger and more zonal (west
to east) jet stream with fewer meanders and
faster west-to-east storm movement. With
Arctic warming becoming strong and tropical warming limited by increased evaporation, the equator-pole temperature gradient
is decreasing, which is leading to greater
meandering of the zonal jet. The larger
waves in the jet stream allow warm moist
air to push farther north and cold, Arctic air
to push farther south, increasing the range
of possible conditions at any given location
and slowing the eastward movement of wet
and dry weather regimes.

For example, in the Great Plains of North
America (the only continent extending
from the subtropics to the Arctic without
an east-west mountain range to separate
cold polar and warm, moist subtropical air

masses), the shifting weather conditions allow for warm, more northward penetration of moist subtropical air, causing, in some years, heavier snowpacks and so increased springtime snowmelt runoff in the northern Great Plains. With its relatively flat geography and a history of less intense weather events, river channels have tended to be shallow. Now, with rapid melting of the heavier snow amounts, major flooding has become more frequent and severe. Thus, not only is the climate changing, but the hydrogeography and land cover of regions are also being affected by the changing weather.

PROJECTIONS OF GLOBAL WARMING THROUGH THE TWENTY-FIRST CENTURY

With global use of coal, petroleum, and natural gas still providing about 80 percent of the world's energy, the atmospheric concentration of CO_2 will continue to increase for some time, even with aggressive attempts to switch to deriving energy from renewable sources. Stopping the ongoing increase in radiative forcing will require CO_2 and other greenhouse gas emissions to decrease to near zero (so not just stopping the growth in emissions); returning to the preindustrial level of radiative forcing would require not only this but also essentially the removal of roughly as much CO_2 from the atmosphere as has been added. That would include the amount that has been taken up by the oceans and biosphere, because much of this carbon would return to the atmosphere as the atmospheric concentration was lowered from its current level.

Computer models of the climate system have been developed to project how the climate will respond to the changing atmospheric concentrations of CO_2 and other climate-altering gases and aerosols (i.e., particles). These models have been built to incorporate both theoretical and empirical understanding of how the climate system functions and how all of the system variables and processes interact and link together to determine the climate that prevails. Thus, the models follow how solar radiation changes through the day and season and is reflected and absorbed by the atmosphere, clouds, land, ice, and oceans. The models also (1) simulate the flow of infrared (heat) radiation and its dependence on temperature and atmospheric composition; (2) calculate the motions of the atmosphere and oceans, both vertically and horizontally; (3) treat the evaporation of water from the surface, the formation of clouds and precipitation, and then accumulation of soil moisture and loss by runoff; (4) treat the buildup and loss of snow cover on land and the expansion and melting of sea ice on the oceans; and more (see Figure 2.3). For the atmospheric component of the system, the model is virtually the same as weather forecast models, but for the oceans, ice, and land surface, the models are constructed to include processes that affect cumulative changes through seasons, years, decades, and longer. Given the influence of small-scale chaotic behavior, the nonlinear governing equations that are the basis of weather forecast models limit the potential predictability of specific large-scale weather situations to only 1–2 weeks. In contrast, model simulations for past multidecadal to multicentennial periods (e.g., the past two centuries) and distant periods in the past have demonstrated that the climate system models, with forcings prescribed as they changed over these periods, reasonably reproduce the climatic conditions that have been reconstructed from observations and geological evidence.

In seeking to project how the climate is likely to evolve in the future, scenarios of the consequences of future societal choices in terms of energy-generated emissions and other factors are used as boundary conditions for simulations using the state-of-the-science climate models. Model results then portray how the climate would be expected

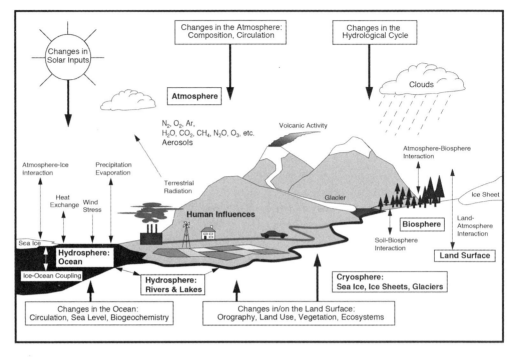

Figure 2.3. Schematic view of the components of the global climate system (bold), their processes and interactions (thin arrows), and some aspects that may change (bold arrows). Presently available models of the atmosphere, oceans, and land, as well as snow and ice surfaces, generally divide each domain into quadrilateral tiles that are about 100 km to 200 km on a side and sliced thin enough to represent vertical variations in each system. Variables such as temperature, water content, and speed of movement are maintained for each tile in each climate system component. Flows and fluxes through the boundaries of every tile and across from one component of the system to another are calculated for each process, then used to update the values of each variable in each tile to a new time using a time step typically measured in minutes. With such a fine time step, climate system models are simulating the evolving "weather" of the atmosphere-ocean-land-ice system and the results are then time- and space-averaged to provide an indication of how the climate is changing in response to natural internal fluctuations and the changing values of external forcings. (Figure 1.1 from IPCC 2001: *Climate Change 2001: The Scientific Basis. Contribution of Working Group I to the Third Assessment Report of the Intergovernmental Panel on Climate Change.*)

to evolve over coming decades and centuries. Because the model results are conditional, being dependent on the scenarios, these results are called projections rather than predictions. For further details, see the

IPCC assessments for a description of global changes and impacts and the most recent US and World Bank assessments for more detailed descriptions in particular regions (Melillo et al. 2014; USGCRP 2017; World Bank 2013, 2014).

As shown in Figure 2.4, the global-scale results indicate that the greater the emissions, the greater the global warming that will occur. These simulations generally indicate that a prolonged reliance on fossil fuels would lead to the global average temperature reaching 4°C above its preindustrial level by 2100 and continuing upward thereafter, whereas a very aggressive shift to reliance on solar, wind, and other renewable energy technologies has the potential to limit the increase in the global average temperature peaking at 2°C–3°C, depending on how rapidly the conversion of the energy system occurs.

Emissions of CO_2 and other greenhouse gases not only cause climate change; many of these gases also contribute to poor air quality, widespread health effects (particularly to women and children as a result of emissions from traditional cooking stoves),

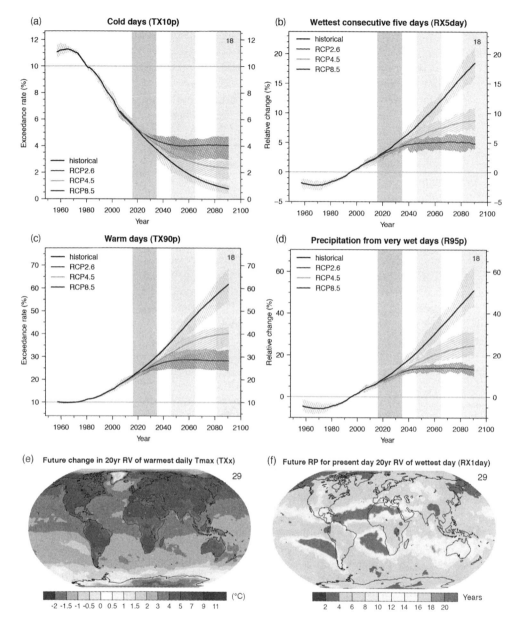

Figure 2.4. Global projections of changes in the likelihood of important occurrences of moderately extreme weather. Graphs on the left show the time history from 1960 to the present (based on observations) and to 2090 (based on projections) of the change in likelihood of what were the days with daily maximum temperature among the (A) coldest and (C) warmest 10% of days during the period 1961–1990. Graphs on the right show the time history from 1960 to present (based on observations) and to 2090 (based on projections) of the change in likelihood of what were (B) the wettest consecutive five days and (D) the wettest 5% of days during the period 1986–2005. The projections are based on scenarios that remain mainly fossil fuel based (black line, RCP8.5), transition slowly away from fossil fuels (light gray, RCP4.5), and transition rapidly to energy technologies that do not emit CO_2 and other greenhouse gases (dark gray). (TFE.9, Figure 1 from Stocker et al. 2013; IPCC Fifth Assessment Report, Working Group I, Technical Summary.)

crop damage, water pollution, glacial melting (due to warming and deposited black carbon) and more. There are thus many reasons to limit climate-changing emissions.

Negotiations to cut emissions of CO_2 and other long-lived greenhouse gases have been the focus of international negotiations through the annual Conference of the Parties (COP) since the UN Framework Convention on Climate Change (UNFCCC) was negotiated in 1992, with the most recent agreement being negotiated in Paris in 2015. Based on the emissions-reduction commitments made in support of the Paris Agreement's goal, the global average temperature is nonetheless still projected to be between 3°C and 4°C above its preindustrial level, which is well above the agreement's aspirational goal of 1.5°C and 2°C; scientific studies suggest that preventing most of the adverse impacts will actually require returning the increase in global average temperature to no more than 0.5°C (equivalent to a CO_2 concentration of 300–350 ppm; e.g., see Hansen et al. 2016).

So, what more can be done? Greatly improving the efficiency in our use of energy (which would result in also reducing emissions of black carbon) and transitioning to renewable energy sources (solar, wind, wave, tidal, water current, etc.) in order to reduce CO_2 emissions are the most environmentally benign ways of reducing human-induced impacts on the climate. Reversing deforestation, even implementing afforestation, and rebuilding soil carbon will also be essential to limiting the rise in the atmospheric CO_2 concentration and later pulling CO_2 back out of the atmosphere. Because the atmospheric lifetime of methane is only about a decade, as compared to the multi-millennial persistence of the human-induced increase in the CO_2 concentration, reducing leakage of methane from the global energy system and its generation by global agriculture can help slow the pace of climate change over the coming few decades (UNEP 2011; Shindell et al. 2012). Reducing emissions of the many halocarbon compounds

can also reduce near-term warming, and the United States and China have taken the lead in doing so. There is thus the potential for nations to make even greater and earlier cuts in climate-warming emissions, including especially emissions of methane, precursors of tropospheric ozone, and black carbon, but there is really very little time for such actions to be taken because of the warming trajectory already created by past emissions.

As serious as the overall warming would be, the unprecedented pace of the warming is at least as serious because it shortens the time for forests and other ecosystems to keep up with the shifting of their preferred climatic zones. In addition to the rapid pace of warming, the accelerating pace of long-term sea-level rise and the lowering of the pH of ocean waters (i.e., acidification) as the CO_2 concentration rises are projected to cause increasing inundation of coastlines and spreading disruption of marine ecosystems, respectively.

BEYOND THE AVERAGE

The focus of international climate change negotiations has primarily been on the projected decade-to-decade changes in global average temperature, but the global average is really only a mathematical construct. Understandably, the multidecadal average change will vary by latitude, season, and local conditions, being greater in high than in low latitudes, greater over land than over the ocean, and greater at night than during the day as a result of greenhouse gases absorbing outgoing longwave radiation re-radiating back toward the surface twenty-four hours a day. In the low latitudes, the temperature increase is smaller than in high latitudes because a greater fraction of the available energy results in evaporation (and evapotranspiration). With the vertical distribution of CO_2-induced radiative forcing exerting a slight tendency to stabilize the troposphere, additional energy from condensation is needed to power convective storms, which leads to the increasing occurrence of intense precipitation. At the same time, warmer oceans are making additional energy available to tropical cyclones (known as typhoons or hurricanes, depending on the region), powering an increase in storm intensity and so in damage, which is then further compounded along vulnerable coastlines because of sea-level rise.

Individual organisms and ecosystems experience the weather at the particular location that they occupy, making changes in the local weather more relevant than changes in long-term average conditions. For example, in areas with variable terrain, intense rains may cause some areas to flood while large-scale, flatter areas end up becoming drier. Considering detailed changes in the weather is thus going to be essential to understanding many biological responses. Useful analyses will need to evaluate changes in the likelihood of various weather types, especially including changes in the frequency, intensity, and duration of extreme conditions (e.g., heat waves, episodes of intense precipitation and drought). Figure 2.5 shows projections for the twenty-first century of changes in the frequency of what were low-likelihood weather events during the mid- to late twentieth century, showing a sharp increase in the occurrence of hot days and wet extremes and a sharp decrease in the occurrence of what were considered extremely cold days. The most recent USGCRP assessment includes substantial new projections of changes in the weather (USGCRP 2017).

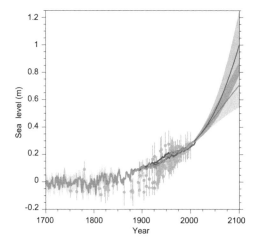

Figure 2.5. Time history of the global average surface air temperature for the period 1900–2100 based on observations for the period 1900–2013 and model projections to 2100 for an emissions scenario that remains strongly fossil fuel based (upper band) or that assumes relatively aggressive conversion to energy technologies that do not release CO_2 and other climate-warming greenhouse gases (lower band). For reference, a warming of 5°C from 1900 to 2100 (and continuing thereafter) would be just less than the warming that occurred from the peak of the last ice age (~20,000 years ago) to the present and the difference between the Cretaceous (>65M years ago) and the present. Even the increase of ~2°C would be very significant, such warm conditions not having occurred globally since well before Pleistocene glacial-interglacial cycling began about 2 million years ago. (Box TS.5, Figure 1 from Field et al. 2014; IPCC Fifth Assessment Report, Working Group II, Technical Summary.)

CLIMATE CHANGE FOR BIOLOGISTS

While policy makers tend to focus on the gradual increase in the global average temperature, biologists will need to consider the full statistical representation of change in atmospheric behavior, including the means, sequences, correlations, higher moments, and more. As change continues to increase, considering only the slow, average changes will provide a less and less adequate sense of the types and nature of impacts on the environment and society. With available ensembles of model projections that have detail comparable to that of weather forecasts, there is the potential for significantly improved environmental impact analyses and more reliable insights for building resilience and avoiding the most significant consequences.

REFERENCES

Arctic Climate Impact Assessment. 2004. "Impacts of a warming Arctic: Arctic climate impact assessment overview report." Cambridge University Press.

Baede, A. P. M., E. Ahlonsou, Y. Ding, and D. Schimel. 2014. "The climate system: An overview." In *Climate Change 2001: The Scientific Basis. Contribution of Working Group I to the Third Assessment Report of the Intergovernmental Panel on Climate Change*, ed. J. T. Houghton, Y. Ding, D. J. Griggs, M. Noguer, P. J. van der Linden, X. Dai, K. Maskell, and C. A. Johnson, 85–98. Cambridge University Press.

Berger, Andre. 2001. "The role of CO_2, sea-level, and vegetation during the Milankovitch forced glacial-interglacial cycles." *Geosphere-Biosphere Interactions and Climate*. Cambridge University Press.

Cramer, Wolfgang, and Gary W. Yohe. 2014. "Detection and attribution of observed impacts." In *Climate Change 2014: Impacts, Adaptation, and Vulnerability: Part A: Global and Sectoral Aspects, Working Group II Contribution to the Fifth Assessment Report of the Intergovernmental Panel on Climate Change*, ed. C. B. Field, V. R. Barros, D. J. Dokken, K. J. Mach, M. D. Mastrandrea, T. E. Bilir, M. Chatterjee, et al., 979–1037. Cambridge University Press.

Field, C. B., V. R. Barros, K. J. Mach, M. D. Mastrandrea, M. van Aalst, W. N. Adger, D. J. Arent, et al. 2014. "Technical summary." In *Climate Change 2014: Impacts, Adaptation, and Vulnerability. Part A: Global and Sectoral Aspects. Contribution of Working Group II to the Fifth Assessment Report of the Intergovernmental Panel on Climate Change*, ed. C. B. Field, V. R. Barros, D. J. Dokken, K. J. Mach, M. D. Mastrandrea, T. E. Bilir, M. Chatterjee, et al., 33–94. Cambridge University Press.

Francis, Jennifer A., and Stephen J. Vavrus. 2012. "Evidence linking Arctic amplification to extreme weather in mid-latitudes." *Geophysical Research Letters* (39). https://doi.org/10.1029/2012GL051000.

Francis, Jennifer A., and Natasa Skific. 2015. "Evidence linking rapid Arctic warming to mid-latitude weather patterns." *Philosophical Transactions of the Royal Society A* 373. https://doi.org/10.1098/rsta.2014.0170.

Hansen, James, Makiko Sato, and Reto Ruedy. 2012. "Perception of climate change." *Proceedings of the National Academy of Sciences* 109. https://doi.org/10.1073/1205276109.

Hansen, James, Makiko Sato, Paul Hearty, Reto Ruedy, et al. 2016. "Ice melt, sea level rise and superstorms: Evidence from paleoclimate data, climate modeling, and modern observations that 2°C warming could be dangerous." *Atmospheric Chemistry and Physics* 16: 3761–3812. https://doi.org10.5194/acp-16-3761-2016.

Intergovernmental Panel on Climate Change. 2013. *Climate Change 2013: The Physical Science*. Cambridge University Press.

Melillo, Jerry M., T. C. Richmond, and Gary W. Yohe, eds. 2014. *Highlights of Climate Change Impacts in the United States: The Third National Climate Assessment*. U.S. Global Change Research Program.

Mysak, Lawrence A. 2008. "Glacial inceptions: Past and future." *Atmosphere-Ocean* 46: 317–341. https://doi.org/10.3137/ao.460303.

Santer, Benjamin D., Susan Solomon, Celine Bonfils, Mark D. Zelinka, Jeffrey F. Painter, Francisco Beltran, John C. Fyfe, et al. 2014. "Observed multi-variable signals of late 20th and early 21st century volcanic activity." *Geophysical Research Letters* 42. https://doi.org/10.1002/2014GL062366.

Santer, Benjamin D., Tom M. L. Wigley, Tim P. Barnett, and Ebby Anyamba. 1995. "Detection of climate change and attribution of causes." In *Climate Change 1995: The Science of Climate Change, Intergovernmental Panel on Climate Change* (IPCC), 407–443. Cambridge University Press.

Shindell, Drew, Johan C. I. Kuylenstierna, Elisabetta Vignati, Rita van Dingenen, Markus Amann, Zbigniew Klimont, Susan C. Anenberg, et al. 2012. "Simultaneously mitigating near-term climate change and improving human health and food security." *Science* 13: 183–189. https://doi.org/10.1126/science.1210026.

Steffen, Will, Jacques Grinevald, Paul Crutzen, and John McNeill. 2011. "The Anthropocene: Conceptual and historical perspectives." *Philosophical Transactions of the Royal Society A* 369: 842–867. https://doi.org/10.1098/rsta.2010.0327.

Stocker, T. F., D. Qin, G.-K. Plattner, L. V. Alexander, S. K. Allen, N. L. Bindoff, F.-M. Bréon, et al. 2013. "Technical summary." In *Climate Change 2013: The Physical Science Basis. Contribution of Working Group I to the Fifth Assessment Report of the Intergovernmental Panel on Climate Change*, ed. T. F. Stocker, D. Qin, G.-K. Plattner, M. Tignor, S. K. Allen, J. Boschung, A. Nauels, Y. Xia, V. Bex, and P. M. Midgley, 31–115. Cambridge University Press.

UN Environment Programme (UNEP) and World Meteorological Organization (WMO). 2011. *Integrated Assessment of Black Carbon and Tropospheric Ozone*. D. Shindell, chair. Nairobi: UNEP.

USGCRP. 2017. *Climate Science Special Report: Fourth National Climate Assessment, Volume I*. Edited by D. J. Wuebbles, D. W. Fahey, K. A. Hibbard, D. J. Dokken, B. C. Stewart, and T. K. Maycock. U.S. Global Change Research Program. https://doi.org/10.7930/J0J964J6.

World Bank Group. 2013. "Turn down the heat: Climate extremes, regional impacts, and the case for resilience." Washington, DC: World Bank.

World Bank Group. 2014. "Turn down the heat: Confronting the new climate normal." Washington, DC: World Bank.

PART II

What Changes Are We Observing?

CHAPTER THREE

Range and Abundance Changes

CAMILLE PARMESAN

INTRODUCTION

Humans have been altering species' distributions and influencing local population abundances for hundreds (perhaps thousands) of years. Hunting pressures and habitat alteration have frequently caused species to disappear locally, often as a prelude to global extinction. Given the magnitude of these and other human activities, it is surprising how many species are undergoing changes in abundances and/or distributions that can be linked in part or entirely to recent anthropogenic climate change.

Detection and attribution of a climate "signal" to observed changes in natural systems has been a challenge for climate change biologists. Further, because climate change science has immediate policy relevance, it is essential that the detection and attribution processes be transparent. As this issue is rarely tackled directly in the literature, I begin with a general discussion of detection and attribution from a biological perspective. From that brief foundation, I proceed to review the impacts that recent climate change has had on species' distributions and local abundances, with an emphasis on meta-analyses that seek out common responses across many hundreds of independent studies. As well as overall patterns of response, I provide a few detailed examples demonstrating the underlying mechanistic links between observed changes in species' distributions and abundances and anthropogenic climate change. I end with an example of the challenges of, as well as potentials for, creative conservation planning for an endangered species already being affected by climate change.

DETECTION AND ATTRIBUTION REVISITED

Detection and attribution remain important issues in climate change science and impacts fields (Cramer et al. 2014). Here, when I refer to *attribution*, I use the definition from Parmesan et al. (2013, 60): "Climate Change Attribution . . . is the process of attributing some significant portion of an observed (detected) biological change to detected trends in climate. Climate here is used broadly to include not only changes in annual means, but also, for example, in patterns of climate variability. . . . Most notably, this type of attribution study relates biological changes to climatic changes regardless of the cause of the climatic changes." Attribution in the natural world is not an easy task. Whereas many wild species have shifted their ranges in concert with regional climate shifts, the leap from correlation to causation is particularly difficult to achieve. Other forces, particularly changes in land use, have clearly affected the distributions of wild species over the twentieth century. It is important, then, to consider the complicating influences of urbanization, conversion of land to agriculture, contaminants, naturally occurring pathogens, overgrazing, and invasion by exotics from other continents.

The effects of each of these factors can rarely be quantified, especially across large scales, so scientists have adopted an inferential approach using multiple lines of evidence to increase scientific rigor. Table 3.1 provides detailed examples of the use of multiple lines of evidence for climate change attribution for two extreme cases: for a single species (*Euphydryas editha*) and for a complex ecosystem (tropical coral reefs). Evidence is broadly derived from the following:

1. A large body of theory that links known regional climate changes to observed changes in the system

2. Known fundamental mechanistic links between thermal and precipi-

tation tolerances, climate variability or climate extremes and the study system

3. Direct observations of the effects on the species or system of specific weather and climate effects from long-term data sets

GLOBALLY COHERENT PATTERNS OF SPECIES DISTRIBUTION SHIFTS

The difficulties of working with data from natural systems have not impeded the emergence of a climate change "signal" in global biological changes. In long-term observational records, complexity of response is found frequently at the population level, but regional- or continental-scale studies generally have found simpler patterns of expected poleward and upward range shifts for those species that have shown changes in distribution.

Thus, conclusions by the scientific community that anthropogenic climate change has affected natural systems derive their robustness from being conducted at global scales. The effects of many factors that are confounded with climate change operate at local or sometimes regional scales but are not consistent over broader scales. At continental to global scales, confounding factors tend to add to noise (i.e., make it more difficult to detect overall climate change responses) but would not be expected to either mimic or bias expected climate change responses. For example, many species have ranges that span much of North America. The destruction of habitat by urbanization might cause localized loss of many species in and near cities, creating a patchwork of "holes" in a species' range. In contrast, increased temperatures would be expected to cause a general poleward shift in that range (or an elevational shift uphill). Thus, observations of poleward and uphill range shifts over continental scales are indicative of response to general warming.

Table 3.1. Attribution and lines of evidence

Lines of evidence	Tropical coral reefs	Euphydryas editha butterfly
Paleo data: document associations between historical climate change and ecological responses	Over the past 490 my, coral reef die-off coincided with increases in CO_2, methane, and/or warm temperatures	NA: Vernon (2008)
Experiments: document a significant role of climate in species' biology	Laboratory experiments show corals bleach under stresses such as warm temperatures, extreme salinities, and high rates of sedimentation: Lesser (1997), Jones et al. (1998), Glynn and D'Croz (1990), Anthony et al. (2007)	Laboratory, greenhouse, and field experiments of temperature manipulations show small increases in temperature increase phenological asynchrony between the butterfly and its host plant, driving increased pre-diapause larval mortality and affecting extinction and colonization dynamics: Weiss et al. (1988), Hellmann (2002), Boughton (1999)
Long-term observations: significant and consistent associations between a climate variable and a species' response	Coral bleaching events consistently follow warm sea surface temperature events (e.g., El Niño): Hoegh-Guldberg (1999)	>50 years of regular (often yearly) censuses across the species' range document multiple population declines and extinctions following drought and increased variability in precipitation: Singer (1971), Singer and Ehrlich (1979), Ehrlich et al. (1980), McLaughlin et al. (2002a), McLaughlin et al. (2002b), Singer and Parmesan (2010). Also early snowmelt (false springs) and unseasonal frost: Thomas et al. (1996)
Fingerprints: responses that uniquely implicate climate change as causal factor	First observations of mass tropical coral bleaching in 1979, concurrent with accelerating SST warming: Hoegh-Guldberg (1999) .	Highest levels of population extinctions along the southern range boundary and lowest levels of population extinctions along northern and high-elevation range boundaries, uniquely consistent with regional warming and not with local or regional habitat degradation and destruction, urbanization, agricultural expansion, or plant invasions: Parmesan (1996), Parmesan (2003), Parmesan (2005). There is a significant downward step in both proportion of population extinctions and snowpack trends above 2,400 m in the Sierra Nevada mountain range: Parmesan (1996), Johnson et al. (1999)

(continued)

Table 3.1.   (continued)

Lines of evidence	Tropical coral reefs	*Euphydryas editha* butterfly
Change in climate variable at relevant scale has been linked to GHG forcing	Ocean warming has been linked to GHG forcing with some GHG projections indicating the Pacific will move toward a more El Niño–like state: Hansen et al. (2006), Latif and Keenlyside (2009), Fedorov and Philander (2000)	Warming and lifting snow lines across western North America have been linked to GHG forcing: Karl et al. (1996), Kapnick and Hall (2012)
Global coherence of responses across taxa and regions	16% of tropical coral reefs lost globally in 1997–1998 El Niño event: Hoegh-Guldberg (1999), Wilkinson (2000)	NA

Note: The stronger the evidence within each line, and the greater the number of different lines of evidence that support a climate change interpretation, the greater is the confidence in attributing an observed change in a particular population, species, or system to observed climate change. Effects of climate change on coral reefs and on a butterfly provide examples of this approach to attribution. (Modified from Parmesan et al. 2013.)

The ever-increasing number of individual studies of climate change impacts in natural systems has allowed the emergence of new studies that merge data from many different publications into "meta-analyses" able to detect common trends when all species are brought together into a single analysis. There are now five major global meta-analyses assessing changes in distributions and abundances of individual species in recent decades (Table 3.2). These analyses showed that about half of species (for which long-term data exist) exhibited significant changes in their distributions over the past 20 to 140 years. These changes are not random but are systematically in the direction expected from regional changes in the climate. One of the major additions to the literature since Parmesan (2005a) are documentations of response to climate change in marine species at a global scale (Poloczanska et al. 2013).

The most remarkable outcome is the consistency of response across diverse species and regions. Responses have now been documented in every taxonomic group for which long-term data exist (e.g., birds, butterflies, trees, mountain flowers, fish, sea

urchins, limpets) and across diverse ecosystems (from temperate terrestrial grasslands to tropical cloud forest, and from coastal estuaries to open ocean). It appears that (unexpectedly) strong responses to current warming trends are swamping other, potentially counteractive forces of global change. Thus, the more visible aspects of human impacts on species' distributions, such as habitat loss, are not masking the impacts of climate change.

Three meta-analyses estimated rate of range boundary changes (Table 3.3). Terrestrial meta-analyses differed somewhat, with Parmesan and Yohe (2003) estimating a 6.1 km/decade shift of poleward and upward range boundaries, and Chen et al. (2011) estimating a 19.2 km/decade shift of poleward range boundaries. The latter number was recalculated by Parmesan from the supplemental data file, taking only data from the poleward range boundary and excluding data on shifts in centroids to make it comparable to Parmesan and Yohe's analysis. But these studies are also fundamentally different in that Chen et al. used the mean shift of all species for a given region as their individual data points (a taxa region mean), whereas

Table 3.2.   Global fingerprints of climate change impacts across wild species

Study	N: species and functional groups	Species in given system (%)	Species or studies focused on distribution or abundance data: % (n) of total in study	For those with long-term distribution or abundance data, % (n) of species showing significant change	Changes consistent with local or regional climate change, % of all species that showed change regardless of type of change
Parmesan and Yohe 2003	1,598	T: 85.2% M: 13.5% F: 1.3%	58% (n = 920)	50% (n = 460/920)	84%[a]
Root et al. 2003	1,468	T: 94% M: 5.4% F: 0.6%	58% (n = 926)	52% (n = 483/926)	82.3%[a]
Rosenzweig et al. 2008	55 studies (~100–200 species)	T: 65%[b] M: 13%[b] F: 22%[b]	33% (n = 18 studies)	—	90%[c]
Poloczanska et al. 2013	857 sp = 1,735 sp × trait combinations[d]	T: 0% M: 100% F: 0%	80% (n = 1,060/1,323 total distributions + abundances)[d]	63% (279/446 total for distributions only)	83%[a]

Note: For species in given system, T = terrestrial, M = marine, and F = freshwater. For changes consistent with local or regional climate change, type of change includes phenological changes. For each data set, a response for an individual species or functional group was classified as (1) no response (no significant change in the measured trait over time), or (2) if a significant change was found, the response was classified as either consistent or not consistent with expectations from local or regional climate trends. Percentages are approximate and estimated for the studies as analyses across each of the studies may differ. The specific metrics of climate change analyzed for associations with biological change vary somewhat across studies, but most use changes in local or regional temperatures (e.g., mean monthly T or mean annual T), with some using precipitation metrics (e.g., total annual rainfall). For example, a consistent response would be poleward range shifts in warming areas. Probability (P) of getting the observed ratio of consistent to not consistent responses by chance was $<10^{-13}$ for Parmesan and Yohe 2003; Root et al. 2003; Root et al. 2005; and Poloczanska et al. 2013; it was <0.001 for Rosenzweig et al. 2008. Test were all binomial tests against $p = 0.5$, performed by Parmesan.

[a]$P < 0.01 \times 10^{-13}$.

[b]Individual species were analyzed by Rosenzweig et al., but data on species not provided in publication—percentages shown are based on numbers of studies.

[c]$P < 0.001$ (from binomial test against random expectation of 50%-50% chance of change in either direction—either consistent or not consistent with local or regional climate change).

[d]For a few species, data were available on multiple traits (e.g., both the leading and the trailing edges of a species' range), such that those species are represented more than once in the analysis. These are the minority, but as numbers given here are based on species × trait combinations, the totals are greater than the total number of species for this analysis.

Parmesan and Yohe used each individual species as data points. Thus the difference in estimate of rates of shift may simply stem from differences in how data were compiled. There are two more likely candidates for these differences: first, Chen et al. (2011) included data from the most recent decade, largely absent from Parmesan and Yohe (2003), and the 2000s was the hottest decade on record; hence, it might be expected that range shifts would have accelerated during this record hot time period; and second, the data from Parmesan and Yohe were distributed more globally than those of Chen et al., with the former including studies through-

out the Northern Hemisphere (most from North America and Europe, but a scattering from Asia and Africa), whereas 76 percent of the species in Chen et al. were from two quite northerly countries (the United Kingdom and Finland), geographically situated in regions that have experienced greater warming than the global average.

More notable is that estimated expansion of leading edges derived from marine systems is several times greater than either terrestrial estimate: at ~72 km/decade (Poloczanska et al. 2013). This may at first seem counterintuitive, because warming has been less than half as much in the oceans as

Table 3.3.   Rates of change in distribution from meta-analyses of marine and terrestrial systems

Study	Observation	Shift (mean ±s.e.m.)	n studies	n	Realm (% studies)	Data criteria
Poloczanska et al. 2013	Leading and trailing edges plus center	30.6 ± 5.2 km/dec	36	360 species × trait combinations	Marine 100%	Single[a] and multispecies studies, climate change inferred
Poloczanska et al. 2013	Trailing edge	15.4 ± 8.7 km/dec	11	106 species	Marine 100%	Single[a] and multispecies studies, climate change inferred
Poloczanska et al. 2013	Leading edge	72.0 ± 13.5 km/dec	27	111 species	Marine 100%	Single[b] and multispecies studies, climate change inferred
Parmesan and Yohe 2003	Leading edge	6.1 ± 2.4 km/dec	4	99 species	Terrestrial 100%	Multispecies studies only, climate change inferred
Chen et al. 2011	Leading edge	19.7 ± 3.7 km/dec[c]	3	16 region × taxon groups	Terrestrial 83%, freshwater 15%, marine 3%	Multispecies studies (>3 species) that infer climate change. Datapoints are each a mean (mean of average response of taxonomic group in a given region), so estimate of shift is a mean of means
Przeslawski et al. 2012	Leading edge	10.6 ± 5.3 km/dec[c]	12	87 species	Marine 100%	Multispecies studies only

Note: The number of studies and number of observations (taxonomic or functional groups) from studies are given, together with a breakdown of studies by realm. The criteria for data inclusion are outlined for each study. Seabirds, anadromous fish, and polar bears were counted as marine. Wading birds were counted as freshwater birds. Minimum time span of observations within studies is 19 years, unless stated otherwise. Multispecies studies include two or more species unless stated otherwise. (Modified from Poloczanska et al. 2013.)

[a]<4% of total number of observations from single-species studies.

[b]<10% of total number of observations from single-species studies.

[c]Recalculated using only leading-edge (cold limit) observations spanning >18 years, and with data after 1990.

on land. But a new measure, called velocity of climate change (VoCC; Loarie et al. 2009), seems to explain this discrepancy. Technically, VoCC is a speed measured as the ratio of the temporal gradient to the spatial gradient, given a particular time frame and emission scenario. So VoCC measures the geographic shift of temperature isotherms through time as the climate warms. A more intuitive way to understand this is that VoCC is the speed required to move across the surface of the Earth if you are trying to maintain a constant temperature under a given level of temperature change. Temperatures in the ocean do not change quickly with geographic distance (i.e., they have a shallow temperature gradient), whereas on land temperatures can change quite quickly over space, especially in mountainous regions. So, for example, if the ocean and land both warm by an equal amount, a fish must travel much farther poleward to maintain the same temperature than would a small mammal on the nearby coast (at the same latitude).

The absolute rate of temperature rise has been about three times greater on land than in the oceans (0.24°C/decade versus 0.7°C/decade, respectively; Burrows et al. 2011), but VoCC gives a different picture. Recent global analyses of VoCC estimates are similar between oceans and land, especially when comparing latitudes where both land and ocean are found (i.e., between 50°S and 80°N), with oceanic isotherms shifting at a rate of 27.5 km/decade and terrestrial isotherms shifting at 27.4 km/decade (Burrows et al. 2011). Further, although absolute temperature change and VoCC are both stronger at the highest latitudes, VoCC is unlike absolute temperature change in being strong in the tropics, particularly in tropical oceans (Burrows et al. 2011). Not surprisingly, VoCC metrics provide a better match to observed range shifts in both marine and terrestrial systems than does absolute temperature change (Burrows et al. 2011; Poloczanska et al. 2013). In regions where VoCC is strongest, indications of whole biome

shifts are emerging. For example, recent large-scale analyses indicate an increasing scrub encroachment into Arctic tundra that is consistent with regional climate change (Serreze et al. 2000; Dial et al. 2016).

DIFFERENTIATING DIAGNOSTIC PATTERNS

Important diagnostic patterns, specific to climate change impacts, helped implicate global warming as the driver of the observed changes in natural systems (defined in Parmesan and Yohe 2003). These patterns include differential responses of cold-adapted and warm-adapted species at the same location and the tracking of decadal temperature swings. For the latter, long time series are essential (>70 years). A typical pattern observed in Britain, Sweden, and Finland was northward shift of the northern range boundaries of birds and butterflies during two twentieth-century warming periods (1930–1945 and 1975–1999), and southward shifts during the intervening cooling period (1950–1970). There were no instances of the opposite pattern (i.e., boundaries shifting southward in warm decades and northward in cool decades). In total, such diagnostic "sign-switching" responses were observed in 294 species by Parmesan and Yohe (2003) and in 24 species by Poloczanska et al. (2013) spread across the globe and ranging from oceanic fish to tropical birds to European butterflies. These specific patterns of biological trends are uniquely predicted by climate trends—no other known driver would cause these specific patterns of response to emerge.

OBSERVED CHANGES IN INDIVIDUAL SPECIES AND IN COMMUNITIES

In spite of the inherent difficulties in detecting climate change impacts in natural systems, studies showing impacts on individual species are numerous (>800 at the time of Parmesan 2006) and increasing every year. The scales of study have varied from local, to regional, to continental, and from 20 years to more than 100 years. Assigning climate change as the cause of the observed biotic changes often has a deeper basis (see Table 3.1), such as a known mechanistic link between climatic variables and biology of the study species (reviewed by Parmesan et al. 2000; Easterling, Evans, et al. 2000; Easterling, Meehl, et al. 2000; Ottersen et al. 2001; Walther et al. 2002). Individual studies are too numerous to review exhaustively, and there are many examples in other chapters of this volume; therefore, I focus here on a few examples that provide insight into mechanistic understanding of the role of climate change on shaping species distributions. (For more examples of distribution and abundance changes, see case study 5 in this volume).

Geographic and taxonomic differences in rates of response have emerged from the relative wealth of phenological data, but long-term observations of the biogeographic dynamics of species are generally too sparse to detect such differences. Nevertheless, a few patterns are emerging. For example, Poloczanska et al. (2013) compared dynamics at the leading and trailing range boundaries for marine species (not always of the same species) and found leading edges to be expanding nearly five times as fast as trailing edges were contracting (see Table 3.3). This asymmetrical pattern of range boundary shifts qualitatively matches that found for European butterflies, in which nearly 67 percent of species expanded their northern range boundaries but <20 percent contracted their southern range boundaries (Parmesan et al. 1999). The reasons for these asymmetrical responses are not clear, but a likely candidate is simply that while range expansion can occur rapidly (in a single year), population extinctions tend to lag behind deterioration of the environment, especially for long-lived species. This process, termed *extinction debt*, is a well-developed concept in the ecological literature (Tilman et al. 1994).

COMPLEXITIES OF CONSERVATION PLANNING WHEN A SYSTEM IS DYNAMIC: AN EXAMPLE FROM *EUPHYDRYAS EDITHA*

E. editha as a Sensitive Indicator of Climate Change Impacts

Having been studied for more than 60 years by dozens of researchers, Edith's checkerspot butterfly presents a model example of how knowledge at many different levels can be integrated to answer complex questions concerning response to climate change. In particular, a great deal is known about the effects of weather and yearly climate variability on individual fitness and on population dynamics, and there is a long history of coupling such long-term observations with experimental manipulations in laboratories, greenhouses, and natural populations. This rich knowledge base allowed for a mechanistic link to be made between large-scale patterns of distribution change in E. *editha* and long-term climate trends (over the twentieth century) (Table 3.1; Parmesan 2005b; Parmesan et al. 2013).

Here, I provide a brief summary of the how researchers have been able to link observed population extinctions to extreme weather and climate events (reviewed in Parmesan et al. 2013). In some populations, phenological asynchrony is the mechanistic basis for climate sensitivity. Field observations and experimental manipulations, performed over many decades, have repeatedly shown that phenological mismatches between E. *editha* and annual host plants routinely cause larvae to starve when hosts senesce before the insects are ready to diapause (Singer 1972; Weiss et al. 1988; Boughton 1999; Hellmann 2002; Singer and McBride 2012). These mismatches existed prior to the onset of current warming but render such populations vulnerable to warming conditions as even a small advancement of plant senescence dramatically increases larval mortality (Singer and Parmesan 2010).

However, shifts in the relative timing of insect development and host senescence are not the only mechanisms related to climate change that have caused population extinctions of this species. Observed responses of different ecotypes of this butterfly living in different habitat types have been highly diverse (Ehrlich et al. 1980; Thomas et al. 1996), and population extinctions have occurred in response to diverse but specific climatic conditions, particularly extreme weather and extreme climate years.

In late winter at low to middle elevations (500 m–1000 m) in the southern half of this species' range (~29°N to 45°N), many populations specialize on annual hosts, seeds normally germinate in response to rainfall, and larvae emerge from their overwintering dormant phase (diapause) in response to a combination of winter chilling and rainfall. There are levels of drought at which the larvae emerge from winter diapause, but the plants do not germinate: this caused several butterfly population extinctions in the 1970s (Ehrlich et al. 1980). At higher elevations (>1,500 m), both seeds and larvae break dormancy at snowmelt, so milder winters and reduced snowpack have advanced both plant and insect activity to the point where both are vulnerable to "false springs." False springs occur when late winter is warm enough that plants and animals respond to this false cue and emerge from winter dormancy, only to be exposed to (and often killed by) later freezing temperatures as typical winter conditions prevail. A series of false springs drove a set of E. editha populations at 2,300 m to extinction in the 1980s and 1990s (Thomas et al. 1996; Boughton 1999).

Anthropogenic climate change has increased the frequencies of false springs at lower montane elevations throughout the world, where snowpack has been gradually declining and melting earlier (Mote et al. 2005; Stewart 2009; Beniston 2012). Specifically, in the Sierra Nevada, which contains the highest elevation populations of E. editha, snowpack dynamics showed a shift at 2,400 m, showing no significant change above 2,400 m but becoming thinner and melting two weeks earlier below 2,400 m (John-son et al. 1999). The regional extinction of a suite of E. editha populations occurred at 2,300 m, caused by frequent false springs consistent with this overall shift toward earlier snowmelt at this elevation.

In sum, three different mechanisms of climate change–caused population extinction were observed in a single butterfly species: shifts in insect and host phenological synchrony caused by hot years, different dormancy responses by plants and insects to drought, and direct mortality of both plants and insects caused by a series of false springs. All three of these extreme climate events either have increased, or are expected to increase, with anthropogenic climate change (Easterling, Evans et al. 2000).

But even though E. editha populations are climate sensitive, this butterfly also, ironically, exhibits traits that appear to make it resilient to changes in climate. First, there is large interpopulation variation in heritable behaviors that have a strong influence on microtemperatures of eggs and early-stage offspring (Bennett et al. 2015). Female E. editha show interpopulation variation in egg placement, exposing their offspring to diverse thermal environments, from 3°C cooler to 20°C hotter than ambient air. Second, the species' single flight season is placed at different times of year across its range, with flight in midsummer in cool climates and diapause through summer in hot climates. Further, whereas both traits are highly heritable, the species contains substantial genetic variation in these traits; therefore, modification of both phenology and egg placement are possible. This ecological and evolutionary flexibility provide options for populations of this species to mitigate future changes in regional climate.

E. editha quino Has Demonstrated Resilience to Climate Change Impacts

The Quino checkerspot, an endangered subspecies of Edith's checkerspot, is currently illustrating remarkable resilience to anthropogenic climate change, despite its

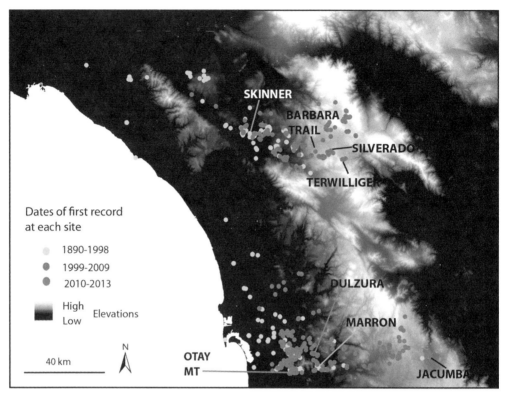

Figure 3.1. Map showing timing of first observation of Quino populations relative to elevation. Topography is depicted by higher elevations having a lighter background color. A small number of the northernmost extinct sites lie outside the map area. (From Parmesan et al. 2015.)

geographical location at the species' equatorial range limit (Parmesan et al. 2015). Quino has recently begun colonizing higher elevations than those historically inhabited by this subspecies, shifting from an average elevation of 360 m throughout most of the twentieth century to an average of 1,164 m by the 2010 decade (Figures 3.1 and 3.2). But, because Quino was historically already at the upper elevational limits of its traditional host plants (principally species of annual *Plantago*), to colonize higher elevations it had to switch its principal host to another species, *Collinsia concolor*, which inhabited the mountain regions around Los Angeles and San Diego. There is no evidence that this host switch was driven by evolutionary changes, and colonies on the novel host

were not well adapted to it (Parmesan et al. 2015). The moral drawn from this is that individuals do not need to be preadapted to novel environments, or to exhibit rapid evolution, in order to colonize beyond historical range boundaries. They merely need to survive.

Short-Term Success, but Long-Term Vulnerability

Quino highlights the difficulty of planning for future climate change impacts. In the near term, an upward range shift has already been encouraged by adding protection to areas into which the butterfly is expanding on its own. However, projected distributions indicate that, even in the medium term (next 40 years), Quino likely will lose all suitable climate space in the region of its current distribution (Figure 3.3). For long-term persistence of Quino, conservation managers will again need to consider assisted colonization. With respect

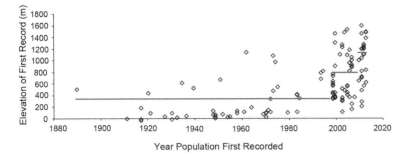

Figure 3.2. Scatter plot of elevations of populations against year that population was first discovered (first recorded). Populations were placed into one of three groups depending on whether they were first discovered in the period 1890–1997, 1998–2009, or 2010–2013. The mean elevation for each grouping is indicated with a bar. Elevation and year of first discovery are highly significantly correlated ($r = 0.60$, $df = 116$, $p < 0.0001$). (From Parmesan et al. 2015.)

to shifting geographic boundaries of species as they attempt to track climate change, the type of stepping-stone approach we advocated for Quino is likely to aid other endangered species as well. True adaptive management, with regular and frequent reassessments of conservation needs and priorities, will become increasingly necessary in a time of rapid climate change. Attention to real-time dynamics, then, can be used to reduce current uncertainty in future projections, both in climate and in species' responses, by informing managers as to which future (projected) pathway appears to be emerging for their species.

SUMMARY

Global coherency in patterns of biological change, a substantial literature linking climate variables with ecological and physiological processes, combined with telltale diagnostic fingerprints allow a causal link to be made between twentieth-century climate change and biological impacts. The patterns of change alone are evidence of a climate change signal (reviewed in Parmesan 2006; Pecl 2017). Poleward range

shifts have been observed in many species, with few shifts toward the equator. Species compositions within communities have generally altered in concert with local temperature rise. At single sites, species from lower latitudes and elevations have tended to increase in abundance as those residing predominately at higher latitudes and elevations suffer local declines. Consistency across studies indicates that local and regional trends reflect global trends.

These types of distributional changes, first documented in the 1990s and continuing to be documented at ever increasing rates, differ idiosyncratically in the exact patterns of colonizations and extinctions across regions. Further, not all species respond to climate change by shifting their distributions. This individualism in species' responses to climate change, resulting in asynchronous range shifts, has led to early signs of altered species interactions as species shift geographically (e.g., the Quino butterfly as a novel herbivore to *Collinsia concolor*). Individual species' range shifts have begun to manifest as shifts in major biomes, particularly in areas of the greatest magnitude of climate change. Traditional conservation models are rapidly being transformed by an accelerating documentation of changes in wild systems that have been linked to climatic trends. Although many effects have been relatively minor, and even benign, the most sensitive species and systems already are being harmed severely (Parmesan 2006). It is clear that the continuation of observed trends is likely to have profound impacts on wild species

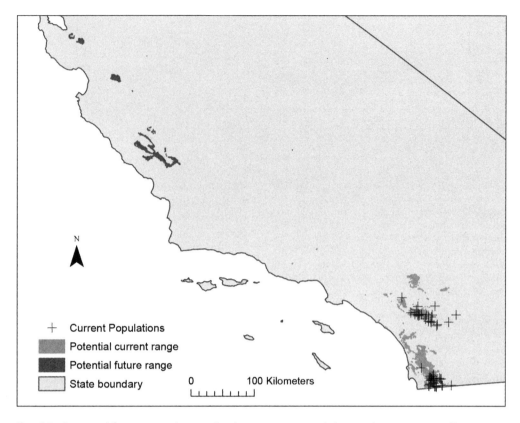

Figure 3.3. Current and future projected species distributions for Quino based on climatic niche models. Known historical records are shown with the plus symbol (+). Current climate projections (light gray) and future climate projections (dark gray) represent an ensemble of all model outputs, with "presence" indicated by light or dark gray coloring if a grid cell was estimated as present in at least 50% of SDM models (i.e., anywhere from 50% to 100% of model agreement). (From Parmesan et al. 2015.)

and consequently on the ways in which the conservation community strives to preserve biodiversity.

REFERENCES

Anthony, K. R. N., S. R. Connolly, and O. Hoegh-Guldberg. 2007. "Bleaching, energetics, and coral mortality risk: Effects of temperature, light, and sediment regime." *Limnology & Oceanography* 52: 716–726.

Beniston, Martin. 2012. "Is snow in the Alps receding or disappearing?" *Wiley Interdisciplinary Reviews—Climate Change* 3 (4): 349–358.

Bennett, Nichole L., Paul M. Severns, Camille Parmesan, and Michael C. Singer. 2015. "Geographic mosaics of phenology, host preference, adult size and microhabitat choice predict butterfly resilience to climate warming." *Oikos* 124 (1). https://doi.org/10.1111/oik.01490.

Boughton, David A. 1999. "Empirical evidence for complex source-sink dynamics with alternative states in a butterfly metapopulation." *Ecology* 80: 2727–2739.

Burrows, Michael T., David S. Schoeman, Lauren B. Buckley, Pippa Moore, Elvira S. Poloczanska, Keith M. Brander, Chris Brown, et al. 2011. "The pace of shifting climate in marine and terrestrial ecosystems." *Science* 334: 652–655.

Chen, I-Ching, Jane K. Hill, Ralf Ohlemüller, David B. Roy, and Chris D. Thomas. 2011. "Rapid range shifts of species associated with high levels of climate warming." *Science* 33: 1024–1026.

Cramer, Wolfgang, Gray W. Yohe, Maximillian Auffhammer, Christian Huggel, Ulf Molau, M. Assunção Faus da Silva Dias, Andrew Solow, Dáithí A. Stone, Lourdes Tibig, Laurens Bouwer, et al. 2014. "Detection and attribution of observed impacts." In *Climate Change 2014: Impacts, Adaptation and Vulnerability. Contribution of Working Group II to the Fifth Assessment Report of the Intergovernmental Panel on Climate Change*, ed. Christopher B. Field, et al., 979–1037. Cambridge University Press.

Dial, Roman J., T. Scott Smeltz, Patrick F. Sullivan, Christina L. Rinas, Katriina Timm, Jason E. Geck, S. Carl Tobin, Trevor S. Golden, and Edward C. Berg. 2016. "Shrubline but not treeline advance matches climate velocity in montane ecosystems of south-central Alaska." *Global Change Biology* 22: 1841–1856. https://doi.org/10.1111/gcb.13207.

Easterling, David R., Jenni L. Evans, P. Ya Groisman, Thomas R. Karl, Kenneth E. Kunkel, and P. Ambenje. 2000. "Observed variability and trends in extreme climate events: A brief review." *Bulletin of the American Meteorological Society* 80: 417–425.

Easterling, David R., Gerald A. Meehl, Camille Parmesan, Stanley A. Chagnon, Thomas R. Karl, and Linda O. Mearns. 2000. "Climate extremes: Observations, modeling, and impacts." *Science* 289: 2068–2074.

Ehrlich, Paul R., Dennis D. Murphy, Michael C. Singer, Caroline B. Sherwood, Ray R. White, and Irene L. Brown. 1980. "Extinction, reduction, stability and increase: The responses of checkerspot butterfly populations to the California drought." *Oecologia* 46: 101–105.

Fedorov, A. V., and S. G. Philander. 2000. "Is El Niño changing?" *Science* 288: 1997–2002.

Glynn, P. W., and L. D'Croz. 1990. "Experimental evidence for high temperature stress as the cause of El Niño-coincident coral mortality." *Coral Reefs* 8: 181–191.

Hansen J., et al. 2006. "Global temperature change." *Philosophical Transactions of the Royal Society of London B* 103 (39): 14288–14293.

Hellmann, Jessica J. 2002. "The effect of an environmental change on mobile butterfly larvae and the nutritional quality of their hosts." *Journal of Animal Ecology* 71: 925–936.

Hoegh-Guldberg, O. 1999. "Climate change, coral bleaching and the future of the world's coral reefs." *Marine Freshwater Research* 50: 839–866.

Jones, R. J., O. Hoegh-Guldberg, A. W. D. Larkum, and U. Schreiber. 1998. "Temperature-induced bleaching of corals begins with impairment of the CO_2 fixation mechanism in zooxanthellae." *Plant Cell & Environment* 21: 1219–1230.

Lesser, M. P. 1997. "Oxidative stress causes coral bleaching during exposure to elevated temperatures." *Coral Reefs* 16: 187–192.

Johnson, Tammy, Jeff Dozier, and Joel Michaelsen. 1999. "Climate change and Sierra Nevada snowpack." In *Interactions between the Cryosphere, Climate and Greenhouse Gases,* ed. Martyn Tranter, Richard Armstrong, Eric Brun, Gerry Jones, Martin Sharp, and Martin Williams, 256: 63–70. International Association of Hydrological Sciences.

Kapnick, S., and A. Hall. 2012. "Causes of recent changes in western North American snowpack." *Climate Dynamics* 38: 1885–1889.

Karl, T. R., R. W. Knight, D. R. Easterling, and R. G. Quayle. 1996. "Indices of climate change for the United States." *Bulletin of the American Meteorological Society* 77: 279–292.

Latif, M., and N. S. Keenlyside. 2009. "El Niño/Southern Oscillation response to global warming." *Philosophical Transactions of the Royal Society of London B* 106 (49): 20578–20583.

Loarie, Scott R., Phillip B. Duffy, Healy Hamilton, Gregory P. Asner, Christopher B. Field, and David D. Ackerly. 2009. "The velocity of climate change." *Nature* 462: 1052–1057.

McLaughlin, J. F., J. J. Hellmann, C. L. Boggs, and P. R. Ehrlich. 2002a. "Climate change hastens population extinctions." *Proceedings of the National Academy of Sciences* 99: 6070–6074.

McLaughlin, J. F., J. J. Hellmann, C. L. Boggs, and P. R. Ehrlich. 2002b. "The route to extinction: Population dynamics of a threatened butterfly." *Oecologia* 132: 538–548.

Mote, Phillip W., Alan F. Hamlet, Martyn P. Clark, and Dennis P. Lettenmaier. 2005. "Declining mountain snowpack in western North America." *Bulletin of the American Meteorological Society* 86 (1): 39–49.

Ottersen, Geir, Benjamin Planque, Andrea Belgrano, Eric Post, Phillip C. Reid, and Nils C. Stenseth. 2001. "Ecological effects of the North Atlantic oscillation." *Oecologia* 128 (1): 1–14.

Parmesan, Camille. 1996. "Climate and species' range." *Nature* 382: 765–766.

Parmesan, Camille. 2003. "Butterflies as bio-indicators of climate change impacts." In *Evolution and Ecology Taking Flight: Butterflies as Model Systems,* edited by C. L. Boggs, W. B. Watt, and P. R. Ehrlich, 541–560. Chicago: University of Chicago Press.

Parmesan, Camille. 2005a. "Biotic response: Range and abundance changes." In *Climate Change and Biodiversity,* ed. T. Lovejoy and L. Hannah, 41–55. Yale University Press.

Parmesan, Camille. 2005b. "Case study: Euphydryas editha." In *Climate Change and Biodiversity,* ed. T. Lovejoy and L. Hannah, 56–60. Yale University Press.

Parmesan, Camille. 2006. "Ecological and evolutionary responses to recent climate change." *Annual Reviews of Ecology, Evolution and Systematics* 37: 637–669.

Parmesan, Camille, Michael T. Burrows, Carlos M. Duarte, Elvira S. Poloczanska, Anthony J. Richardson, David S. Schoeman, and Michael C. Singer. 2013. "Beyond climate change attribution in ecology and conservation research." *Ecology Letters* 16 (S1): 58–71. https://doi.org/10.1111/ele.12098.

Parmesan, Camille, Terry L. Root, and Michael R. Willig. 2000. "Impacts of extreme weather and climate on terrestrial biota." *Bulletin of the American Meteorological Society* 81: 443–450.

Parmesan, Camille, Nils Ryrholm, Constantí Stefanescu, Jane K. Hill, Chris D. Thomas, Henri Descimon, Brian Huntley, Lauri Kaila, Jaakko Kullberg, Toomas Tammaru, W. John Tennent, Jeremey A. Thomas, and

Martin Warren. 1999. "Poleward shifts in geographical ranges of butterfly species associated with regional warming." *Nature* 399: 579–583.

Parmesan, Camille, Alison Williams-Anderson, Matthew Mikheyev, Alexander S. Moskwik, and Michael C. Singer. 2015. "Endangered Quino checkerspot butterfly and climate change: Short-term success but long-term vulnerability?" *Journal of Insect Conservation* 19 (2): 185–204.

Parmesan, Camille, and Gray W. Yohe. 2003. "A globally coherent fingerprint of climate change impacts across natural systems." *Nature* 421: 37–42.

Pecl, Gretta T., Miguel B. Araújo, Johann D. Bell, Julia Blanchard, Timothy C. Bonebrake, I-Ching Chen, Timothy D. Clark, Robert K. Colwell, Finn Danielsen, Birgitta Evengård, et al. 2017. "Biodiversity redistribution under climate change: Impacts on ecosystems and human well-being." *Science* 355 (6332): art. eaai9214. https://doi.org/10.1126/science.aai9214.

Poloczanska, Elvira S., Christopher J. Brown, William J. Sydeman, Wolfgang Kiessling, David S. Schoeman, Pippa J. Moore, Keith Brander, et al. 2013. "Global imprint of climate change on marine life." *Nature Climate Change* 3 (10): 919–925.

Przeslawski, Rachel, Inke Falkner, Michael B. Ashcroft, and Pat Hutchings. 2012. "Using rigorous selection criteria to investigate marine range shifts." *Estuarine, Coastal and Shelf Science* 113: 205–212.

Root, Terry L., Jeff T. Price, Kimberly R. Hall, Stephen H. Schneider, Cynthia Rosenzweig, and J. Allan Pounds. 2003. "Fingerprints of global warming on wild animals and plants." *Nature* 421: 57–60.

Serreze, M. C., J. E. Walsh, F. Stuart Chapin III, T. Osterkamp, M. Dyurgerov, V. Romanovsky, W. C. Oechel, J. Morison, T. Zhang, and R. G. Barry. 2000. "Observational evidence of recent change in the northern high-latitude environment." *Climatic Change* 46: 159–207.

Singer, Michael C. 1972. "Complex components of habitat suitability within a butterfly colony." *Science* 176: 75–77.

Singer, M. C., and P. R. Ehrlich. 1979. "Population dynamics of the checkerspot butterfly *Euphydryas editha*." *Fortschritte der Zoologie* 25: 53–60.

Singer, Michael C., and Carolyn S. McBride. 2012. "Geographic mosaics of species' association: A definition and an example driven by plant–insect phenological synchrony." *Ecology* 93 (12): 2658–2673.

Singer, Michael C., and Camille Parmesan. 2010. "Phenological asynchrony between herbivorous insects and their hosts: A naturally-evolved starting point for climate change impacts?" *Philosophical Transactions of the Royal Society of London* 365 (155): 3161–3176.

Stewart, Iris T. 2009. "Changes in snowpack and snowmelt runoff for key mountain regions." *Hydrological Processes* 23 (1): 78–94.

Thomas, Chris D., Michael C. Singer, and David A. Boughton. 1996. "Catastrophic extinction of population sources in a butterfly metapopulation." *American Naturalist* 148: 957–975.

Tilman, David, Robert M. May, Clarence L. Lehman, and Martin A. Nowak. 1994. "Habitat destruction and the extinction debt." *Nature* 371: 65–66. https://doi.org/10.1038/371065a0.

Vernon, J. E. N. 2008. "Mass extinctions and ocean acidification: Biological constraints on geological dilemmas." *Coral Reefs* 27: 459.

Walther, Gian-Reto, Eric Post, Peter Convey, Annette Menzel, Camille Parmesan, Trevor J. C. Beebee, Jean-Marc Fromentin, Ove Hoegh-Guldberg, and Franz Bairlein. 2002. "Ecological responses to recent climate change." *Nature* 416 (6879): 389–395.

Weiss, Stuart B., Dennis D. Murphy, and Raymond R. White. 1988. "Sun, slope and butterflies: Topographic determinants of habitat quality for *Euphydryas editha*." *Ecology* 69 (5): 1486–1496.

Wilkinson, C. R., ed. 2000. *Global Coral Reef Monitoring Network: Status of Coral Reefs of the World in 2000*. Townsville: Australian Institute of Marine Sciences.

The Bering Sea and Climate Change

Lee W. Cooper

The Bering and Chukchi Seas encompass an extensive continental shelf to the north and east of the deep Bering Sea, which is oceanographically an extension of the North Pacific. Sea-ice cover extends typically south to the boundary between the shallow shelf and the deeper basin in the Bering Sea from November to June. The Chukchi Sea has much more extensive ice cover, which melts back to a minimum in September. In recent years, the retreat of seasonal sea ice in late summer has often receded so that there is ice only over the deep Arctic Ocean (Figure CS1.1).

In the Bering Sea, the shallow shelf can be divided into two ecosystems. To the south, a fish-dominated pelagic system persists that is a globally important fishery, including Alaskan pollock (*Gadus chalcogrammus*); snow, tanner, and king crab (*Chionoecetes opilio, Chionoecetes tanneri,* and *Paralithodes camtschaticus*); as well as Pacific salmon (five species in the genus *Oncorhynchus*). To the north and east, cold-water temperatures (~0°C) occur year-round, influenced by winter sea-ice formation, and these temperatures inhibit reproduction of most fish species found farther south. A key ecological organizing principle for the northern Bering Sea and the adjoining Chukchi Sea north of Bering Strait is that the shallow productive waters lead to strong pelagic-benthic coupling to the seafloor and deposition of fresh chlorophyll coinciding with the spring bloom. This northern ecosystem is dominated by marine macroinvertebrates (e.g., clams, polychaetes, sipunculids, crustaceans) that feed on the high production deposited rapidly to the seafloor, which in turn serve as food resources for diving birds and mammals such as gray whales (*Eschrichtius robustus*), bearded seals (*Erignathus barbatus*), eiders (genus *Somateria*), and walruses (*Odobenus rosmarus*). North of the Bering Strait in the Chukchi Sea, and between St. Lawrence Island and Bering Strait, the persistence of seasonal sea ice has significantly declined over the past three decades, although be-

tween St. Matthew Island and St. Lawrence Island no significant decreases or, in some cases, modest increases in sea-ice persistence have been observed. It is thought that where sea ice has diminished, such as in the northern Bering Sea and the Chukchi Sea, major ecological shifts to a more pelagic ecosystem can be expected (Grebmeier et al. 2006), as illustrated by Figure CS1.2. For example, walrus feed on abundant clam populations in the Chukchi Sea shelf in the summer and use remnant sea ice as a resting offshore platform, but sea ice is now often persisting only over deeper waters of the Arctic Ocean in late summer, beyond the diving capacity of walruses and at depths where no significant food is present on the seafloor. Once ice has retreated so that it is only over deep water, walrus must come ashore in Alaska and Russia and swim much farther distances to offshore feeding grounds (Figure CS1.2).

Changes to the northern Bering Sea and Chukchi Sea benthos has been well studied over the past several decades (Grebmeier et al. 2006), and it is clear that dynamic biomass changes are occurring and that decreases in biomass could be having an impact on marine mammal and seabird populations. In addition to changes in sea temperature and the spatial distribution of

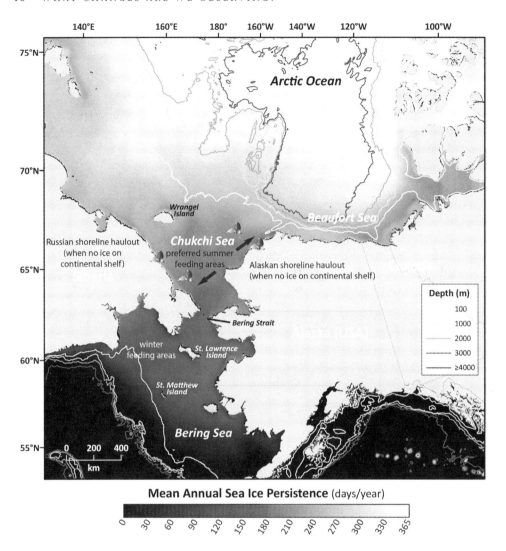

Mean Annual Sea Ice Persistence (days/year)

Figure CS1.1. Persistence of sea ice (in days per year) for the Bering and Chukchi Seas, 2003–2010. Presence of sea ice is defined as more than a 15% sea-ice concentration threshold in each pixel analyzed, using the Advanced Microwave Scanning Radiometer (AMSR-E) satellite time series. (Analysis and figure courtesy of Karen Frey, Graduate School of Geography, Clark University. Prepared by Lee Cooper.)

the cold-water pool that inhibits northward migration of many fish, already-seasonally-persistent low pH in parts of the northern Bering Sea are likely to be further affected by ocean acidification due to cold temperatures and freshwater melt. This may be influencing the replacement of clam popula-

tions southwest of St. Lawrence Island with polychaetes that may be less vulnerable to seasonal acidification. The area south of St. Lawrence Island is known to be important foraging areas for spectacled eiders (*Somateria fischeri*) and walruses, which heavily depend on bivalves for food.

The possibility that commercial fisheries in the southeastern Bering Sea, which are the largest by biomass in the United States, could eventually move northward onto the shallow northern shelf now dominated by soft, bottom benthic animals is understood by fisheries managers. For example, the North Pacific Fisheries Management Council,

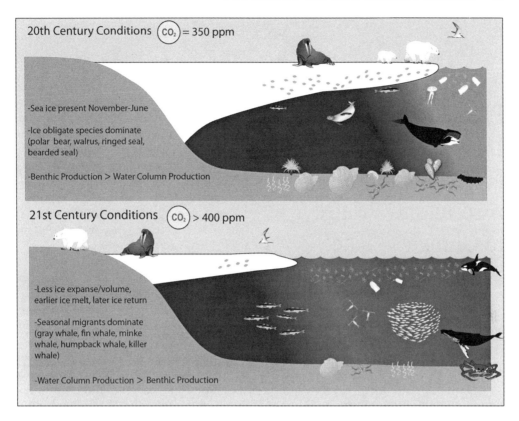

Figure CS1.2. It is thought that where sea ice has diminished, as in the northern Bering Sea and Chukchi Sea, major ecological shifts can be expected (Grebmeier et al. 2006), as shown in the twin diagrams. Less extensive sea ice will lead to lower deposition of sea ice and spring-bloom algae (*bottom*), and a lower biomass benthic community. Fish and zooplankton will increase in importance, and pelagic feeding whales such as orcas and humpbacks will become more important than sea-ice-associated bowheads and benthic feeding pinnipeds. (Prepared by Lee Cooper.)

which regulates fisheries in the Bering Sea, has restricted potentially destructive trawling of the bottom and other activities north and south of St. Lawrence Island in recognition of the dependence of predators such as walruses and spectacled eiders on healthy populations of organisms on the seafloor. The two Siberian Yupik communities on St. Lawrence Island, Savoonga and Gambell, are in turn dependent on subsistence harvests of walruses and bearded seals, in addition to other marine mammals. These subsistence-based communities look warily toward the

possibility of fisheries expansion as the climate warms, even though limited scientific trawling and sampling of the bottom by US and Japanese scientists continues to monitor for possible shifts in commercially harvestable fish and shellfish populations.

REFERENCE

Grebmeier, Jacqueline M., Lee W. Cooper, Howard M. Feder, and Boris I. Sirenko. 2006. "Ecosystem dynamics of the Pacific-influenced Northern Bering and Chukchi Seas in the Amerasian Arctic." *Progress in Oceanography* 71 (2–4): 331–361. http://doi.org/10.1016/j.pocean.2006.10.001.

CHAPTER FOUR

Phenological Dynamics in Pollinator–Plant Associations Related to Climate Change

ERIC POST AND MICHAEL AVERY

INTRODUCTION

Organisms that inhabit seasonal environments—and some that inhabit aseasonal environments—typically display life-history traits characterized by the importance of the timing of their expression. Examples include the annual timing of migration, nesting, and egg laying in birds; emergence, calling, and egg laying in frogs and toads; emergence, flowering, and fruit or seed set in plants; emergence and ground or aerial activity in insects; and migration and offspring production in some mammals.

The timing of expression of such life-history events in any given season or annual period of production is an inherently individual-level phenomenon governed by the interaction between abiotic and biotic constraints and drivers acting as selective forces on the evolution of life-history traits (Post 2013). On an individual basis, seasonal changes in biological activity in individual organisms may be cued by changes in diel light-to-dark ratios, daily maximum or minimum temperatures, or resource availability. In a simplistic sense, then, the timing of life-history events concerned with, for example, offspring production by individuals in seasonal environments can be viewed as balancing the costs of failed reproduction due to adverse environmental conditions with the benefits of reproducing earlier in the season than conspecifics with which competition for resources will increase later in the season (Iwasa and Levin 1995; Post et al. 2001). Despite the obvious fitness relevance of this perspective to the individual, seasonal and interannual variations in the timing of life-history events are commonly studied or observed at the population level and comprise the discipline of phenology (Schwartz 2003; Post 2013). Hence, phenology may be regarded as the seasonal timing of expression of life-history traits concerned with growth, reproduction, and survival.

RECENT TRENDS IN PHENOLOGY RELATED TO CLIMATE CHANGE

The most intuitive biological indicators of climate change are trends in phenology. Examples of earlier (i.e., advancing) timing of spring events are abundant, but those most commonly reported concern the timing of plant phenology, including green-up and flowering. Estimates of rates of plant phenological advance through time, denoted as *trends*, and rates of plant phenological advance in response to temperature increase or changes in other abiotic factors such as timing of snowmelt, denoted as *climatic sensitivity*, vary greatly over scales of observation and levels of biological organization (Parmesan 2007). Observations deriving from satellite imagery covering vast spatial scales and combining multiple species can produce widely differing rates of phenological advance, even when obtained for a single region. For instance, MODIS satellite observations over the Svalbard archipelago in High Arctic Norway revealed no trend in the onset of the plant-growing season between 2000 and 2013 (Karlsen et al. 2014). During the same period, MODIS satellite data covering four subzones of the Yamal Peninsula in Arctic Russia revealed variation in trends in the start of the plant-growing season among these subzones from an advance of −5.3 days per decade to a delay of 18.9 days per decade (Zeng, Jia, and Forbes 2013). Over the entire Northern Hemisphere, the timing of onset of the annual plant-growing season as detected by satellite observations advanced by −5.2 days per decade between the years 1982 and 1999, but this rate slowed to −0.2 days per decade between 2000 and 2008 (Jeong et al. 2011).

In contrast to regional-scale satellite observations that are blind to individual species, studies conducted at local sites have in some instances documented substantially greater trends in plant phenology. Among 24 species of perennial plants occurring at a study site in Mediterranean Spain, the timing of leaf unfolding advanced between 1943 and 2003 at a mean rate of −4.8 days per decade (Gordo and Sanz 2009). However, rates of advance in 20 percent of these species were as high as −7 to −7.5 days per decade (Gordo and Sanz 2009). Among 232 plant species monitored over a 27-year period on the island of Guernsey in the English Channel, annual first-flowering date advanced on average by −5.2 days per decade, but this rate was as high as −13.7 days per decade for the subset of species comprising evergreens (Bock et al. 2014). Similarly, climatic sensitivity in this study averaged −9.7 days of advance in first flowering per degree of warming among all species but was greatest for evergreen species, at −10.9 days of advance per degree of warming (Bock et al. 2014).

Trends in first-flowering dates since the early 1930s for 24 species at a site near Mohonk Lake, New York, in the northeastern United States, varied from as low as −0.05 days per decade to −1.5 days per decade, with some species, such as painted trillium (*Trillium undulatum*), displaying significant delays in the onset of flowering by up to 1.6 days per decade (Cook et al. 2008). In contrast to these modest rates of advance, or in some instances delays, record-warm spring temperatures in Massachusetts, also in northeastern United States, and in Wisconsin, in the northern Midwest United States, were accompanied by record-early flowering in some species of plants at long-term phenological monitoring sites in both locations (Ellwood et al. 2013). In Massachusetts, flowering occurred 21 days earlier during the record-warm spring of 2010 than during the mid-nineteenth century, whereas in Wisconsin, flowering occurred 24 days earlier in the record-warm spring of 2012 than during the mid 1930s (Ell-wood et al. 2013).

The greatest community-averaged rates of phenological advance yet reported derive from multiannual studies in the Arctic. From 1996 to 2005, the average timing of flowering at the community level at Zackenberg in High Arctic Greenland advanced by −13.7 days per decade (Høye et al. 2007). More recently, the estimated rate of advance of flowering at this site was reduced to −8.7 days per decade with the inclusion of data through 2011 (Iler, Høye, et al. 2013). At a Low Arctic site near Kangerlussuaq, Greenland, the onset of the plant-growing season at the community level advanced between 2002 and 2013 at a rate of −24 days per decade (Post et al. 2016). At both sites, species displayed highly individualistic rates of phenological advance. For instance, at Zackenberg, rates of advance in the timing of flowering varied from insignificant in *Saxifraga oppositifolia* to approximately −30 days per decade in *Silene acaulis* (Høye et al. 2007). In Kangerlussuaq, rates of advance in the timing of onset of spring growth varied from insignificant in *Salix glauca* to approximately −29 days per decade in *Pyrola grandiflora* (Post et al. 2016). In such instances, when species undergo highly idiosyncratic rates of phenological advance, the phenological community may become reorganized, with species that are more responsive to the alleviation of abiotic constraints by climate change assuming earlier positions in the sequence of emergence or flowering (Post 2013; Post et al. 2016). Such shifts in plant phenology related to climate change that alter the timing, sequence, and overlap of co-occurring plant species may have consequences for reproduction and survival of consumers dependent on the predictability in time and space of plant resource availability. Next, we provide a brief overview of consequences of differential phenological shifts at adjacent trophic levels for consumer-resource interactions before proceeding with a review of such interactions in pollinator-plant systems.

PHENOLOGICAL DYNAMICS IN CONSUMER-RESOURCE INTERACTIONS

When species interact in a consumer-resource relationship, the seasonal timing of activity or seasonal timing of the local presence of the interacting species may be of critical importance to the survival and reproductive success of both consumer and resource. If the seasonal timing of life-history events in consumer and resource species responds differently to abiotic cues that covary with climate change, such as local seasonal or mean annual temperature, then the phenological dynamics of consumer and resource may become uncoupled. In such instances, phenology is studied as trophic match-mismatch (Stenseth and Mysterud 2002; Miller-Rushing et al. 2010).

Empirical investigations of trophic match-mismatch in terrestrial systems have typically focused on situations in which the seasonal timing of reproduction by consumers is cued by factors that differ from those that cue the seasonal timing of activity in resource species (Visser and Both 2005). Such studies have revealed that trophic mismatch associated with recent climate change tends to arise when resource phenology advances while consumer phenology does not, or when it advances at a greater rate than that of consumer phenology (Both, Bijlsma, and Visser 2005; Both et al. 2009; Plard et al. 2014; Post and Forchhammer 2008; Kerby, Wilmers, and Post 2013; Miller-Rushing et al. 2010). Migratory birds such as the pied flycatcher (*Ficedula hypoleuca*) that breed in Northern Europe, for instance, have not advanced their breeding dates at the same rate at which the emergence phenology of their caterpillar prey have advanced (Both et al. 2006; Both, Bijlsma, and Visser 2005), a mismatch that has resulted in declines in fledging success and population abundance in the affected populations (Both et al. 2006). Similarly, Western roe deer (*Capreolus capreolus*) in a population in France have not shifted to earlier annual offspring production while the phenology of their forage plants has advanced (Plard et al. 2014). This has contributed to a decline in reproductive success and individual fitness over the longer term in the focal roe deer population (Plard et al. 2014).

In cases such as these, there is potential for declines in populations of consumer species driven by their inability to adjust their reproductive phenology apace with rapid advances in the phenology of their resource species in response to climate change. In mutualistic interactions, however, both species function as consumers of and resources for the other, and they may depend on each other for survival and reproduction. The potential for phenological mismatches in such associations may have implications for the persistence of both species involved, with magnified consequences for diversity loss. Hence, the main focus for the rest of this chapter is on phenological responses to climate change in mutualistic species interactions typified by pollinator-plant associations.

IMPLICATIONS OF DIFFERENTIAL PHENOLOGICAL ADVANCE IN POLLINATOR-PLANT ASSOCIATIONS

Recent declines in abundance and species richness of insect pollinators, in particular wild and domestic bees (Potts et al. 2010; Ollerton et al. 2014; Bartomeus et al. 2013), have raised concerns about the implications for ecosystem services they provide, both natural and agricultural (Thomann et al. 2013; Kudo 2014; Burkle and Alarcon 2011). One such implication that has attracted considerable attention is the potential for pollinator declines and extinctions to increase pollen limitation in flowering plants, thereby reducing reproductive success of plants, their abundance, or their role in the productivity of natural and agricultural systems (Thomann et al. 2013; Hegland et al. 2009). As a result, coupled pollinator-plant associations face dual biodiversity implica-

tions when one species or group of species in the association undergoes a decline that, in turn, has the potential to drive declines in the associated species or species group.

General patterns in insect pollinator declines and their causes are elusive, but candidate drivers include changes in agricultural practices, use of pesticides, pathogens, and climate change (Potts et al. 2010; Williams, Araújo, and Rasmont 2007; Hegland et al. 2009). In this chapter, we avoid presenting arguments that favor one potential driver of pollinator declines over others. Rather, we review the evidence for a role of climate change in phenological dynamics of pollinator-plant associations. The intent of this review is to stimulate thinking about the biodiversity implications of climate change for pollinator-plant associations through disruption of phenological associations in these mutualisms.

One of the most taxonomically and temporally comprehensive analyses of declines in insect pollinators examined trends in abundance of bees in the northeastern United States, analyzing trends and drivers of them in 438 species over a 140-year span (Bartomeus et al. 2013). Of the ecological drivers of changes in abundance included in the analysis, narrow dietary breadth and narrow phenological breadth (indexed as the species-specific annual number of days of adult activity) best predicted species-level declines in abundance (Bartomeus et al. 2013). As well, the timing of peak seasonal activity in pollinator species groups may influence their vulnerability to extreme climatic events. Bumblebees, for instance, may be particularly vulnerable to heat waves, which typically occur late in summer when bumblebee activity is greatest (Rasmont and Iserbyt 2012). Despite widespread pollinator declines and a basis for associating such declines with climate change–driven disruption of phenological associations between plants and pollinators, there is limited evidence to date in support of a direct role of climate change in either pollinator declines or declines in pollinator-dependent plant species (Potts et al. 2010). In part, this may be attributed to a scarcity of case studies, because currently only a few investigations have focused on testing this hypothesis (Rafferty et al. 2013). Nonetheless, evidence of differential phenological shifts in pollinator versus flowering-plant phenology indicates the potential for climate change to disrupt such associations.

In many specialized plant-pollinator associations, the flowering phenology of the focal plant species typically precedes that of the timing of pollinator activity on a seasonal basis, and plant-flowering phenology advances more quickly in response to warming than does pollinator phenology (Forrest and Thomson 2011; Solga, Harmon, and Ganguli 2014; Kudo and Ida 2013). In parts of Northern Europe, specialists and sedentary species of bee and syrphid fly pollinators have declined since 1980, but generalists and highly mobile species have not (Potts et al. 2010). Among plants in the same areas, insect-pollinated species have undergone greater declines than have wind- and self-pollinated species (Potts et al. 2010). Whether or not recent climate change has contributed to these patterns, they indicate the consequences of phenological niche conservatism or narrow phenological niche breadth for species unable to adjust to rapid shifts in environmental conditions (*sensu* Bartomeus et al. 2013).

Some data indicate that specialized pollinators are more likely than generalists to suffer resource loss from phenological mismatch with their plant mutualists if they do not keep pace with advancing timing of flowering in response to climate change (Hegland et al. 2009). For example, an observational study based on 14 years of data documented a decline in abundance of pollinator species of muscid flies and chironomid midges associated with a shorter flowering season at a High Arctic site in Greenland (Høye et al. 2013). The decline in muscid fly abundance at the site is particularly troublesome considering that they are the key pollinator group for at least one common species of flowering plant in the High Arctic, mountain avens (*Dryas* sp.) (Tiusanen et al. 2016). Seasonal overlap between the phenology of bee pollinators and their flowering-plant mutualists was also disrupted during an anomalously warm spring in a mid-latitude alpine system (Kudo 2014). In the United Kingdom, species of bumblebees currently in decline are those presumed to have narrow climatic niches (Williams, Araújo, and Rasmont 2007; Williams, Colla, and Xie 2009), whereas, generally, pollinator-plant associations expected to persist under climate change are those that display generalized adaptations to climatic conditions and species associations (Burkle and Alarcon 2011).

A recent and comprehensive examination of the consequences for pollinator species' richness of changes in floral plant phenology focused on associations involving 665 species of insect pollinators and the phenological dynamics of 132 species of their plant hosts in a Mediterranean community near Athens, Greece (Petanidou et al. 2014). This study classified floral resources according to few- versus many-flowered plant species and annual versus perennial plant species, and it quantified the numbers of insect pollinators visiting each plant species over 4 years. Pollinators comprised bees, muscoid flies, hoverflies, beetles, wasps, butterflies, and true bugs. The numbers of pollinator species observed visiting plants were compared to the annual deviation in plant flowering phenology from the 4-year mean to analyze consequences for pollinator richness of advances or delays in flowering phenology.

This analysis revealed, at the scale of the entire plant community, a positive association between onset of flowering and pollinator richness but a negative association between flowering duration and pollinator richness (Figure 4.1).

These relationships suggest that earlier onset of flowering, and prolonged flowering across the community, were associated with visits by fewer species of insect polli-

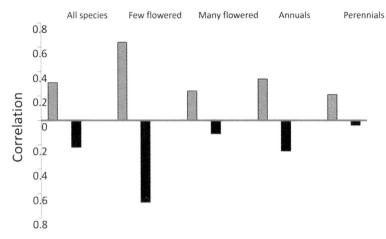

Figure 4.1. Magnitude and direction of correlations between insect pollinator richness and the onset (gray columns) and duration (black columns) of flowering across all plant species, and individually among few-flowered species, many-flowered species, annuals, and perennials in a Mediterranean community near Athens, Greece. (Adapted from Petanidou et al. 2014.)

nators. The same associations were evident for few-flowered species, many-flowered species, annuals, and perennials individually (Figure 4.1). However, the associations were strongest for few-flowered species, and the negative correlation between flowering duration and pollinator richness was not significant for perennials (Petanidou et al. 2014). Because onset and duration of flowering were negatively correlated for the entire community and each category of flowering plants individually, the relationships between flowering duration and pollinator species richness may actually be driven by the negative response of pollinator richness to earlier onset of flowering (Petanidou et al. 2014).

Evidence deriving primarily from observational studies illustrates the potential for development of phenological mismatches between insect pollinators and flowering plants as a result of climate change. Experimental studies on this topic have, by contrast, been few to date and primarily "temporal transplants" (Forrest 2015). Such experiments have focused on advancing the phenology of plants in growth cham-

bers and then transferring them to outdoor plots with plants in an ambient phenological state (Forrest 2015). Results of such experiments have been mixed, with observations of both increased (Rafferty and Ives 2011) and reduced (Parsche, Frund, and Tscharntke 2011) pollinator visitation of experimentally advanced plants compared to phenologically ambient plants. Multiple observational studies have demonstrated more rapid phenological advance in flowering plants than in their insect mutualists in response to warming (Iler, Inouye, et al. 2013; Kudo and Ida 2013; Solga, Harmon, and Ganguli 2014; Forrest and Thomson 2011; Rafferty and Ives 2012; Warren, Bahn, and Bradford 2011). Reduced pollinator effectiveness may be expected to result from increasing phenological mismatch involving specialists in such cases (Rafferty et al. 2013; Rafferty and Ives 2011; Kudo and Ida 2013). However, integrity of pollination services to plants can be maintained despite phenological shifts in response to climate change in associations involving generalist pollinators or pollinators that advance their timing apace with changes in flowering phenology (Cleland et al. 2012; Bartomeus et al. 2011; Iler, Inouye, et al. 2013; Solga, Harmon, and Ganguli 2014; Gilman et al. 2012).

At least one observational study has shown that temporal overlap between insect pollinators and their floral resources may remain intact even when species at each trophic level respond differentially to climate change. This analysis used a 20-year record of timing of flowering by nine plant species and timing of emergence at the community level by syrphid flies comprising at least 15 flower-visiting species at an alpine site at Rocky Mountain Biological Station in Colorado (Iler, Inouye, et al. 2013). Flowering phenology at this site was associated strongly with the timing of snowmelt and generally preceded timing of emergence by syrphid pollinators. As a consequence, earlier annual snowmelt was accompanied by earlier flowering across plant species, and flowering phenology advanced at a higher rate in association with earlier snowmelt than did fly emergence (Figure 4.2). Similarly, the termination of the flowering season advanced more strongly with earlier snowmelt than did the end of the annual period of fly activity (Figure 4.2).

Because syrphid flies at the study site generally terminated activity prior to the end of the annual flowering season, these differential rates of phenological advance in response to earlier snowmelt resulted in greater overall phenological overlap between flowers and syrphid flies in years with early compared to late snowmelt (Iler, Inouye, et al. 2013). Across the range of dates observed in this 20-year study, the number of days of phenological overlap between

flowering plants and syrphid fly pollinators varied by more than twofold between the earliest and latest dates of annual snowmelt (Iler, Inouye, et al. 2013).

As noted earlier, Høye et al. (2013) documented declines in the abundance of insect pollinators resulting from phenological mismatch with advancement and spatial compression of flowering by the plants they visit. Importantly, plants themselves may suffer adverse consequences of advancing flowering phenology that outpaces shifts in the timing of activity by their insect pollinators. A review of six studies of plant phenological dynamics comprising between 19 and 385 species each, and of four studies of phenological dynamics of insect pollinators comprising between 10 and 35 species each, concluded that species-level responses to climate change are highly individualistic, with few general patterns (Solga, Harmon, and Ganguli 2014).

This megareview highlighted three examples of plant species that experienced ad-

Figure 4.2. Model coefficients quantifying the rate of phenological advance in flowering onset, onset of syrphid fly activity, end of flowering, and end of syrphid fly activity in relation to the annual timing of snowmelt over a 20-year observational study at Rocky Mountain Biological Laboratory, Colorado, United States. Asterisks indicate statistical significance ($p < 0.05$) of taxa × snowmelt coefficients. (Adapted from Iler, Inouye, et al. 2013b.)

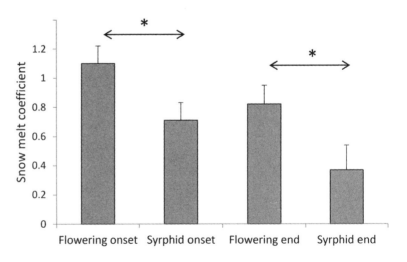

verse consequences of the loss of pollinator services as a result of mismatches deriving from differential phenological responses to climate change. A species of glacier lily (*Erythronium grandiflorum*) in a subalpine meadow in Colorado suffered pollen limitation early in the flowering season due to an absence of active bumblebee pollinators that early in the season (Thomson 2010). Similarly, a species of star-of-Bethlehem (*Gagea lutea*) in Japan advanced its flowering phenology in response to warmer spring temperatures to an extent that rendered it unavailable to its bee pollinator (Kudo et al. 2004). Finally, the flowering plant *yan hu suo* (*Corydalis ambigua*) exhibited reduced seed set at a site in Japan following an advancement of its flowering phenology in response to springtime warming because its bee pollinator did not undergo an advancement of its emergence phenology (Kudo et al. 2004). Although these examples apply to flowering plants with specialist pollinators or plant species that are visited primarily by a single species of pollinator, generalist pollinators can apparently replace other pollinators whose phenology does not advance in nonspecialist associations (Solga, Harmon, and Ganguli 2014). Similarly, generalist pollinators that show preference for some plant species may switch their preference to other species if they become phenologically mismatched from their preferred resources (Forrest and Thomson 2011).

As is evident in the examples here, phenological mismatch is commonly studied as an outcome of differential rates of advance in springtime events between floral resources and pollinators in general or between pollinators classified as specialists or generalists. In our final example, we review a study of species-specific phenological dynamics in paired pollinator-plant associations to examine interactions at the species level.

Bartomeus et al. (2011) analyzed data on the emergence phenology of 10 species of generalist bees in the northeastern United States and southeastern Canada spanning 130 years and compared the rates of advance in springtime activity by those species to published estimates of rates of advance in flowering dates by 38 species of plants from the same region. Particularly noteworthy features of this analysis include the length of the time series for both taxonomic groups, the temporal and spatial coherence among the pollinator and flowering plant data sets, and the restriction of this analysis to plant species visited by the species of pollinators included in the study (Bartomeus et al. 2011). Across the entire time series, the 10 species of bees in this study advanced their emergence phenology by approximately −0.8 days per decade, but the rate of advance has nearly doubled since 1970, indicating a response to accelerated warming (Bartomeus et al. 2011).

Analyses of pooled data spanning the entire time series revealed that the rates of advance of springtime activity by bees and the timing of flowering by plants in this region have largely paralleled each other, suggesting an absence of evidence of developing mismatch (Bartomeus et al. 2011). Furthermore, the plant species in this comparison exhibited greater rates of advancement in flowering phenology since 1970, and for both bees and plants, early-active species displayed more rapid phenological advancement than those active later in the season (Bartomeus et al. 2011). These observations suggest that a phenological mismatch between bee pollinators and flowering plants in the same general region has not yet developed, but the authors caution that such a mismatch could develop with additional warming in the future (Bartomeus et al. 2011).

That a mismatch is likely for at least some of the species in this study is suggested by a comparison of rates of advance by individual bee species and the plant species they visit (Figure 4.3). We plotted rates of advance in springtime activity by each of the 10 bee species in this study versus rates of advance in flowering phenology only for those species visited by each of them in the

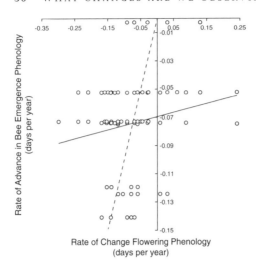

Figure 4.3. Rates of advance in emergence phenology of 10 species of bee pollinators plotted against rates of advance in flowering phenology of the species they visit in northeastern United States and southeastern Canada. Rates are expressed as days per year and were estimated using linear regression of dates of emergence or flowering against year. The solid trend line is for the pooled scatter across all species pairs. The dashed line represents parity (i.e., equal rates of advance by bee and plant species pairs). (Plotted from coefficients reported in Bartomeus et al. 2011.)

region where they occur. The slope of the pooled scatter plot (Figure 4.3, solid line) deviates substantially from parity (Figure 4.3, dashed line), which suggests that the bee species have lagged behind rates of phenological advance in the plant species.

By contrast, when we plot pairwise differences in the rates of advance by individual plant species and the rates of advance by individual species of bees visiting them, there is considerable variation within each bee species in the extent to which its rate of phenological advance has lagged, exceeded, or matched that of the plant species it uses (Figure 4.4). Among species of bees in the genus Andrena, rates of advance in emergence phenology lagged behind rates of advance in flowering phenology in 13 species pairs, whereas rates of bee advancement exceeded those of flowering advancement in 9 species pairs (Figure 4.4, top panel). For species of bees in the genus Bombus, advances in emer-

gence phenology lagged behind advances in flowering phenology in 14 species pairs, whereas rates of bee advancement exceeded rates of flowering advancement in 7 species pairs (Figure 4.4, middle panel). In contrast, bee species belonging to the genera Osmia and Colletes displayed rates of phenological advance that largely exceeded those of the plant species they visit, with 10 species pairs showing greater rates of advance in bees than in plants and 4 species pairs indicating the opposite (Figure 4.4, bottom panel). For purposes of this comparison, we considered a rate differential equal to or less than 0.20 as indicative of approximate parity in rates of advance between bee species and their associated plant species. Three of 25 (12 percent) species pairs showed equivalent rates of advance for the genus Andrena, 5 of 26 (19 percent) species pairs approximated parity in rates of advance for the genus Bombus, whereas 5 of 19 (26 percent) species pairs in the genera Osmia and Colletes exhibited rates of phenological advance equivalent to those of the plant species they pollinate. Together these results suggest that these three species of Andrena bees may be at greatest risk of suffering mismatch with floral resources under future warming, but plant species visited by the species of Osmia and Colletes in this study may have the greatest prospects for avoiding loss of pollinator services with future warming.

CONCLUSIONS

In applying insights developed from this review, we can arrive at several conclusions regarding the biodiversity implications of climate change for pollinator-plant associations. There appears to be considerable potential for mismatch to develop in pollinator-plant mutualisms, or to become exacerbated where it already exists, as Earth's climate continues to warm. This would seem to be most likely in associations involving species that become active early in the season and in those involving species with narrow phe-

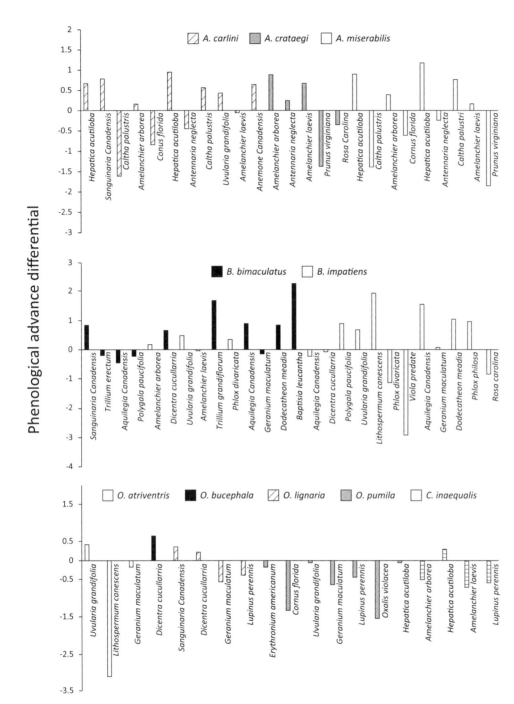

Figure 4.4. Differences in the rate of phenological advance between bees and the plants they visit, expressed as the rate of advance in flowering (days per decade) minus the rate of advance in bee emergence (days per decade). Bee genera include *Andrena* (top), *Bombus* (middle), and *Colletes* and *Osmia* (bottom). Positive values indicate bees lagged behind flowers, whereas negative values indicate bee emergence phenology advanced more rapidly than flowering phenology for any species pair. (Estimated using slopes reported in Bartomeus et al. 2011.)

nological niches. Phenological niche specialization, whether relative to dependence on species interactions or to abiotic conditions, is likely to result in a loss of existing pollinator-plant associations if advances in the timing of abiotic conditions favorable for reproductive activity are not matched by sufficient shifts in the timing of pollinator activity. Importantly, this may occur even if mean abiotic conditions themselves, such as seasonal or annual temperature, do not change but the timing of critical thermal thresholds does change. As well, advances in the timing of pollinator activity, even if these keep pace with advances in flowering phenology, may not ensure maintenance of existing pollinator-plant associations if, for instance, pollinator advances increase their early-season exposure to natural enemies or adverse temperature fluctuations. Finally, in cases of pollinator loss due to phenological mismatches resulting from more rapid advancement in the timing of flowering compared to the timing of pollinator activity, persistence of the plant species involved in existing pollinator-plant mutualisms may depend on replacement by pollinator species with broader, more flexible, and hence less specialized phenological niches. Complementary measures to conserve both native floral and pollinator diversity should be undertaken where lacking and maintained where already in effect, as strategies to buffer both taxonomic groups against phenological shifts by the species upon which they depend. Programs to foster the proliferation of urban green spaces and pollinator gardens may also offer particularly affordable and achievable local actions, with the ancillary benefit of raising public awareness of the importance of pollinator conservation for ecosystem services (Banaszak-Cibicka, Ratynska, and Dylewski 2016).

REFERENCES

Banaszak-Cibicka, W., H. Ratynska, and L. Dylewski. 2016. "Features of urban green space favourable for large and diverse bee populations (Hymenoptera: Apoidea: Apiformes)." *Urban Forestry & Urban Greening* no. 20: 448–452. https://doi.org/10.1016/j.ufug.2016.10.015.

Bartomeus, Ignasi, John S. Ascher, Jason Gibbs, Bryan N. Danforth, David L. Wagner, Shannon M. Hedtke, and Rachael Winfree. 2013. "Historical changes in northeastern US bee pollinators related to shared ecological traits." *Proceedings of the National Academy of Sciences* 110 (12): 4656–4660. https://doi.org/10.1073/pnas.1218503110.

Bartomeus, Ignasi, John S. Ascher, David Wagner, Bryan N. Danforth, Sheila Colla, Sarah Kornbluth, and Rachael Winfree. 2011. "Climate-associated phenological advances in bee pollinators and bee-pollinated plants." *Proceedings of the National Academy of Sciences* 51 (108): 20645–20649. https://doi.org/10.1073/pnas.1115559108.

Bock, A., T. H. Sparks, N. Estrella, N. Jee, A. Casebow, C. Schunk, M. Leuchner, and A. Menzel. 2014. "Changes in first flowering dates and flowering duration of 232 plant species on the island of Guernsey." *Global Change Biology* 11 (20): 3508–3519. https://doi.org/10.1111/gcb.12579.

Both, C., R. G. Bijlsma, and M. E. Visser. 2005. "Climatic effects on timing of spring migration and breeding in a long-distance migrant, the pied flycatcher Ficedula hypoleuca." *Journal of Avian Biology* 5 (36): 368–373.

Both, C., S. Bouwhuis, C. M. Lessells, and M. E. Visser. 2006. "Climate change and population declines in a long-distance migratory bird." *Nature* 441 (7089): 81–83.

Both, C., M. van Asch, R. G. Bijlsma, A. B. van den Burg, and M. E. Visser. 2009. "Climate change and unequal phenological changes across four trophic levels: Constraints or adaptations?" *Journal of Animal Ecology* 1 (78): 73–83. https://doi.org/10.1111/j.1365-2656.2008.01458.x.

Burkle, Laura A., and Ruben Alarcon. 2011. "The future of plant-pollinator diversity: Understanding interaction networks across time, space, and global change." *American Journal of Botany* 3 (98): 528–538. https://doi.org/10.3732/ajb.1000391.

Cleland, Elsa E., Jenica M. Allen, Theresa M. Crimmins, Jennifer A. Dunne, Stephanie Pau, Steven E. Travers, Erika S. Zavaleta, and Elizabeth M. Wolkovich. 2012. "Phenological tracking enables positive species responses to climate change." *Ecology* 8 (93): 1765–1771.

Cook, B. I., E. R. Cook, P. C. Huth, J. E. Thompson, A. Forster, and D. Smiley. 2008. "A cross-taxa phenological dataset from Mohonk Lake, NY and its relationship to climate." *International Journal of Climatology* 10 (28): 1369–1383. https://doi.org/10.1002/joc.1629.

Ellwood, E. R., S. A. Temple, R. B. Primack, N. L. Bradley, and C. C. Davis. 2013. "Record-breaking early flowering in the eastern United States." *PLOS One* 1 (8). https://doi.org/10.1371/journal.pone.0053788.

Forrest, J. R. K. 2015. "Plant-pollinator interactions and phenological change: what can we learn about climate impacts from experiments and observations?" *Oikos* 1 (124): 4–13. https://doi.org/10.1111/oik.01386.

Forrest, Jessica R. K., and James D. Thomson. 2011. "An examination of synchrony between insect emergence and flowering in Rocky Mountain meadows." *Ecological Monographs* 3 (81): 469–491. https://doi.org/10.1890/10-1885.1.

Gilman, R. Tucker, Nicholas S. Fabina, Karen C. Abbott, and Nicole E. Rafferty. 2012. "Evolution of plant-pollinator mutualisms in response to climate change." *Evolutionary Applications* 1 (5): 2–16. https://doi.org/10.1111/j.1752-4571.2011.00202.x.

Gordo, O., and J. J. Sanz. 2009. "Long-term temporal changes of plant phenology in the Western Mediterranean." *Global Change Biology* 8 (15): 1930–1948. https://doi.org/10.1111/j.1365-2486.2009.01851.x.

Hegland, Stein Joar, Anders Nielsen, Amparo Lazaro, Anne-Line Bjerknes, and Orjan Totland. 2009. "How does climate warming affect plant-pollinator interactions?" *Ecology Letters* 2 (12): 184–195. https://doi.org/10.1111/j.1461-0248.2008.01269.x.

Høye, T. T., E. Post, H. Meltofte, N. M. Schmidt, and M. C. Forchhammer. 2007. "Rapid advancement of spring in the High Arctic." *Current Biology* 12 (17): R449–R451.

Høye, T. T., E. Post, N. M. Schmidt, K. Trojelsgaard, and M. C. Forchhammer. 2013. "Shorter flowering seasons and declining abundance of flower visitors in a warmer Arctic." *Nature Climate Change* 8 (3): 759–763. https://doi.org/10.1038/nclimate1909.

Iler, A. M., T. T. Høye, D. W. Inouye, and N. M. Schmidt. 2013. "Long-term trends mask variation in the direction and magnitude of short-term phenological shifts." *American Journal of Botany* 7 (100): 1398–1406. https://doi.org/10.3732/ajb.1200490.

Iler, A. M., D. W. Inouye, T. T. Høye, A. J. Miller-Rushing, L. A. Burkle, and E. B. Johnston. 2013. "Maintenance of temporal synchrony between syrphid flies and floral resources despite differential phenological responses to climate." *Global Change Biology* 8 (19): 2348–2359. https://doi.org/10.1111/gcb.12246.

Iwasa, Y., and S. A. Levin. 1995. "The timing of life history events." *Journal of Theoretical Biology* 1 (172): 33–42. https://doi.org/10.1006/jtbi.1995.0003.

Jeong, S. J., C. H. Ho, H. J. Gim, and M. E. Brown. 2011. "Phenology shifts at start vs. end of growing season in temperate vegetation over the Northern Hemisphere for the period 1982–2008." *Global Change Biology* 7 (17): 2385–2399. https://doi.org/10.1111/j.1365-2486.2011.02397.x.

Karlsen, S. R., A. Elvebakk, K. A. Hogda, and T. Grydeland. 2014. "Spatial and temporal variability in the onset of the growing season on Svalbard, Arctic Norway—Measured by MODIS-NDVI satellite data." *Remote Sensing* 9 (6): 8088–8106. https://doi.org/10.3390/rs6098088.

Kerby, J. T., C. C. Wilmers, and E. Post. 2013. "Climate change, phenology, and the nature of consumer-resource interactions: advancing the match/mismatch hypothesis." In *Trait-Mediated Indirect Interactions: Ecological and Evolutionary Perspectives*, ed. T. Ohgushi, O. J. Schmitz, and R. D. Holt, 508–525. Cambridge University Press.

Kudo, Gaku. 2014. "Vulnerability of phenological synchrony between plants and pollinators in an alpine ecosystem." *Ecological Research* 4 (29): 571–581. https://doi.org/10.1007/s11284-013-1108-z.

Kudo, Gaku, and Takashi Y. Ida. 2013. "Early onset of spring increases the phenological mismatch between plants and pollinators." *Ecology* 10 (94): 2311–2320. https://doi.org/10.1890/12-2003.1.

Kudo, G., Y. Nishikawa, T. Kasagi, and S. Kosuge. 2004. "Does seed production of spring ephemerals decrease when spring comes early?" *Ecological Research* 2 (19): 255–259. https://doi.org/10.1111/j.1440-1703.2003.00630.x.

Miller-Rushing, A. J., T. T. Hoye, D. W. Inouye, and E. Post. 2010. "The effects of phenological mismatches on demography." *Philosophical Transactions of the Royal Society B—Biological Sciences* 1555 (365): 3177–3186. https://doi.org/10.1098/rstb.2010.0148.

Ollerton, J., H. Erenler, M. Edwards, and R. Crockett. 2014. "Extinctions of aculeate pollinators in Britain and the role of large-scale agricultural changes." *Science* no. 346: 1360–1362.

Parmesan, Camille. 2007. "Influences of species, latitudes and methodologies on estimates of phenological response to global warming." *Global Change Biology* 9 (13): 1860–1872. https://doi.org/10.1111/j.1365-2486.2007.01404.x.

Parsche, S., J. Frund, and T. Tscharntke. 2011. "Experimental environmental change and mutualistic vs. antagonistic plant flower-visitor interactions." *Perspectives in Plant Ecology Evolution and Systematics* 1 (13): 27–35. https://doi.org/10.1016/j.ppees.2010.12.001.

Petanidou, Theodora, Athanasios S. Kallimanis, Stefanos P. Sgardelis, Antonios D. Mazaris, John D. Pantis, and Nickolas M. Waser. 2014. "Variable flowering phenology and pollinator use in a community suggest future phenological mismatch." *Acta Oecologica/International Journal of Ecology* no. 59: 104–111. https://doi.org/10.1016/j.actao.2014.06.001.

Plard, F., J. M. Gaillard, T. Coulson, A. J. M. Hewison, D. Delorme, C. Warnant, and C. Bonenfant. 2014. "Mismatch between birth date and vegetation phenology slows the demography of roe deer." *PLOS Biology* 4 (12). https://doi.org/10.1371/journal.pbio.1001828.

Post, E. 2013. *Ecology of Climate Change: The Importance of Biotic Interactions*. Monographs in Population Biology 52. Princeton University Press.

Post, E., and M. C. Forchhammer. 2008. "Climate change reduces reproductive success of an arctic herbivore through trophic mismatch." *Philosophical Transactions of the Royal Society of London, Series B* (363): 2369–2375.

Post, E., J. Kerby, C. Pedersen, and H. Steltzer. 2016. "Highly individualistic rates of plant phenological advance associated with arctic sea ice dynamics."

Biology Letters (12): 20160332. https://doi.org/10.1098/rsbl.2016.0332.

Post, E., S. A. Levin, Y. Iwasa, and N. C. Stenseth. 2001. "Reproductive asynchrony increases with environmental disturbance." *Evolution* 55 (4): 830–834.

Potts, Simon G., Jacobus C. Biesmeijer, Claire Kremen, Peter Neumann, Oliver Schweiger, and William E. Kunin. 2010. "Global pollinator declines: trends, impacts and drivers." *Trends in Ecology & Evolution* 25 (6): 345–353. https://doi.org/10.1016/j.tree.2010.01.007.

Rafferty, N. E., P. J. CaraDonna, L. A. Burkle, A. M. Iler, and J. L. Bronstein. 2013. "Phenological overlap of interacting species in a changing climate: An assessment of available approaches." *Ecology and Evolution* 3 (9): 3183–3193. https://doi.org/10.1002/ece3.668.

Rafferty, Nicole E., and Anthony R. Ives. 2011. "Effects of experimental shifts in flowering phenology on plant-pollinator interactions." *Ecology Letters* 14 (1): 69–74. https://doi.org/10.1111/j.1461–0248.2010.01557.x.

Rafferty, Nicole E., and Anthony R. Ives. 2012. "Pollinator effectiveness varies with experimental shifts in flowering time." *Ecology* 93 (4): 803–814.

Rasmont, P., and S. Iserbyt. 2012. "The bumblebees scarcity syndrome: Are heat waves leading to local extinctions of bumblebees (Hymenoptera: Apidae: Bombus)?" *Annales de la société entomologique de France* 48 (3–4): 275–280.

Schwartz, M. D. 2003. *Phenology: An Integrative Environmental Science.* Kluwer Academic Publishers.

Solga, Michelle J., Jason P. Harmon, and Amy C. Ganguli. 2014. "Timing is everything: An overview of phenological changes to plants and their pollinators." *Natural Areas Journal* 34 (2): 227–234.

Stenseth, N. C., and A. Mysterud. 2002. "Climate, changing phenology, and other life history and traits: Nonlinearity and match-mismatch to the environment." *Proceedings of the National Academy of Sciences* 21 (99): 13379–13381. https://doi.org/10.1073/pnas.212519399.

Thomann, Michel, Eric Imbert, Celine Devaux, and Pierre-Olivier Cheptou. 2013. "Flowering plants under global pollinator decline." *Trends in Plant Science* 18 (7): 353–359. https://doi.org/10.1016/j.tplants.2013.04.002.

Thomson, James D. 2010. "Flowering phenology, fruiting success and progressive deterioration of pollination in an early-flowering geophyte." *Philosophical Transactions of the Royal Society B—Biological Sciences* 1555 (365): 3187–3199. https://doi.org/10.1098/rstb.2010.0115.

Tiusanen, M., P. D. N. Hebert, N. M. Schmidt, and T. Roslin. 2016. "One fly to rule them all—muscid flies are the key pollinators in the Arctic." *Proceedings of the Royal Society B—Biological Sciences* 1839 (283). https://doi.org/10.1098/rspb.2016.1271.

Visser, M. E., and C. Both. 2005. "Shifts in phenology due to global climate change: The need for a yardstick." *Proceedings of the Royal Society B-Biological Sciences* 1581 (272): 2561–2569.

Warren, Robert J., II, Volker Bahn, and Mark A. Bradford. 2011. "Temperature cues phenological synchrony in ant-mediated seed dispersal." *Global Change Biology* 7 (17): 2444–2454. https://doi.org/10.1111/j.1365-2486.2010.02386.x.

Williams, P. H., M. B. Araújo, and P. Rasmont. 2007. "Can vulnerability among British bumblebee (*Bombus*) species be explained by niche position and breadth?" *Biological Conservation* 138 (3–4): 493–505. https://doi.org/10.1016/j.biocon.2007.06.001.

Williams, Paul, Sheila Colla, and Zhenghua Xie. 2009. "Bumblebee vulnerability: Common correlates of winners and losers across three continents." *Conservation Biology* 23 (4): 931–940. https://doi.org/10.1111/j.1523-1739.2009.01176.x.

Zeng, H. Q., G. S. Jia, and B. C. Forbes. 2013. "Shifts in Arctic phenology in response to climate and anthropogenic factors as detected from multiple satellite time series." *Environmental Research Letters* 8 (3). https://doi.org/10.1088/1748-9326/8/3/035036.

CHAPTER FIVE

Coral Reefs: Megadiversity Meets Unprecedented Environmental Change

OVE HOEGH-GULDBERG

The most distinguishing feature of our planet is the presence of liquid water covering 71 percent of its surface. This ocean nurtured the beginnings of life, and today it is essential for its maintenance and future. The ocean performs a vital regulatory role, stabilizing planetary temperature, regulating the gas content of the atmosphere, and determining the climate and weather systems on which life depends. In many ways, the ocean represents the heart and lungs of our planet, and life as we know it would not be possible without its presence.

Unfortunately, the latest science tells us that we are starting to change the ocean in fundamental ways that threaten its ability to provide the ecosystem goods and services on which much of humanity lives. These changes have been highlighted in the latest Intergovernmental Panel on Climate Change (IPCC), which included several new chapters on the potential response of the ocean to rapid changes in global climate (Hoegh-Guldberg et al. 2014; Pörtner et al. 2014). These chapters have established a consensus that concludes that unprecedented and rapid changes are occurring in ocean temperature, volume, structure, chemistry, and therefore life itself.

The ocean plays a crucial role in stabilizing planetary conditions, mainly due to its size and enormous capacity to absorb carbon dioxide (CO_2) and heat with minimal change in temperature. As a result, the ocean has absorbed 93 percent of the extra energy from the enhanced greenhouse effect, and approximately 30 percent of the CO_2 generated by human activities. These changes have been essential in reducing the rate at which warming has manifested itself over the 150 years since the Industrial Revolution, when human activities began to drive ever-increasing amounts of CO_2 and other greenhouse gases into the atmosphere.

Absorbing this extra heat and CO_2, however, has had an impact on the ocean. Ocean temperatures have increased, leading to rapid sea-level rise, decrease in sea-ice extent, changes in ocean structure and gas content, and impacts on the planet's hydrological cycle. At the same time, ocean pH has dropped rapidly, along with changes in the concentration of key ion species such as carbonate and bicarbonate.

Not surprisingly, life in the ocean has responded to these fundamental changes. More than 80 percent of the studies lengthy enough to detect discernible changes in the distribution and abundance of species have reported rapid changes in the direction expected by climate change (e.g., shifts to higher latitudes as warming occurs). Other studies reveal that the combination of local and global factors drive down the oxygen content of deeper areas of the ocean, to such an extent that oxygen-dependent life-forms are beginning to disappear from many deeper-water regions. These are now being referred to as "dead zones." In other parts of the ocean, rates of primary productivity are either decreasing or increasing, driven by conditions such as retreating sea ice, warming, and other factors such as changing winds and nutrient concentrations (Hoegh-Guldberg et al. 2014; Pörtner et al. 2014).

Some ecosystems shift more quickly and are showing greater changes than others. In these cases, ecosystems are fundamentally changing in terms of their distribution and abundance. In the case of coral reefs, this contraction poses a serious threat to the biological diversity of the ocean, given that coral reefs provide habitat for at least 25 percent of all marine species on the planet. In turn, the loss of coral reefs presents a major challenge to the 850 million people who live within 100 km of tropical coastlines and derive a number of key benefits, such as food and livelihood from them, as well as a range of other ecosystem goods and services (Burke et al. 2011).

Coral reefs provide a stark example of the connection among climate change, biodiversity, changing ecosystems, and human well-being. The present chapter focuses on coral reefs and describes the physiological, ecological, and economic challenges of a rapidly warming and acidifying ocean, with many of the messages being transferable to other ecosystems and regions.

PHYSICAL AND CHEMICAL CHANGES IN THE OCEAN

Earth's ocean is a vast region that stretches from the high-tide mark to the deepest ocean trench (11,030 m) and occupies a volume of 1.3 billion km^3. Most of the ocean is difficult to visit, requiring advanced technologies to venture anywhere below 40 m from the surface. Consequently, most of the ocean is unexplored, with the common assertion that we know more about the surfaces of Mars or the Moon than we do about the deep ocean. Until recently, an order of magnitude more papers were being published on land-based change as opposed to that occurring in the ocean (Hoegh-Guldberg and Bruno 2010). Knowing so little about the ocean is a major vulnerability in terms of our understanding of planetary systems and how they are likely to change as the average global surface temperature and ocean pH change.

Many early studies concluded that climate change was unlikely to influence the ocean by much given its size and thermal inertia (i.e., it takes a lot of heat to change ocean temperature). The ocean is highly structured, however, and as a consequence, the upper layers of the ocean (a relatively thin layer of several hundred meters) acts somewhat independently of the rest of the ocean. As a consequence, changes in the temperature of the surface layer of the ocean are not too far behind that of the surface temperature of the planet as a whole (IPCC 2013). While average global temperature has increased by approximately 0.7°C since 1950, average global average sea surface temperatures (SSTs) in the Indian, Atlantic, and Pacific Oceans have increased by 0.65°C, 0.41°C, and 0.31°C, respectively (Hoegh-Guldberg et al. 2014). Naturally, changes in average sea surface temperature vary geographically and may have higher or lower values on average, depending on the local oceanographic dynamics as well as the relative influence of long-term patterns of variability (Hoegh-Guldberg et al. 2014, table 30-1).

At the same time that the ocean has been warming, ocean chemistry has changed in response to the unprecedented inward flux of CO_2. These changes arise from the fact that CO_2 reacts with water upon entering the ocean, forming dilute carbonic acid (Figure 5.1A). Carbonic acid molecules subsequently disassociate, releasing protons, which causes a decrease in pH as well as the conversion of carbonate ions into bicarbonate (Caldeira and Wickett 2003; Hoegh-Guldberg et al. 2007; Kleypas, Buddemeier, et al. 1999). These changes have considerable implications for a range of physiological systems, from the gas exchange of fish to the calcifying abilities of corals and many other organisms that make shells and skeletons by precipitating calcium carbonate from seawater (Gattuso et al. 2014; Kroeker et al. 2013; Pörtner et al. 2014). The increase of anthropogenic CO_2 has lowered ocean pH by 0.1 units—equivalent to a 26 percent

increase in the concentration of protons (Caldeira and Wickett 2003). The current rate at which CO_2, and hence ocean chemistry, is changing is faster than at any other time in 65 million years, if not 300 million years (Hönisch et al. 2012; Pachauri et al. 2014). As several authors have observed, these changes are pushing organisms and ecosystems like coral reefs outside the envelope of environmental conditions under which they have evolved (Figure 5.1B).

Figure 5.1. (A) Linkages between the buildup of atmospheric CO_2 and the slowing of coral calcification due to ocean acidification. Approximately 25% of CO_2 emitted by humans in the period 2000 to 2006 was taken up by the ocean where it combined with water to produce carbonic acid, which releases a proton that combines with a carbonate ion. This decreases the concentration of carbonate, making it unavailable to marine calcifiers such as corals. (B) Temperature, $[CO_2]_{atm}$, and carbonate-ion concentrations reconstructed for the past 420,000 years. Carbonate concentrations were calculated from $CO_{2\ atm}$ and temperature deviations from today's conditions with the Vostok Ice Core data set, assuming constant salinity (34 parts per trillion), mean sea temperature (25°C), and total alkalinity (2300 mmol kg^{-1}). Further details of these calculations are in the SOM. Acidity of the ocean varies by ±0.1 pH units over the past 420,000 years (individual values not shown). The thresholds for major changes to coral communities are indicated for thermal stress (+2°C) and carbonate-ion concentrations ([carbonate] = 200 µmol kg^{-1}, approximate aragonite saturation $\sim\Omega_{aragonite}$ = 3.3; $[CO_2]_{atm}$ = 480 ppm). Coral Reef Scenarios CRS-A, CRS-B, and CRS-C are indicated as A, B, and C, respectively, with analogs from extant reefs depicted in Figure 5.5. Arrows pointing progressively toward the right-hand top square indicate the pathway that is being followed toward $[CO_2]_{atm}$ of more than 500 ppm.

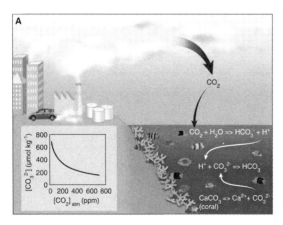

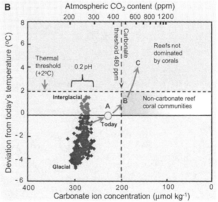

BIOLOGICAL RESPONSES TO RAPID OCEAN CHANGE

To some, the changes in temperature and chemistry might seem too small to significantly affect organisms and ecosystems at first. There is now abundant evidence, however, that life in the ocean is undergoing a series of fundamental changes in response to the changes in temperature and chemistry that have occurred already and that will occur in the future. Poloczanska and colleagues (Poloczanska et al. 2014; Poloczanska et al. 2013) examined a global database of 208 peer-reviewed papers, recording observed responses from more than 800 species or assemblages across a range of regions and taxonomic groups, from phytoplankton to marine mammals, in the distribution and abundance of marine or-

ganisms. The results are startling by any measure. Approximately 80 percent of the papers examined exhibited trends among marine organisms consistent with climate change, suggesting that life in the ocean is already responding to the changes in ocean temperature, structure, pH, and carbonate chemistry (Poloczanska et al. 2014; Poloc-

Figure 5.2. Ranges of change in distribution (km/decade) for marine taxonomic groups, measured at the leading edges (circles) and trailing edges (triangles). Average distribution shifts were calculated using all data, regardless of range location, and are shown as squares. Distribution shifts have been square-root transformed; standard errors may be asymmetric as a result. Positive distribution changes are consistent with warming (into previously cooler waters, generally poleward). Means ± standard error are shown, along with number of observations. Non-bony fishes include sharks, rays, lampreys, and hagfish. (From Poloczanska et al. 2013.)

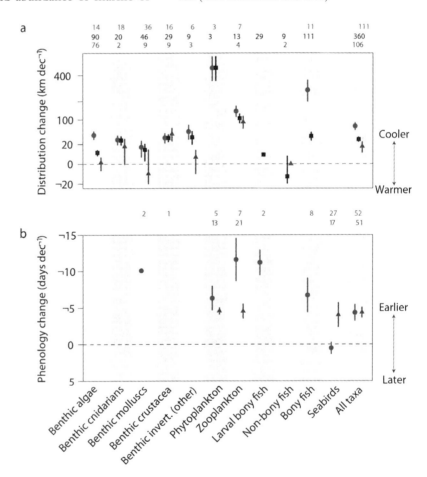

zanska et al. 2013). In some cases, highly mobile creatures such as phytoplankton and zooplankton have been moving at the rates of 10 km–40 km per year toward higher latitudes as the waters have been warming (Figure 5.2). Less mobile organisms such as mollusks, benthic algae, and cnidarians have been traveling at rates between 0 km and 2 km per year.

These differential changes in the location of particular organisms as a function of their relative mobility toward warming higher latitudes suggests that substantial changes are likely in the composition of ecosystems (Hoegh-Guldberg and Bruno 2010; Hoegh-Guldberg et al. 2014; Poloczanska et al. 2014; Poloczanska et al. 2013). In basic terms, less mobile organisms will be able to keep up with more mobile ones, leading to novel ecosystems looms (e.g., coral fish with corals) and creating challenges in terms of managing rapidly moving organisms and ecosystems. Although we have some insight into the sorts of changes that are likely, it is hard to anticipate how these changes are likely to play out over the coming decades and century.

Biological responses are not restricted only to changes in sea temperature per se. Changes in ocean temperature drive a range of other phenomena, from changes in the volume and height of the ocean to changes in the relative stability of the ocean water column (IPCC 2013). At the same time, many ocean organisms and ecosystems are sensitive to recent and projected changes in pH and carbonate chemistry (Gattuso et al. 2014; Kroeker et al. 2013; Pörtner et al. 2014). Changes to pH, CO_2, and carbonate ion concentration have been found to be important to the developing stages of some organisms, as well as the ability to form shells and skeletons in others (Kroeker et al. 2013; Pörtner et al. 2014). At more fundamental levels, there is evidence that neural systems of some organisms such as fish, as well as the respiratory systems of organisms such as cephalopods, can be seriously affected by changing pH and related aspects of ocean chemistry (Pörtner et al. 2014). We are only beginning to understand the full ramifications of these types of changes on organisms, and we fully appreciated the impact of rising levels of CO_2 on marine systems as little as only 15 years ago (Kleypas, Buddemeier, et al. 1999).

CORAL REEF ECOSYSTEMS: MEGADIVERSITY UNDER PRESSURE

Although coral reefs only occupy less than 0.1 percent of the Earth's surface, they provide habitat for one in every four species of marine organisms. Although a large number of species remain to be discovered, studies have attempted to estimate the biological diversity of coral reefs on the basis of the observed rate of discovery and assumptions about the drivers of biodiversity. These types of analysis suggest that anywhere from 1 million to 9 million species live in and around coral reefs (Reaka-Kudla 1997). Other studies such as the Census of Marine Life (Costello et al. 2010) may yield even higher estimates when they are complete. While this megadiversity on its own is an important legacy of evolution, it has significant, mostly unexplored benefits to humanity (e.g., pharmaceuticals). Coral reefs provide a range of other benefits to humanity in the form of food, income (e.g., tourism, fisheries), cultural services, recreation, and coastal protection (i.e., coral reefs build coastal barrier systems that protect entire coastlines) (Cinner et al. 2009; Costanza et al. 2014; Burke et al. 2011; Moberg and Folke 1999).

Coral reefs have long been a prominent feature of coastal ecosystems and are typified by the presence of large numbers of Scleractinian or other reef-building corals. The Scleractinian corals first appeared around 237 million years ago, about 40 million years after the Permian mass extinction event (Stanley and Fautin 2001). The presence of Scleractinian corals within the fossil record is characterized by periods in which carbonate reef systems were present and others, referred to as "gaps," where coral reefs were absent as significant reef builders for millions of years (Veron 2008). Notably, reef gaps were typically associated with the five major mass extinction events linked to high CO_2 concentrations that built up over thousands, if not millions, of years in the Earth's atmosphere (Berner and Kothavala 2001). As a result of these changes, Scleractinian corals and many other organisms underwent significant decreases in diversity and abundance (Veron 2008).

Coral reefs are dominated by Scleractinian corals that form extensive ecosystems in tropical regions (warm-water coral reefs), as well as ecosystems in deeper, low-light habitats (so-called mesophotic coral reefs (Bongaerts et al. 2011) and in deep (down to 3000 m) locations (cold-water coral reefs; Freiwald et al. 2004). Below 200 m depth there is so little light that photosynthesis is no longer possible, and hence reefs are made of corals that do not form symbioses with single-celled dinoflagellates. All of these coral communities, however, attract and provide habitat for significant ecosystems, many of which are very important for fisheries and human livelihoods. Given space constraints, the rest of this chapter focuses on warm-water coral reefs.

Today, most warm-water coral reefs are found in sunlit and low-nutrient coastal waters between the latitudes 30°N and 33°S. Tropical and subtropical waters that accommodate warm-water coral reefs are also saturated with respect to the calcium and carbonate ions necessary for skeletal formation by corals and other organisms. The ratio of calcium carbonate to the saturation constant of aragonite (the principal form of calcium carbonate in coral skeletons) is a critical variable (Caldeira and Wickett 2003; Kleypas, Buddemeier, et al. 1999). If too low, calcium carbonate deposition (calcification) is unable to keep up with calcium carbonate erosion and dissolution (decalcification). The ratio of these two broad ecological processes ultimately determines

whether or not carbonate coral reefs exist. Areas such as the eastern Pacific, where upwelling waters at the equator are rich in CO_2 and low in pH and carbonate, do not have significant carbonate reef development (Manzello et al. 2008). However, aragonite saturation states of 3.3 or higher are generally associated with coral reef communities that can form significant calcium carbonate structures (Hoegh-Guldberg et al. 2007; Kleypas, McManus, et al. 1999).

The contribution of Scleractinian corals to reef building is strongly dependent on an ancient symbiosis with dinoflagellate protists from the genus *Symbiodinium* (Muscatine 1990; Muscatine and Porter 1977; Stanley and Fautin 2001). The dinoflagellates are located within the gastrodermal tissues of corals, where they continue to photosynthesize (as any free-living dinoflagellate would), providing as much as 95 percent of the organic carbon produced by the host. In return, the coral host supplies inorganic nitrogen and phosphorus to the dinoflagellates, compounds that are usually at low concentrations in the clear tropical and subtropical waters in which corals and their dinoflagellates exist (Muscatine and Porter 1977). This mutualistic symbiosis is highly efficient and leads to the internal recycling of inorganic compounds between primary producer and consumer. Corals also feed on particles suspended in the water column, adding yet another source of nutrients and energy to the otherwise oligotrophic waters typical of most coral reefs. The capacity for polytrophic behavior has been important to the success of corals, and hence coral reefs, over tens of millions of years of existence (Stanley and Fautin 2001).

As a result of access to large quantities of organic carbon and energy, corals are able to precipitate large amounts of calcium carbonate and build up significant skeletal structures. These skeletal structures remain after corals have died and are glued together by the calcifying activities of other organisms such as encrusting calcareous red algae. Other organisms, such as calcifying invertebrates, plankton, and plants, also contribute significant amounts of calcium carbonate to reef systems as they develop. Over time, these limestone-like deposits build up three-dimensional reef structures and, over the longer term, form islands and coastal barrier systems. The three-dimensional nature of coral reefs provides a complex habitat for the tens of thousands of species that live in and around typical coral reefs.

LOCAL AND GLOBAL OCEAN WARMING AND ACIDIFICATION

Coral reefs have experienced relatively small shifts in temperature and seawater chemistry over the past 420,000 years at least (Hoegh-Guldberg et al. 2007) (Figure 5.1B). When compared to the current rapid pace of change in a number of environmental parameters, even the shifts between ice age and interglacial periods resulted in relatively small changes over much longer periods of time. As a consequence, the world's largest continuous reef system, the Great Barrier Reef, has waxed and waned over thousands of years as sea levels have risen and fallen. These changes are not directly comparable to the changes we are seeing today because they occurred over much longer periods of time, whereas current change is thought to be occurring faster than at any other time in the past 65 million years, if not 300 million years (Hönisch et al. 2012; IPCC 2013; Pachauri et al. 2014). These observations go for both local (e.g., pollution, overexploitation of key organisms; Burke et al. 2011) and global (e.g., ocean warming and acidification) sources (Hoegh-Guldberg 2014; Hoegh-Guldberg et al. 2014).

As human populations have increased, coastal areas have been modified, and the demand for local resources has grown exponentially. As much as 75 percent of coral reefs are threatened, with as much as 95 percent in danger of being lost by 2050 (Burke et al. 2011; Hoegh-Guldberg 1999;

Hoegh-Guldberg et al. 2007). These projections are borne out by a growing list of long-term studies that are reporting a serious contraction in reef-building corals over the past 30–50 years (Bruno and Selig 2007; De'ath et al. 2012; Gardner et al. 2003). Given the central role of massive corals to reefs, this is also associated with the loss of the three-dimensional structure of coral reefs. In addition, mesocosm studies like that of Dove et al. (2013) reveal rapid loss of three-dimensional structure as corals disappear under the pressure of future concentrations of atmospheric CO_2.

There is little doubt to many experts that the greatest threat to coral reefs and the hundreds of millions of people that depend on them is rapid ocean warming and acidification, due to the vulnerability of the mutualistic symbiosis between Symbiodinium and reef-building corals (Gattuso et al. 2014; Hoegh-Guldberg et al. 2014; Pörtner et al. 2014). When rapid changes occur, or conditions are pushed outside the environmental envelope in which corals normally exist, the symbiosis begins the breakdown (Glynn 1993, 2012; Hoegh-Guldberg 1999; Hoegh-Guldberg and Smith 1989). This results in mass exodus of the brown dinoflagellates out of the coral host's tissues, leaving the coral host alive but bleached white in appearance (hence the term "coral bleaching"; see Plate 2).

Around 1980, populations of corals began to bleach across large areas of the world's tropical regions, with no precedent in the scientific literature. Over time, information began to accumulate that suggested that these large-scale changes to coral reefs are associated with short periods of time when sea temperatures were 1°C–2°C above the long-term summer maxima for a particular region (Hoegh-Guldberg 1999). Since that time, the frequency and intensity of coral bleaching has increased, with the phenomenon being reported in most tropical and subtropical regions of the world. Although some coral reefs have recovered from mass coral bleaching and mortality, many reefs have not (Baker et al. 2008). These types of studies have confirmed the conclusions of modeling studies based on the thermal sensitivity of coral reefs and underlying mechanisms (Donner et al. 2007; Donner et al. 2005; Frieler et al. 2013; Hoegh-Guldberg 1999). Studies of bleaching and recovery show that even with aggressive action on CO_2 emissions from the burning of fossil fuels and deforestation (e.g., the Paris Agreement in December 2015; Davenport 2015; Schellnhuber et al. 2016), only 10 percent of today's corals will survive (Donner et al. 2005; Frieler et al. 2013; Hoegh-Guldberg 1999; Hoegh-Guldberg et al. 2014). However, given the importance of coral reefs in terms of biological diversity, ecosystem services, and human well-being, saving the remaining 10 percent of coral reefs will deliver benefits in terms of the regeneration of coral reefs under a stabilized climate, and hence should be a priority of leaders and policy makers (Pachauri et al. 2014).

It is very important to appreciate the challenge ahead. In 2012, a group of scientists at the Australian Institute of Marine Science published the results of a long-term study that rigorously investigated changes in the distribution and abundance of reef-building corals on the Great Barrier Reef. Their conclusions were startling. Not only had the growth rates of at least one group of reef-building corals across the Great Barrier Reef decreased significantly since 1990 (De'ath et al. 2009); the amount of coral on the Great Barrier Reef had declined by 50 percent since 1983 (De'ath et al. 2012). A range of factors were included, but it is clear when considering the three main factors (i.e., cyclones, starfish predation, and ocean warming) that the demise of the Great Barrier Reef is a complex combination of both local and global factors. Central to this observation is that the ecological changes under way on the Great Barrier Reef probably involve synergies and antagonisms that interact to produce an ecosystem that is losing resilience even while experiencing a greater number of impacts. Like any prizefighter,

the Great Barrier Reef can probably take a few heavy punches, but if the ability of its corals to calcify and grow is compromised, then it is not surprising that the impact of disturbances (i.e., loss of corals) will grow over time. These negative trends are occurring even though the Great Barrier Reef is considered to have very low human pressures on it as compared to reefs in other regions, and it has been managed by a well-resourced and modern marine park system that has often been held up as the bastion of good practice.

These heavy punches keep coming, though. As 2015 finished—the warmest year on record—tropical waters across the planet began to exhibit exceptional warming trends once again. This included the Great Barrier Reef, which exhibited sea surface temperatures between 1°C and 3°C higher than the long-term summer average in the first few months of 2016. By late March, large sections of the northern Great Barrier Reef were exhibiting mass coral bleaching on a scale never seen before (Hoegh-Guldberg and Ridgway 2016; Hughes et al. 2016). In some cases, more than 90 percent of corals on reefs were bleached; in another case, less bleaching was reported. There was a clear gradient in heat stress and the extent of mass coral bleaching over the 2,300 km range of the Great Barrier Reef, with reefs in the southern third seeing much lower temperatures and hence less mass coral bleaching. The resulting mortality of corals was also exceptional. By October and November, the northern sector of the Great Barrier Reef (about 800 km) was exhibiting an average loss of 67 percent of coral cover on sample of 60 reefs (Hughes et al. 2016).

These results indicate that projections made by scientists almost two decades ago are coming to pass, with the clear likelihood that more frequent and more intense bleaching events not only are a reality but also clearly threaten the very existence of the world's largest continuous coral reef. Following on from events in Pacific, this coral bleaching has affected a wide range of other regions, including the Maldives, Hawaii, Japan, Southeast Asia, and many other countries, resembling the events associated with the 1998 El Niño (Hoegh-Guldberg 1999; Hoegh-Guldberg and Ridgway 2016), albeit with greater intensity and mortality.

CONCLUSION: REDUCING THE RATE OF GLOBAL CHANGE TO ZERO AND ACTING DECISIVELY ON LOCAL FACTORS

There is no doubt about the importance and interconnectedness of the ocean to all aspects of life on our planet. Without the ocean functioning as it does today, life would be very different, if not impossible. Nonetheless, we are in the process of playing catchup in terms of understanding the impacts of global change on the ocean, whether from local or global sources. With the latest scientific consensus revealing that the ocean is undergoing changes that are unprecedented in 65 million years, if not 300 million years (Hönisch et al. 2012; IPCC 2013), there must be equally unprecedented action at a global level to solve the serious problems that face our watery world. In many ways, we know what the solutions are; it is a pathway to action that we now need to define and act on, and quickly.

In terms of strong action on this issue, it comes down to two major issues. The first is that we need to rapidly stabilize global conditions, whether they be rising temperature, changing ocean chemistry, or other aspects of climate. Without stabilization, the ability of organisms and ecosystems to adapt will be severely compromised, if not entirely impossible. As has been discussed elsewhere, the ability for adaptation to play a role in minimizing the impacts of a rapidly changing climate comes down to the stabilization of conditions, so that genetic processes can catch up with the rate of exceptional environmental change (Hoegh-Guldberg 2012). To stabilize the climate conditions sufficiently for megadiverse ecosystems such as coral reefs to meet this challenge, atmospheric CO_2 must be constrained well below the 450 ppm/2°C limit (Frieler et al. 2013), which requires meeting the goals of the Paris Agreement as quickly as possible (Davenport 2015; Schellnhuber et al. 2016). As emphasized by the events of 2016 on the Great Barrier Reef and the world's other coral reefs, we have probably underestimated the sensitivity of the world's biological diversity and ecosystems to climate change, further emphasizing the need for us to move even more rapidly to the goals of the Paris Agreement.

The second major action that we need is to rapidly reduce the influence of local stresses such as declining water quality, overexploitation, pollution, and the physical destruction of marine ecosystems such as coral reefs. In addition to the direct effects of these factors on the distribution and abundance of healthy coral reefs, there is the significant contribution that these local factors make to reducing the resilience of coral-dominated ecosystems to impacts and disturbances. Therefore, it is very important that we confront and solve issues such as unsustainable coastal deforestation and agriculture, as well as reforming the management of other stressors from the physical destruction of coral reefs to the management and reduction of overexploitation of fisheries and other reef uses. Direct action of this type has the potential to build the resilience of coral reefs and to buy important time for the global community as it struggles to rapidly reduce emissions of CO_2 and other dangerous greenhouse gases.

REFERENCES

Baker, Andrew C., Peter W. Glynn, and Bernhard Riegl. 2008. "Climate change and coral reef bleaching: An ecological assessment of long-term impacts, recovery trends and future outlook." *Estuarine Coastal and Shelf Science* 80 (4): 435–471.

Berner, Robert A., and Zavareth Kothavala. 2001. "GEO-CARB III: A revised model of atmospheric CO_2 over

Phanerozoic time." *American Journal of Science* 301 (2): 182–204.

Bongaerts, P., T. C. L. Bridge, D. Kline, P. Muir, C. Wallace, R. Beaman, and O. Hoegh-Guldberg. 2011. "Mesophotic coral ecosystems on the walls of Coral Sea atolls." *Coral Reefs* 30 no. 2: 335.

Bruno, John F., and Elizabeth R. Selig. 2007. "Regional decline of coral cover in the Indo-Pacific: timing, extent, and subregional comparisons." *PLOS One* 2 (8): e711.

Burke, Lauretta, Kathleen Reytar, Mark Spalding, and Allison Perry. 2011. *Reefs at Risk Revisited*. World Resources Institute.

Caldeira, Ken, and Michael E. Wickett. 2003. "Oceanography: Anthropogenic carbon and ocean pH." *Nature* 425 (6956): 365.

Cinner, Joshua E., Timothy R. McClanahan, Tim M. Daw, Nicholas A. Graham, Joseph Maina, Shaun K. Wilson, and Terence P. Hughes. 2009. "Linking social and ecological systems to sustain coral reef fisheries." *Current Biology* 19 (3): 206–212.

Costanza, Robert, Rudolf de Groot, Paul Sutton, Sander van der Ploeg, Sharolyn J. Anderson, Ida Kubiszewski, Stephen Farber, and R. Kerry Turner. 2014. "Changes in the global value of ecosystem services." *Global Environmental Change* 2: 152–158.

Costello, Mark John, Marta Coll, Roberto Danovaro, Pat Halpin, Henn Ojaveer, and Patricia Miloslavich. 2010. "A census of marine biodiversity knowledge, resources, and future challenges." *PLOS One* 5 (8).

Davenport, Coral. 2015. "Nations approve landmark climate accord in Paris." *New York Times*, December 12.

De'ath, Glenn, Katharina E. Fabricius, Hugh Sweatman, and Marji Puotinen. 2012. "The 27-year decline of coral cover on the Great Barrier Reef and its causes." *Proceedings of the National Academy of Sciences* 109 (44): 17995–17999.

De'ath, Glenn, Janice M. Lough, and Katharina E. Fabricius. 2009. "Declining coral calcification on the Great Barrier Reef." *Science* 323 (5910): 116–119.

Donner, Simon D., Thomas R. Knutson, and Michael Oppenheimer. 2007. "Model-based assessment of the role of human-induced climate change in the 2005 Caribbean coral bleaching event." *Proceedings of the National Academy of Sciences* 104 (13): 5483–5488.

Donner, Simon D., William J. Skirving, Christopher M. Little, Michael Oppenheimer, and Ove Hoegh-Guldberg. 2005. "Global assessment of coral bleaching and required rates of adaptation under climate change." *Global Change Biology* 11 (12): 2251–2265.

Dove, Sophie G., David I. Kline, Olga Pantos, Florent E. Angly, Gene W. Tyson, and Ove Hoegh-Guldberg. 2013. "Future reef decalcification under a business-as-usual CO_2 emission scenario." *Proceedings of the National Academy of Sciences* 110 (38): 15342–15347.

Freiwald, André, Jane Heldge Fosså, Anthony Grehan, Tony Koslow, and J. Murray Roberts. 2004. *Cold-Water Coral Reefs: Out of Sight—No Longer out of Mind*. Cambridge: UNEP-WCMC.

Frieler, K., M. Meinshausen, A. Golly, M. Mengel, K. Lebek, S. D. Donner, and O. Hoegh-Guldberg. 2013. "Limiting global warming to 2 degrees C is unlikely to save most coral reefs." *Nature Climate Change* 3 (2): 165–170.

Gardner, Toby A., Isabelle M. Côté, Jennifer A. Gill, Alastair Grant, and Andrew R. Watkinson. 2003. "Long-term region-wide declines in Caribbean corals." *Science* 301 (5635): 958–960.

Gattuso, J. P., P. G. Brewer, O. Hoegh-Guldberg, J. A. Kleypas, H. O. Pörtner, and D. N. Schmidt. 2014. "Cross-chapter box on ocean acidification." In *Climate Change 2014: Impacts, Adaptation, and Vulnerability. Part A: Global and Sectoral Aspects. Contribution of Working Group II to the Fifth Assessment Report of the Intergovernmental Panel of Climate Change*, ed. C. B. Field, et al., 129–131. Cambridge University Press.

Glynn, P. W. 1993. "Coral reef bleaching: Ecological perspectives." *Coral Reef* 12 (1): 1–17.

Glynn, Peter W. 2012. "Global warming and widespread coral mortality: Evidence of first coral reef extinctions." In *Saving a Million Species*, ed. Lee Hannah, 103–120. Island Press/Center for Resource Economics.

Hoegh-Guldberg, Ove. 1999. "Coral bleaching, climate change and the future of the world's coral reefs." *Marine and Freshwater Research* 50 (8): 839–866.

Hoegh-Guldberg, Ove. 2012. "The adaptation of coral reefs to climate change: Is the Red Queen being outpaced?" *Scientia Marina* 76 (2): 403–408.

Hoegh-Guldberg, Ove. 2014. "Coral reefs in the Anthropocene: Persistence or the end of the line?" *Geological Society, London, Special Publications* 395 (1): 167–183.

Hoegh-Guldberg, Ove, and John F. Bruno. 2010. "The impact of climate change on the world's marine ecosystems." *Science* 328 (5985): 1523–1528.

Hoegh-Guldberg, Ove, Peter J. Mumby, Anthony J. Hooten, Robert S. Steneck, Paul Greenfield, Edgardo Gomez, C. Drew Harvell, et al. 2007. "Coral reefs under rapid climate change and ocean acidification." *Science* 318 (5857): 1737–1742.

Hoegh-Guldberg, O., and T. Ridgway. 2016. "Coral bleaching comes to the Great Barrier Reef as record-breaking global temperatures continue." *The Conversation*, March 21, 2016.

Hoegh-Guldberg, Ove, and G. Jason Smith. 1989. "The effect of sudden changes in temperature, light and salinity on the population density and export of zooxanthellae from the reef corals Stylophora pistillata and Seriatopra hystrix." *Journal of Experimental Marine Biology and Ecology* 129 (3): 279–303.

Hoegh-Guldberg, O., et al. 2014. "The ocean." In *Climate Change 2014: Impacts, Adaptation, and Vulnerability. Part B: Regional Aspects. Contribution of Working Group II to the Fifth Assessment Report of the Intergovernmental Panel on Climate Change*, ed. V. R. Barros, et al., 1655–1731. Cambridge University Press.

Hönisch, Bärbel, Andy Ridgwell, Daniela N. Schmidt, Ellen Thomas, Samantha J. Gibbs, Appy Sluijs, Richard

Zeebe, et al. 2012. "The geological record of ocean acidification." *Science* 335 (6072): 1058–1063.

Hughes, T. P., B. Schaffelke, and J. Kerry. 2016. "How much coral has died in the Great Barrier Reef's worst bleaching event?" *The Conversation*, November 29, 2016.

IPCC. 2013. *Climate Change 2013: The Physical Science Basis. Contribution of Working Group I to the Fifth Assessment Report of the Intergovernmental Panel on Climate Change.* Cambridge University Press.

Kleypas, Joan A., Robert W. Buddemeier, David Archer, Jean-Pierre Gattuso, Chris Langdon, and Bradley N. Opdyke. 1999. "Geochemical consequences of increased atmospheric carbon dioxide on coral reefs." *Science* 284, no: 5411: 118–120.

Kleypas, Joan A., John W. McManus, and Lambert A. B. Meñez. 1999. "Environmental limits to coral reef development: Where do we draw the line?" *American Zoologist* 39 (1): 146–159.

Kroeker, Kristy J., Rebecca L. Kordas, Ryan Crim, Iris E. Hendriks, Laura Ramajo, Gerald S. Singh, Carlos M. Duarte, and Jean-Pierre Gattuso. 2013. "Impacts of ocean acidification on marine organisms: Quantifying sensitivities and interaction with warming." *Global Change Biology* 19 (6): 1884–1896.

Manzello, Derek P., Joan A. Kleypas, David A. Budd, C. Mark Eakin, Peter W. Glynn, and Chris Langdon. 2008. "Poorly cemented coral reefs of the eastern tropical Pacific: Possible insights into reef development in a high-CO(2) world." *Proceedings of the National Academy of Sciences* 105 (30): 10450–10455.

Moberg, Fredrik, and Carl Folke. 1999. "Ecological goods and services of coral reef ecosystems." *Ecological Economics* 29 (2): 215–233.

Muscatine, L. 1990. "The role of symbiotic algae in carbon and energy flux in reef corals." *Ecosystems of the World* 25: 75–87.

Muscatine, Leonard, and James W. Porter. 1977. "Reef corals: Mutualistic symbioses adapted to nutrient-poor environments." *Bioscience* 27 (7): 454–460.

Pachauri, Rajendra K., Myles R. Allen, Vicente R. Barros, John Broome, Wolfgang Cramer, Renate Christ, John A. Church, et al. 2014. *Climate Change 2014: Synthesis Report. Contribution of Working Groups I, II and III to the Fifth Assessment Report of the Intergovernmental Panel on Climate Change.* IPCC.

Poloczanska, Elvira S., Christopher J. Brown, William J. Sydeman, Wolfgang Kiessling, David S. Schoeman, Pippa J. Moore, Keith Brander, et al. 2013. "Global imprint of climate change on marine life." *Nature Climate Change* 3 (10): 919–925.

Poloczanska, E. S., O. Hoegh-Guldberg, W. Cheung, H. O. Pörtner, and M. Burrows. 2014. "Cross-chapter box on observed global responses of marine biogeography, abundance, and phenology to climate change." In *Climate Change 2014: Impacts, Adaptation, and Vulnerability. Part A: Global and Sectoral Aspects. Contribution of Working Group II to the Fifth Assessment Report of the Intergovernmental Panel of Climate Change,* ed. C. B. Field, et al., 123–127. Cambridge University Press.

Pörtner, Hans-Otto, David M. Karl, Philip W. Boyd, William Cheung, Salvador E. Lluch-Cota, Yukihiro Nojiri, Daniela N. Schmidt, et al. 2014. "Ocean systems." In *Climate Change 2014: Impacts, Adaptation, and Vulnerability. Part A: Global and Sectoral Aspects. Contribution of Working Group II to the Fifth Assessment Report of the Intergovernmental Panel on Climate Change,* 411–484. Cambridge University Press.

Reaka-Kudla, Marjorie L. 1997. "Global biodiversity of coral reefs: A comparison with rainforests." In *Biodiversity II: Understanding and Protecting Our Biological Resources,* ed. M. L. Reaka-Kudla, and D. E. Wilson, 551. Joseph Henry Press.

Schellnhuber, Hans Joachim, Stefan Rahmstorf, and Ricarda Winkelmann. 2016. "Why the right climate target was agreed in Paris." *Nature Climate Change* 6 (7): 649–653.

Stanley, George D., and Daphne G. Fautin. 2001. "The origins of modern corals." *Science* 291 (5510): 1913–1914.

Veron, J. E. N. 2008. *A Reef in Time: The Great Barrier Reef from Beginning to End.* Harvard University Press.

CHAPTER SIX

Genetic Signatures of Historical and Contemporary Responses to Climate Change

BRETT R. RIDDLE

Biologists have long been interested in the responses of lineages and biotas to grand changes in climates. Wallace (1880) associated the cold and warm cycles of "glacial epochs" to patterns of distributional change and the extinction of mammals in Europe and North America. Adams (1905) expanded on this theme by envisioning whole biotas in North America expanding their distributions from five ice-age "biotic preserves" ("refugia" in modern terminology) as continental glaciers receded. Haffer (1969) envisioned avian diversification in the Amazonian region during Pleistocene cycles of forest expansion and contraction.

Of course, we now know that climate change is the "norm" at different temporal and spatial scales, including multiple glacial-interglacial cycles over the past 2.5 million years (Ma) and a progressive shift from a generally warmer to cooler Earth beginning about 12 mya associated with Antarctic reglaciation (Zachos et al. 2001). Extant populations, species, and higher taxa represent the survivors of a long history of distributional, diversification, and adaptive responses to shifting climates—and the signatures of any or all of these histories are conceptually decipherable in the genetic variation recorded in different partitions of a genome.

Whereas the earliest attempts to decipher genetic variation within and among natural populations relied on sampling variation in the nuclear genome with techniques such as variation in chromosome morphology (Dobzhansky et al. 1966) and protein electrophoresis (e.g., Lewontin and Hubby 1966), the ability to reconstruct the genetic signatures of historical responses increased markedly over three decades ago, when phylogeography was introduced as a discipline that bridged a theoretical and empirical gap between phylogenetics above the species level and population genetics below it (Avise 2000). The sorts of responses that subsequently have been documented include distributional dynamics (range contraction and expansion, range shifting) and population subdivision, resulting in restrictions in gene flow (genetically differentiated populations) or cessation of gene flow (diversification of a single ancestral lineage into two or more evolutionarily distinct lineages). Phylogeography originally sampled genetic variation in natural populations by bypassing the complex mixture of selected, neutral, and relatively slowly evolving variation in the nuclear genome in favor of much smaller and simpler organelle genomes—mitochondrial DNA (mtDNA) in animals and fungi and chloroplast DNA (cpDNA) in plants (Avise 2000). Organelle genomes sampled mostly sex-biased (maternal) gene flow, and ancestral lineages could be "captured" through interspecific gene flow, but these disadvantages were somewhat countered by the advantageous quality of evolving and "fixing" new mutations in populations more rapidly than did much of the nuclear genome. Although early studies tended to be ad hoc and not statistically rigorous, changes in these attributes elevated the discipline to one that tested alternative hypotheses (e.g., zero, one, or multiple Pleistocene refugia) within a statistically defensible context (Hickerson et al. 2010). A smaller set of phylogeography studies are concerned with responses of groups of codistributed species, whether they simply occur together in a community or region or form more intimate biological and ecological associations (e.g., pollinators and plant hosts). These kinds of studies fall under the umbrella of comparative phylogeography (Avise et al. 2016), and they offer the potential to track the responses of partial or whole biotas to climate change.

Biologists have recognized that genetic diversity could be a product of gene flow and selection within and among populations on contemporary landscapes, influenced more by ecological attributes such as habitat connectivity or fragmentation, metapopulation structure, and recent adaptive evolution than by historical dynamics. This recognition generated the discipline of landscape genetics (Manel et al. 2003) and has expanded subsequently with access to genetic markers that evolve much more quickly than even the organelle markers that motivated the growth of phylogeography (e.g., microsatellites; restriction site associated DNA sequencing, or RADSeq; and other single nucleotide polymorphisms [SNPs] methods). It is not uncommon to recognize that one must understand population structure in terms of both historical and contemporary forces (Epps and Keyghobadi 2015; Rissler 2016), yet differences between phylogeography and landscape genetics have not been obvious in all studies, although attempts have been made to delineate the two in terms of spatiotemporal scales (Figure 6.1), optimal genetic markers, analytical approaches, and ecological versus evolutionary arenas (Epps and Keyghobadi 2015; Rissler 2016).

Phylogeography and landscape genetics alike have entered new phases with incorporation of next-generation sequencing (NGS) platforms (Thomson et al. 2010). On the one hand, these approaches offer new opportunities for phylogeographers to add a nuclear DNA component to studies in the form of phylogenomics (DeSalle and Rosenfeld 2013) that matches in resolution that of organelle DNA and to fish out genes involved with divergence and a history of adaptive evolution (e.g., Brewer et al. 2014; Jezkova et al. 2015; Schield et al. 2015). On the other hand, perhaps the most interesting form of contemporary response, adaptive evolution to current landscapes and climates, can now be investigated at a candidate gene level through the emerging potential of landscape genomics (e.g., Sork et al. 2016). Additionally, several newer techniques are increasing capacity to add an ancient DNA component to studies

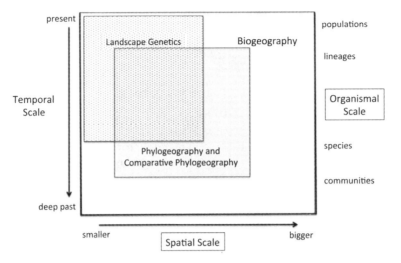

Figure 6.1. A depiction of the generally perceived relationship between phylogeography or comparative phylogeography and landscape genetics within the overarching realm of biogeographic research along three scales—temporal, spatial, and organismal. (From Rissler 2016.)

by broadly sampling across the mitochondrial and nuclear genomes of fecal samples and museum specimens (McCormack et al. 2016; van der Valk et al. 2017).

Rissler (2016) argued that with the increasing power of genetics to blur lines between spatiotemporal scales, nonadaptive versus adaptive variation, and ecological versus evolutionary contexts, phylogeography and landscape genetics represents a continuum of questions within biogeography (Table 6.1). Nevertheless, in the remainder of this chapter I concentrate largely on the phylogeographic component of this continuum because it is the approach that captures an integrated signature of responses to a history of responses to climate change, both within species and among closely related species—at spatial and temporal scales that often are commensurate with delineation of biotic reserves, biodiversity hotspots, and large-scale range dynamics. This scale is receiving increasing recognition as critical for establishing a spatial and temporal evolutionary framework within which to design studies at the landscape scale (Epps and Keyghobadi

2015). I close with several additional observations regarding the potential of landscape genomics, perhaps in combination with phylogeography, to address the potential of populations and species to respond to climate change through ongoing population-level processes, including adaptive evolution. I concentrate primarily on nonmarine biodiversity, although a number of the issues discussed are relevant in the marine realm as well (e.g., Bowen 2016).

THE RELEVANCE OF A PHYLOGEOGRAPHIC PERSPECTIVE

Phylogeographic studies increasingly have incorporated data from nDNA for good theoretical and empirical reasons (e.g., Edwards and Bensch 2009)—often along with organelle DNA, for a multiple genome perspective. Organelle-DNA-based studies are still employed frequently, to good advantage, in many phylogeography studies, for reasons including ease of preparation and analysis and relatively low cost, which allows for the maximization of numbers of populations and individuals that can be sampled (Riddle 2016); although technical advances continue to lower the cost of nDNA approaches such as RADSeq (Andrews et al. 2016) and other SNPs methods (Leaché et al. 2017). Moreover, the signal

Table 6.1. Temporally scaled continua among several topics and questions associated with historical biotic responses to climate change, addressed with genetic approaches

Time frame (years before present)	Forcing events	Dominant geographic processes	Biogeographic context	Analytical realm	Overarching question
10,000–1,000,000	Multiple glacial-interglacial cycles	Inter- and intraspecific range dynamics; refugia; community or biota assembly-disassembly	Phylogeography	Inter- and intraspecific phylogenetics	How do species and assemblages of species become isolated, connected, and interact at contact zones; what patterns of biotic diversification result from glacial-interglacial climate cycles?
1,000–10,000	Holocene warming-cooling episodes; anthropogenically modified landscapes	Intraspecific range dynamics; community assembly-disassembly	Phylogeography; landscape genetics	Intraspecific phylogenetics; population genetics	How do recent landscapes interact with paleoclimates to isolate populations and facilitate or restrict gene flow among them?
0–1,000	Ongoing warm-cool phases; anthropogenically modified climates and landscapes	Shifting population structure; gene flow dynamics	Landscape genetics	Population genetics	How do current and recent landscapes interact with climate, etc., to facilitate or restrict gene flow among populations?

(at least in animal mtDNA) of variation is usually stronger than it is for most nDNA data sets, allowing for establishment of a first approximation of genetic architecture that can be treated as an hypothesis to be addressed with nDNA for concordance across genomes, episodes of past hybridization, and so on (Barrowclough and Zink 2009). However, increasing access to NGS approaches to sampling the nuclear genome could motivate the transition of phylogeography from predominantly mtDNA and cpDNA to a more inclusive genomic framework in the foreseeable future (Edwards et al. 2015).

Within the broader purview of conservation genetics, phylogeography is one component of a triad of past, present, and future priorities that have different objectives and use different approaches (Bowen 2016). Under Bowen's (2016) framework, the past is explored in a phylogenetic context—how many species or lineages, how different they are from one another, and the like (often the units that conservation biology has used, rather contentiously, to delineate evolutionarily significant units; Casacci et al. 2014); the present is concerned with the ecology of population genetic variation, herein discussed as "landscape genetics" (again, often considered in conservation biology as management units); and the future is the realm of topics such as dispersal potential among geographically distributed populations and range shifting under changing climates, explored within a phylogeographic context. Although I agree in general with Bowen's (2016) perspective, I view the role of phylogeography in conservation genetics as incorporating an historical framework in two ways. First, this is an approach that often has revealed "cryptic lineages" that are the product of past divergence, unknowable prior to phylogeographic-scale studies that are then made available for assessment through phylogenetics analyses (Riddle and Hafner 2006). Second, much of empirical phylogeographic research has been concerned with the reconstruction of past range dynamics, for example, into and out of refugia, which forms an important basis for trying to predict responses to future climate change (Sanchez-Ramirez et al. 2015).

How relevant has phylogeography been within the context of climate change and biodiversity research? A topics search of the Web of Science database on June 12, 2016, using "phylogeograph*" returned 21,036 citations, confirming the overall popularity of this approach; while a search using both "phylogeograph*" and "climate change*" returned 1,131; another using "phylogeograph*" and "biodiversity" returned 1,475; and another using "phylogeograph*," "biodiversity," and "climate change*" yielded 235 citations. In the same search sequence as above, if "phylogeograph*" is replaced with "comparative phylogeograph*" the numbers returned are 151, 171, and 36, respectively, indicating a detectable interest in the topic at a codistributed lineages scale.

GEOGRAPHIC RESPONSES TO PAST CLIMATE CHANGE

A large body of phylogeographic work is focused temporally on the responses of lineages and biotas to climate change during the latest transition from glacial to interglacial climates (Riddle 2016), often referencing the latest Quaternary—that is, since the Late Glacial Maximum, LGM, about 25 kya (thousands of years ago). However, a focus only on this single episode of climate transition, though pronounced, would be an oversimplification in considering the influence of historical climate change on patterns of distribution, diversification, and adaptive evolution. For example, in a review of comparative phylogeography in continental biotas (Riddle 2016), a substantial portion of studies focused on Neogene, or both Neogene and Quaternary time frames, thereby not only covering that portion of Earth history trending from generally warmer to cooler climates over the past few millions of years (Zachos et al. 2001) but also within that expanded time frame, requiring attention to Neogene geological transformations (e.g., plateau and mountain uplift, closure of the Isthmus of Panama) that are coupled with large-scale climate changes.

Even a relatively short interval of time, the Pleistocene-Holocene transition, should not be interpreted as a single event with stable climates before and after. The Younger Dryas represented an episode of recooling following several thousand years of warming, starting about 12.9 kya (Renssen et al. 2015) and ending about 11.5 kya. Other Holocene anomalies include a cooling event 8.2 kya (Morrill and Jacobsen 2005) and Mid-Holocene "warm period" in the Northern Hemisphere about 6 kya (Bartlein et al. 2011). Smaller and more recent events such as the Medieval Warm Period of the ninth to thirteenth centuries and the Little Ice Age of the fifteenth to nineteenth centuries illustrate the point that lineages and biotas must frequently respond to variable climates, although responses to these smaller and temporally much more recent events likely would be different in kind and at smaller spatial scales and therefore are not likely to register through a typical phylogeographic approach. Rather, landscape genetics (Epps and Keyghobadi 2015;

Rissler 2016) might become a more useful framework for measuring response within these more recent time frames and typically smaller spatial scales.

Frequently, populations and species have avoided extinction through changes in geographic distribution that tracked shifting habitats. Often, this sort of geographic dynamic is envisioned as a range contraction into a refugial phase (Haffer 1969; Arenas et al. 2012), followed by range expansion as preferred habitats expand within a different climatic regime (Hewitt 2004; Mona et al. 2014). Should an ancestral lineage be subdivided into more than a single refuge, the microevolutionary processes of genetic drift and selection could act as drivers of diversification, from the production of genetic structure among populations to the generation of evolutionarily distinct lineages and perhaps even new species (Damasceno et al. 2014). Genes may be exchanged again once previously isolated populations come into contact at contact zones (if envisioned as happening between multiple taxa, called suture zones; Remington 1968). Contact zones are places that have been investigated for evidence that complete reproductive isolation was established during refugial isolation. For example, Schield et al. (2015) used a combination of mtDNA and nDNA RADSeq in the western diamondback rattlesnake (*Crotalus atrox*) to corroborate earlier findings that populations were historically isolated into eastern and western populations long enough to be recognizable as distinct mtDNA evolutionary lineages. By contrast, whereas the nDNA data supported an episode of historical isolation, they demonstrated ongoing gene flow following secondary contact. Should such a process of refugial isolation or range expansion occur across multiple glacial-interglacial cycles, it has the potential to be Pleistocene "species pumps," generators of new biodiversity detectable at the level of distinct evolutionary lineages, an idea that harkens back to the tropical forest refugia model of Haffer (1969) but was subsequently expanded to

include, for example, montane habitat islands (Toussaint et al. 2013), land-bridge islands (Papadopoulou and Knowles 2015), and deserts (Riddle and Hafner 2006). Indeed, the same contact zone investigated in detail for western diamondback rattlesnake dynamics is also a point of contact for multiple arid-adapted species (Riddle 2016), suggesting a biota-wide pattern of responses to historical climate change with the potential to elucidate taxonomic, phylogenetic, and ecological attributes contributing to how a particular lineage has responded in the past (Pyron and Burbrink 2007).

Nevertheless, the most immediate impacts from current and future climate change likely will occur over shorter time frames than generally thought to be necessary to foster generation of new evolutionary lineages and species. Therefore, it seems appropriate to emphasize the lessons from history that focus more directly on impacts from the redistribution of populations, species, and assemblages of species on the landscape (range contraction, perhaps into refugia, and expansion; range shifting). Indeed, these sorts of responses have been popular topics for investigation—a topics search of the Web of Science database on June 28, 2016, using "phylogeograph*," "pleistocene," and "refug*" returned 1,250 studies. A search replacing the first term with "comparative phylogeograph*" returned 185 studies, suggesting interest in the question of whether codistributed lineages respond in a concerted fashion by sharing refugia (Hewitt 2004), which does not always occur in a simple way (Stewart et al. 2010; Pelletier et al. 2015).

Variants of this common theme are postulated refugial distribution during the Last Interglacial (LIG) rather than glacial (LGM) stage, predicted by habitat preferences (Stewart et al. 2010; Latinne et al. 2015); consequences of microrefugia on population structure, diversification, and range dynamics under climate change (Hannah et al. 2014; Mee et al. 2014); relative distributional stasis during the LGM (Jezkova

et al. 2011); and predictions that stable refugial areas represent genetic diversity
hotspots, perhaps worthy of focal attention
as high-priority conservation management
areas (Wood et al. 2013). Most of the recent
such studies employ a species distribution
modeling (SDM) procedure to create an
independent model of predicted range dynamics that is then available for evaluation
through phylogeographic analysis. For example, Pelletier et al. (2015) used hindcast
models of predicted species distribution at
21 kya in three congeneric species of salamander from northwestern North America
to support an inference that the geographic
range of only one of the three species likely
contracted into multiple glacial-age refugia,
driving evolution of distinct lineages detectable genetically among extant populations.
Scaling up of such studies could be used at a
codistributed species scale to identify range
dynamics associated with putative biodiversity hotspots.

A generalized consensus on historical
distributional responses to climate change
has not appeared, but it is entirely possible
that each "system" (e.g., lineage or codistributed lineages, biome, landscape, biogeographic region) requires independent evaluation to establish a baseline for predicting
future distributional responses. Nevertheless, refined analytical tools are demonstrating promise in the ongoing growth of approaches to modeling the historical ranges
of intraspecific populations and locations of
refugia (Rosauer et al. 2015).

In what ways does ecology matter in
determining a particular response to past
climate change and in explaining idiosyncratic responses to the same underlying
climatic transitions? Angert and Schemske
(2005) addressed this question experimentally with reciprocal transplants of two
species of monkey flower (Mimulus cardinalis
and M. lewisii) and presented evidence for reduced fitness in each species at range margins. Jezkova et al. (2015) took advantage of
a postulated post-LGM route of northward
range expansion of arid-adapted vertebrates

and invertebrates into the western Great Basin, United States, from historically late glacial "stable" (or refugial) populations to the
south in the Mojave Desert. They contrasted
mtDNA and RADseq sequencing to produce
nDNA SNP data sets for two codistributed
kangaroo rat species, one a sand substrate
specialist with patchy distributions (Dipodomys deserti) and the other a substrate generalist with more connected populations (D.
merriami). On the basis of well-established
models that predicted that there would be
a loss of genetic diversity in recently expanded populations (Hewitt 2004), with
loss more extreme in D. deserti due to smaller
effective populations with less gene flow
between them. First, SDM modeling supported the prediction that the ranges of
both species were contracted into a southern (Mojave Desert) distribution during the
LGM (perhaps not surprising given that
much of their expanded ranges were part
of a large Pleistocene lake at that time). Second, the more rapid and extreme loss of genetic diversity in the substrate specialist was
confirmed for mtDNA. Third, loss of nDNA
variation mirrored mtDNA qualitatively but
was quantitatively not as extreme, perhaps
because either effective population sizes of
organelle DNA are smaller or because of
sex-biased dispersal. Thus, organelle DNA
may be a more sensitive marker recording
range expansions out of Pleistocene refugia,
but it may not accurately reflect adaptive
niche evolution in the nuclear genome.

NICHE EVOLUTION UNDER
CLIMATE CHANGE

Do niches evolve as populations and species respond to climate change? Tradition-
ally, biogeographers and evolutionary biologists have framed this question through
the dichotomy of "stay at home and evolve
new niches" or "move and retain niches"
as habitats shift across space (Weeks et al.
2014). Further, much thinking about climatically driven niche evolution has fo-

cused on an interspecific or higher taxon scale, often at time frames deeper than the latest Pleistocene to Holocene shifts (Cooney et al. 2016). However, of greater relevance to the issue of responses to ongoing and future climate change would be a focus on niche evolution, or stasis, as intraspecific populations shift distributions (Wüest et al. 2015; Lancaster et al. 2016). A relevant question here is this: how rapidly might we expect populations to undergo adaptive evolutionary responses to newly encountered climatic regimes, whether they stay at home or move? Modeling can contribute to establishing a theoretical framework, exploring parameters including landscape structure (smooth gradient vs. patchy), genetic architecture (many loci with small phenotypic effect vs. few with large effect), average dispersal distances, and interactions among all these variables (Schiffers et al. 2014).

Empirical evidence will continue to come from genomic studies that can identify genes under selection, but if focused on the last major glacial (LGM) to interglacial transition, we may not expect to have sufficient resolution in traditional Sanger DNA sequencing techniques to be able to find that gene or suite of genes driven to mutation and fixation of new genotypes under selection across such a short time frame. New opportunities arise, however, through the use of NGS approaches as applied to both DNA and RNA (transcriptome) assessments of genome evolution (Brewer et al. 2014). Focusing on a phylogeographic context, larger amounts of genome-wide data than provided through popular subsampling procedures such as RADSeq might be required to detect selection in natural populations, particularly if linkage disequilibrium is low (Edwards et al. 2015). One such study identified SNPs from transcriptome data and the mtDNA *NADH* gene in the American pika (*Ochotona princeps*), a focal taxon for investigating the effects of climate change in western North America, suggesting local adaptation to different thermal

and respiratory conditions along elevation gradients (Lemay et al. 2013).

RESPONSES TO CONTEMPORARY CLIMATE CHANGE

Climate change during the Anthropocene is the broad time frame of focal interest in attempting to predict consequences of human-mediated activities for biodiversity. Yet developing an empirical context for the role of contemporary landscapes on genetic structure within and between populations can be challenging owing to the inherent lag between the time of a relatively recent but historical change in landscape configuration—either through climate change alone or with some combination of physical habitat alteration. Given the recency of many of these changes, extant measures of genetic diversity across a landscape may not reflect current configurations but rather retain a time-lag signature of past conditions. One important emphasis in the ongoing development of a landscape genetics that has relevance for conservation management (Keller et al. 2015) is the development of ways to sort historical from contemporary influences on genetic structure and gene flow, and the building of a cross-referencing conceptual and analytical environment between phylogeography and landscape genetics will be important for doing so (Epps and Keyghobadi 2015; Rissler 2016).

As landscape genetics progresses into an NGS genomics framework, opportunities to investigate more explicitly the relationship between adaptive evolution and local selection regimes should increase. Sork et al. (2016) summarize work on climate-associated candidate genes in the California endemic oak *Quercus lobata*, including those associated with bud burst and flowering, growth, osmotic stress, and temperature stress; 10 out of 40 candidate genes indicated spatially divergent selection. Extensions of this work should include much larger samples of SNPs drawn from NGS approaches to sampling the genome (Sork et al. 2016).

LESSONS: EXTINCTION RESILIENCE AND EXTINCTION RISK

An important issue with regard to future biodiversity is whether and with what degree of accuracy can inferences about historical and contemporary responses to climate change be used as a guide to predict responses under rapidly changing climates. Herein, I provided an overview of several ways in which phylogeographers have developed capacity to decipher genetic signatures of what can be called *extinction resilience*—those distributional, demographic, and adaptive properties that have buffered populations, species, and biotas from extinction within time frames that have witnessed profoundly changing climates. What is required is the ongoing development of connections between these insights and an emerging framework for being able to predict extinction risk as populations, species, and biotas are required to track changing climates into the future. A combined effort that incorporates, for example, NGS and more traditional genomics, SDMs, explicit linkages between landscapes and genetics, and connections between genotypic and phenotypic evolution in an explicit climatic niche context—at population, species, and biotic scales—promises to advance understanding of the future of biodiversity under climate change.

REFERENCES

Adams, Chas C. 1905. "The postglacial dispersal of the North American biota." *Biological Bulletin* 9 (1): 53–71.

Andrews, Kimberley R., Jeffrey M. Good, Michael R. Miller, Gordon Luikart, and Paul A. Hohenlohe. 2016. "Harnessing the power of RADSeq for ecological and evolutionary genomics." *Nature Reviews Genetics* 17: 81–92.

Angert, A. L., and D. W. Schemske. 2005. "The evolution of species' distributions: Reciprocal transplants across the elevation ranges of *Mimulus cardinalis* and *M. lewisii*." *Evolution* 59 (8): 1671–1684.

Arenas Miguel, Nicholas Ray, Mattias Currat, and Laurent Excoffier. 2011. "Consequences of range contractions and range shifts on molecular diversity." *Molecular Biology and Evolution* 29 (1): 207–218.

Avise, John C. 2000. *Phylogeography: The History and Formation of Species.* Harvard University Press.

Avise, John C., Brian W. Bowen, and Francisco J. Ayala. 2016. "In light of evolution X: Comparative phylogeography." *Proceedings of the National Academy of Sciences* 113 (29): 7957–7961.

Barrowclough, George F., and Robert M. Zink. 2009. "Funds enough, and time: mtDNA, nuDNA and the discovery of divergence." *Molecular Ecology* 18 (14): 2934–2936.

Bartlein, P. J., S. P. Harrison, S. Brewer, S. Connor, B. A. S. Davis, K. Gajewski, J. Guiot, T. I. Harrison-Prentice, A. Henderson, O. Peyron, I. C. Prentice, M. Scholze, H. Seppa, B. Shuman, S. Sugita, R. S. Thompson, A. E. Viau, J. Williams, and H. Wu. 2011. "Pollen-based continental climate reconstructions at 6 and 21 ka: A global synthesis." *Climate Dynamics* 37 (3–4): 775–802.

Bowen, Brian W. 2016. "The three domains of conservation genetics: Case histories from Hawaiian waters." *Journal of Heredity* 107 (4): 309–317.

Brewer, Michael S., Darko D. Cotoras, Peter J. P. Croucher, and Rosemary G. Gillespie. 2014. "New sequencing technologies, the development of genomics tools, and their applications in evolutionary arachnology." *Journal of Arachnology* 42 (1): 1–15.

Casacci, L. P., F. Barbero, and E. Balletto. 2014. "The 'evolutionarily significant unit' concept and its applicability in biological conservation." *Italian Journal of Zoology* 81 (2): 182–193.

Cooney, Christopher R., Nathalie Seddon, and Joseph A. Tobias. 2016. "Widespread correlations between climatic niche evolution and species diversification in birds." *Journal of Animal Ecology* 85 (4): 869–878.

Damasceno, Roberto, Maria L. Strangas, Ana C. Carnaval, Miguel T. Rodrigues, and Craig Moritz. 2014. "Revisiting the vanishing refuge model of diversification." *Frontiers in Genetics* 5: 1–12.

DeSalle, Rob, and Jeffrey A. Rosenfeld. 2013. *Phylogenomics: A Primer.* Garland Science.

Dobzhansky, T., W. W. Anderson, and O. Pavlovsky. 1966. "Genetics of natural populations: XXXVIII. Continuity and change in populations of *Drosophila pseudoobscura* in western United States." *Evolution* 20 (3): 418–427.

Edwards, Scott, and Staffan Bensch. 2009. "Looking forwards or looking backwards in avian phylogeography? A comment on Zink and Barrowclough 2008." *Molecular Ecology* 18 (14): 2930–2933.

Edwards, Scott V., Allison J. Shultz, and Shane Campbell-Staton. 2015. "Next-generation sequencing and the expanding domain of phylogeography." *Folia Zoologica* 64: 187–206.

Epps, Clinton W., and Nusha Keyghobadi. 2015. "Landscape genetics in a changing world: Disentangling historical and contemporary influences and inferring change." *Molecular Ecology* 24 (24): 6021–6040.

Haffer, Jürgen. 1969. "Speciation in Amazonian forest birds." *Science* 165 (3889): 131–137.

Hannah, Lee, Lorraine Flint, Alexandra D. Syphard, Max A. Moritz, Lauren B. Buckley, and Ian M. McCullough. 2014. "Fine-grain modeling of species' response to climate change: Holdouts, stepping-stones, and microrefugia." *Trends in Ecology & Evolution* 29 (7): 390–397.

Hewitt, G. M. 2004. "Genetic consequences of climatic oscillations in the Quaternary." *Philosophical Transactions of the Royal Society of London Series B—Biological Sciences* 359 (1442): 183–195.

Hickerson, M. J., B. C. Carstens, J. Cavender-Bares, K. A. Crandall, C. H. Graham, J. B. Johnson, L. Rissler, P. F. Victoriano, and A. D. Yoder. 2010. "Phylogeography's past, present, and future: 10 years after." *Molecular Phylogenetics and Evolution* 54 (1): 291–301.

Jezkova, Tereza, Viktoria Olah-Hemmings, and Brett R. Riddle. 2011. "Niche shifting in response to warming climate after the last glacial maximum: inference from genetic data and niche assessments in the chisel-toothed kangaroo rat (Dipodomys microps)." *Global Change Biology* 17 (11): 3486–3502.

Jezkova, Tereza, Brett R. Riddle, Daren C. Card, Drew R. Schield, Mallory E. Eckstut, and Todd A. Castoe. 2015. "Genetic consequences of postglacial range expansion in two codistributed rodents (genus Dipodomys) depend on ecology and genetic locus." *Molecular Ecology* 24 (1): 83–97.

Keller, Daniela, Rolf Holderegger, Maarten J. van Strien, and Janine Bolliger. 2015. "How to make landscape genetics beneficial for conservation management?" *Conservation Genetics* 16 (3): 503–512.

Lancaster, Lesley T., Rachael Y. Dudaniec, Pallavi Chauhan, Maren Wellenreuther, Erik I. Svensson, and Bengt Hansson. 2016. "Gene expression under thermal stress varies across a geographical range expansion front." *Molecular Ecology* 25 (5): 1141–1156.

Leaché, Adam D., and Jamie R. Oaks. 2017. "The utility of single nucleotide polymorphism (SNP) data in phylogenetics." *Annual Review of Ecology, Evolution, and Systematics* 48: 69–84.

Lemay, Matthew A., Philippe Henry, Clayton T. Lamb, Kelsey M. Robson, and Michael A. Russello. 2013. "Novel genomic resources for a climate change sensitive mammal: characterization of the American pika transcriptome." *BMC Genomics* 14 (1): 311. https://doi.org/10.1186/1471-2164-14-311.

Lewontin, Richard C., and Jack L. Hubby. 1966. "A molecular approach to the study of genic heterozygosity in natural populations. II. Amount of variation and degree of heterozygosity in natural populations of Drosophila pseudoobscura." *Genetics* 54 (2): 595–609.

McCormack, John E., Whitney L. E. Tsai, and Brant C. Faircloth. 2016. "Sequence capture of ultraconserved elements from bird museum specimens." *Molecular Ecology Resources* 16: 1189–1203.

Mee, Jonathan A., and Jean-Sébastien Moore. 2014. "The ecological and evolutionary implications of microrefugia." *Journal of Biogeography* 41 (5): 837–841.

Manel, Stéphanie, Michael K. Schwartz, Gordon Luikart, and Pierre Taberlet. 2003. "Landscape genetics: Combining landscape ecology and population genetics." *Trends in Ecology & Evolution* 18 (4): 189–197.

Mona, Stefano, Nicolas Ray, Miguel Arenas, and Laurent Excoffier. 2014. "Genetic consequences of habitat fragmentation during a range expansion." *Heredity* 112 (3): 291–299.

Morrill, Carrie, and Robert M. Jacobsen. 2005. "How widespread were climate anomalies 8200 years ago?" *Geophysical Research Letters* 32 (19): L19701.

Papadopoulou, Anna, and L. Lacey Knowles. 2015. "Genomic tests of the species-pump hypothesis: Recent island connectivity cycles drive population divergence but not speciation in Caribbean crickets across the Virgin Islands." *Evolution* 69 (6): 1501–1517.

Pelletier, Tara A., Charlie Crisafulli, Steve Wagner, Amanda J. Zellmer, and Bryan C. Carstens. 2015. "Historical species distribution models predict species limits in western Plethodon salamanders." *Systematic Biology* 64 (6): 909–925.

Pyron, R. Alexander, and Frank T. Burbrink. 2007. "Hard and soft allopatry: Physically and ecologically mediated modes of geographic speciation." *Journal of Biogeography* 37 (10): 2005–2015.

Remington, Charles L. 1968. "Suture-zones of hybrid interaction between recently joined biotas." In *Evolutionary Biology*, ed. T. Dobzhansky, M. K. Hecht, and W. C. Steere, 321–428. New York: Plenum Press.

Renssen, Hans, Aurélien Mairesse, Hugues Goosse, Pierre Mathiot, Oliver Heiri, Didier M. Roche, Kerim H. Nisancioglu, and Paul J. Valdes. 2015. "Multiple causes of the Younger Dryas cold period." *Nature Geoscience* 8 (12): 946–980.

Riddle, Brett R. 2016. "Comparative phylogeography clarifies the complexity and problems that drove A. R. Wallace to favor islands." *Proceedings of the National Academy of Sciences* 113 (29): 7970–7977.

Riddle, Brett R., and David J. Hafner. 2006. "A step-wise approach to integrating phylogeographic and phylogenetic biogeographic perspectives on the history of a core North American warm deserts biota." *Journal of Arid Environments* 66 (3): 435–461.

Rissler, Leslie J. 2016. "Union of phylogeography and landscape genetics." *Proceedings of the National Academy of Sciences* 113 (29): 8079–8086.

Rosauer, Dan F., Renee A. Catullo, Jeremy VanDerWal, Adnan Moussalli, and Craig Moritz. 2015. "Lineage range estimation method reveals fine-scale endemism linked to Pleistocene stability in Australian rainforest herpetofauna." *PLOS One* 10 (5): e0126274.

Schield, Drew R., Daren C. Card, Richard H. Adams, Tereza Jezkova, Jacobo Reyes-Velasco, F. Nicole Proctor,

Carol L. Spencer, Hans-Werner Herrmann, Stephen P. Mackessy, and Todd A. Castoe. 2015. "Incipient speciation with biased gene flow between two lineages of the western diamondback rattlesnake (Crotalus atrox)." Molecular Phylogenetics and Evolution 83: 213–223.

Schiffers, Katja, Frank M. Schurr, Justin M. J. Travis, Anne Duputié, Vincent M. Eckhart, Sébastien Lavergne, Greg McInerny, et al. 2014. "Landscape structure and genetic architecture jointly impact rates of niche evolution." Ecography 37 (12): 1218–1229.

Sork, Victoria L., Kevin Squire, Paul F. Gugger, Stephanie E. Steele, Eric D. Levy, and Andrew J. Eckert. 2016. "Landscape genomic analysis of candidate genes for climate adaptation in a California endemic oak, Quercus lobata." American Journal of Botany 103 (1): 33–46.

Stewart, John R., Adrian M. Lister, Ian Barnes, and Love Dalen. 2010. "Refugia revisited: Individualistic responses of species in space and time." Proceedings of the Royal Society B—Biological Sciences 277 (1682): 661–671.

Thomson, Robert C., Ian J. Wang, and Jarrett R. Johnson. "Genome-enabled development of DNA markers for ecology, evolution and conservation." Molecular Ecology 19 (11): 2184–2195.

Toussaint, Emmanuel F. A., Katayo Sagata, Suriani Surbakti, Lars Hendrich, and Michael Balke. 2013. "Australasian sky islands act as a diversity pump facilitating peripheral speciation and complex reversal from narrow endemic to widespread ecological supertramp." Ecology and Evolution 3 (4): 1031–1049.

Van der Valk, Tom, Frida Lona Durazo, Love Dalén, and Katerina Guschanski. 2017. "Whole mitochondrial genome capture from faecal samples and museum-preserved specimens." Molecular Ecology Resources 17: e111–e121.

Wallace, A. R. 1880. Island Life, or the Phenomena and Causes of Insular Faunas and Floras. Macmillan.

Weeks, Andrea, Felipe Zapata, Susan K. Pell, Douglas C. Daly, John D. Mitchell, and Paul V. A. Fine. 2014. "To move or to evolve: Contrasting patterns of intercontinental connectivity and climatic niche evolution in 'Terebinthaceae' (Anacardiaceae and Burseraceae)." Frontiers in Genetics 5: 1–20.

Wood, Dustin A., Amy G. Vandergast, Kelly R. Barr, Rich D. Inman, Todd C. Esque, Kenneth E. Nussear, and Robert N. Fisher. 2013. "Comparative phylogeography reveals deep lineages and regional evolutionary hotspots in the Mojave and Sonoran Deserts." Diversity and Distributions 19 (7): 722–737.

Wüest, Rafael O., Alexandre Antonelli, Niklaus E. Zimmermann, and H. Peter Linder. 2015. "Available climate regimes drive niche diversification during range expansion." American Naturalist 185 (5): 640–652.

Zachos, James, Mark Pagani, Lisa Sloan, Ellen Thomas, and Katharina Billups. 2001. "Trends, rhythms, and aberrations in global climate 65 Ma to present." Science 292 (5517): 686–693.

Climate Change and Salmon Populations

Donald J. Noakes

Natural populations of salmon (*Oncorhychus* and *Salmo*) are widely distributed throughout the Northern Hemisphere. In addition to their ecological, social, and cultural importance, a number of Atlantic and Pacific salmon stocks support significant commercial, recreational, and Aboriginal (First Nation) fisheries. Consequently, reasonable long-term records of catch or abundance exist for many salmon populations, and these provide convincing evidence that most populations have fluctuated between periods of high and low abundance for hundreds and perhaps thousands of years. Although some geographic differences are evident, the widespread synchronous changes in salmon abundance between periods of low and high productivity appear to be closely associated with regional or global changes in climate (Noakes and Beamish 2011). However, each species has exhibited distinct responses to climate change, with some species thriving while others struggle to survive and avoid extinction within the same climate regime. This apparent paradox highlights the unique nature of salmon, as they are available in almost unlimited numbers through aquaculture or via hatchery production while at the same time listed as an endangered species within the same geographic region or even within the same river system. The current state of global warming is no different, with some species and stocks of salmon in the North Pacific exhibiting very high abundance and other stocks threatened with extinction.

Salmon are anadromous, so climate change could affect survival and/or growth or other factors (biological or ecological) in either the freshwater or the marine stages of their life cycle. While large-scale hatchery programs for many species of salmon have made it more difficult to assess and quantify the effects of climate change on salmon populations, there is convincing evidence that climate change has affected both wild and hatchery salmon (Noakes and Beamish 2011). Hatchery programs have also caused unanticipated consequences, both biological and ecological, and the additional complexity associated with climate change means that more risk-averse approaches to salmon management and conservation will be required to address the goals, objectives, and expectations of various competing user groups and stakeholders.

A number of regional indexes have been developed to examine the impact of climate change on salmon. Typically, these involve either sea-level pressure as an index of ocean mixing and/or upwelling productivity or sea surface temperature as a measure of ocean warming (Noakes and Beamish 2011). Regionally, the various climate indexes tend to be significantly correlated, but the correlation is less between the Atlantic and the Pacific. In the Pacific, regime shifts (changes in the climate and associated changes in salmon abundance) were identified in 1925, 1947, 1977, 1989, and 2000. Similar regime shifts are evident in the Atlantic about 1925 and 1977, as well as a change in the late 1980s or early 1990s. These regime shifts represent changes in the physical environment in which the salmon live, and although the direction (positive or negative) of the change is suggestive of how salmon populations will respond, the magnitude of the environmental change is not necessarily proportional to the change observed in salmon abundance or survival.

The six commercial species of salmon are all anadromous, and each species has been affected differently by climate change. Pink (*O. gorbuscha*) and chum (*O. keta*) salmon spend relatively little time in freshwater, and wild populations of each are heavily supplemented by extensive hatchery programs throughout the Pacific. Coho (*O. kisutch*), Chinook (*O. tshawytscha*), and Atlantic (*S. salar*) typically spend two or more years in their natal rivers or streams (or hatchery) before migrating to the ocean, where they will rear for another year or more before returning to spawn. Sockeye (*O. nerka*) typically spend two years in rearing lakes or streams before migrating to sea to live for another two years. While spawning channels are occasionally used to enhance sockeye production, there are very few hatcheries releasing artificially reared sockeye, which is a significant difference from the other commercially harvested salmon species.

There are many factors affecting the survival and growth of salmon. A substantial increase in Pacific salmon production occurred in 1977, coincident with a shift in regional climate indexes reflective of higher productivity (Figure CS2.1). The increase was especially apparent for pink and chum salmon, although some of the increase could be attributed to large-scale enhancement efforts for these species. The increase for sockeye was more modest, with larger increases in production associated with more northerly stocks. Production for these three species has remained at high levels (nearly 1 million tons) since 1977 (perhaps due to hatchery production for pink and chum), although sockeye salmon production has decreased since the early 1990s (Figure CS2.1). The same is not true for Atlantic, Chinook, and Coho salmon. The abundance (and/or survival) of Atlantic and Chinook has declined almost steadily since the late 1970s, from about 900,000 Atlantic and 700,000 Chinook to about 100,000 Atlantic and 25,000 Chinook salmon (Figure CS2.1), despite very significant salmon enhancement efforts, including the use of hatchery fish. While freshwater factors

could influence production for these species, the large number of hatchery fish produced and released for these species should (and is argued or assumed to) compensate for any such losses. There is some evidence to support a decrease in marine survival for these species as the reason for their decline in abundance, although factors associated with the fitness and survival of hatchery-produced fish may also contribute to the observed decline.

DISCUSSION

Some salmon species and populations (particularly more northerly stocks) have thrived as a result of the recent changes in climate.

Figure CS2.1. All-nation abundance of sockeye, pink, and chum salmon in the North Pacific recent catches are nearly 1 million tons. Prefishery abundance of North American 2SW (sea winter) Atlantic salmon (solid line) and Chinook salmon catches in the Strait of Georgia, British Columbia. Atlantic salmon abundance and Strait of Georgia Chinook salmon catches have each decreased from several hundred thousand fish in the late 1970s to about 100,000 and 25,000 fish, respectively, in 2013. *Sources:* ICES (2007) Report of the Working Group on North Atlantic Salmon (WGNAS), 11–20 April 2007, Copenhagen, Denmark. ICES CM 2007/ACFM-13, 378p and the Pacific Salmon Commission (2014) Joint Chinook Technical Committee Annual Report of Catch and Escapement for 2013, TCCHINOOK (14)-2: 239p. North Pacific Anadromous Fish Commission, http://www.npafc.org/new/science_statistics.html. (Courtesy of Donald Noakes.)

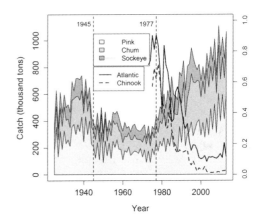

Others have undergone very significant declines and remain at low levels of abundance. With our limited understanding of complex ecosystem-level changes, a more careful approach to fisheries management and conservation will be required in the future. There is a higher public expectation with respect to salmon, because important stocks are often located in close proximity to large population centers. Salmon are an iconic and ecological and economically important species worthy of protection for future generations.

REFERENCE

Noakes, Donald J., and Richard J. Beamish. 2011. "Shifting the balance: Towards sustainable salmon populations and fisheries of the future." In *Sustainable Fisheries: Multi-Level Approaches to a Global Problem*, ed. W. W. Taylor, A. J. Lynch, and M. G. Schechter, 23–50. Bethesda, MD: American Fisheries Society.

CHAPTER SEVEN

Rapid Broad-Scale Ecosystem Changes and Their Consequences for Biodiversity

DAVID D. BRESHEARS, JASON P. FIELD,
DARIN J. LAW, JUAN C. VILLEGAS,
CRAIG D. ALLEN, NEIL S. COBB, AND
JOHN B. BRADFORD

INTRODUCTION

Biodiversity contributes to and depends on ecosystem structure and associated function. Ecosystem structure, such as the amount and type of tree cover, influences fundamental abiotic variables such as near-ground incoming solar radiation (e.g., Royer et al. 2011; Villegas et al. 2017), which in turn affects species and associated biodiversity (e.g., Trotter et al. 2008). In many systems, foundational, dominant, or keystone species (or species groups) are important in determining biodiversity, often because of their role in determining ecosystem structure. At spatial scales ranging from ecosystems to regions and larger, structural characteristics of vegetation or other structurally dominant organisms such as corals can influence species diversity, whether focused on alpha diversity (mean species diversity at the habitat level), beta diversity (differentiation among habitats), or gamma diversity (total species diversity across a landscape; Whittaker 1960).

Climate change is already fundamentally altering ecosystems at broad scales, and these changes are projected to increase (IPCC 2014). Such ecosystem changes can occur rapidly in response to extreme events such as droughts, floods, and hurricanes (IPCC 2012). Consequently, rapid broad-scale changes in ecosystems are of increasing concern. Several rapid ecological changes have occurred at spatial scales that are sufficiently broad to represent biome changes (Gonzalez et al. 2010, Settele et al. 2014; Figure 7.1). Rapid broad-scale changes differ from other patterns of vegetation dynamics in that they result in a "crash" in one or more populations (Breshears et al. 2008) over large areas of the affected region. Rapid broad-scale changes triggered by climate can include megafires, drought-triggered tree die-off, and associated pest and pathogen outbreaks (Logan et al. 2003; Breshears et al. 2005; Safranyik et al. 2007; Berner et al. 2017), and hurricanes and wind-throw events (IPCC 2012, 2014). These rapid broad-scale changes can rapidly alter other factors such as microclimate (Royer et al. 2011; Villegas et al. 2017; Zemp et al. 2017), which in turn can affect numerous other species and associated biodiversity. Many examples of broad-scale changes are documented in the paleoecology literature (Settele et al. 2014), although the temporal resolution at which those events can be resolved is relatively coarse (often centuries or longer). Contemporary events have highlighted that broad-scale changes can occur rapidly (years or less; Figure 7.1; see Breshears et al. 2005, Gonzalez et al. 2010; Stephens et al. 2014; Settele et al. 2014; Hartmann et al. 2018). These rapid broad-scale changes will have important consequences for biodiversity beyond the more commonly considered direct impacts of climate change (e.g., species physiology, phenology, and distribution; see Chapters 1, 3, and 4) or indirect effects, by which changes in one species can also affect other species (Cahill et al. 2013) up to the point of causing species coextinctions (Koh et al. 2004). The focus of this chapter is to alert readers to recent and projected rapid ecosystem changes that are expected to have potential consequences for biodiversity at ecosystem, landscape, and regional scales.

EXAMPLE SYSTEMS WITH VULNERABILITY TO RAPID BROAD-SCALE CHANGE

Climate change is already shifting the distribution of biomes across elevations and latitudes throughout the world, especially in temperate, tropical, and boreal areas (IPCC 2014). Predicted changes in climate and modeled vegetation projections indicate that up to one-half of the global land area could be highly vulnerable to biome change (Gonzalez et al. 2010; Settele et al. 2014). Projected vegetation changes suggest the following: the temperate mixed forest biome and the boreal conifer forest biome show the highest vulnerability as a fraction of total biome area; the tundra and alpine biome and the boreal conifer forest biome are most vulnerable in total land area; and tropical evergreen broadleaf forest and desert biomes show the lowest vulnerability in most cases (Gonzalez et al. 2010).

Many types of rapid broad-scale changes of sufficient magnitude to affect biome boundaries have already occurred recently across the globe (Figure 7.1; Gonzalez et al. 2010; Settele et al. 2014). For terrestrial ecosystems, the combined effects of drought and heat and their associated effects on pests and pathogens (e.g., Logan et al. 2003) have resulted in some of the most pronounced broad-scale ecosystem changes. Regional drought, which periodically affects most systems, under the hotter conditions associated with climate change (together referred to as "global-change-type drought" [Breshears et al. 2005] or more simply "hotter drought" [Allen et al. 2015]), can result in increased plant water stress that leads to widespread mortality, particularly for trees (Allen et al. 2010; Figure 7.1). The time to mortality during drought for tree species continues to increase with temperature, with lethal drought events becoming ever more frequent (Adams et al. 2017). Increased water stress in trees, especially in dominant and codominant species, can result in greater vulnerability to megafires (Millar and Stephenson 2015), as well as increasing probability of pest and pathogen outbreaks (e.g., bark beetles; Logan et al. 2003) and of mortality directly from the drought itself (Breshears et al. 2005; Allen et al. 2015). Such regional-scale droughts are of particular concern in the Amazon, where tree die-off and fire feedbacks, if sufficient in magnitude, could have profound impacts on biodiversity (Brando et

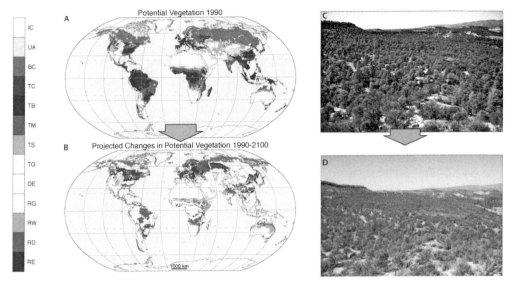

Figure 7.1. Climate-driven projected changes in vegetation often are represented as maps of spatial change (A, B). Implicit and potentially overlooked in such presentations is the underlying widespread tree mortality (C, D) that necessarily would occur with many such projected changes to produce biome-level shifts in vegetation. Left panels of projected vegetation shifts are from Gonzalez et al. (2010), using the MC1 dynamic global vegetation model. (A) Modeled potential vegetation under observed 1961–1990 climate. (B) Modeled potential vegetation under projected 2071–2100 climate, where any of nine climate scenario combinations of general circulation model and emissions pathway drives projected vegetation change. Displayed biomes, in panels A and B, from poles to equator: ice (IC), tundra and alpine (UA), boreal conifer forest (BC), temperate conifer forest (TC), temperate broadleaf forest (TB), temperate mixed forest (TM), temperate shrubland (TS), temperate grassland (TG), desert (DE), tropical grassland (RG), tropical woodland (RW), tropical deciduous broadleaf forest (RD), tropical evergreen broadleaf forest (RE). Right panel photographs show (C) *Pinus edulis* mortality under way in a southwestern US woodland (October 2002), and (D) the same view after dead *P. edulis* trees have dropped needles but trunks remain standing (May 2004). (Photos by C. D. Allen from Breshears et al. 2009.)

al. 2014). Megafires—crown fires at scales unprecedented historically (and perhaps largely unprecedented prehistorically)—can kill trees within hours to days and can drastically alter landscape structure at broad scales (Stephens et al. 2014). In upland tropical systems where biodiversity is still high, warming climate can change

fog-belt locations and characteristics, putting fog-dependent species at risk and leaving them with limited migration options (Foster 2001). Warming is altering fundamental structural characteristics in boreal and tundra systems, where loss of permafrost changes ecosystem structure (Settele et al. 2014). Presented here are examples of systems that are vulnerable to rapid broadscale changes triggered by climate change.

Evergreen Temperate, Semiarid, Coniferous Woodlands

Piñon pine (*Pinus edulis*) mortality occurred across the US Southwest in response to a hot drought and concomitant bark beetle (*Ips confusus*) outbreaks at the start of the millennium (2002–2004; Breshears et al. 2005; Figure 7.1 and Plate 3). Piñon pine is a foundational species that provides resources either directly or indirectly to a diverse set of individual species, populations, and communities (e.g., Trotter et al. 2008). Biodiversity at multiple scales is likely to be greatly reduced in conjunction with major reductions in such foundational species. Experiments with potted plants in glasshouses or in the field with in situ plants, as well as correlations between climate conditions and tree mortality, all highlight the vulnerability of piñon pine to warmer temperature and associated increases in vapor pressure deficit during drought (Allen et al. 2015). Even though piñon-juniper (*Juniperus monosperma*) woodlands can have a relatively sparse amount of tree cover premortality, loss of just one of the two codominant tree species is sufficient to greatly alter site microclimate (Royer et al. 2011). Piñon mortality across the southwestern United States in 2002 led to conversion of piñon-juniper woodland to a juniper woodland savanna (Clifford et al. 2011), and for the 12 years after that mortality event, piñon recruitment has been low enough to suggest that the woodland vegetation could be replaced by savanna or grassland vegetation (Redmond et al. 2015).

Evergreen Temperate, Coniferous, Montane Forests

A mountain pine beetle outbreak, which began in the mid-1990s, caused widespread mortality in lodgepole pine (*Pinus contorta*) stands throughout the western United States and Canada (Raffa et al. 2008). Lodgepole pine, which is broadly distributed in western North America from the Yukon Territory to Baja, California (Safranyik et al. 2007), is a classic fire-dependent species with high levels of cone serotiny (Minckley et al. 2011). Mountain pine beetles (*Dendroctonus ponderosae*) are a well-recognized and pervasive disturbance agent in lodgepole pine forests (Safranyik et al. 2007). The mid-1990s outbreak appears to be unprecedented over at least the past century (Raffa et al. 2008), likely due to unusually hot and dry conditions that induce stress in the host trees, similar to the stress driving change in piñon-juniper woodlands (Allen et al. 2015). The severity of the recent outbreak may have also been influenced by abnormally high and extensive host abundance, possibly as a result of several decades of fire suppression (Raffa et al. 2008). Fire and insect outbreaks can drive negative feedbacks in which young, recently burned stands are

less susceptible to beetle invasion (Kulakowski et al. 2012) and in which stands experiencing substantial mortality due to beetles are less likely to experience crown fires (Simard et al. 2010). However, although paleoecological evidence suggests disturbance influences lodgepole pine ranges, patterns of moisture availability also play a key role (Minckley et al. 2011) and are likely to shift in response to climate change (Bradford et al. 2014).

Boreal-Tundra Ecosystems

Boreal and tundra ecosystems are vast, and although their biodiversity can be relatively low, they are potentially vulnerable to rapid broad-scale changes triggered by climate. Climate models predict that boreal and tundra ecosystems will change dramatically and rapidly as temperatures rise via changes in permafrost thaw rates, increased fire frequency, and increased shrub cover (IPCC 2012, 2014). Interactions among permafrost thaw, wildfires, and shrub expansion are expected to collectively trigger rapid broad-scale changes (Figure 7.2; Settele et al. 2014). Reductions in biodiversity are expected in at least some cases for tundra (Callaghan, Tweedie, et al. 2011). For example, warmer temperatures on Disko Island, Central West Greenland were associated with net species loss at a fell-field, even though warmer temperatures had little effect on species losses and gains in a nearby herb-slope community (Callaghan, Christensen, et al. 2011). Boreal species diversity increased in Yukon, Canada during a 42-year period over which temperatures increased by 2°C (Danby et al. 2011), highlighting the potential for rapid changes as temperatures continue to warm. Mosses and lichens may be replaced by vascular plants and their litter (Callaghan, Tweedie, et al. 2011), although some studies suggest they may not respond to warming and under some conditions bryophyte biomass could actually increase (Hudson and Henry 2009). Projected climate change scenarios raise concern that transitions in

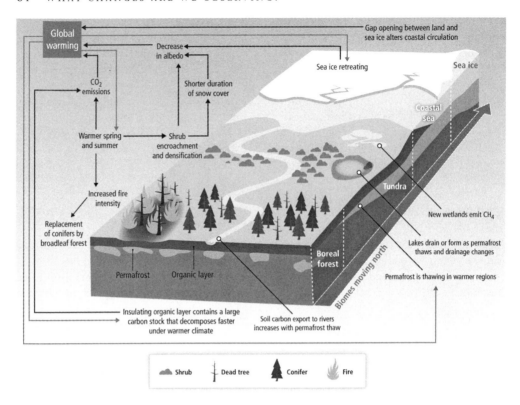

Figure 7.2. Tundra-boreal biome shift. Earth-system models predict a northward shift of Arctic vegetation with climate warming, as the boreal biome migrates into what is currently tundra. Observations of shrub expansion in tundra, increased tree growth at the tundra-forest transition, and tree mortality at the southern extent of the boreal forest in recent decades are consistent with model projections. Vegetation changes associated with a biome shift, which is facilitated by intensification of the fire regime, will modify surface energy budgets, and net ecosystem carbon balance, permafrost thawing, and methane emissions, with net feedbacks to additional climate change. (Figure 4-10 and figure 4-8 from IPCC 2014.)

boreal-tundra ecosystems and decline in their associated biodiversity will accelerate as conditions continue to warm, potentially triggering changes even more rapidly than have occurred to date (Settele et al. 2014).

Lowland Tropical Forests—The Amazon Region

Large areas of wet tropical forests, including the Amazon Region, have undergone significant changes linked to processes re-

lated to climate and global change, such as drought, fire, and land use (Brando et al. 2014). Experimental results indicate that the interaction between drought and fire could lead to a near-future dieback of the Amazon Region forest, with associated biodiversity loss in one of the most species-rich areas on the planet (see Case Study 6, Figure CS6.1). Hot droughts in association with human-caused fires are predicted to significantly reduce Amazon forest area in this century (Bush et al. 2008), possibly even in coming decades (Settele et al. 2014). Projected changes to climate are expected to exacerbate forests already degraded by land-use change, which has already had a significant impact on biodiversity. These changes include grasses replacing woody plants and potentially a rapid transition to a savanna state that supports significantly less species (Silvério et al. 2013). Feedbacks between vegetation and the atmosphere suggest frequent extreme drought events could potentially destabilize significant parts of forest in the Amazon region (Zemp et al. 2017).

Tropical Montane Ecosystems

Tropical montane ecosystems are uniquely diverse, containing extremes in topography with pronounced climatic gradients (Bruijnzeel 2004). Steep climatic gradients in tropical montane ecosystems typically constrain species distribution; numerous species occur within relatively narrow altitudinal bands. Sensitive to small changes in climate, tropical montane ecosystems are prone to rapid broad-scale changes. Species within narrow elevational bands are vulnerable to changes in land use that include transformation of natural ecosystems into, for example, agricultural lands. These changes can lead to shifts in species distributions, constrain migration, and exacerbate biodiversity loss (Magrin et al. 2014). For example, many species thrive in cloud forests, depend on fog interception, and are particularly vulnerable to climate change. Permanent or seasonal contact with fog is necessary for survival, and small changes in temperature can dramatically alter humidity, fog, and cloud density threatening survival (Villegas et al. 2008). Consequently, climate change and related disturbances to hydroclimatic variables that determine the presence of fog can have significant implications for biodiversity in these systems (Bruijnzeel 2004); and consequently, increased rates of species extinction are expected (Magrin et al. 2014).

INSIGHTS, RESEARCH NEEDS, AND MANAGEMENT CHALLENGES

Of the many ways in which climate change can and will have an impact on biodiversity, rapid, broad-scale changes in ecosystems pose a particular challenge as to whether species can track changes in climate. Species that cannot adapt to local changes will be required to track the pace of changing climate conditions, or "climate velocity," in rates of reproduction, dispersal area, and establishment to take advantage of transient windows of suitable habitat (IPCC 2014; Figure 7.3). Tracking climate velocity depends on reproduction, dispersal, germination, and/or establishment rates, which themselves depend on species-specific characteristics such as generation intervals and seed size. Many plant species and communities are unlikely to be able to track the climate velocity (Settele et al. 2014), particularly, for example, in flatter—rather than steeper—terrains, where the dispersal rate needed to track changes in climate is higher than expected for many species. Gradients in elevation (mountains) have been used as a proxy for expected changes over long latitudinal distances of flat terrain (Jump et al. 2009; Settele et al. 2014), although larger distances make this proxy problematic (Jump et al. 2009).

The impacts on biodiversity are potentially greater when ecosystem changes are broad scale because species' abilities to track climate can be overwhelmed. Consequently, rapid broad-scale ecosystem changes require consideration of not only local impacts of climate change but also broad-scale impacts across a region. The consequences to biodiversity and associated ecosystem services also depend on how patchy the remnant areas of an impacted ecosystem are in response to changing climate (López-Hoffman et al. 2013). Additionally, biodiversity in one region could potentially be affected by broad-scale ecosystem changes in another, nonadjacent region through "ecoclimatic teleconnections," whereby local vegetation change in one location, such as drought-induced tree die-off, results in climatic effects not only locally but also as transmitted through atmospheric circulation, potentially affecting vegetation elsewhere (Garcia et al. 2016; Stark et al. 2016; Swann et al. 2018). The scope of types of impacts also needs to be expanded when considering the biodiversity consequences of rapid broad-scale changes in ecosystems.

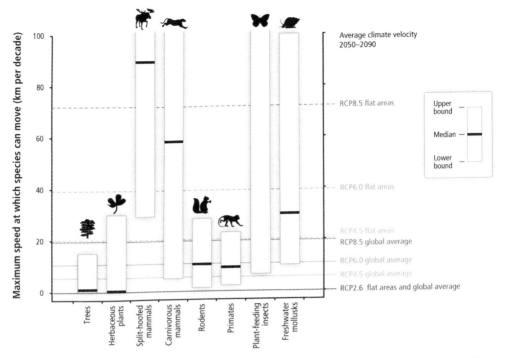

Figure 7.3. Maximum speeds at which species can move across landscapes (based on observations and models; vertical axis on left), compared with speeds at which temperatures are projected to move across landscapes (climate velocities for temperature; vertical axis on right). Human interventions, such as transport or habitat fragmentation, can greatly increase or decrease speeds of movement. White boxes with black bars indicate ranges and medians of maximum movement speeds for trees, plants, mammals, plant-feeding insects (median not estimated), and freshwater mollusks. For RCP2.6, 4.5, 6.0, and 8.5 for 2050–2090, horizontal lines show climate velocity for the global-land-area average and for large flat regions. Species with maximum speeds below each line are expected to be unable to track warming in the absence of human intervention. (Figure SPM from IPCC 2014.)

When species groups that greatly influence ecosystem structure are negatively affected, other factors such as consequences of changing microclimate must be considered in addition to the direct impacts of a climate event or trend.

Management of biodiversity will be challenged given that the capacity for ecosystems to adapt to the projected rates and magnitudes of climate change in the twenty-first century may be insufficient in

some cases to avoid substantial loss of species and ecosystem services at broad scales (Settele et al. 2014; Allen et al. 2015; Millar and Stephenson 2015). Management actions, such as assisted migration, can increase the inherent capacity of ecosystems and their constituent organisms to adapt to a different climate and can reduce the risk of local extinctions and undesirable impacts on ecosystem function (Settele et al. 2014). In many cases, species will have to migrate across unfavorable habitats to reach suitable climates or areas of lower land-use pressure, including protected areas (IPCC 2014). When habitat fragmentation affects the migration of key species or when the rate of climate change is faster than natural migration rates, management actions such as assisted migration may increase the ability of the system to maintain its structure and function by tolerating such changes (Hoegh-Guldberg et al. 2008). Management actions for assisted migration include maintaining or improving existing habitat, maintaining or improving migration corridors, and directly translocating species or genetically distinct populations within a species (Hoegh-Guldberg et al. 2008). Keystone species such as pollinators and other dominant species such as trees are important to manage under changing climate to preserve biodiversity and to maintain the provision of ecosystem services (Zarnetske et al. 2012). Climate change will present many natural resource management challenges that will require land managers and policy makers to work together to combat the impacts of a changing climate on ecosystem function and biodiversity.

The ability to anticipate, predict, and manage biodiversity under rapid broad-scale ecological change overlaps with more general considerations about biodiversity responses to climate change and requires considering the following. First, there is need for greater recognition of the risks of rapid broad-scale change and their associated consequences. For example, western US forests were generally not viewed as highly vulnerable in the early 1990s, but within the past two decades have undergone enormous broad-scale rapid change through forest die-off and wildfire (Allen et al. 2015). Second, an improved understanding is needed of the secondary impacts of broad-scale ecosystem change, such as forest die-off effects on microclimate change being propagated elsewhere through ecoclimatic teleconnections (Garcia et al. 2016; Stark et al. 2016; Swann et al. 2018). Third, to account for these impacts, conservation strategies are needed that allow for "bet hedging" for protected areas (Davison et al. 2011). Fourth, improved methods of assessing broad-scale change are needed during early stages to enable feasible rapid responses (Hartmann et al. 2018). Fifth, broad-scale changes may particularly amplify the need to consider assisted migration (Hoegh-Guldberg et al. 2008). Sixth, depending on the patchiness of the change, scientists may wish to consider whether microclimate refugia can be established within an impacted area. For example, the watering of iconic groves of ancient giant sequoias (*Sequoiadendron giganteum*) in western US national parks might become a priority under extreme drought even though current land management practices would generally preclude that option (Grant et al. 2013). In conclusion, land managers and policy makers should consider the suite of potential impacts on biodiversity and associated ecosystem services, rather than single species-by-species cases, when attempting to preserve biodiversity and maintain the provision of ecosystem services, especially under rapid broad-scale climate-driven changes that have already begun to occur and are expected to increase in the future.

ACKNOWLEDGMENTS

This publication was supported by the National Science Foundation through Macrosystems Biology (NSF EF-1340624, EF-1550756 and EF-1550756) and Critical Zone Observatories (Santa Catalina Mountains and Jemez River Basin; NSF EAR-1331408), Philecology Foundation, Fort Worth for Biosphere 2, Arizona Agricultural Experiment Station, the U.S. Geological Survey's Ecosystems and Climate and Land Use Change mission areas, through the USGS Western Mountain Initiative project; and Colciencias (Programa de investigación en gestión del riesgo asociado a cambio climático y ambiental en cuencas hidrográficas—convocatoria 543/2011). Any use of trade, product, or firm names is for descriptive purposes only and does not imply endorsement by the U.S. Government.

REFERENCES

Adams, H. D., G. A. Barron-Gafford, R. L. Minor, A. A. Gardea, L. Patrick Bentley, D. J. Law, D. D. Breshears, N. G. McDowell, and T. E. Huxman. 2017. "Temperature response surfaces for mortality risk of tree species with future drought." *Environmental Research Letters* 12: 115014.

Allen, Craig D., David D. Breshears, and Nate G. McDowell. 2015. "On underestimation of global vulnerability to tree mortality and forest die-off from hotter drought in the Anthropocene." *Ecosphere* 6 (8): art. 129.

Allen, Craig D., Alison K. Macalady, Haroun Chenchouni, Dominique Bachelet, Nate McDowell, Michel Vennetier, Thomas Kitzberger, et al. 2010. "A global overview of drought and heat-induced tree mortality reveals emerging climate change risks for forests." *Forest Ecology and Management* 259 (4): 660–684.

Bradford, John B., Daniel Schlaepfer, and William K. Lauenroth. 2014. "Ecohydrology of adjacent sagebrush and lodgepole pine ecosystems: The consequences of climate change and disturbance." *Ecosystems* 17 (4): 590–605.

Brando, Paulo Monteiro, Jennifer K. Balch, Daniel C. Nepstad, Douglas C. Morton, Francis E. Putz, Michael T. Coe, Divino Silvério, et al. 2014. "Abrupt increases in Amazonian tree mortality due to drought-fire interactions." *Proceedings of the National Academy of Sciences* 111 (17): 6347–6352.

Breshears, David D., Neil S. Cobb, Paul M. Rich, Kevin P. Price, Craig D. Allen, Randy G. Balice, William H. Romme, et al. 2005. "Regional vegetation die-off in response to global-change-type drought." *Proceedings of the National Academy of Sciences of the United States of America* 102 (42): 15144–15148.

Breshears, David D., Travis E. Huxman, Henry D. Adams, Chris B. Zou, and Jennifer E. Davison. 2008. "Vegetation synchronously leans upslope as climate warms." *Proceedings of the National Academy of Sciences* 105 (33): 11591–11592.

Bruijnzeel, Leendert Adriaan. 2004. "Hydrological functions of tropical forests: Not seeing the soil for the trees?" *Agriculture, Ecosystems & Environment* 104 (1): 185–228.

Bush, M. B., M. R. Silman, C. McMichael, and S. Saatchi. 2008. "Fire, climate change and biodiversity in Amazonia: A late-Holocene perspective." *Philosophical Transactions of the Royal Society B—Biological Sciences* 363 (1498): 1795–1802.

Cahill, Abigail E., Matthew E. Aiello-Lammens, M. Caitlin Fisher-Reid, Xia Hua, Caitlin J. Karanewsky, Hae Yeong Ryu, Gena C. Sbeglia, et al. 2012. "How does climate change cause extinction?" In *Proceedings of the Royal Society B*: rspb20121890.

Callaghan, Terry V., Torben R. Christensen, and Elin J. Jantze. 2011. "Plant and vegetation dynamics on Disko Island, West Greenland: Snapshots separated by over 40 years." *AMBIO: A Journal of the Human Environment* 40 (6): 624–637.

Callaghan, Terry V., Craig E. Tweedie, Jonas Åkerman, Christopher Andrews, Johan Bergstedt, Malcolm G. Butler, Torben R. Christensen, et al. 2011. "Multidecadal changes in tundra environments and ecosystems: synthesis of the International Polar Year-Back to the Future Project (IPY-BTF)." *AMBIO: A Journal of the Human Environment* 40 (6): 705–716.

Clifford, Michael J., Neil S. Cobb, and Michaela Buenemann. 2011. "Long-term tree cover dynamics in a pinyon-juniper woodland: Climate-change-type drought resets successional clock." *Ecosystems* 14 (6): 949–962.

Danby, Ryan K., Saewan Koh, David S. Hik, and Larry W. Price. 2011. "Four decades of plant community change in the alpine tundra of southwest Yukon, Canada." *AMBIO: A Journal of the Human Environment* 40 (6): 660–671.

Davison, Jennifer E., David D. Breshears, Willem J. D. Van Leeuwen, and Grant M. Casady. 2011. "Remotely sensed vegetation phenology and productivity along a climatic gradient: On the value of incorporating the dimension of woody plant cover." *Global Ecology and Biogeography* 20 (1): 101–113.

Foster, Pru. 2001. "The potential negative impacts of global climate change on tropical montane cloud forests." *Earth-Science Reviews* 55 (1): 73–106.

Garcia, E. S., A. L. S. Swann, J. C. Villegas, D. D. Breshears, D. J. Law, S. R. Saleska, S. C. Stark. 2016. "Synergistic ecoclimate teleconnections from forest loss in different regions structure global ecological responses." *PLOS One* 11: e0165042.

Gonzalez, Patrick, Ronald P. Neilson, James M. Lenihan, and Raymond J. Drapek. 2010. "Global patterns in the vulnerability of ecosystems to vegetation shifts due to climate change." *Global Ecology and Biogeography* 19 (6): 755–768.

Grant, Gordon E., Christina L. Tague, and Craig D. Allen. 2013. "Watering the forest for the trees: An emerging priority for managing water in forest landscapes." *Frontiers in Ecology and the Environment* 11 (6): 314–321.

Hartmann, Henrik, Catarina F. Moura, William R. L. Anderegg, Nadine K. Ruehr, Yann Salmon, Craig D. Allen, Stefan K. Arndt, David D. Breshears, Hendrik Davi, David Galbraith, et al. 2018. "Research frontiers for improving our understanding of drought-induced tree and forest mortality." *New Phytologist* 218 (1): 15–28.

Hoegh-Guldberg, Ove, Lesley Hughes, Sue McIntyre, D. B. Lindenmayer, C. Parmesan, Hugh P. Possingham, and C. D. Thomas. 2008. "Assisted colonization and rapid climate change." *Science* 321: 325–326.

Hudson, James M. G., and Greg H. R. Henry. 2009. "Increased plant biomass in a High Arctic heath community from 1981 to 2008." *Ecology* 90 (10): 2657–2663.

IPCC. 2012. *Managing the Risks of Extreme Events and Disasters to Advance Climate Change Adaptation: A Special Report of Working Groups I and II of the Intergovernmental Panel on Climate Change*, ed. C. B. Field, V. Barros, T. F. Stocker, D. Qin, D. J. Dokken, K. L. Ebi, M. D. Mastrandrea, K. J. Mach, G.-K. Plattner, S. K. Allen, M. Tignor, and P. M. Midgley. Cambridge University Press.

IPCC. 2014. "Climate change 2014: Impacts, adaptation, and vulnerability. Part A: Global and sectoral aspects." In *Contribution of Working Group II to the Fifth Assessment Report of the Intergovernmental Panel on Climate Change*, ed. C. B. Field, V. R. Barros, D. J. Dokken, K. J. Mach, M. D. Mastrandrea, T. E. Bilir, M. Chatterjee, et al. Cambridge University Press. https://doi.org/10.1017/CBO9781107415379.

Jump, Alistair S., Csaba Mátyás, and Josep Peñuelas. 2009. "The altitude-for-latitude disparity in the range retractions of woody species." *Trends in Ecology & Evolution* 24 (12): 694–701.

Koh, Lian Pin, Robert R. Dunn, Navjot S. Sodhi, Robert K. Colwell, Heather C. Proctor, and Vincent S. Smith. 2004. "Species coextinctions and the biodiversity crisis." *Science* 305 (5690): 1632–1634.

Kulakowski, Dominik, Daniel Jarvis, Thomas T. Veblen, and Jeremy Smith. 2012. "Stand-replacing fires reduce susceptibility of lodgepole pine to mountain pine beetle outbreaks in Colorado." *Journal of Biogeography* 39 (11): 2052–2060.

Logan, Jesse A., Jacques Régnière, and James A. Powell. 2003. "Assessing the impacts of global warming on forest pest dynamics." *Frontiers in Ecology and the Environment* 1 (3): 130–137.

López-Hoffman, Laura, David D. Breshears, Craig D. Allen, and Marc L. Miller. 2013. "Key landscape ecology metrics for assessing climate change adaptation options: Rate of change and patchiness of impacts." *Ecosphere* 4 (8): 1–18.

Magrin, G. O., J. A. Marengo, J. P. Boulanger, M. S. Buckeridge, E. Castellanos, G. Poveda, F. R. Scarano, and S. Vicuña. 2014. "Central and South America." *Climate Change*: 1499–1566.

Millar, Constance I., and Nathan L. Stephenson. 2015. "Temperate forest health in an era of emerging megadisturbance." *Science* 349 (6250): 823–826.

Minckley, Thomas A., Robert K. Shriver, and B. Shuman. 2012. "Resilience and regime change in a southern Rocky Mountain ecosystem during the past 17000 years." *Ecological Monographs* 82 (1): 49–68.

Raffa, Kenneth F., Brian H. Aukema, Barbara J. Bentz, Allan L. Carroll, Jeffrey A. Hicke, Monica G. Turner, and William H. Romme. 2008. "Cross-scale drivers of natural disturbances prone to anthropogenic amplification: the dynamics of bark beetle eruptions." *AIBS Bulletin* 58 (6): 501–517.

Redmond, Miranda D., Neil S. Cobb, Michael J. Clifford, and Nichole N. Barger. 2015. "Woodland recovery following drought-induced tree mortality across an environmental stress gradient." *Global Change Biology* 21 (10): 3685–3695.

Royer, Patrick D., Neil S. Cobb, Michael J. Clifford, Cho-Ying Huang, David D. Breshears, Henry D. Adams, and Juan Camilo Villegas. 2011. "Extreme climatic event-triggered overstorey vegetation loss increases understorey solar input regionally: Primary and secondary ecological implications." *Journal of Ecology* 99 (3): 714–723.

Safranyik, Les, and Allan L. Carroll. 2006. "The biology and epidemiology of the mountain pine beetle in lodgepole pine forests." In *The Mountain Pine Beetle: A Synthesis of Biology, Management, and Impacts on Lodgepole Pine*, ed. L. Safranyik and W. R. Wilson, 3–66. Victoria, BC: Natural Resources Canada, Canadian Forest Service, and Pacific Forestry Centre.

Settele, J., R. Scholes, R. Betts, S. Bunn, P. Leadley, D. Nepstad, J. T. Overpeck, and M. A. Taboada. 2014. "Terrestrial and inland water systems." In *Climate Change 2014: Impacts, Adaptation, and Vulnerability. Part A: Global and Sectoral Aspects. Contribution of Working Group II to the Fifth Assessment Report of the Intergovernmental Panel on Climate Change*, ed. C. B. Field, V. R. Barros, D. J. Dokken, K. J. Mach, M. D. Mastrandrea, T. E. Bilir, M. Chatterjee, K. L. Ebi, Y. O. Estrada, R. C. Genova, B. Girma, E. S. Kissel, A. N. Levy, S. MacCracken, P. R. Mastrandrea, and L. L. White, 271–359. Cambridge University Press.

Silvério, Divino V., Paulo M. Brando, Jennifer K. Balch, Francis E. Putz, Daniel C. Nepstad, Claudinei Oliveira-Santos, and Mercedes M. C. Bustamante. 2013. "Testing the Amazon savannization hypothesis: Fire effects on invasion of a neotropical forest by native cerrado and exotic pasture grasses." *Philosophical Transactions of the Royal Society of London B: Biological Sciences* 368 (1619): 20120427.

Simard, Martin, William H. Romme, Jacob M. Griffin, and Monica G. Turner. 2011. "Do mountain pine beetle outbreaks change the probability of active crown fire in lodgepole pine forests?" *Ecological Monographs* 81 (1): 3–24.

Stark, Scott C., David D. Breshears, Elizabeth S. Garcia, Darin J. Law, David M. Minor, Scott R. Saleska, Abigail L. S. Swann, et al. 2016. "Toward accounting for ecoclimate teleconnections: Intra- and inter-continental consequences of altered energy balance after vegetation change." *Landscape Ecology* 31 (1): 181–194.

Stephens, Scott L., Neil Burrows, Alexander Buyantuyev, Robert W. Gray, Robert E. Keane, Rick Kubian, Shirong Liu, et al. 2014. "Temperate and boreal forest mega-fires: Characteristics and challenges." *Frontiers in Ecology and the Environment* 12 (2): 115–122.

Swann, A. L. S., M. Laguë, E. S. Garcia, D. D. Breshears, J. P. Field, D. J. P. Moore, S. R. Saleska, S. C. Stark, J. C. Villegas, D. J. Law, and D. M. Minor. 2018. "Continental-scale consequences of tree die-offs in North America: Identifying where forest loss matters most." *Environmental Research Letters*.

Trotter, R. T., N. S. Cobb, and T. G. Whitham. 2008. "Arthropod community diversity and trophic structure: a comparison between extremes of plant stress." *EcologicaEntomology* 33: 1–11.

Villegas, J. C., D. J. Law, S. C. Stark, D. M. Minor, D. D. Breshears, S. R. Saleska, A. L. S. Swann, E. S. Garcia, E. M. Bella, J. M. Morton, N. S. Cobb, G. A. Barron-Gafford, M. E. Litvak, and T. E. Kolb. 2017. "Prototype campaign assessment of disturbance-induced tree-loss effects on surface properties for atmospheric modeling." *Ecosphere* 8 (3): e01698.

Villegas, Juan Camilo, Conrado Tobón, and David D. Breshears. 2008. "Fog interception by non-vascular epiphytes in tropical montane cloud forests: Dependencies on gauge type and meteorological conditions." *Hydrological Processes* 22 (14): 2484–2492.

Whittaker, Robert Harding. 1960. "Vegetation of the Siskiyou mountains, Oregon and California." *Ecological Monographs* 30 (3): 279–338.

Zarnetske, Phoebe L., David K. Skelly, and Mark C. Urban. 2012. "Biotic multipliers of climate change." *Science* 336 (6088): 1516–1518.

Zemp, Delphine Clara, Carl-Friedrich Schleussner, Henrique M. J. Barbosa, Marina Hirota, Vincente Montade, Gilvan Sampaio, Arie Staal, Lan Wang-Erlandsson, and Anja Rammig. 2017. "Self-amplified Amazon forest loss due to vegetation-atmosphere feedbacks." *Nature Communications* 8: 14681.

Rapidly Diverging Population Trends of Adélie Penguins Reveal Limits to a Flexible Species' Adaptability to Anthropogenic Climate Change

Grant Ballard and David Ainley

Adélie penguins (*Pygoscelis adeliae*) thrive in some of the most dramatically variable habitats and weather on Earth. They spend much of their life at sea, more than 40 m underwater, often under sea ice, where they find their food. They also spend roughly 3 months per year mostly on land, living in dense colonies for breeding. They regularly contend with the transition between open and frozen ocean, and with terrain alternately blanketed in snow and ice, then swept clear by high winds. Adélie penguins are one of only a handful of species to be able to survive extended periods in subzero air temperatures, out of the relatively warm water that harbors most of the biodiversity to be found at the highest latitudes of the Southern Ocean. They require both ice-free terrain to nest and nearby open water to forage. The combination of these two things is rare in Antarctica, but where they are found, so, too, are Adélies.

Over the past 12 million–15 million years, Adélie penguins have contended with a wide range of climates and consequent impacts to their habitat. Ice sheets have repeatedly expanded and retreated hundreds of kilometers, destroying or creating nesting habitat, and Adélie populations have grown and shrunk corresponding to interglacial and glacial periods, respectively (Li et al. 2014). The comings and goings of Adélies through geologic time, as determined by dating subfossil bones, have been used to validate the dates of ice-sheet advances and retreats. A warm period 2,000–4,000 years BP is even known as the "penguin optimum" for the widespread extent of ancient penguin colonies found from this time period—including places where they have not yet reoccupied (Baroni and Orombelli 1997).

As a result of recent and ongoing changes in climate and sea ice, Adélies are now encountering changes of magnitudes previously inferred from the geologic and genetic record, but at a faster pace. Between 1979 and 2010, the period of time during which the sea-ice field in Antarctica expands, known as the sea-ice season (Stammerjohn et al. 2012; see Plate 5), has declined by 3 months in the Arctic and in the Antarctic Peninsula region. This change in persistence coincides with dramatic reductions in the overall extent of sea ice in both those regions. The opposite is true in East Antarctica and in the Ross Sea region, where the sea-ice season has extended by two months over the same period. These changes are among the largest phenological shifts so far associated with anthropogenic climate change, and along with warming temperatures and increasing precipitation, they have profound implications for sea-ice ecosystems (Sailley et al. 2013). In response, breeding populations of Adélie penguins are retreating southward from much of the Antarctic Peninsula—a place continuously occupied by Adélies for 500–800 years (Emslie 2001)—but they are slowly increasing as sea ice loosens farther south in that region (Lynch et al. 2012). Concomitantly, Adélie populations in the southern Ross Sea are expanding rapidly (Lyver et al. 2014; see Plate 5), in some cases exploiting nesting habitat recently exposed by retreating ice sheets (LaRue et al. 2013).

The mechanisms driving these changes also relate to the specific adaptations of Adélie penguins to the sea-ice environment. During summer months, the presence of 6 percent–15 percent sea-ice cover is optimal; more or less ice and foraging trips grow longer, with less food delivered to chicks (Ballard, Dugger, et al. 2010). The sea ice itself is a substrate for diatom growth and subsequent grazing by krill and copepods, which in turn are food for small fish (especially sea-ice-dwelling silverfish—*Pleuragramma antarcticum*) and penguins. Silverfish are also major prey for Adélie penguins, particularly in the Ross Sea (Ainley et al. 2003). It is possibly the loss of silverfish that explains the decrease in Adélie penguins in the northern Antarctic Peninsula. Without sea ice, the food web in the Southern Ocean is far less complex and less suited to higher trophic level predators like penguins, as it is dominated by algae (*Phaeocystis antarctica*) with limited grazing by pteropods, which penguins and fish do not appear to consume. Sea ice is also a platform for Adélies to rest upon, molt, and seek shelter from predators, and can provide a source of freshwater when snow accumulates on its surface.

Sea-ice variability has also substantially impacted Adélie penguins' migratory patterns through time, with today's populations ranging from essentially nonmigratory at the more northerly parts of their range to long-distance migratory at the southernmost portions—with annual journeys of up to ~18,000 km (round trip)—an astounding feat for a flightless animal (Ballard, Toniolo, et al. 2010). Their apparent need for some amount of daylight during all phases of their annual cycle appears to limit the potential range of wintering and migration to north of 72.7°S—the latitude of zero midwinter twilight. It is likely that, as a result of climate change, more suitable breeding habitat will become available farther south than it currently exists (as glaciers retreat and free up more coastline for nesting), but it may not be possible for Adélies to migrate much farther than they already do and still have sufficient time to raise young, given the short breeding season of high-latitude ecosystems. In this way, Adélies are caught between astronomically imposed limits and anthropogenic climate change.

The diverging Adélie penguin population trends described here are projected to continue during the next few decades, after which sea ice, as warming reaches 2°C above preindustrial levels, is predicted to decrease everywhere in the Southern Ocean (Ainley et al. 2010). With the retreat of sea ice, Adélie penguins and the other three truly ice-obligate seabird species (Antarctic petrel—*Thalassoica antarctica*; snow petrel—*Pagodrama nivea*; and emperor penguin—*Aptenodytes forsteri*) will eventually disappear (Ainley et al. 2010; Jenouvrier et al. 2014), to be replaced by ice-tolerant and ice-avoiding species.

ACKNOWLEDGMENTS

We are grateful for support from the National Science Foundation Grants OPP-0944411 and 1543498, and to Ian Gaffney for production of Plate 5. Point Blue Conservation Science Contribution #2113.

REFERENCES

Ainley, David, Joellen Russell, Stephanie Jenouvrier, Eric Woehler, Philip O'B. Lyver, William R. Fraser, and Gerald L. Kooyman. 2010. "Antarctic penguin response to habitat change as Earth's troposphere reaches 2°C above preindustrial levels." *Ecological Monographs* 80 (1): 49–66.

Ballard, Grant, Katie M. Dugger, Nadav Nur, and David G. Ainley. 2010. "Foraging strategies of Adélie penguins: Adjusting body condition to cope with environmental variability." *Marine Ecology Progress Series* 405: 287–302.

Ballard, Grant, Viola Toniolo, David G. Ainley, Claire L. Parkinson, Kevin R. Arrigo, and Phil N. Trathan. 2010. "Responding to climate change: Adélie penguins confront astronomical and ocean boundaries." *Ecology* 91 (7): 2056–2069.

Baroni, Carlo, and Giuseppe Orombelli. 1994. "Abandoned penguin rookeries as Holocene paleoclimatic indicators in Antarctica." *Geology* 22 (1): 23–26.

Emslie, Steven D. 2001. "Radiocarbon dates from abandoned penguin colonies in the Antarctic Peninsula region." *Antarctic Science* 13 (3): 289–295.

Jenouvrier, Stéphanie, Marika Holland, Julienne Stroeve, Mark Serreze, Christophe Barbraud, Henri Weimerskirch, and Hal Caswell. 2014. "Projected continent-wide declines of the emperor penguin under climate change." *Nature Climate Change* 4 (8): 715–718. https://doi.org/10.1038/NCLIMATE2280.

LaRue, Michelle A., David G. Ainley, Matt Swanson, Katie M. Dugger, O. Phil, B. Lyver, Kerry Barton, and Grant Ballard. 2013. "Climate change winners: Receding ice fields facilitate colony expansion and altered dynamics in an Adélie penguin metapopulation." *PLOS One* 8 (4): e60568. https://doi.org/10.1371/journal.pone.0060568.

Li, Cai, Yong Zhang, Jianwen Li, Lesheng Kong, Haofu Hu, Hailin Pan, Luohao Xu, et al. 2014. "Two Antarctic penguin genomes reveal insights into their evolutionary history and molecular changes related to the Antarctic environment." *GigaScience* 3 (1): 27.

Lynch, Heather J., Ron Naveen, Philip N. Trathan, and William F. Fagan. 2012. "Spatially integrated assessment reveals widespread changes in penguin populations on the Antarctic Peninsula." *Ecology* 93 (6): 1367–1377.

Lyver, Phil O'B., Mandy Barron, Kerry J. Barton, David G. Ainley, Annie Pollard, Shulamit Gordon, Stephen McNeill, Grant Ballard, and Peter R. Wilson. 2014. "Trends in the breeding population of Adélie penguins in the Ross Sea, 1981–2012: A coincidence of climate and resource extraction effects." *PLOS One* 9 (3): e91188.

Sailley, Sévrine F., Hugh W. Ducklow, Holly V. Moeller, William R. Fraser, Oscar M. E. Schofield, Deborah K. Steinberg, Lori M. Garzio, and Scott C. Doney. 2013. "Carbon fluxes and pelagic ecosystem dynamics near two western Antarctic Peninsula Adélie penguin colonies: An inverse model approach." *Marine Ecology Progress Series* 492: 253–272. https://doi.org/10.3354/meps10534.

Stammerjohn, Sharon, Robert Massom, David Rind, and Douglas Martinson. 2012. "Regions of rapid sea ice change: An inter-hemispheric seasonal comparison." *Geophysical Research Letters* 39 (6). https://doi.org/10.1029/2012GL050874.

PART III

What Does the Past Tell Us?

CHAPTER EIGHT

A Paleoecological Perspective on Sudden Climate Change and Biodiversity Crises

JEFFREY PARK

INTRODUCTION

Understanding past climate change is central to understanding possible future impacts on biodiversity. Future environmental change will differ from past disruptions in the Earth system, but the biosphere will respond with mechanisms familiar from the geologic past. Potential rates of future environmental responses can be inferred from the study of past changes.

Sepkoski (1996) argued for five "mass-extinction" events in Earth's preserved fossil record, based on a threshold percentage (39 percent) of marine genera disappearing from Earth's sediments nearly simultaneously. Barnosky et al. (2011) recalibrated the mass-extinction threshold to 75 percent of all species in order to compare these past events with modern extinction rates. They found that, at least for terrestrial vertebrates, present-day rates of biodiversity loss are comparable to past crises, if extrapolated for several centuries. It is reasonable to ask whether human activity is generating a sixth mass extinction (e.g., Dirzo et al. 2014). For many people, especially nonscientists, it is difficult to credit our species with the power to destroy the biosphere as we know it. Against this comfortable skepticism, however, Zeebe et al. (2016) estimate that humans contribute carbon-based greenhouse gases (carbon dioxide and methane) into Earth's atmosphere at 10 times the rate of any natural source since the extinction of the dinosaurs 66 million years ago (Figure 8.1). More ominously, Bond and Grasby (2017) argue that most global "biotic crises" in Earth history correlate with sudden changes in atmospheric CO_2. The famous mass extinction of the dinosaurs (Schulte et

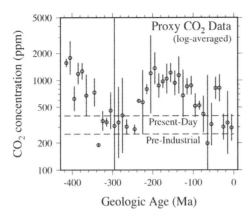

Figure 8.1. Compilation of atmospheric CO_2 levels estimated from a variety of geologic proxy data, from the data set of Park and Royer (2011). Individual proxy data were log-averaged within 10 Ma intervals of estimated geologic age. The error bars are computed from the sample variances of these sums. For reference purposes, horizontal dashed lines mark the preindustrial and present-day concentrations of atmospheric CO_2.

Eon	Era	Period	Age (Ma)
			541
Proterozoic	Neo-proterozoic	Ediacaran	635
		Cryogenian	850
		Tonian	1000
	Meso-proterozoic	Stenian	1200
		Ectasian	1400
		Calymmian	1600
	Paleo-proterozoic	Statherian	1800
		Orosirian	2050
		Rhyacian	2300
		Siderian	2500
Archean	Neoarchean		2800
	Mesoarchean		3200
	Paleoarchean		3600
	Eoarchean		4000
	Hadean		4560

(Precambrian spans Proterozoic and Archean)

al. 2010) is the only mass-extinction event that correlates persuasively with a meteor strike. Are we unwittingly replicating the conditions for past biodiversity crises?

Nineteenth-century geologists divided Earth history into discrete intervals, often based on correlated changes in the types of sediments, but more often based on abrupt changes in the mix of fossil species within the sediments (Rudwick 1985; Figure 8.2). The best-documented biotic transitions in Earth history mark the boundaries of three eras within the Phanerozoic Eon, the Paleozoic (541.0 Ma–252.2 Ma; Figure 8.3),[1] Mesozoic (252.2 Ma–66.0 Ma), and Cenozoic (66.0 Ma–present; Figure 8.4). Smaller subdivisions of time within the geologic time scale (Gradstein and Ogg 2012) include periods, such as the Cambrian (541.0 Ma–485.4 Ma) and the Cretaceous (145.0 Ma–66.0 Ma); epochs, such as the Eocene (56.0 Ma–33.9 Ma) and the Pennsylvanian (323.2 Ma–298.9 Ma); and age or stage, such as the Danian (66.0 Ma–61.6 Ma) and the Frasnian (382.7 Ma–372.2 Ma). At each boundary between time intervals, multiple fossil species may go extinct; the appearance of new spe-

Figure 8.2. Geologic time scale, after Gradstein and Ogg (2012), covering the first ~4 billion years of Earth history. Within this broad time interval known as the Precambrian, evidence for biologic activity spans both the Proterozoic and Archean eons, primarily via layered stromatolites and trace fossils, as well as chemical and isotopic signatures. Within the Neoproterozoic era, heavy dashed lines border the Cryogenic period, an interval of extreme cold climates during which molecular clocks place the basal diversification of all metazoan life forms.

cies often occurs more gradually (D'Hondt 2005; Hull et al. 2011).

Stratigraphers have identified a new epoch of geologic history, called the Anthro-

Eon	Era	Period	Epoch	Stage	Age (Ma)
Phanerozoic	Paleozoic	Permian	Lopingian	Changhsingian	252.2
					254.2
				Wuchiapingian	259.8
			Guadalupian	Capitanian	265.1
				Wordian	268.8
				Roadian	272.3
			Cisuralian	Kungurian	279.3
				Artinskian	290.1
				Sakmarian	295.5
				Asselian	298.9
		Carboniferous	Pennsylvanian	Gzhelian	303.7
				Kasimovian	307.0
				Moscovian	315.2
				Bashkirian	323.2
			Mississippian	Serpukhovian	330.9
				Visean	346.7
				Tournaisian	358.9
		Devonian	Upper	Famennian	372.2
				Frasnian	382.7
			Middle	Givetian	387.7
				Eifelian	393.3
			Lower	Emsian	407.6
				Pragian	410.8
				Lochkovian	419.2
		Silurian	Pridoli	—	423.0
			Ludlow	Ludfordian	425.6
				Gorstian	427.4
			Wenlock	Homerian	430.5
				Sheinwoodian	433.4
			Llandovery	Telychian	438.5
				Aeronian	440.8
				Rhuddanian	443.8
		Ordovicin	Upper	Hirnantian	445.2
				Katian	453.0
				Sandbian	458.4
			Middle	Darriwilian	467.3
				Dapingian	470.0
			Lower	Floian	477.7
				Tremadocian	485.4
		Cambrian	Furongian	Stage 10	489.5
				Jiangshanian	494
				Paibian	497
			Series 3	Guzhangian	500.5
				Drumian	504.5
				Stage 5	509
			Series 2	Stage 4	514
				Stage 3	521
			Terreneuvian	Stage 2	529
				Fortunian	541.0

Figure 8.3. Geologic time scale, after Gradstein and Ogg (2012), covering the Paleozoic Era of Earth history (541.0–252.2 Ma). Life with hard parts evolved rapidly in the Cambrian, leaving an abundant fossil record. Within the Paleozoic lie most of the severe biodiversity crises in the history of life. Heavy dashed lines mark such crises that have been identified by McGhee et al. (2013), culminating in the end-Permian mass extinction at 252.2 Ma, associated with the volcanic degassing associated with the Siberian Traps (Cui and Kump 2015).

pocene (Steffen et al. 2011; Zalasiewicz et al. 2011), which spans the impact of human activity on the sedimentary record from the start of the Industrial Revolution. Sediment-accumulation patterns of the previous Holocene (11.8 ka to 0.25 ka)[2] and Pleistocene (2.6 Ma to 11.8 ka) epochs have been disrupted by the land-use practices of human societies, as have chemical compositions, isotopic tracers, and fossil assemblages within the sediments. We can compare Anthropocene climate projections with the natural climate variations of the Holocene and Pleistocene, through which Earth's species had sufficient evolutionary resilience to produce our present-day biodiversity (Svenning et al. 2015). How likely are anthropogenic climate changes to exceed the natural climate variations and possibly threaten this resilience?

THE EARTH SYSTEM AND ITS CARBON CYCLE

What can we gain from knowledge of Earth's climate history? The first lessons are conceptual. Geologists view Earth as a system of interlocking processes, dominated by geochemical cycles (e.g., Berner et al. 1983; Raymo et al. 1988; White and Blum 1995). Every chemical element essential to life follows a loop-transit around, and within, our planet. Earth is habitable because plate tectonics recycles water, carbon, sulfur, and other elements from its surface into its interior and back out again. Atmospheric carbon dioxide (CO_2) forms a tiny portion of Earth's carbon budget. However, the CO_2 greenhouse effect controls Earth's

Era	Period	Epoch	Stage	Age(Ma)
Cenozoic	Quaternary	Holocene	—	0.0118
Cenozoic	Quaternary	Pleistocene	Upper	0.126
Cenozoic	Quaternary	Pleistocene	Ionian	0.781
Cenozoic	Quaternary	Pleistocene	Calabrian	1.806
Cenozoic	Quaternary	Pleistocene	Gelasian	2.588
Cenozoic	Neogene	Pliocene	Piacenzian	3.600
Cenozoic	Neogene	Pliocene	Zanclean	5.333
Cenozoic	Neogene	Miocene	Messinian	7.246
Cenozoic	Neogene	Miocene	Tortonian	11.63
Cenozoic	Neogene	Miocene	Serravalian	13.82
Cenozoic	Neogene	Miocene	Langhian	15.97
Cenozoic	Neogene	Miocene	Burdigalian	20.44
Cenozoic	Neogene	Miocene	Aquitanian	23.03
Cenozoic	Paleogene	Oligocene	Chattian	28.1
Cenozoic	Paleogene	Oligocene	Rupelian	33.9
Cenozoic	Paleogene	Eocene	Priabonian	37.8
Cenozoic	Paleogene	Eocene	Bartonian	41.2
Cenozoic	Paleogene	Eocene	Lutetian	47.8
Cenozoic	Paleogene	Eocene	Ypresian	56.0
Cenozoic	Paleogene	Paleocene	Thanetian	59.2
Cenozoic	Paleogene	Paleocene	Selandian	61.6
Cenozoic	Paleogene	Paleocene	Danian	66.0
Mesozoic	Cretaceous	Upper	Maastrichtian	72.1
Mesozoic	Cretaceous	Upper	Campanian	83.6
Mesozoic	Cretaceous	Upper	Santonian	86.3
Mesozoic	Cretaceous	Upper	Coniacian	89.8
Mesozoic	Cretaceous	Upper	Turonian	93.9
Mesozoic	Cretaceous	Upper	Cenomanian	100.5
Mesozoic	Cretaceous		Albian -->	113.0
Mesozoic	Cretaceous		Aptian -->	126.3
Mesozoic	Cretaceous	Lower	Barremian	130.8
Mesozoic	Cretaceous	Lower	Hauterivian	133.9
Mesozoic	Cretaceous	Lower	Valanginian	139.4
Mesozoic	Cretaceous	Lower	Berriasian	145.0
Mesozoic	Jurassic	Upper	Tithonian	152.1
Mesozoic	Jurassic	Upper	Kimmeridgian	157.3
Mesozoic	Jurassic	Upper	Oxfordian	163.5
Mesozoic	Jurassic	Middle	Callovian	166.1
Mesozoic	Jurassic	Middle	Bathonian	168.3
Mesozoic	Jurassic	Middle	Bajocian	170.3
Mesozoic	Jurassic	Middle	Aalenian	174.1
Mesozoic	Jurassic	Lower	Toarcian	182.7
Mesozoic	Jurassic	Lower	Pliensbachian	190.8
Mesozoic	Jurassic	Lower	Sinemurian	199.3
Mesozoic	Jurassic	Lower	Hettangian	201.3
Mesozoic	Triassic	Upper	Rhaetian	209.5
Mesozoic	Triassic	Upper	Norian	228.4
Mesozoic	Triassic	Upper	Carnian	237.0
Mesozoic	Triassic	Middle	Ladinian	241.5
Mesozoic	Triassic	Middle	Anisian	247.1
Mesozoic	Triassic	Lower	Olenekian	250.0
Mesozoic	Triassic	Lower	Induan	252.2

Figure 8.4. Geologic time scale, after Gradstein and Ogg (2012), covering the Mesozoic and Cenozoic eras of Earth history (252.2 Ma to present). Terrestrial vertebrates and carbonate-secreting plankton evolved and diversified during this interval. Although the former species group is more entertaining in museum displays, the latter group may be responsible for buffering the impacts of extreme greenhouse gas fluctuations. Heavy dashed lines mark three mass extinctions of the five listed by Sepkoski (1996) and Barnosky et al. (2011): the end-Permian, end-Triassic, and end-Cretaceous events. Light-dashed lines mark environmental crises that have been associated with ocean anoxic events (OAEs) by Jenkyns (2010), as well as the Paleocene-Eocene Thermal Maximum (PETM) at 56.0 Ma and key transitions into sustained glacial conditions at 33.9 Ma (Oligocene glaciation) and at 0.781 Ma (large-amplitude 100 kyr Pleistocene glacial cycles).

average surface temperature, as well as the vigor of Earth's climate system (Alley 2012; Houghton 2015). By contrast, water vapor has a powerful greenhouse effect, but H_2O quickly condenses and precipitates in the form of rain or snow. A warmer atmosphere can maintain a higher steady-state humidity, so water vapor is properly treated as a CO_2 feedback (Lacis et al. 2010; Stevens et al. 2016).

Reconstructions from the deep past suggest that the most sudden shifts in Earth's past climate have left fingerprints of greenhouse gas disruption, such as ocean anoxia or acidification (Kidder and Worsley 2010; McGhee et al. 2013; Self et al. 2014; Molina 2015). Methane (CH_4) is both a greenhouse gas and an important by-product of the biosphere and carbon cycle, and it has been identified as a potential trigger for abrupt climate change in Earth's past (e.g., Zeebe et al. 2009). However, methane rapidly oxidizes to carbon dioxide in the ocean or in the atmosphere (Dickens 2011). Even if CH_4 pulls the trigger of abrupt climate change, it is CO_2 that empties the ammunition clip.

The Earth-system paradigm is relevant to efforts to shield biodiversity from anthropogenic climate change. The climate that we were once familiar with, characteristic of the mid-twentieth century, was at best a steady-state balance of interactions between competing geological and biological processes. It was not a stable equilibrium point. If we could somehow "geo-engineer" the climate system to the global-average temperature of either 1950 or 1750, the details of the steady-state would be new. Geographic shifts in heat absorption and emission would alter the regional patterns of climate (Robock et al. 2009; Irvine et al. 2010).

Biodiversity is key to climate resiliency. Earth's ultimate buffer against a runaway greenhouse effect, such as occurs on our sister planet Venus, is the consumption of CO_2 by chemical weathering of silicate rocks (Gislason et al. 2009; Beaulieu et al. 2010). During 4.54 billion years of Earth's history, extreme climate states, and the biodiversity crises associated with them, have become less common as the tree of life grew more branches. This suggests that biological processes have evolved that can react either more quickly, or more efficiently, to environmental changes. There is no need to invoke a Gaia superorganism to explain such adaptation (Levin 1998) if the segments of DNA that facilitate climate resilience, if not the species themselves, survive to face the next disruption. The key adaptive gene for mitigating human-made climate change could be lurking in an unappreciated organism.

The following sections of this chapter elaborate on the above concepts. The chapter first outlines several of the many tools that geologists use to reconstruct past environments and offers a synoptic view of climate history to place anthropogenic global warming in proper context. The chapter then discusses the successes and shortcomings of the Earth-system concept in explaining the diverse stable environments in Earth's past and the often-disruptive transitions between them. Finally, the chapter focuses more directly on environmental changes that can be projected as the climate system catches up to the CO_2 concentrations that mankind has imposed, or will soon impose, on it.

BIODIVERSITY CRISES IN EARTH HISTORY

The biotic catastrophes that mark mass extinctions arose from extreme conditions that are challenging to reconstruct millions of years afterward. Even the meteor strike that ended the Cretaceous period needed some unlucky feedbacks to elevate its lethality (D'Hondt 2005; Schulte et al. 2010). Because greenhouse gasses are key environmental factors, it is not surprising that sudden CO_2 extremes play important roles in most mass extinctions (Bond and Grasby 2017; Ernst and Youbi 2017). Geologists are developing narratives of past biodiversity crises that uncomfortably parallel some of the geologic processes that human civilization has unleashed (Figure 8.5). We describe the unique features of three such crises—Snowball Earth (Hoffman et al. 1998), as well as the end-Permian and the end-Cretaceous events—and seek common features and trends among the others.

Life as we know it roots its evolution in the most sustained climate extremes of Earth history. DNA clocks place the origin and diversification of most multicellular life forms (metazoans) during the climate crises of the Cryogenian period (850 Ma–635 Ma) of the Neoproterozoic Era (Erwin 2015). There were two so-called Snowball Earth episodes of sustained glacial conditions at both high and low latitudes. The Sturtian glaciation (ca. 700 Ma) was longer, with duration estimated to be 57 Myr (Rooney et al. 2015), and was followed by the shorter Marinoan glaciation (ca. 650 Ma) with duration >4 Myr (Prave et al. 2016). These two extreme events lasted at least 10 times longer than any global-scale glaciation since 635 Ma (e.g., Finnegan et al. 2011).

We do not know the scope of whatever DNA diversity was lost in the biotic crises of the Cryogenian. The microbial metabolisms and life strategies of its extinct biodiversity did not generate climate feedbacks to buffer the Earth system sufficiently to prevent near-permanent glaciation. Persistent Snow-

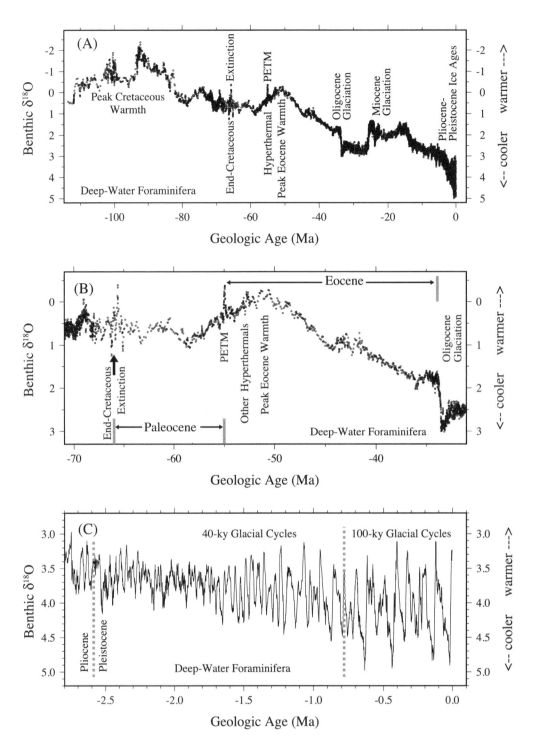

Figure 8.5. (A) Climate change since 114 Ma, as recorded by the oxygen-isotope ratio δ¹⁸O recorded in the carbonate tests of benthic foraminifera, microfauna of the deep-ocean (Zachos et al. 2008; Friedrich et al. 2012). Variations in δ¹⁸O depend on ambient water temperature and global ice volume. Both cooling temperatures and increased ice volumes cause δ¹⁸O to become "heavier," so the y-axis is reversed to match the sense of temperature change. From a plateau during the warm ice-free conditions of the mid-Cretaceous period, values of δ¹⁸O decline as the deep waters of the global ocean become cool from "greenhouse" to "icehouse" climate conditions. Near the start of the Oligocene epoch (33.9 Ma), short-term fluctuations of δ¹⁸O record the cyclic advance and retreat of continental ice sheets, largely correlated with the timing of Earth's orbital cycles. (B) Climate changes from the end-Cretaceous mass extinction (high variability reflecting climate and environmental instability), then warming to the peak Eocene warmth, punctuated by transient episodes of warmth, such as the Paleocene-Eocene Thermal Maximum (PETM), then gradual cooling until an abrupt cooling at the Eocene-Oligocene transition. (C) Fluctuations in benthic δ¹⁸O associated largely with Pleistocene glacial-interglacial cycles. Initially paced by the 41-kyr obliquity cycle, the glacial cycles since ~0.8 Ma have been larger and longer, reflecting nonlinear feedbacks in the Earth system.

ball Earth conditions have not developed in the more diverse ecosystems of the Phanerozoic, although several extinction events are marked by severe but more transient glaciations. Metazoans evolved to consume photosynthesizing organisms for food, perhaps limiting the ability of plants, phytoplankton, and algae to extract CO_2 from the atmosphere. At greenhouse extremes, innovations afforded by plant evolution may have increased the buffering resilience of photosynthesis. Franks et al. (2014) argue from leaf fossils that sustained atmospheric CO_2 levels above 1,000 ppm have not occurred since the evolution of forests in the Devonian (419.2 Ma–358.9 Ma).

McGhee et al. (2013) identify 11 distinct biodiversity crises in the Phanerozoic Eon, a set that includes the big-five mass extinctions of Sepkoski (1996). With few exceptions, these crises correlate with the eruption of large igneous provinces (LIPs) on land (Bond and Wignall 2014), usually in the form of flood basalts. The end-Permian event was extreme partly because the coeval Siberian Traps were the largest of these LIPs. Environmental stresses associated with massive disruptions of the carbon cycle lie behind most "rebuilding episodes" of the biosphere, including many extinction events too small to qualify for the list compiled by McGhee et al. (2013). LIPs can cause extremes of both hot and cold climates. Eruptions release CO_2 that is dissolved in the magma, sometimes adding to the greenhouse release by volatilizing carbon-rich strata near the surface. Extensive fresh exposures of flood basalt, however, will consume CO_2 via chemical weathering in a delayed feedback and can overcompensate for the CO_2 release, diminishing greenhouse warming enough to induce widespread glaciation.

The largest extinction event occurred at the close of the Permian period, in a process that spans the time between dated strata at 251.941 ±0.037 Ma and 251.880 ±0.031 Ma (Burgess et al. 2014). The coeval eruption of the Siberian Traps, the largest continental flood-basalt event of the Phanerozoic, is coupled with extreme fluctuations in $\delta^{13}C$, pointing to a massive degasing of carbon dioxide from Earth's mantle, as well as the volatilization of carbon-rich sediments in contact with magmas that intruded into subsurface sills (Burgess et al. 2017). From $\delta^{18}O$ measurements from apatite within fossil conodonts, Joachimski et al. (2012) and Sun et al. (2012) infer rapid temperature increases of 8°C and more during the end-Permian extinction interval. Cui and Kump (2015) review evidence from different studies to conclude that CO_2 levels increased by roughly three times, consistent with a climate sensitivity of 5°C–6°C for doubled CO_2 during the end Permian.

Knoll et al. (2007) compared an array of environmental impacts from rapid CO_2 release. Oxygen depletion in warm waters led to asphyxiation. Hydrogen-sulfide influx from volcanism poisoned eukaryotic microbes. A shift of primary production to prokaryotic cyanobacteria would render the base of the food chain less nutritious for marine vertebrates. Anoxia persisted over wide areas as microbial photosynthesis continued, even as higher organisms went extinct. Ocean acidification from dissolved CO_2 would increase the extinction rate of species that secreted calcium carbonate, particularly species that secreted $CaCO_3$ in open seawater to form reefs. Suffocation by dissolved CO_2, known as hypercapnia, is less spectacular as a kill agent. Nevertheless, Knoll et al. (2007) argued that hypercapnia explains best the extreme lethality of the end-Permian event.

Nearly all past biotic crises are coincident with a transient drop in $\delta^{13}C$ consistent with a release of carbon of some kind, but their kill factors are not uniform. Evidence for widespread hypercapnia is unique to the end-Permian event. Oxygen isotopes from biodiversity-crisis carbonates often display evidence of extreme global warming as an ecosystem disruptor. For biodiversity crises that terminated the Ordovician (485.4 Ma–443.6 Ma) and Devonian (419.2 Ma–358.9 Ma) periods, geologic evidence points to widespread glaciation following LIP eruption (Isaacson et al. 2008; Finnegan et al. 2011; Retallack 2015; Jones et al. 2017). At least one major kill factor for marine ecosystems, at the end of the Givetian epoch (387.7 Ma–382.7 Ma) of the Devonian period, has been associated with rapid sea-level rise, possibly an indirect effect of greenhouse warming, which drowned reefs and opened marine passageways for invasive species (McGhee et al. 2013).

The possible exception to this rule is the well-known end-Cretaceous mass extinction. The Deccan Traps LIP eruption in India was roughly coeval with the Chicxulub impact and has fueled controversy over the meteor hypothesis for decades (e.g., Chenet et al. 2007). Henehan et al. (2016) compares the lethal impacts of the Deccan Traps eruption with those of the meteor strike, using both sedimentary markers and a carbon-cycle model. Improvements in radiometric dating now place the largest Deccan eruption roughly 250 kyr prior to the Chicxulub bolide impact and end-Cretaceous mass extinction (Schoene et al. 2015; but see Renne et al. 2015). During the preimpact interval, carbonate microfossils accumulated at the seafloor in poor condition, suggesting degradation by added acidity from CO_2 emitted from the Deccan Traps. However, only 2°C–3°C greenhouse warming is estimated from isotopic data, along with disruption in species habitat ranges, but not widespread preimpact extinction (Thibault and Gardin 2010). Henehan et al. (2016) hypothesize that widespread carbonate dissolution on the seafloor, sourced from carbonate plankton in the photic zone, acted to buffer the CO_2-induced acidity of the ocean.

HAS BIODIVERSITY EVOLVED RESISTANCE TO CLIMATE DISRUPTION?

The end-Cretaceous extinction began as a volcanism-related carbon-cycle disruption, but its status as the last of Sepkoski's mass extinctions required a severe shock of a different kind. A hopeful interpretation, at least as far as a potential Anthropocene mass extinction is concerned, is that the gradual diversification of life into new ecological niches has made Earth's biosphere more resilient to exogenous carbon-cycle disruption. Several of the biodiversity crises of the Paleozoic, for instance, are notable for the extinction of whole classes of reef-building organisms (McGhee et al. 2013). This suggests that reefs, which formed large test beds for evolutionary innovation and also were among Earth's principal sinks for carbonate, lacked resiliency to carbon-cycle shocks in the Paleozoic and early Mesozoic. Either the evolution of corals provided better resiliency, or the evolution of other carbonate-secreting organisms created an effective "crumple zone" for reef ecosystems, similar to how a well-engineered automobile can absorb the impact of a traffic collision.

Large-scale environmental disruptions and sudden biotic turnover did not cease in the later Mesozoic and Cenozoic eras. Many marine environments in the Mesozoic experienced episodic anoxia, often paced by Earth's orbital cycles (Lanci et al. 2010; Eldrett et al. 2015). These cyclic variations have scarce impact on overall extinction rates, although preserved microfossil abundances varied. These regular cycles are punctuated at intervals of 10 Myr to 60 Myr with clusters of orbital layers that contain greatly elevated levels of organic carbon (Jenkyns 2010; Laurin et al. 2016). When the sedimentary evidence can be correlated globally, geologists term these clusters of orbital cycles the ocean anoxic events, or OAEs (Erba 2004). Jenkyns (2010) reviews global OAEs in detail at 183 Ma (Toarcian stage of the Jurassic period), 120 Ma (Selli event), 111 Ma (Paquier event), and 93 Ma (Bonarelli event), and suggests extending the OAE classification to the Early Cenozoic Paleocene-Eocene Thermal Maximum (PETM) at 56.0 Ma. In some cases the OAEs are correlated with large igneous-province

(LIP) eruptions, such as the Karoo-Farrar LIP for the Toarcian OAE (Caruthers et al. 2013) and the undersea eruption of the Ontong Java Plateau with the Selli event (Bottini et al. 2012). The associated ecosystem disruption of the Toarcian OAE (183 Ma) has been described as a "mass extinction" by some (Gomez and Goy 2011; Caruthers et al. 2013; Krencker et al. 2014), but its impact was significantly smaller than the biodiversity crises reviewed by McGhee et al. (2013).

Jenkyns (2010) notes that all OAEs of this time period share evidence for a sudden injection of CO_2 or CH_4 into Earth's atmosphere. Oxygen isotopes within carbonate fossils typically suggest increases to ocean temperatures of 5°C–7°C (see also Gomez and Goy 2011; Krencker et al. 2014) but also evidence from strontium or osmium isotopes that continental weathering accelerated to limit the CO_2 at least partially. The clustering of orbital cycles within OAEs indicates that enhanced-greenhouse conditions persisted for 50 kyr–400 kyr, during which marine biologic activity is strong and the relative absence of carbonate microfossils suggest an acidic deep ocean. Positive excursions in the carbon-isotope ratio $\delta^{13}C$ of marine carbonates are consistent with enhanced burial of organic carbon. Negative shifts in $\delta^{13}C$ also occur within the duration of most OAEs, which suggests the injection of isotopically-light carbon from volcanism, methane emissions, or other mechanisms for volatizing organic-carbon deposits. The detailed stratigraphy of geochemical markers within OAEs suggests complex feedbacks within the Earth system over 100-kyr intervals. OAEs typically conclude with a return to the previous climate conditions. Extinctions impact biodiversity during OAEs, but ecosystem functions appear to recover quickly, at least from the proxy studies thus far.

During the Cenozoic, Earth's carbon cycle has been tested multiple times by sudden inputs or extractions of atmospheric CO_2 or CH_4, but no biodiversity crises have oc-

curred. A dramatic greenhouse "hyperthermal" that marks the stratigraphic boundary between the Paleocene and Eocene epochs has been proposed as an analogue for the Anthropocene by Zeebe and Zachos (2013). The Paleocene-Eocene Thermal Maximum (PETM) elevated Earth's surface temperatures by 5°C–8°C for 170 kyr–200 kyr via an injection of either CH_4 or CO_2 at rates that are sudden (e.g., 3,000 petagrams of carbon in 5 kyr–10 kyr) relative to baseline geologic processes, but far slower than atmospheric change since the Industrial Revolution (perhaps 3,500–5,000 petagrams of carbon in 500 years). Several hyperthermal events have been identified in the isotope fluctuations of benthic foraminifera of the early Cenozoic, with repeat timing that correlated with 100-kyr orbital eccentricity cycles (Nicolo et al. 2007; Sexton et al. 2011; Laurin et al. 2016).

An abrupt cooling at 34 Ma (Liu et al. 2009; Hren et al. 2013) accompanied the extinction of many planktonic marine foraminifera species at the Eocene-Oligocene epoch transition (Zachos et al. 1999; Ivany et al. 2000; Pearson et al. 2008), after which the Antarctic ice sheet becomes an important factor in climate. After this transition Earth's orbital cycles modulated the growth of ice sheets in an icehouse climate, rather than governed the release of buried carbon in a greenhouse climate. Waxing and waning of the ice sheets before the mid-Pleistocene (0.8 Ma) appear to have responded in a simple linear manner to high-latitude insolation (Imbrie et al. 1992), but the most recent ice-age cycles have the time scale of Earth's eccentricity cycle (100 kyr), the insolation signal of which is too small to govern ice volume directly (Imbrie et al. 1993). Ambient air preserved in ice cores indicates that during the glacial periods atmospheric CO_2 was at least 100 ppm lower than during warm interglacial times (Jouzel 2013). The CO_2 variations validate dramatically the greenhouse gas control of Earth's climate change but pose a problem that is not yet solved. What caused 100-kyr cycles in atmospheric CO_2 during the late Pleistocene epoch?

A popular Earth-system paradigm for the 100-kyr ice-age cycle is the oceanic "conveyor belt for salt," which is facilitated by the production of deep-ocean water in the high-latitude North Atlantic (Broecker 1990). North Atlantic deep water (NADW) balances the transport of salt against the transport of water, via the atmosphere, from the net-evaporative Atlantic to the net-precipitative Pacific. In the conveyor-belt model, the NADW supplies the deep ocean with greenhouse gasses absorbed from Earth's atmosphere at the sea surface (Thornalley et al. 2011). Fluctuations in $\delta^{13}C$ within deep-sea sediment cores indicate that NADW formation is unstable, strong during interglacial intervals similar to the present day and weak during glacial intervals. Maasch and Saltzmann (1990) proposed a simple nonlinear ice-age climate model in which NADW formation stockpiles atmospheric CO_2 in the deep ocean, decreasing greenhouse gas concentrations in the shallow ocean and atmosphere, and leading to glacial conditions worldwide. Given the estimated time scales of chemical transport in the ocean (Broecker and Peng 1987), this nonlinear model predicted an alternation between glacial and interglacial states, which is consistent with paleoclimate data.

Subsequent research has revised the original conveyor-belt model significantly. Keeling and Stephens (2001) proposed that deepwater sources in the Antarctic, not in the North Atlantic, govern the supply of cold dense deep water to the world ocean. Motivated by strong late-Pleistocene correlations between atmospheric CO_2 and Antarctic temperature proxies, Stephens and Keeling (2000) argued that fluctuations in the sea ice bordering Antarctica, not the glacial meltwater from the North American ice sheet, govern the principal drawdown mechanism of CO_2 from Earth's atmosphere into the deep ocean. Recent oceanographic models for glacial-interglacial variability

(Allen et al. 2015; Howe et al. 2016; Yu et al. 2016; Jaccard et al. 2016) focus on changes in the relative volumes of NADW and Antarctic bottom water (AABW) in the abyssal ocean, with CO_2 storage in glacial periods associated with the Antarctic water mass. There is observational evidence that dissolved CO_2 trapped in the deep Southern Ocean was vented to the atmosphere during the last transition out of glacial conditions, at roughly 12 ka (Skinner et al. 2010; Basak et al. 2018).

An emphasis on the physical transport of CO_2 by water masses, however, ignores key chemical and biological carbon-cycle interactions in the ocean. A stockpile of dissolved CO_2 in the deep ocean would lead to acidification. Although ocean acidification impedes carbonate secretion by marine organisms (Doney et al. 2009, 2014), the long coexistence of high CO_2 levels and carbonate deposition in Earth history tells us that the biosphere has strategies to compensate, if given enough time to adjust. Broecker and Peng (1987) hypothesized the carbonate compensation mechanism to explain glacial-to-interglacial transitions. It buffers the deep ocean, after a time delay of a few millennia, via a reorganization of dissolved-nutrient transport and ecosystem function. Geologic proxies for carbonate-anion (CO_3^{-2}) in the world ocean suggest that the glacial and interglacial states in the deep Pacific and Indian Oceans had similar values (Honisch et al. 2008; Yu et al. 2014), in contrast to strong differences in the surface-water proxies. These studies indicate that global patterns of biological carbon consumption, either through photosynthesis or $CaCO_3$ secretion, differed greatly during glacial times, relative to the Holocene.

Gottschalk et al. (2016) isolated transient upward fluctuations in atmospheric CO_2 during the most-recent interval of glacial climate, which occurred at irregular intervals of 5–10 kyr between 25 ka and 65 ka, and correlated these with indicators of decreased primary biological production in the Southern Ocean. These fluctuations are strong evidence of a biological carbon pump during glacial periods, which helps the deep ocean store dissolved CO_2 without excessive acidification. Despite its importance, the precise biological actors for this buffering are not yet certain (Marchitto et al. 2005; Yu et al. 2010, 2013, 2014). The glacial and interglacial extremes of the late Pleistocene ice ages are maintained by biochemical steady states, as well as by dynamic ocean-circulation and ice-sheet balances (e.g., Brovkin et al. 2007; Buchanan et al. 2016).

Extinction rates in the marine realm have been low overall during the "icehouse" Earth climate interval since the Eocene-Oligocene boundary (Thomas 2007; Moritz and Agudo 2013). Interestingly, subdued extinction rates characterized the late-Paleozoic icehouse-climate interval as well. After the Serpukhovian biodiversity crisis, caused by sudden global cooling (McGhee et al. 2014), both extinction and speciation occurred at subdued rates through the later Carboniferous period (Stanley and Powell 2003). Milankovitch cycles have been identified in organic-rich cyclothems (Fielding et al. 2008; Davydov et al. 2010) deposited by swampy Carboniferous forests, suggesting that excessive carbon burial periodically depleted Earth's greenhouse blanket and led to glaciation. The geographical pattern of glaciation changed over the 50-My duration of the late-Paleozoic ice ages (Montanez and Poulsen 2013), suggesting, as with the Pleistocene, multiple steady-state balances in Earth's past carbon cycle between ocean circulation, surface weathering, plant growth, and marine calcification.

The concept of multiple steady climate states, mediated by separate biological interactions with Earth's carbon cycle, might apply to enhanced-greenhouse episodes as well. Zeebe et al. (2009) argue from modeling experiments that the initial burst of greenhouse gases that started the Paleocene-Eocene Thermal Maximum (PETM) were insufficient to explain the 170–200 kyr duration of the event. Natural sinks for atmospheric CO_2 would have scrubbed even

3,000 Pg C from the climate system in a far-shorter time. Their modeling experiments suggest that a persistent input of carbon to the atmosphere maintained the PETM warmth. Zeebe et al. (2009) model this steady input as exogenous, but it could also arise from a new balance of carbon utilization within the existing range of biodiversity. The same species assemblage (for the most part), but with different abundances, might recycle atmospheric CO_2 to maintain elevated concentrations. As the Anthropocene proceeds, a global ecological reorganization could create a new, warmer "normal" that will persist longer than simple nonbiotic models predict.

LESS THAN CATASTROPHE: BIODIVERSITY IN A CHANGEABLE CLIMATE

Extinction is only one of the potential impacts to biodiversity by ice-age cycles or an enhanced-greenhouse excursion. Human civilization has developed within a stable Holocene climate, tempting us to interpret modern-day ecosystems as stable and mature. Both DNA evidence and comparisons between theoretical and real ecosystems have revealed that many modern ecosystems had not reached a stable state before the impacts of modern civilization could be felt. Terrestrial ecosystems disappear or transform as ice sheets wax and wane, but biodiversity at the species and genus level is preserved in climate refugia, isolated locations where warm-adapted organisms can survive a glacial interval, or where cold-adapted species can survive a interglacial interval (Svenning et al. 2008; Woodruff 2010). Using DNA to separate species lineages, Hewitt (2004) argues that many modern high-latitude Northern Hemisphere ecosystems were tethered to specific refugia during the past glacial intervals and have remained distinct during the interglacials. Sandel et al. (2011) defines climate-change velocity as the rate at which a climate-specialized organism must migrate to follow the environmental conditions to which it has adapted. Ordonez and Svenning (2015) argue from distributions of plant species in Europe that many species cannot migrate quickly enough for some ecosystems to stabilize during a typical 10–20 kyr interglacial interval. Ordonez and Svenning (2016) argue that cyclic or intermittent climate transitions impair "functional diversity," a metric of how efficiently an ecosystem utilizes the trophic opportunities of its ambient environment.

The lesson offered by cyclic ice-age biodiversity disruptions is that we should not expect natural ecosystems to adjust quickly and completely to the Anthropocene greenhouse excursion, at least not within centuries or millennia. Ordonez et al. (2016) use species information and climate projections to identify locations that will experience novel climate conditions, regions where environmental gradients will migrate most rapidly, and locations where temperature and precipitation trends will diverge, so that an existing species has no migration path that could follow its optimal environment. The bumpy path that ecosystems must follow to their new locations is reflected in the sedimentary record of fires during the most recent glacial-to-interglacial transition (Marlon et al. 2009, 2013). Intervals of rapid climate change, such as near the onset and conclusion of the Younger Dryas cooling interval (12.9–11.7 ka), are marked by large increases of fire activity.

Nearly all the large biodiversity crises in last 541 million years of Phanerozoic Earth history have involved a severe transient disruption to the carbon cycle and to the atmospheric concentration of carbon dioxide, the most important terrestrial greenhouse gas. Most large biodiversity crises correlate in time with surface eruptions of large igneous provinces (LIPs), which release large volumes of CO_2 that were dissolved in their magma. Isotopic evidence for large sudden increases in ocean temperature is common during crises, consistent with an enhanced greenhouse.

In some cases evidence for widespread glaciation is found instead, consistent with CO_2 consumption via chemical weathering of the exposed LIP eruption, which in some cases can overcompensate for the released CO_2.

Over geologic time the extinction events associated with LIP volcanism and other methods of rapid carbon release have weakened, although impacts often were substantial in particular ecosystems. The ocean anoxic events (OAEs) of the Mesozoic era and the hyperthermal events of the Eocene epoch elevated temperatures and CO_2 levels for 20–200 kyr, longer than one might expect a single CO_2 emission to persist in the atmosphere. We hypothesize that the climate system found steady-state balances between carbon fluxes, ocean circulations, and radiation balances during these enhanced greenhouse intervals, possibly containing biotic feedbacks that buffered the severity of the climate change on overall biodiversity. The glacial and interglacial climate extremes of the recent late-Pleistocene ice ages have been modeled to represent two separate such climate-biosphere steady-state balances.

In the Anthropocene, the environmental impacts of anthropogenic greenhouse gas release are exacerbated by other ecosystem disruptions from humans, such as chemical pollution and habitat reduction and fragmentation. We cannot feel assured that the apparent resilience of biodiversity during the Cenozoic to sudden climate changes will prevail in the context of these additional impacts. Anthropogenic greenhouse gases are playing the role of the Large Igneous Provinces in past crises, but we humans may also be playing the role of the end-Cretaceous meteor. There is much we still do not understand about the detailed balances within Earth's carbon cycle. We had better learn them quickly.

NOTES

1. Ages of geologic events and materials are denoted by ka, Ma, and Ga. These correspond to thousands, millions, and bil-

lions of years ago, respectively. Intervals of time are denoted by kyr, Myr, and Gyr (e.g., if two meteorites struck the Earth at 10 Ma and 13 Ma, their impacts would be 3 Myr apart). The commonly accepted value of Earth's age (4.54 Ga) is bracketed by the age of meteorites in the Solar System (Dalrymple 2001), and the oldest surviving terrestrial material (Wilde et al. 2001).

2. If we choose 0.25 ka for the onset of the Anthropocene, it coincides roughly with the 1769 patent for James Watt's steam engine.

REFERENCES

Allen, K. A., E. L. Sikes, B. Hönisch, A. C. Elmore, T. P. Guilderson, Y. Rosenthal, and R. F. Anderson. 2015. "Southwest Pacific deep water carbonate chemistry linked to high southern latitude climate and atmospheric CO_2 during the Last Glacial Termination." Quaternary Science Review 122: 180–191. https://doi.org/10.1016/j.quascirev.2015.05.007.

Alley, R. 2012. The Climate Control Knob, National Climate Seminar Series. Island Press.

Barnosky, A. D., N. Matzke, S. Tomiy, G. O. U. Wogan, B. Swartz, T. B. Quental, C. Marshall, J. L. McGuire, E. L. Lindsey, K. C. Maguire, B. Mersey, and E. A. Ferrer. 2011. "Has the Earth's sixth mass extinction already arrived?" Nature 471: 51–57. https://doi.org/10.1038/nature09678.

Basak, C., H. Fröllje, F. Lamy, R. Gersonde, V. Benz, R. F. Anderson, M. Molina-Kescher, and K. Pahnke. 2018. "Breakup of last glacial deep stratification in the South Pacific." Science 359: 900–904. https://doi.org/10.1126/science.aao2473.

Beaulieu, E., Y. Goddéris, D. Labat, C. Roelandt, P. Oliva, and B. Guerrero. 2010. "Impact of atmospheric CO_2 levels on continental silicate weathering." Geochemistry, Geophysics, Geosystems 11: Q07007. https://doi.org/10.1029/2010GC003078.

Berner, R. A., A. C. Lasaga, and R. M. Garrels. 1983. "The carbonate-silicate geochemical cycle and its effect on atmospheric carbon-dioxide over the past 100 million years." American Journal of Science 283: 641–683. https://doi.org/10.2475/ajs.283.7.641.

Bond, D. P., and P. B. Wignall. 2014. "Large igneous provinces and mass extinctions: An update." In Volcanism, Impacts, and Mass Extinctions: Causes and Effects, ed. G. Keller and A. C. Kerr, Geological Society of America Special Papers, 505: 29–55. https://doi.org/10.1130/2014.2505(02).

Bond, D. P., and S. E. Grasby. 2017. "On the causes of mass extinctions." Palaeogeography, Palaeoclimatology, Palaeoecology 478: 3–29. https://doi.org/10.1016/j.palaeo.2016.11.005.

Bottini, C., A. S. Cohen, E. Erba, H. C. Jenkyns, and A. L. Coe. 2012. "Osmium-isotope evidence for volcanism, weathering, and ocean mixing during the early Aptian OAE 1a." Geology 40: 583–586. https://doi.org/10.1130/G33140.1.

Broecker, W. S., and T. H. Peng. 1987. "The role of $CaCO_3$ compensation in the glacial to interglacial atmo-spheric CO_2 change." Global Biogeochemical Cycles 1: 15–29. https://doi.org/10.1029/GB001i001p00015.

Broecker, W. S. 1990. The Great Ocean Conveyor: Discovering the Trigger for Abrupt Climate Change. Princeton University Press.

Brovkin, V., A. Ganopolski, D. Archer, and S. Rahmstorf. 2007. "Lowering of glacial atmospheric CO_2 in response to changes in oceanic circulation and marine biogeochemistry." Paleoceanography 22: PA4202. https://doi.org/10.1029/2006PA001380.

Buchanan, P. J., R. J. Matear, A. Lenton, S. J. Phipps, Z. Chase, and D. M. Etheridge. 2016. "The simulated climate of the Last Glacial Maximum and insights into the global marine carbon cycle." Climate of the Past 12: 2271–2295. https://doi.org/10.5194/cp-12-2271-2016.

Burgess, S. D., S. Bowring, and S. Z. Shen. 2014. "High-precision timeline for Earth's most severe extinction." Proceedings of the National Academy of Sciences 111: 3316–3321. https://doi.org/10.1073/pnas.1317692111.

Burgess, S. D., J. D. Muirhead, and S. Bowring. 2017. "Initial pulse of Siberian Traps sills as the trigger of the end-Permian mass extinction." Nature Communications 8: 164. https://doi.org/10.1038/s41467-017-00083-9.

Caruthers, A. H., P. L. Smith, and D. R. Gröcke. 2013. "The Pliensbachian–Toarcian (Early Jurassic) extinction, a global multi-phased event." Palaeogeography, Palaeoclimatology, Palaeoecology 386: 104–118. https://doi.org/10.1016/j.palaeo.2013.05.010.

Chenet, A. L., X. Quidelleur, F. Fluteau, V. Courtillot, and S. Bajpai. 2007. "$^{40}K–^{40}Ar$ dating of the Main Deccan large igneous province: Further evidence of KTB age and short duration." Earth and Planetary Science Letters 263: 1–15. https://doi.org/10.1016/j.epsl.2007.07.011.

Cui, Y., and L. R. Kump. 2015. "Global warming and the end-Permian extinction event: Proxy and modeling perspectives." Earth-Science Reviews 149: 5–22. https://doi.org/10.1016/j.earscirev.2014.04.007.

Dalrymple, G. B. 2001. "The age of the Earth in the twentieth century: A problem (mostly) solved." In The Age of the Earth: From 4004BC to AD 2002, ed. C. L. E. Lewis and S. J. Knell, 190, 205–221. Geological Society, London. https://doi.org/10.1144/GSL.SP.2001.190.01.14.

Davydov, V. I., J. L. Crowley, M. D. Schmitz, and V. I. Poletaev. 2010. "High-precision U-Pb zircon age calibration of the global Carboniferous time scale and Milankovitch band cyclicity in the Donets Basin, eastern Ukraine." Geochemistry, Geophysics, Geosystems 11: Q0AA04. https://doi.org/10.1029/2009GC002736.

D'Hondt, S. 2005. "Consequences of the Cretaceous/Paleogene mass extinction for marine ecosystems." Annual Review of Ecology, Evolution, and Systematics, 36: 295–317, https://doi.org/10.1146/annurev.ecolsys.35.021103.105715.

Dickens, G. R. 2011. "Down the rabbit hole: Toward appropriate discussion of methane release from gas hydrate systems during the Paleocene-Eocene thermal maximum and other past hyperthermal events." Climate of the Past 7: 831–846. https://doi.org/10.5194/cp-7-831-2011.

Dirzo, R., H. S. Young, M. Galetti, G. Ceballos, N. J. Isaac, and B. Collen. 2014. "Defaunation in the Anthropocene." *Science* 345: 401–406. https://doi.org/10.1126/science.1251817.

Doney, S. C., V. J. Fabry, R. A. Feely, and J. A. Kleypas. 2009. "Ocean acidification: The other CO_2 problem." *Annual Review of Marine Science* 1: 169–192. https://doi.org/10.1146/annurev.marine.010908.163834.

Doney, S. C., L. Bopp, and M. C. Long. 2014. "Historical and future trends in ocean climate and biogeochemistry." *Oceanography* 27: 108–119. https://doi.org/10.5670/oceanog.2014.14.

Eldrett, J. S., C. Ma, S. C. Bergman, A. Ozkan, D. Minisini, B. Lutz, S. J. Jackett, C. Macaulay, and A. E. Kelly. 2015. "Origin of limestone–marlstone cycles: Astronomic forcing of organic-rich sedimentary rocks from the Cenomanian to early Coniacian of the Cretaceous Western Interior Seaway, USA." *Earth and Planetary Science Letters* 423: 98–113. https://doi.org/10.1016/j.epsl.2015.04.026.

Erba, E. 2004. "Calcareous nannofossils and Mesozoic oceanic anoxic events." *Marine Micropaleontology* 52: 85–106. https://doi.org/10.1016/j.marmicro.2004.04.007.

Ernst, R. E., and N. Youbi. 2017. "How Large Igneous Provinces affect global climate, sometimes cause mass extinctions, and represent natural markers in the geological record." *Palaeogeography, Palaeoclimatology, Palaeoecology* 478: 30–52. https://doi.org/10.1016/j.palaeo.2017.03.014.

Erwin, D. H. 2015. "Early metazoan life: Divergence, environment and ecology." *Philosophical Transactions of the Royal Society (Series B)* 370: 20150036. https://doi.org/10.1098/rstb.2015.0036.

Fielding, C. R., T. D. Frank, and J. L. Isbell. 2008. "The late Paleozoic ice age—A review of current understanding and synthesis of global climate patterns." In *Resolving the Late Paleozoic Ice Age in Time and Space*, ed. C. R. Fielding, T. D. Frank, and J. L. Isbell, *Geological Society of America Special Papers*, 441: 343–354. https://doi.org/10.1130/2008.2441(24).

Finnegan, S., K. Bergmann, J. M. Eiler, D. S. Jones, D. A. Fike, I. Eisenman, N. C. Hughes, A. K. Tripati, and W. W. Fischer. 2011. "The magnitude and duration of Late Ordovician–Early Silurian glaciation." *Science* 331: 903–906. https://doi.org/10.1126/science.1200803.

Franks, P. J., D. L. Royer, D. J. Beerling, P. K. Van de Water, D. J. Cantrill, M. M. Barbour, and J. A. Berry. 2014. "New constraints on atmospheric CO_2 concentration for the Phanerozoic." *Geophysical Research Letters* 41: 4685–4694. https://doi.org/10.1002/2014GL060457.

Friedrich, O., R. D. Norris, and J. Erbacher. 2012. "Evolution of middle to Late Cretaceous oceans: A 55 m.y. record of Earth's temperature and carbon cycle." *Geology* 40: 107–110. https://doi.org/10.1130/G32701.1.

Gislason, S. R., E. H. Oelkers, E. S. Eiriksdottir, M. I. Kardjilov, G. Gisladottir, B. Sigfusson, A. Snorrason, S. Elefsen, J. Hardardottir, P. Torssander, and N. Oskarsson. 2009. "Direct evidence of the feedback between climate and weathering." *Earth and Planetary Science Letters* 277: 213–222. https://doi.org/10.1016/j.epsl.2008.10.018.

Gómez, J. J., and A. Goy. 2011. "Warming-driven mass extinction in the Early Toarcian (Early Jurassic) of northern and central Spain: Correlation with other time-equivalent European sections." *Palaeogeography, Palaeoclimatology, Palaeoecology* 306: 176–195. https://doi.org/10.1016/j.palaeo.2011.04.018.

Gottschalk, J., L. C. Skinner, J. Lippold, H. Vogel, N. Frank, S. L. Jaccard, and C. Waelbroeck. 2016. "Biological and physical controls in the Southern Ocean on past millennial-scale atmospheric CO_2 changes." *Nature Communications* 7: 11539. https://doi.org/10.1038/ncomms11539.

Gradstein, F. M., and J. G. Ogg. 2012. "The chronographic scale." In *The Geologic Time Scale 2012*, ed. F. M. Gradstein, J. G. Ogg, M. Schmitz, and G. Ogg, 31–42. Elsevier. https://doi.org/10.1016/B978-0-444-59425-9.00002-0.

Henehan, M. J., P. M. Hull, D. E. Penman, J. W. Rae, and D. N. Schmidt. 2016. "Biogeochemical significance of pelagic ecosystem function: An end-Cretaceous case study." *Philosophical Transactions of the Royal Society (Series B)* 371: 20150510. https://doi.org/10.1098/rstb.2015.0510.

Hewitt, G. M. 2004. "Genetic consequences of climatic oscillations in the Quaternary." *Philosophical Transactions of the Royal Society (Series B)* 359: 183–195. https://doi.org/10.1098/rstb.2003.1388.

Hoffman, P. F., A. J. Kaufman, G. P. Halverson, and D. P. Schrag. 1998. "A Neoproterozoic snowball earth." *Science* 281: 1342–1346. https://doi.org/10.1126/science.281.5381.1342.

Hönisch, B., T. Bickert, and N. G. Hemming. 2008. "Modern and Pleistocene boron isotope composition of the benthic foraminifer Cibicidoides wuellerstorfi." *Earth and Planetary Science Letters* 272: 309–318. https://doi.org/10.1016/j.epsl.2008.04.047.

Houghton, J. 2015. *Global Warming: The Complete Briefing*. 5th ed. Cambridge University Press.

Howe, J. N. W., A. M. Piotrowski, T. L. Noble, S. Mulitza, C. M. Chiessi, and G. Bayon. 2016. "North Atlantic deep water production during the last glacial maximum." *Nature Communications* 7: 11765. https://doi.org/10.1038/ncomms11765.

Hren, M. T., N. D. Sheldon, S. T. Grimes, M. E. Collinson, J. J. Hooker, M. Bugler, and K. C. Lohmann. 2013. "Terrestrial cooling in Northern Europe during the Eocene–Oligocene transition." *Proceedings of the National Academy of Sciences* 110: 7562–7567. https://doi.org/10.1073/pnas.1210930110.

Hull, P. M., R. D. Norris, T. J. Bralower, and J. D. Schueth. 2011. "A role for chance in marine recovery from the end-Cretaceous extinction." *Nature Geoscience* 4: 856–860. https://doi.org/10.1038/NGEO1302.

Imbrie, J., E. A. Boyle, S. C. Clemens, A. Duffy, W. R. Howard, G. Kukla, J. Kutzbach, D. G. Martinson, A.

McIntyre, A. C. Mix, B. Molfino, J. J. Morley, L. C. Peterson, N. G. Pisias, W. L. Prell, M. E. Raymo, N. J. Shackleton, and J. R. Toggweiler. 1992. "On the structure and origin of major glaciation cycles 1: Linear responses to Milankovitch forcing." *Paleoceanography* 7: 701–738. https://doi.org/10.1029/92PA02253.

Imbrie, J., A. Berger, E. A. Boyle, S. C. Clemens, A. Duffy, W. R. Howard, G. Kukla, J. Kutzbach, D. G. Martinson, A. Mcintyre, A. C. Mix, B. Molfino, J. J. Morley, L. C. Peterson, N. G. Pisias, W. L. Prell, M. E. Raymo, N. J. Shackleton, and J. R. Toggweiler. 1993. "On the structure and origin of major glaciation cycles 2: The 100,000-year cycle." *Paleoceanography* 8: 699–735. https://doi.org/10.1029/93PA02751.

Irvine, P. J., A. Ridgwell, and D. J. Lunt. 2010. "Assessing the regional disparities in geoengineering impacts." *Geophysical Research Letters* 37: L18702. https://doi.org/10.1029/2010GL044447.

Isaacson, P. E., E. Diaz-Martinez, G. W. Grader, J. Kalvoda, O. Babek, and F. X. Devuyst. 2008. "Late Devonian–earliest Mississippian glaciation in Gondwanaland and its biogeographic consequences." *Palaeogeography, Palaeoclimatology, Palaeoecology* 268: 126–142. https://doi.org/10.1016/j.palaeo.2008.03.047.

Ivany, L. C., W. P. Patterson, and K. C. Lohmann. 2000. "Cooler winters as a possible cause of mass extinctions at the Eocene/Oligocene boundary." *Nature* 407: 887–890. https://doi.org/10.1038/35038044.

Jaccard, S. L., E. D. Galbraith, A. Martínez-García, and R. F. Anderson. 2016. "Covariation of deep southern-ocean oxygenation and atmospheric CO_2 through the last ice age." *Nature* 530: 207–210. https://doi.org/10.1038/nature16514.

Jenkyns, H. C. 2010. "Geochemistry of oceanic anoxic events." *Geochemistry, Geophysics, Geosystems* 11: Q03004, https://doi.org/10.1029/2009GC002788.

Joachimski, M. M., X. Lai, S. Shen, H. Jiang, G. Luo, B. Chen, J. Chen, and Y. Sun. 2012. "Climate warming in the latest Permian and the Permian–Triassic mass extinction." *Geology* 40: 195–198. https://doi.org/10.1130/G32707.1.

Jones, D. S., A. M. Martini, D. A. Fike, and K. Kaiho. 2017. "A volcanic trigger for the Late Ordovician mass extinction? Mercury data from south China and Laurentia." *Geology* 45: 631–634. https://doi.org/10.1130/G38940.1.

Jouzel, J. 2013. "A brief history of ice core science over the last 50 yr." *Climate of the Past* 9: 2525–2547. https://doi.org/10.5194/cp-9-2525-2013.

Keeling, R. F., and B. B. Stephens. 2001. "Antarctic sea ice and the control of Pleistocene climate instability." *Paleoceanography* 16: 112–131. https://doi.org/10.1029/2000PA000529.

Kidder, D. L., and T. R. Worsley. 2010. "Phanerozoic large igneous provinces (LIPs), HEATT (haline euxinic acidic thermal transgression) episodes, and mass extinctions." *Palaeogeography, Palaeoclimatology, Palaeoecology* 295: 162–191. https://doi.org/10.1016/j.palaeo.2010.05.036.

Knoll, A. H., R. K. Bambach, J. L. Payne, S. Pruss, and W. W. Fischer. 2007. "Paleophysiology and end-Permian mass extinction." *Earth and Planetary Science Letters* 256: 295–313. https://doi.org/10.1016/j.epsl.2007.02.018.

Krencker, F. N., S. Bodin, R. Hoffmann, G. Suan, E. Mattioli, L. Kabiri, K. B. Föllmi, and A. Immenhauser. 2014. "The middle Toarcian cold snap: Trigger of mass extinction and carbonate factory demise." *Global and Planetary Change* 117: 64–78. https://doi.org/10.1016/j.gloplacha.2014.03.008.

Lacis, A. A., G. A. Schmidt, D. Rind, and R. A. Ruedy. 2010. "Atmospheric CO_2: Principal control knob governing Earth's temperature." *Science* 330: 356–359. https://doi.org/10.1126/science.1190653.

Lanci, L., G. Muttoni, and E. Erba. 2010. "Astronomical tuning of the Cenomanian Scaglia Bianca Formation at Furlo, Italy." *Earth and Planetary Science Letters* 292: 231–237. https://doi.org/10.1016/j.epsl.2010.01.041.

Laurin, J., S. R. Meyers, S. Galeotti, and L. Lanci. 2016. "Frequency modulation reveals the phasing of orbital eccentricity during Cretaceous Oceanic Anoxic Event II and the Eocene hyperthermals." *Earth and Planetary Science Letters* 442: 143–156. https://doi.org/10.1016/j.epsl.2016.02.047.

Levin, S. A. 1998. "Ecosystems and the biosphere as complex adaptive systems." *Ecosystems* 1: 431–436. https://doi.org/10.1007/s100219900037.

Liu, Z., M. Pagani, D. Zinniker, R. DeConto, M. Huber, H. Brinkhuis, S. R. Shah, R. M. Leckie, and A. Pearson. 2009. "Global cooling during the Eocene-Oligocene climate transition." *Science* 323: 1187–1190. https://doi.org/10.1126/science.1166368.

Maasch, K. A., and B. Saltzman. 1990. "A low-order dynamical model of global climatic variability over the full Pleistocene." *Journal of Geophysical Research* 95: 1955–1963. https://doi.org/10.1029/JD095iD02p01955.

Marchitto, T. M., J. Lynch-Stieglitz, and S. R. Hemming. 2005. "Deep Pacific $CaCO_3$ compensation and glacial–interglacial atmospheric CO_2." *Earth and Planetary Science Letters* 231: 317–336. https://doi.org/10.1016/j.epsl.2004.12.024.

Marlon, J. R., P. J. Bartlein, M. K. Walsh, S. P. Harrison, K. J. Brown, M. E. Edwards, P. E. Higuera, M. J. Power, R. S. Anderson, C. Briles, A. Brunelle, C. Carcaillet, M. Daniels, F. S. Hu, M. Lavoie, C. Long, T. Minckley, P. J. H. Richard, A. C. Scott, D. S. Shafer, W. Tinner, C. E. Umbanhowar Jr., and C. Whitlock. 2009. "Wildfire responses to abrupt climate change in North America." *Proceedings of the National Academy of Sciences* 106: 2519–2524. https://doi.org/10.1073/pnas.0808212106.

Marlon, J. R., P. J. Bartlein, A. L. Daniau, S. P. Harrison, S. Y. Maezumi, M. J. Power, W. Tinner, and B. Vannière. 2013. "Global biomass burning: A synthesis and review of Holocene paleofire records and their controls." *Quaternary Science Reviews* 65: 5–25. https://doi.org/10.1016/j.quascirev.2012.11.029.

McGhee, G. R., M. E. Clapham, P. M. Sheehan, D. J. Bottjer, and M. L. Droser. 2013. "A new ecological-severity ranking of major Phanerozoic biodiversity crises." *Palaeogeography, Palaeoclimatology, Palaeoecology* 370: 260–270. https://doi.org/10.1016/j.palaeo.2012.12.019.

McGhee, G. R., P. M. Sheehan, D. J. Bottjer, and M. L. Droser. 2014. "Ecological ranking of Phanerozoic biodiversity crises: The Serpukhovian (early Carboniferous) crisis had a greater ecological impact than the end-Ordovician." *Geology* 40: 147–150. https://doi.org/10.1130/G32679.1.

Molina, E. 2015. "Evidence and causes of the main extinction events in the Paleogene based on extinction and survival patterns of foraminifera." *Earth-Science Reviews* 140: 166–181. https://doi.org/10.1016/j.earscirev.2014.11.008.

Montañez, I. P., and C. J. Poulsen. 2013. "The Late Paleozoic ice age: An evolving paradigm." *Annual Review of Earth and Planetary Sciences* 41: 629–656. https://doi.org/10.1146/annurev.earth.031208.100118.

Moritz, C., and R. Agudo. 2013. "The future of species under climate change: Resilience or decline?" *Science* 341: 504–508. https://doi.org/10.1126/science.1237190.

Nicolo, M. J., G. R. Dickens, C. J. Hollis, and J. C. Zachos. 2007. "Multiple early Eocene hyperthermals: Their sedimentary expression on the New Zealand continental margin and in the deep sea." *Geology* 35: 699–702. https://doi.org/10.1130/G23648A.1.

Ordonez, A., and J. C. Svenning. 2015. "Geographic patterns in functional diversity deficits are linked to glacial-interglacial climate stability and accessibility." *Global Ecology and Biogeography* 24: 826–837. https://doi.org/10.1111/geb.12324.

Ordonez, A., and J. C. Svenning. 2016. "Strong paleoclimatic legacies in current plant functional diversity patterns across Europe." *Ecology and Evolution* 6: 3405–3416. https://doi.org/10.1002/ece3.2131.

Ordonez, A., J. W. Williams, and J. C. Svenning. 2016. "Mapping climatic mechanisms likely to favour the emergence of novel communities." *Nature Climate Change* 6: 1104–1109. https://doi.org/10.1038/NCLIMATE3127.

Park, J., and D. L. Royer. 2011. "Geologic constraints on the glacial amplification of Phanerozoic climate sensitivity." *American Journal of Science* 311: 1–26. https://doi.org/10.2475/01.2011.01.

Pearson, P. N., I. K. McMillan, B. S. Wade, T. D. Jones, H. K. Coxall, P. R. Bown, and C. H. Lear. 2008. "Extinction and environmental change across the Eocene-Oligocene boundary in Tanzania." *Geology* 36: 179–182. https://doi.org/10.1130/G24308A.1.

Prave, A. R., D. J. Condon, K. H. Hoffmann, S. Tapster, and A. E. Fallick. 2016. "Duration and nature of the end-Cryogenian (Marinoan) glaciation." *Geology* 44: 631–634. https://doi.org/10.1130/G38089.1.

Raymo, M. E., W. F. Ruddiman, and P. N. Froelich. 1988. "Influence of late Cenozoic mountain building on ocean geochemical cycles." *Geology* 16: 649–653. https://doi.org/10.1130/0091-7613(1988)016<0649:IOLCMB>2.3.CO;2.

Renne, P. R., C. J. Sprain, M. A. Richards, S. Self, L. Vanderkluysen, and K. Pande. 2015. "State shift in Deccan volcanism at the Cretaceous-Paleogene boundary, possibly induced by impact." *Science* 350: 76–78. https://doi.org/10.1126/science.aac7549.

Retallack, G. J. 2015. "Late Ordovician glaciation initiated by early land plant evolution and punctuated by greenhouse mass extinctions." *Journal of Geology* 123: 509–538. https://doi.org/10.1086/683663.

Robock, A., A. Marquardt, B. Kravitz, and G. Stenchikov. 2009. "Benefits, risks, and costs of stratospheric geoengineering." *Geophysical Research Letters* 36: L19703. https://doi.org/10.1029/2009GL039209.

Rooney, A. D., J. V. Strauss, A. D. Brandon, and F. A. Macdonald. 2015. "A Cryogenian chronology: Two long-lasting synchronous Neoproterozoic glaciations." *Geology* 43: 459–462. https://doi.org/10.1130/G36511.1.

Rudwick, M. J. S. 1985. *The Great Devonian Controversy: The Shaping of Scientific Knowledge among Gentlemanly Specialists.* University of Chicago.

Sandel, B., L. Arge, B. Dalsgaard, R. G. Davies, K. J. Gaston, W. J. Sutherland, and J. C. Svenning. 2011. "The influence of Late Quaternary climate-change velocity on species endemism." *Science* 334: 660–664. https://doi.org/10.1126/science.1210173.

Schoene, B., K. M. Samperton, M. P. Eddy, G. Keller, T. Adatte, S. A. Bowring, S. F. Khadri, and B. Gertsch. 2015. "U-Pb geochronology of the Deccan Traps and relation to the end-Cretaceous mass extinction." *Science* 347: 182–184, https://doi.org/10.1126/science.aaa0118.

Schulte, P., L. Alegret, I. Arenillas, J. A. Arz, P. J. Barton, P. R. Bown, T. J. Bralower, G. L. Christeson, P. Claeys, C. S. Cockell, et al. 2010. "The Chicxulub asteroid impact and mass extinction at the Cretaceous-Paleogene boundary." *Science* 327: 1214–1218. https://doi.org/10.1126/science.1177265.

Self, S., A. Schmidt, and T. A. Mather. 2014. "Emplacement characteristics, time scales, and volcanic gas release rates of continental flood basalt eruptions on Earth." In *Volcanism, Impacts, and Mass Extinctions: Causes and Effects,* ed. G. Keller and A. C. Kerr, *Geological Society of America Special Paper* 505: 319–337. https://doi.org/10.1130/2014.2505(16).

Sepkoski, J. J., Jr. 1996. "Patterns of Phanerozoic extinction: A perspective from global data bases." In *Global Events and Event Stratigraphy in the Phanerozoic,* ed. O. H. Walliser, 35–51. Springer. https://doi.org/10.1007/978-3-642-79634-0_4.

Sexton, P. F., R. D. Norris, P. A. Wilson, H. Pälike, T. Westerhold, U. Röhl, C. T. Bolton, and S. Gibbs. 2011. "Eocene global warming events driven by ventilation of oceanic dissolved organic carbon." *Nature* 471: 349–352. https://doi.org/10.1038/nature09826.

Skinner, L. C., S. Fallon, C. Waelbroeck, E. Michel, and S. Barker. 2010. "Ventilation of the deep Southern Ocean and deglacial CO$_2$ rise." *Science* 328: 1147–1151. https://doi.org/10.1126/science.1183627.

Stanley, S. M., and M. G. Powell. 2003. "Depressed rates of origination and extinction during the late Paleozoic ice age: A new state for the global marine ecosystem." *Geology* 31: 877–880. https://doi.org/10.1130/G19654R.1.

Steffen, W., J. Grinevald, P. Crutzen, and J. McNeill. 2011. "The Anthropocene: Conceptual and historical perspectives." *Philosophical Transactions of the Royal Society (Series A)* 369: 842–867. https://doi.org/10.1098/rsta.2010.0327.

Stephens, B. B., and R. F. Keeling. 2000. "The influence of Antarctic sea ice on glacial-interglacial CO$_2$ variations." *Nature* 404: 171–174. https://doi.org/10.1038/35004556.

Stevens, B., S. C. Sherwood, S. Bony, and M. J. Webb. 2016. "Prospects for narrowing bounds on Earth's equilibrium climate sensitivity." *Earth's Future* 4: 512–522. https://doi.org/10.1002/2016EF000376.

Sun, Y., M. M. Joachimski, P. B. Wignall, C. Yan, Y. Chen, H. Jiang, L. Wang, and X. Lai. 2012. "Lethally hot temperatures during the Early Triassic greenhouse." *Science* 338: 366–370. https://doi.org/10.1126/science.1224126.

Svenning, J. C., S. Normand, and M. Kageyama. 2008. "Glacial refugia of temperate trees in Europe: Insights from species distribution modelling." *Journal of Ecology* 96: 1117–1127. https://doi.org/10.1111/j.1365-2745.2008.01422.x.

Svenning, J. C., W. L. Eiserhardt, S. Normand, A. Ordonez, and B. Sandel. 2015. "The influence of paleoclimate on present-day patterns in biodiversity and ecosystems." *Annual Review of Ecology, Evolution, and Systematics* 46: 551–572. https://doi.org/10.1146/annurev-ecolsys-112414-054314.

Thibault, N., and S. Gardin. 2010. "The calcareous nannofossil response to the end-Cretaceous warm event in the Tropical Pacific." *Palaeogeography, Palaeoclimatology, Palaeoecology* 291: 239–252. https://doi.org/10.1016/j.palaeo.2010.02.036.

Thomas, E. 2007. "Cenozoic mass extinctions in the deep sea: What perturbs the largest habitat on Earth?" In *Large Ecosystem Perturbations: Causes and Consequences*, ed. S. Monechi, R. Coccioni, and M. R. Rampino, *Geological Society of America Special Paper* 424: 1–23. https://doi.org/10.1130/2007.2424(01).

Thornalley, D. J. R., S. Barker, W. S. Broecker, H. Elderfield, and N. McCave. 2011. "The deglacial evolution of North Atlantic deep convection." *Science* 331: 202–205. https://doi.org/10.1126/science.1196812.

White, A. F., and A. E. Blum. 1995. "Effects of climate on chemical weathering in watersheds." *Geochimica et Cosmochimica Acta* 59: 1729–1747. https://doi.org/10.1016/0016-7037(95)00078-E.

Wilde, S. A., J. W. Valley, W. H. Peck, and C. M. Graham. 2001. "Evidence from detrital zircons for the existence of continental crust and oceans on the Earth 4.4 Gyr ago." *Nature* 409: 175–178. https://doi.org/10.1038/35051550.

Woodruff, D. S. 2010. "Biogeography and conservation in Southeast Asia: How 2.7 million years of repeated environmental fluctuations affect today's patterns and the future of the remaining refugial-phase biodiversity." *Biodiversity and Conservation* 19: 919–941. https://doi.org/10.1007/s10531-010-9783-3.

Yu, J., G. L. Foster, H. Elderfield, W. S. Broecker, and E. Clark. 2010. "An evaluation of benthic foraminiferal B/Ca and δ^{11}B for deep ocean carbonate ion and pH reconstructions." *Earth and Planetary Science Letters* 293: 114–120. https://doi.org/10.1016/j.epsl.2010.02.029.

Yu, J., R. F. Anderson, Z. Jin, J. W. Rae, B. N. Opdyke, and S. M. Eggins. 2013. "Responses of the deep ocean carbonate system to carbon reorganization during the Last Glacial–interglacial cycle." *Quaternary Science Reviews* 76: 39–52. https://doi.org/10.1016/j.quascirev.2013.06.020.

Yu, J., R. F. Anderson, and E. J. Rohling. 2014. "Deep ocean carbonate chemistry and glacial-interglacial atmospheric CO-changes." *Oceanography* 27: 16–25. https://doi.org/10.5670/oceanog.2014.04.

Yu, J., L. Menviel, Z. D. Jin, D. J. R. Thornalley, S. Barker, G. Marino, E. J. Rohling, Y. Cai, F. Zhang, X. Wang, Y. Dai, P. Chen, and W. S. Broecker. 2016. "Sequestration of carbon in the deep Atlantic during the last glaciation." *Nature Geoscience* 9: 319–324. https://doi.org/10.1038/ngeo2657.

Zachos, J. C., B. N. Opdyke, T. M. Quinn, C. E. Jones, and A. N. Halliday. 1999. "Early Cenozoic glaciation, Antarctic weathering, and seawater ^{87}Sr/^{86}Sr: Is there a link?" *Chemical Geology* 161: 165–180. https://doi.org/10.1016/S0009-2541(99)00085-6.

Zachos, J. C., G. R. Dickens, and R. E. Zeebe. 2008. "An early Cenozoic perspective on greenhouse warming and carbon-cycle dynamics." *Nature* 451: 279–283. https://doi.org/10.1038/nature06588.

Zalasiewicz, J., M. Williams, A. Haywood, and M. Ellis. 2011. "The Anthropocene: A new epoch of geological time?" *Philosophical Transactions of the Royal Society (Series A)* 369: 835–841. https://doi.org/10.1098/rsta.2010.0339.

Zeebe, R. E., and J. C. Zachos. 2013. "Long-term legacy of massive carbon input to the Earth system: Anthropocene versus Eocene." *Philosophical Transactions of the Royal Society (Series A)* 371: 20120006. https://doi.org/10.1098/rsta.2012.0006.

Zeebe, R. E., J. C. Zachos, and G. R. Dickens. 2009. "Carbon dioxide forcing alone insufficient to explain Palaeocene–Eocene Thermal Maximum warming." *Nature Geoscience* 2: 576–580. https://doi.org/10.1038/NGEO578.

Zeebe, R. E., A. Ridgwell, and J. C. Zachos. 2016. "Anthropogenic carbon release rate unprecedented during the past 66 million years." *Nature Geoscience* 9: 325–329. https://doi.org/10.1038/NGEO2681.

CHAPTER NINE

Climate Change, Conservation, and the Metaphor of Deep Time

RICHARD B. ARONSON

INTRODUCTION

Climate change is surpassing overharvesting as the greatest threat to marine life. Rising temperatures and ocean acidification—both consequences of artificially high atmospheric levels of carbon dioxide—are acting at scales from the global to the local in the marine environment. There is as yet no evidence of elevated extinction rates in marine systems as a result of contemporary climate change, but climate change is clearly altering ecological structure and function by driving shifts in the behaviors, phenologies, and geographic and bathymetric ranges of species (Walther et al. 2002; Norris et al. 2013; Molinos et al. 2015). The fossil record in deep time—geological time spanning millions to hundreds of millions of years—displays the evolutionary and biogeographic impacts of past climatic changes, providing a baseline for evaluating extinction risk (Finnegan et al. 2015). Moreover, the history of marine systems in deep time makes it possible to link current and future climate change with community structure and function.

The goal of this chapter is to develop a paleobiological framework for characterizing marine communities in a biosphere increasingly afflicted by human activities. This framework is built on the metaphor of the time stamp: characterizing a living community as structurally and functionally resembling a community from another geologic interval. Climate change, overharvesting, and other anthropic insults are interpreted as driving marine communities in one temporal direction or another, with the goal of formulating effective management strategies. The destruction of coral reefs and the impacts of overfishing in temperate marine environments illustrate the heuristic value of the time stamp. Climate change is also shifting the time stamp of benthic communities in Antarctica. Because Antarctica is one of the last places on Earth to remain in near-pristine condition, the time-shifting impacts of climate and the policy options available are instructive and cautionary.

MARINE COMMUNITIES IN DEEP TIME

The Phanerozoic eon, which extends from about 541 Ma (millions of years ago) to the present (Box 9.1), is the interval during which multicellular life as we know it has dominated the global biota. The Phanerozoic succeeded the Proterozoic eon (the "Precambrian"), which ran from 2,500–541 Ma. This brief summary of the Phanerozoic history of marine life draws from reviews by Sepkoski (1991) and Sheehan (2001). The focus is on nearshore, shallow-marine, soft-bottom habitats, which have a fossil record that is abundant, detailed, and well studied.

Sepkoski partitioned marine diversity into three broad groups, or "evolutionary faunas," that were globally dominant at successive times during the Phanerozoic: a Cambrian fauna, a Paleozoic fauna, and

---

Box 9.1   **The Geologic Time Scale**

Geologic time is divided hierarchically into eons, eras, periods, epochs, and ages. The abridged table below lists the names and time spans of the divisions relevant to this chapter. Ma denotes millions of years ago, with dates based on the US Geological Survey's 2012 stratigraphic chart. The Anthropocene, a developing concept for an epoch of geologically discernible human influence, began in the nineteenth century or earlier.

Eon	Era	Period	Epoch	Time span (Ma)
Phanerozoic				540–0
	Cenozoic			66.0–0
		Neogene		23.0–0
			Anthropocene	~0–?
			Holocene	0.01–0
			Pleistocene	2.6–0.01
			Pliocene	5.3–2.6
			Miocene	23.0–5.3
		Paleogene		66.0–23.0
			Oligocene	33.9–23.0
			Eocene	56.0–33.9
			Paleocene	66.0–56.0
	Mesozoic			252–66
		Cretaceous		145–66
		Jurassic		201–145
		Triassic		252–201
	Paleozoic			540–252
		Permian		299–252
		Carboniferous		359–299
		Devonian		419–359
		Silurian		444–419
		Ordovician		485–444
		Cambrian		541–485
Proterozoic				2500–540

Modern fauna. The Cambrian fauna diversified during the Cambrian period (541–485 Ma) and was dominated by trilobites, which consumed organic matter at the sedimentary surface. Other important elements included low-lying suspension feeders, such as inarticulate brachiopods (lamp shells) (suspension feeders are animals that feed on organic particles, including plankton, bacteria, and detritus, in the water column). By the end of the Paleozoic, 252 Ma, the Cambrian fauna was reduced to a minor component of marine diversity, and trilobites were extinct.

The Paleozoic fauna initially diversified in the Ordovician Period (485–444 Ma) and was dominated by epifaunal suspension-feeders that fed higher above the seafloor. Particularly abundant were articulate brachiopods (lamp shells designated as rhynchonelliforms), stalked crinoids (sea lilies), stromatoporoids (an archaic group of sponges), rugose and tabulate corals (the extinct horn and table corals), bryozoans (moss animals), and ophiuroids (brittlestars). The primary predators were asteroids (sea stars), polychaetes (marine worms), shelled cephalopods (similar to today's nautilus), placoderms (ancient armored fish), and primitive cartilaginous fish (relatives of modern sharks and rays). Burrowing organisms, or bioturbators, expanded downward into the sediments.

The Modern fauna diversified during the Mesozoic (252–66 Ma) and Cenozoic (66 Ma–present). Gastropods (snails), infaunal bivalves (burrowing clams), and echinoids (sea urchins and sand dollars) are prominent constituents. Skeleton-crushing predators diversified and became the trophic dominants. These "durophagous" predators include teleosts (modern bony fish), neoselachians (modern sharks and rays), reptant decapods (large, walking crustaceans such as crabs and lobsters), coleoids (shell-less cephalopods, including octopus and squid), marine mammals, and now-extinct marine reptiles such as mosasaurs. Bioturbators burrowed deeper and more actively than their Paleozoic counterparts.

Evolutionary novelties, notably including the modern, durophagous predators, evolved primarily in nearshore, coastal habitats and then expanded offshore to outer-shelf and bathyal (upper-deep-sea) environments (Jablonski et al. 1983). The Modern fauna progressively displaced many taxa of the Paleozoic fauna, especially stalked crinoids and other epifaunal suspension feeders, to offshore, deepwater environments.

Today, the deep sea is the refuge of Paleozoic-type communities and "living fossils" such as the coelacanth Latimeria, the nautilus, and stalked crinoids. Dense populations of epifaunal ophiuroids occur in outer-shelf and bathyal environments (Metaxas and Griffin 2004; Aronson 2017). "Brittlestar beds" also persist in a few coastal locations characterized by low levels of durophagous predation, either because predators are naturally absent or as a likely consequence of centuries of overfishing (Aronson 1989). The most prevalent predators within those ophiuroid populations are seastars and other Paleozoic-grade invertebrates.

The increase in predation that began in the Mesozoic modernized benthic communities in nearshore environments. New, durophagous predators were the primary drivers of several important trends: (1) the evolution of increased defensive architecture in mollusks and other skeletonized invertebrates; (2) the loss of dense populations of epifaunal suspension feeders from nearshore, shallow, soft-bottom environments; and (3) the transition from epifaunal to infaunal life modes in bivalves. This "Mesozoic marine revolution" (Vermeij 1977) transcended the mass extinction at the end of the Cretaceous and accelerated during the Cenozoic. Although it is a top-down construct, its macroevolutionary dynamics were fueled from the bottom up by an increased supply of energy from plankton in the water column (Knoll and Follows 2016). These trends are expressed to varying degrees and at different times in different geographic regions (Monarrez et al. 2017; Whittle et al. 2018).

CLIMATE AND BIODIVERSITY
IN DEEP TIME

Direct impacts of climate change on marine biodiversity have been difficult to discern over deep time. Some episodes of elevated extinction have been causally linked to trends of heating or cooling. Examples, in temporal order, include cooling in the Late Ordovician mass extinction, circa 450 Ma; the late Paleozoic ice age, which began in the Mississippian, circa 330 Ma; high-temperature conditions 252–250 Ma associated with the end-Permian mass extinction; cooling at the Eocene–Oligocene transition, 34 Ma; and cooling at the Pliocene–Pleistocene transition, 2.6 Ma. At other times, biological patterns have been driven by other physical changes or have been complicated by the ecology of the taxa on which climate is acting. Ocean acidification, for example, has played a critical role in the history of marine biodiversity (Norris et al. 2013).

Glaciations that occurred 850–635 Ma, near the end of the Proterozoic, were the most intense in Earth's history. These Varangian glaciations, which were followed by climatic warming, may have delayed—or perhaps accelerated!—the diversification of animal life in the Cambrian by causing mass extinctions of the existing faunas, although rising oxygen levels, burial of organic carbon, and tectonic shifts probably played a more central role in the timing of the "Cambrian explosion." In contrast, the Pleistocene glaciations did not drive mass extinctions in the ocean, but they altered the geographic ranges of marine species and, therefore, changed the composition of marine communities (Roy and Pandolfi 2005). Likewise, although benthic foraminiferans (bottom-dwelling, shelled amoebas) experienced high rates of extinction during the warming trend of the Paleocene–Eocene Thermal Maximum (PETM; ca. 56 Ma), other groups of small marine organisms responded primarily by shifting their geographic ranges (Norris et al. 2013). The different responses of biotas at different times reflect the temporal scales over which the temperature changes occurred (along with the associated changes in sea level, acidification, and other factors), the departures of the thermal shifts from background conditions, the temporal resolution of the fossil record in different time frames, the prevailing composition of the biosphere, and other events at the time. The variability of climatic impacts in the past notwithstanding, contemporary climate change is altering living communities in ways the fossil record can illuminate.

THE METAPHORICAL TIME STAMP

The idea of time stamping marine communities is founded on three broad generalizations. The first is the primacy of top-down control of marine food webs. Strong, cascading, predatory impacts are readily apparent in many or most nearshore marine ecosystems (Estes et al. 2011). The second is that top predators, being the largest, most energetic, and least abundant members of marine communities, are more vulnerable to direct exploitation by humans than are lower-level consumers and producers (Pauly et al. 1998; Estes et al. 2011). The third is that trophic levels have been added sequentially to marine-benthic communities through Phanerozoic time, as we have already seen. Combining these generalizations, we can view the cascading impacts of predation through the lens of paleobiology. Artificially reducing the abundance of top predators through exploitation, climate change, and other impacts can metaphorically force a community back in time to an anachronistic state with a retrograde structure. Alternatively, perturbation can create new, no-analogue communities (Williams and Jackson 2007; Jackson 2008), which, maintaining the metaphor, could be considered in some sense futuristic as opposed to anachronistic. The metaphor can be extended to changes in the abundance of strongly interacting consumers at lower trophic levels, as well as to nontrophic classes of strong interactors such as foundation species, or ecosystem engineers.

Stamping a living community with its time equivalent from the fossil record should not be taken to mean that the living community fully replicates a particular paleocommunity from the distant past. That would be impossible after millions to hundreds of millions of years of change in the ocean. Rather, the time stamp denotes some commonality of pattern and process, yielding clues to anthropogenic causation and suggesting remedial actions. The intent is to draw a revealing caricature, not one that is overly exaggerated. As in politics and life generally, absurd caricatures are not so useful for identifying the features that need improvement.

The time stamp is implicit in interpreting the Pleistocene extinctions of terrestrial megafauna. Neotropical fruiting plants with large seeds, bereft of the megafaunal dispersers with which they coevolved, now live in a no-analogue world but find some temporal redress in the form of large, domesticated animals such as cattle and horses (Janzen and Martin 1982; Guimarães et al. 2008). Levin et al. (2002) likewise argued that grazing by feral horses at least partially re-creates the impacts of Pleistocene megaherbivores in North American salt marshes. Rewilding (Donlan et al. 2006) is an explicit management proposal to restore North American ecosystems to some semblance of their prehuman states by importing living megafauna from other continents.

The degradation of coral reefs provides an example from the marine realm. Climate change, overharvesting, and nutrient loading all kill corals and promote the growth of macroalgae (seaweeds) on reefs. Some ecologists predict that reef communities will ultimately degrade to a "seabed of slime," dominated by expansive mats of cyanobacteria (i.e., blue-green algae) reminiscent of the Proterozoic seas (Pandolfi et al. 2003; Jackson 2008). Cyanobacterial mats formed hardened, layered stromatolites,

which were the basis of the earliest living reefs until they were outmoded by the rapid diversification of mobile consumers in the Cambrian. Modern, stromatolite-dominated seascapes are likewise characterized by low levels of grazing. Moreover, anachronistic communities dominated by microbially built structures reappeared at various times in the Phanerozoic following the clearance of ecospace by extinction events (Sheehan and Harris 2004). Whether coral reefs are degrading to a Proterozoic state and on what schedule remain contentious issues. The complementary prediction for pelagic systems is that gelatinous zooplankton will inherit the water column after all the fish are gone. The hypothesis of retrogradation to a Proterozoic-style "ocean of jelly" has found only partial support (Condon et al. 2012).

OVERFISHING AND POST-MODERN COMMUNITIES

The increased abundance of epifaunal ophiuroids in overfished coastal habitats illustrates how removing top predators can drive benthic communities in a Paleozoic direction. Human exploitation of predators such as fish and sea otters can also lead to explosive increases in sea urchins. Sea urchins, which belong to Sepkoski's Modern fauna, are potent herbivores. When released from predation pressure they can defoliate expansive beds of seagrasses and macroalgae in shallow-water environments. Aronson (1990) distinguished modern, temperate-zone communities, which are naturally dominated by predators and macroalgae, from overfished, "post-modern" communities dominated by humans and sea urchins. On tropical reefs, in contrast, cultivating artificially dense, post-modern populations of sea urchins might be a useful management tool to combat the displacement of corals by macroalgae and cyanobacterial mats (Precht and Aronson 2006).

Aronson (1990) broadened the onshore-offshore model to incorporate human exploitation. Fishing activity by *Homo sapiens*, an evolutionary novelty that originated in nearshore environments, sequentially alters community structure along the onshore-to-offshore gradient. Food webs progressively become depleted from their top predators downward (Pauly et al. 1998) and from coastal habitats to the edge of the continental shelf and into the bathyal zone. The onshore-offshore trend is reflected in projections that bottom fisheries will continue to expand to the slow-growing, highly vulnerable stocks that live in deep-sea environments, with obvious implications for marine policy and conservation planning.

The post-modern time stamp also applies to overfished demersal and pelagic assemblages in which coleoid cephalopods are increasing. Like sea urchins, coleoids are components of the Modern fauna, and they are remarkably convergent on fish in their structure, function, and ecology (Packard 1972). Catches of octopus, squid, and cuttlefish have increased recently, in large part because the teleosts that are their predators and competitors have been overfished by humans (Doubleday et al. 2016).

Jackson (2008) expressed the same dark sentiment that marine communities have entered post-modern times with his dystopian reference to the brave new ocean. Whether we view perturbed marine communities as retrograde fragments or temporal chimeras, we can best understand the bleak future of life in the sea as a warped and shattered reflection of the distant past. Antarctica provides a vivid case study of the value of the metaphorical time stamp for managing a highly vulnerable marine ecosystem that is being altered by climate change.

EOCENE CLIMATE CHANGE AND ANTARCTIC MARINE COMMUNITIES

Antarctic ecosystems are extraordinarily sensitive to human impacts, and a high degree of endemism aggravates the threat of climate change (Aronson et al. 2011; Chown et al. 2015). Apart from endemism per se, marine-benthic communities on the continental shelves surrounding Antarctica are unique in their structure and function. Declining temperatures over the last 40 million years have placed a Paleozoic time stamp on the living benthic communities, but now rapidly warming seas are poised to reverse that dynamic.

The living bottom-fauna in Antarctica lacks the durophagous teleosts, neoselachians, and reptant decapods that generally exert top-down control at temperate, tropical, and Arctic latitudes. As a result, shallow-benthic communities in Antarctica resemble benthic communities of the Paleozoic marine world, as well as contemporary offshore or deep-sea communities (Aronson et al. 2007; Aronson 2017). Living macrofaunal assemblages on soft bottoms in Antarctica are dominated by ophiuroids, crinoids, anthozoans (sea anemones and their relatives), sponges, and other epifaunal suspension feeders. Brachiopods are common on rocky outcrops in some localities. The primary predators are slow moving, Paleozoic-grade invertebrates, including asteroids, giant nemerteans (ribbon worms), and giant pycnogonids (sea spiders). The thin, undefended shells of Antarctic gastropods are more vulnerable to shell-crushing predators than their counterparts in the temperate zones and the tropics, a latitudinal analogue of the Mesozoic marine revolution (Vermeij 1978; Watson et al. 2017). Other retrograde, Paleozoic features include a phylogenetically archaic bivalve fauna and, related, an impoverished infauna (Crame 2014).

Climatic cooling began in Antarctica in the late Eocene with the opening of the Drake Passage and establishment of the Antarctic Circumpolar Current (ACC). The cold and isolation created by the ACC drove the trend toward anachronistic community structure in the Antarctic shelf fauna. Geochemical analysis of bivalve shells from the Eocene La Meseta Formation at Seymour Island, in the Weddell Sea off the eastern coast of the Antarctic Peninsula, suggest a temperature drop around 41 Ma (Douglas et al. 2014). That event was the first pulse of the long-term trend that led to glaciation of the continent and its polar climate as we know it today.

The La Meseta Formation, which spans ~55 Ma to 33.5 Ma—most of the Eocene—is a highly fossiliferous, uplifted channel-fill that preserves faunas from nearshore, shallow-marine environments. The benthic communities were dominated by gastropods and bivalves, typical of Cenozoic nearshore communities worldwide. Far from typical, however, was the abrupt appearance of dense populations of epifaunal ophiuroids and crinoids following the cooling pulse 41 Ma. Other changes across the cooling event included a decline of sharks, a decline of defensive architecture in gastropods, and an increase of epifaunal, suspension-feeding bivalves (Long 1992; Stilwell and Zinsmeister 1992). Durophagous teleosts, sharks, and decapods went extinct in Antarctica after the Eocene-Oligocene boundary. The notothenioid fish, which survive in Antarctica by producing antifreeze glycoproteins, radiated as early as the late Eocene in response to cooling (Near et al. 2012). They are the only remaining teleosts in shelf environments in Antarctica, and they never evolved to feed on hard-shelled prey. This functional idiosyncrasy is an accident of phylogeny not shared with Arctic teleosts, many of which produce antifreeze compounds and are also shell crushers (Aronson et al. 2007).

The faunal response to cooling 41 Ma—retrogradation to a quasi-Paleozoic state—was an initial exemplar of later, widespread ecological changes to the Antarctic benthos. Fossil evidence suggests that subsequent warming and cooling were accompanied by fluctuations in the time stamp of community structure. At some point, however, sea temperatures were uniformly cold enough that retrograde communities became the norm. Rapidly warming seas now appear to be erasing the Paleozoic time stamp and

remodernizing the Antarctic benthos by re-admitting durophagous predators.

THE FUTURE OF ANTARCTIC MARINE LIFE

Much has been written about the ecological impacts of ocean acidification at polar latitudes, but warming seas pose a more immediate threat to the continued survival of benthic species in Antarctica (Gutt et al. 2015). Because water temperatures are virtually aseasonal, marine ectotherms in Antarctica have evolved physiologies that are cold stenothermal: they tolerate a narrow range of low temperatures. Warming directly threatens Antarctic ectotherms with extinction through physiological stress (Pörtner 2006; see also Griffiths et al. 2017). Rising temperatures also act indirectly, accelerating biological invasions and range expansions from warmer environments (e.g., Walther et al. 2002). As a consequence, durophagous crustaceans appear to be on the verge of returning to shelf communities off the western Antarctic Peninsula (WAP).

The continental shelf of the WAP is the marine region of Antarctica most susceptible to biological invasion for several reasons. First, the WAP is close to source populations of potential invaders from southern South America. A vagrant crab from South America, likely transported in the ballast of a tourist ship, was recently discovered at Deception Island, and crab larvae are drifting southward into the Southern Ocean in eddies coming off the ACC (Aronson, Frederich, et al. 2015). Second, sea surface temperatures over the shelf adjacent to the WAP have risen approximately 1.5°C during the past 50 years, which is considerably more rapid than the global average (Schmidtko et al. 2014). Third, the near-bottom waters over the continental slope of the WAP are slightly warmer than the shallow waters over the inner shelf. This oceanographic peculiarity arises from downwelling within

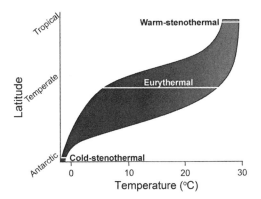

Figure 9.1. Annual temperature range within shallow-marine habitats as a function of latitude. The lack of thermal seasonality has driven the evolution of cold stenothermy in Antarctic ectotherms. Ectotherms living near the equator tend to be warm stenothermal, again because of reduced seasonality. Temperate-marine ectotherms are eurythermal in comparison. Extinction risks are elevated in Antarctica from the physiological stresses of climate change. (Modified from Aronson et al. 2011.)

the ACC and upwelling off the WAP. Predatory king crabs (anomurans in the family Lithodidae) are already living on the continental slope but are excluded from colder, nearshore habitats. They have the potential to expand shoreward as temperatures rise.

Lithodids are among the most cold tolerant of the reptant decapods (Hall and Thatje 2011), and they are widespread along the slope off the WAP (Figure 9.1). Large, reproductively viable populations of one species, *Paralomis birsteini*, are currently living at a minimum temperature of ~0.4°C, centered in a depth range of 1100–1500 m (Aronson, Smith, et al. 2015) and extending as shallow as 721 m. In 2010, C. Smith et al. (2012) documented a viable population of another lithodid, *Neolithodes yaldwyni*, at slope depths in Palmer Deep, a glacially carved trough in the shelf. Like *Paralomis*, *Neolithodes* are normally found on the continental slope, so at some point they presumably came over the shelf break at 500 m—as either adults or demersal (bottom-associated) larvae—to migrate into the trough. When lithodids initially established populations in Antarctica is unknown.

Lithodids prey on invertebrates, including echinoderms and mollusks. They and their prey exhibit complementary depth distributions in Palmer Deep and on the continental slope where they occur at high abundance (C. Smith et al. 2012; K. Smith et al. 2017). Nearshore waters are still too cold for lithodids to survive, but the outer shelf, at 400–600 m depth, is already warm enough. At current rates of warming, no physiological barriers will prevent lithodids from expanding shoreward across the shelf over the next few decades. The abundant, naïve, undefended echinoderms, mollusks, and other invertebrates will be easy pickings, compromising the Paleozoic character of the Antarctic shelf faunas.

SUMMARY AND CONCLUSIONS

Deep time provides a framework for understanding the structure of marine-benthic communities under varying levels of human exploitation. Overexploitation of top predators has tended to reconfigure benthic assemblages so they come to resemble Paleozoic communities, or alternatively "postmodern" communities if humans are added to the scenario. This metaphorical retrogradation (or progradation) has generally proceeded from nearshore habitats to offshore and deep-sea environments. The value of the temporal metaphor lies beyond mere academic interest: the time stamp of a community can provide concrete direction to the formulation of policies and action plans in a world increasingly driven by climate change. In Antarctica, climatic warming, by altering benthic food webs, is on the verge of remodernizing the anachronistic communities living on the continental shelf.

One need look no further than Antarctica's iconic penguins to see the danger of climate change. Warming along the WAP is reducing sea ice, which is reducing the productivity of krill, a primary food source of penguins. The penguins are declining. The story is more complicated than that (Trivelpiece et al. 2011), but climatically induced reductions in krill are having a significant impact on the abundance of penguins. There is only so far south along the WAP that penguins and other marine species can shift their ranges before running into the continent and facing extinction.

The direct climatic threat to marine biodiversity on a regional level within Antarctica is magnified by the prospect of intercontinental homogenization of community structure: the loss of diversity among biogeographic provinces (Aronson et al. 2007; Aronson, Frederich, et al. 2015; Aronson, Smith, et al. 2015). Warming seas and increasing ship traffic are placing the Antarctic fauna at risk from climatically mediated biological invasions and range expansions of marine taxa that include durophagous predators. Crabs will be the first shell-crushing predators to arrive in benthic-shelf communities, but durophagous fish will follow (Aronson et al. 2007). Remodernization of the shelf benthos would destroy the endemic, retrograde character of Antarctic marine life.

Apart from limiting direct environmental damage from pollution and fishing, two clear prescriptions are to control carbon emissions globally and regulate ship traffic in the Southern Ocean (Aronson et al. 2011). The Ross Sea Marine Protected Area, which was created in 2016, will protect Antarctic fish stocks and other resources. Like many other marine protected areas, however, this MPA—the world's largest at 1.57 million km²—will rapidly be rendered obsolete if climate change is not halted and reversed. As with the broader environmental crisis, the question is whether the international community has the political will to rise to the occasion. The conceptual framework provided by the metaphor of the time stamp, including the allied concept of no-analogue communities, hopefully will help align practitioners and academics as they work together on desperate problems of monumental proportion.

ACKNOWLEDGMENTS

I thank J. Eastman, K. Heck, L. Ivany, J. McClintock, R. Moody, W. Precht, K. Smith, and S. Thatje for helpful advice and discussion. Comments from the editors and several anonymous reviewers greatly improved the manuscript. Contribution 140 from the Institute for Research on Global Climate Change at Florida Tech, this work was supported by the National Science Foundation through Grants ANT-1141877 and OCE-1535007.

REFERENCES

Aronson, R. B. 1990. "Onshore-offshore patterns of human fishing activity." *Palaios* 5: 88–93.

Aronson, R. B. 2017. "Metaphor, inference, and prediction in paleoecology: Climate change and the Antarctic bottom fauna." In *Conservation Paleobiology: Science and Practice*, ed. G. P. Dietl and K. W. Flessa, 183–200. University of Chicago Press.

Aronson, R. B., M. Frederich, R. Price, and S. Thatje. 2015. "Prospects for the return of shell-crushing crabs to Antarctica." *Journal of Biogeography* 42: 1–7.

Aronson, R. B., K. E. Smith, S. C. Vos, J. B. McClintock, M. O. Amsler, P.-O. Moksnes, D. S. Ellis, et al. 2015. "No barrier to emergence of bathyal king crabs on the Antarctic shelf." *Proceedings of the National Academy of Sciences* 112: 12997–13002.

Aronson, R. B., S. Thatje, A. Clarke, L. S. Peck, D. B. Blake, C. D. Wilga, and B. A. Siebel. 2007. "Climate change and invasibility of the Antarctic benthos." *Annual Review of Ecology, Evolution, and Systematics* 38: 129–154.

Aronson, R. B., S. Thatje, J. B. McClintock, and K. A. Hughes. 2011. "Anthropogenic impacts on marine ecosystems in Antarctica: The Yearbook in Ecology and Conservation Biology." *Annals of the New York Academy of Sciences* 1223: 82–107.

Chown, S. L., A. Clarke, C. I. Fraser, S. C. Cary, K. L. Moon, and M. A. McGeoch. 2015. "The changing form of Antarctic biodiversity." *Nature* 522: 431–438.

Condon, R. H., W. M. Graham, C. M. Duarte, K. A. Pitt, K. H. Lucas, S. H. D. Haddock, K. R. Sutherland, et al. 2012. "Questioning the rise of gelatinous zooplankton in the world's oceans." *BioScience* 62: 160–169.

Crame, J. A. 2014. "Evolutionary setting." In *Biogeographic Atlas of the Southern Ocean*, ed. C. De Broye and P. Koubbi, 32–35. Scientific Committee on Antarctic Research, Cambridge.

Donlan, C. J., J. Berger, C. E. Bock, J. H. Bock, D. A. Burney, J. A. Estes, D. Forman, et al. 2006. "Pleistocene rewilding: An optimistic agenda for twenty-first century conservation." *American Naturalist* 168: 660–681.

Doubleday, Z. A., T. A. A. Prowse, A. Arkhipkin, G. J. Pierce, J. Semmens, M. Steer, S. C. Leporati, S. Lourenço, A. Quetglas, W. Sauer, and B. M. Gillanders. 2016. "Global proliferation of cephalopods." *Current Biology* 26: R406–R407.

Douglas, P. M. J., H. P. Affek, L. C. Ivany, A. J. P. Houben, W. P. Sijp, A. Sluijs, S. Schouten, and M. Pagani. 2014. "Pronounced zonal heterogeneity in Eocene southern high-latitude sea surface temperatures." *Proceedings of the National Academy of Sciences* 111: 6582–6587.

Estes, J. A., J. Terborgh, J. S. Bashares, M. E. Power, J. Berger, W. J. Bond, S. R. Carpenter, et al. 2011. "Trophic downgrading of Planet Earth." *Science* 333: 301–306.

Finnegan, S., S. C. Anderson, P. G. Harnik, C. Simpson, D. P. Tittensor, J. E. Byrnes, Z. V. Finkel, et al. 2015. "Paleontological baselines for evaluating extinction risk in the modern oceans." *Science* 348: 567–570.

Griffiths, H. J., A. J. S. Meijers, and T. J. Bracegirdle. 2017. "More losers than winners in a century of future Southern Ocean seafloor warming." *Nature Climate Change* 7: 749–754.

Griffiths, H. J., R. J. Whittle, S. J. Roberts, M. Belchier, and K. Linse. 2013. "Antarctic crabs: Invasion or endurance?" *PLOS One* 8: e66981.

Guimarães, P. R., M. Galetti, and P. Jordano. 2008. "Seed dispersal anachronisms: Rethinking the fruits the gomphotheres ate." *PLOS One* 3: e1745.

Gutt, J., N. Bertler, T. J. Bracegirdle, A. Buschmann, J. Comiso, G. Hosie, E. Isla, et al. 2015. "The Southern Ocean ecosystem under multiple climate change stresses—An integrated circumpolar assessment." *Global Change Biology* 21: 1434–1453.

Hall, S., and S. Thatje. 2011. "Temperature-driven biogeography of the deep-sea family Lithodidae (Crustacea: Decapoda: Anomura) in the Southern Ocean." *Polar Biology* 34: 363–370.

Jablonski, D., J. J. Sepkoski Jr., D. J. Bottjer, and P. M. Sheehan. 1983. "Onshore-offshore patterns in the evolution of Phanerozoic shelf communities." *Science* 222: 1123–1125.

Jackson, J. B. C. 2008. "Ecological extinction and evolution in the brave new ocean." *Proceedings of the National Academy of Sciences* 105: 11458–11465.

Janzen, D. H., and P. S. Martin. 1982. "Neotropical anachronisms: The fruits the gomphotheres ate." *Science* 215: 19–27.

Knoll, A. H., and M. J. Follows. 2016. "A bottom-up perspective on ecosystem change in Mesozoic oceans." *Proceedings of the Royal Society, Series B* 283: 20161755.

Levin, P. S., J. Ellis, R. Petrik, and M. E. Hay. 2002. "Indirect effects of feral horses on estuarine communities." *Conservation Biology* 16: 1364–1371.

Long, D. J. 1992. "Paleoecology of Eocene Antarctic sharks." In *The Antarctic Paleoenvironment: A Perspective on Global Change, Part 1. Antarctic Research Series* 56, ed. J. P.

Kennett and D. A. Warnke, 131–139. American Geophysical Union.

Metaxas, A., and B. Griffin. 2004. "Dense beds of the ophiuroid *Ophiacantha abyssicola* on the continental slope off Nova Scotia, Canada." *Deep-Sea Research I* 51: 1307–1317.

Molinos, J.-G., B. S. Halpern, D. S. Schoeman, C. J. Brown, W. Kiessling, P. J. Moore, J. M. Pandolfi, E. S. Poloczanska, A. J. Richardson, and M. T. Burrows. 2015. "Climate velocity and the future global redistribution of marine biodiversity." *Nature Climate Change* 6: 83–88.

Monarrez, P. M., M. Aberhan, and S. M. Holland. 2017. "Regional and environmental variation in escalatory ecological trends: A western Tethys hotspot for escalation?" *Paleobiology* 43: 569–586.

Near, T. J., A. Dornburg, K. L. Kuhn, J. T. Eastman, J. N. Pennington, T. Patarnello, L. Zane, D. A. Fernández, and C. D. Jones. 2012. "Ancient climate change, antifreeze, and the evolutionary diversification of Antarctic fishes." *Proceedings of the National Academy of Sciences* 109: 3434–3439.

Norris, R. D., S. K. Turner, P. M. Hull, and A. Ridgwell. 2013. "Marine ecosystem responses to Cenozoic climate change." *Science* 341: 492–498.

Packard, A. 1972. "Cephalopods and fish: The limits of convergence." *Biological Reviews of the Cambridge Philosophical Society* 47: 241–307.

Pandolfi, J. M., J. B. C. Jackson, N. Baron, R. H. Bradbury, H. M. Guzman, T. P. Hughes, C. V. Kappel, F. Micheli, J. C. Ogden, H. P. Possingham, and E. Sala. 2005. "Are U.S. coral reefs on the slippery slope to slime?" *Science* 307: 1725–1726.

Pauly, D., V. Christiansen, J. Dalsgaard, R. Froese, and F. C. Torres Jr. 1998. "Fishing down marine food webs." *Science* 279: 860–863.

Pörtner, H. O. 2006. "Climate-dependent evolution of Antarctic ectotherms: An integrative analysis." *Deep-Sea Research II* 53: 1071–1104.

Precht, W. F., and R. B. Aronson. 2006. "Death and resurrection of Caribbean coral reefs: A paleoecological perspective." In *Coral Reef Conservation*, ed. I. Côté and J. Reynolds, 40–77. Cambridge University Press.

Roy, K., and J. M. Pandolfi. 2005. "Responses of species and ecosystems to past climate change." In *Climate Change and Biodiversity*, ed. T. E. Lovejoy and L. Hannah, 160–175. Yale University Press.

Schmidtko, S., K. J. Heywood, A. F. Thompson, and S. Aoki. 2014. "Multidecadal warming of Antarctic waters." *Science* 346: 1227–1230.

Sepkoski, J. J., Jr. 1991. "Diversity in the Phanerozoic oceans: A partisan view." In *The Unity of Evolutionary Biology: Proceedings of the Fourth International Congress of Systematic and Evolutionary Biology (Volume 1)*, ed. E. C. Dudley, 210–236. Dioscorides Press.

Sheehan, P. M. 2001. "History of marine diversity." *Geological Journal* 36: 231–249.

Sheehan, P. M., and M. T. Harris. 2004. "Microbialite resurgence after the Late Ordovician extinction." *Nature* 430: 75–78.

Smith, C. R., L. J. Grange, D. L. Honig, L. Naudts, B. Huber, L. Guidi, and E. Domack. 2012. "A large population of king crabs in Palmer Deep on the west Antarctic Peninsula shelf and potential invasive impacts." *Proceedings of the Royal Society, Series B* 279: 1017–1026.

Smith, K. E., R. B. Aronson, B. V. Steffel, M. O. Amsler, S. Thatje, H. Singh, J. Anderson, et al. 2017. "Climate change and the threat of novel marine predators in Antarctica." *Ecosphere* 8: e02017.

Stilwell, J. D., and W. J. Zinsmeister. 1992. "Molluscan systematics and biostratigraphy: Lower Tertiary La Meseta Formation, Seymour Island, Antarctic Peninsula." *Antarctic Research Series* 55. American Geophysical Union.

Trivelpiece, W. Z., J. T. Hinke, A. K. Miller, C. S. Reiss, S. G. Trivelpiece, and G. M. Watters. 2011. "Variability in krill biomass links harvesting and climate warming to penguin population changes in Antarctica." *Proceedings of the National Academy of Sciences* 108: 7625–7628.

Vermeij, G. J. 1977. "The Mesozoic marine revolution: Evidence from snails, predators and grazers." *Paleobiology* 3: 245–258.

Vermeij, G. J. 1978. *Biogeography and Adaptation: Patterns of Marine Life.* Harvard University Press.

Walther, G.-R., E. Post, P. Convey, A. Menzel, C. Parmesan, T. J. C. Beebee, J.-M. Fromentin, O. Hoegh-Guldberg, and F. Bairlein. 2002. "Ecological responses to recent climate change." *Nature* 416: 389–395.

Watson, S.-A., S. A. Morley, and L. S. Peck. 2017. "Latitudinal trends in shell production cost from the tropics to the poles." *Science Advances* 3: e1701362.

Whittle, R. J., A. W. Hunter, D. J. Cantrill, and K. J. McNamara. 2018. "Globally discordant Isocrinida (Crinoidea) migration confirms asynchronous Marine Mesozoic Revolution." *Communications Biology* 1: 46. https://doi.org/10.1038/s42003-018-0048-0.

Williams, J. W., and S. T. Jackson. 2007. "Novel climates, no-analog communities, and ecological surprises." *Frontiers in Ecology and the Environment* 5: 475–482.

The Effects of Sea-Level Rise on Habitats and Species

Céline Bellard, Camille Leclerc, and Franck Courchamp

Based on "Impact of sea level rise on the 10 insular biodiversity hotspots," Bellard et al. (2014). All the sources cited here, if not listed at the end of the case study, can be found in this key reference.

Sea-level rise is one of the main components of climate change. In agreement with climate models, projections showed that sea level will likely continue to rise in more than 95 percent of the ocean area during the coming decades. The current estimates give an increase of 0.26 to 0.98 m for the period 2081–2100 (Stocker et al. 2013). Small islands and other low-lying coastal regions are expected to be the most affected by sea-level rise. There are about 180,000 islands worldwide that are often characterized by a high proportion of endemic species, particularly vulnerable to threats. Several studies have investigated how sea-level rise may affect biodiversity. For instance, sea-level rise may affect marine ecosystems by inducing changes of parameters such as available light, salinity, and temperature. The long-term persistence of freshwater-dependent ecosystems can also be affected by the increase of sea level (Traill et al. 2011). In addition, species in terrestrial ecosystems that might be permanently inundated could be locally extinct as a result of sea-level rise. Besides, the economic cost of sea-level rise is also important. For example, the average global flood losses are estimated to be approximately 5 percent of global gross domestic product in 2005 across all cities (Hanson et al. 2010). As a result, sea-level change will cause habitat change and loss of coastal areas, both of which will have important consequences for society and biodiversity. Because sea-level rise is predicted to continue to increase in the next decades, a number of studies have investigated its future effects on insular biodiversity. For example, Bellard et al. (2014) investigated the potential effects of sea-level rise on the 10 insular biodiversity hotspots following an increase in sea level by 1 m to 6 m. These hotspots contain more than 50,000 endemic plant species and about 670 endemic mammals. Because insular hotspots are densely populated islands (31.8 percent of all humanity) the implications of sea-level rise could be considerable for human societies. In this study, they estimated the number of islands partially and entirely submerged (Figure CS4.1) following sea-level rise. They showed that 6 percent–19 percent of the 4,447 islands studied could be submerged under an increase of sea-level rise of 1 m–6 m.

In particular, three biodiversity hotspots were identified as the most vulnerable to sea-level rise. The Caribbean Islands hotspot was the most threatened, with 63 of its islands entirely submerged under a scenario of sea-level rise by 1 m. Sundaland and the Philippines were also predicted to be particularly affected, with a potential loss of 61 and 48 islands, respectively (Figure CS4.1). In addition, 11.4 percent of all islands may lose at least 50 percent of their habitat with an increase of sea level of 1 m. These islands harbor a high number of endemic species that are of high concern for conservation. Consequently, about 20 to 300 vascular plant species have been predicted at risk of extinction due to sea-level rise. In addition, species already listed as threatened by the International Union for Conservation of Nature (i.e., endemic plants and reptiles from New Caledonia) were more vulnerable to sea-level rise. Focusing on the Pacific and Southeast Asia (over 12,900 islands), Wetzel et al. (2012) found similar results (i.e., between 15 percent and 62 percent of islands would

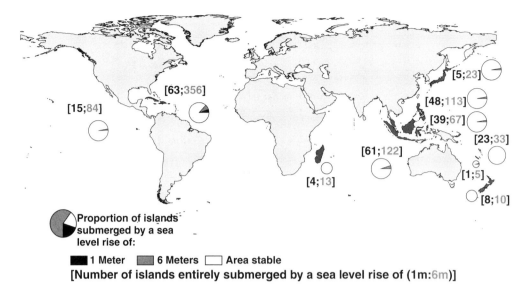

Figure CS4.1. Percentage of islands included in the 10 insular biodiversity hotspots predicted to be submerged from 1 m to 6 m by sea-level rise scenarios.

be completely inundated under scenarios of sea-level rise by 1 m to 6 m), highlighting the extinction risk for 54 mammals of the Indo-Malaysian islands. Using the hotspot of the forest of East Australia as a case study, Bellard et al. (2016) also illustrate that nearly 860 km² could be entirely submerged with an increase of sea level by 1 m, with water intrusion reaching up to 35 km inland. This inundated area mainly includes natural areas such as national parks that harbor high species richness (about 150 species per pixel) and low protected areas. They also demonstrated that inundated areas might lead to the displacement of 20,600 people, especially around Graton, Port Macquarie, and Taree cities in Australia. Consequently, the future cost of sea-level rise for society is expected to be important. Hallegatte et al. (2013) calculated that average global flood losses could reach US$52 billion by 2050 in the 136 largest coastal cities. Without adaptation to mitigate the effect of rising sea levels, Hinkel et al. (2014) estimate that 0.2 percent–4.6 percent of global population is expected to be flooded annually in 2100, with expected annual losses of 0.3 percent–9.3 percent to global gross domestic product.

It should be noted that the majority of the studies that investigated the effect of sea-level rise at a regional or global scale used simplified scenarios. For instance, they simulated a uniform increase in sea-level rise worldwide. In addition, most of them did not consider the potential effects of lateral erosion, intrusion of saline waters, decadal and centennial tides, and eventually secondary impacts like displacement of human populations and agricultural lands. Besides, they do not include the dynamic response of islands. Yet Webb and Kench (2010) highlighted the dynamic nature of sea-level rise response of islets in four atolls of the Central Pacific. Over periods of 19 to 61 years, 43 percent of islands showed an increase (and stability), whereas only 14 percent showed a decrease.

Finally, there are a variety of models, from a simple drowning model to a process-based model, that have been recently developed with their own strengths and weaknesses to investigate the effect of sea-level rise. These techniques are suited for different management objectives regarding sea-level rise (Runting, Wilson, and Rhodes 2013). Therefore, there is much room to improve the current techniques and knowledge of the likely effects of sea-level rise.

To conclude, although the magnitude of sea-level rise and its potential effects are difficult to predict with certainty, it is established that sea-level rise will occur and threaten a large and unique part of biodiversity. Therefore, vulnerability assessments and current conservation programs should include this threat among others criteria of prioritization schemes.

REFERENCES

Bellard, C., C. Leclerc, and F. Courchamp. 2014. "Impact of sea level rise on the 10 insular biodiversity hotspots." *Global Ecology and Biogeography* 23: 203–212. https://doi.org/10.1111/geb.12093.

Bellard, C., C. Leclerc, B. D. Hoffmann, and F. Courchamp. 2016. "Vulnerability to climate change and sea-level rise of the 35th biodiversity hotspot, the Forests of East Australia." *Environmental Conservation* 43: 79–89. https://doi.org/10.1017/S037689291500020X.

Hallegatte, S., C. Green, R. J. Nicholls, and J. Corfee-Morlot. 2013. "Future flood losses in major coastal cities." *Nature Climate Change* 3: 802–806. https://doi.org/10.1038/nclimate1979.

Hinkel, J., D. Lincke, A. T. Vafeidis, M. Perrette, R. J. Nicholls, R. S. J. Tol, et al. 2014. "Coastal flood damage and adaptation costs under 21st century sea-level rise." *Proceedings of the National Academy of Sciences* 111: 3293–3297. https://doi.org/10.1073/pnas.1222469111.

Runting, R. K., K. A. Wilson, and J. R. Rhodes. 2013. "Does more mean less? The value of information for conservation planning under sea level rise." *Global Change Biology* 19: 352–363. https://doi.org/10.1111/gcb.12064.

Stocker, T. F., D. Qin, G.-K. Plattner, L. V. Alexander, S. K. Allen, N. L. Bindoff, F.-M. Bréon, et al. 2013. "Technical summary." In *Climate Change 2013: The Physical Science Basis. Contribution of Working Group I to the Fifth Assessment Report of the Intergovernmental Panel on Climate Change*, ed. T. F. Stocker, D. Qin, G.-K. Plattner, M. Tignor, S. K. Allen, J. Boschung, A. Nauels, Y. Xia, V. Bex, and P. M. Midgley, 33–115. Cambridge University Press. https://doi.org/10.1017/CBO9781107415324.005.

CHAPTER TEN

Past Abrupt Changes in Climate and Terrestrial Ecosystems

JOHN W. WILLIAMS AND KEVIN D. BURKE

INTRODUCTION

Managing ecosystems in a quickly changing world can be distilled to the challenge
of managing rates of change. We must slow
the rates of processes deemed harmful to
species of concern, ecosystem health, or
ecosystem services. Via adaption, we seek
to accelerate adaptive processes and, via car-
bon-mitigation strategies, buy time for species to adapt. We must prepare for abrupt
ecological responses triggered by climatic
extreme events and by nonlinear responses
and feedbacks of ecological systems to more
gradual forcings (National Research Council 2013) and for the ecological surprises
caused by the rapid transition of the climate
system to a state potentially without precedent in human history.

Ecological processes operate across a
wide range of timescales, and the ecological effects of recent climatic trends may not
manifest for years or even centuries (Svenning and Sandel 2013). Hence, the geological record is essential to understand how
abrupt dynamics manifest in both climate
and ecological systems (National Research
Council 2013). Past climate changes were
often large and abrupt (Chapter 8), with
some rates of change occurring as fast or
even faster than those expected for this century (Table 10.1). Paleoclimatic and paleoecological proxies offer a rich record of past
climatic changes and the population, species, community, and ecosystem responses
and feedbacks to those changes. High-resolution records enable annual- to decadal-
scale studies of past ecological dynamics
(Tinner and Lotter 2001), while networks
of paleoecological records enable study of
climate-driven range shifts and other spatiotemporal processes at timescales of cen-

turies to millennia (Williams et al. 2004), as well as ecological responses to large but infrequent climatic extreme events. Thus, the resilience of ecological systems to rapid climate change and rates of ecological response can be studied. Species have four options in a changing climate—move, adapt, persist, or die—and all four have happened in the past.

No geological event is a perfect analog for the climatic and ecological changes expected for this century, but many offer instructive model systems (Williams et al. 2013). Here we review several such events, each illuminating a different aspect of biological dynamics during periods of rapid change. We focus on terrestrial vegetation, because plants are well documented in the fossil record, are fundamental to terrestrial ecosystem function, and are at risk of both abrupt threshold-type responses and adaptive lags (Aitken et al. 2008; Williams et al. 2011; Svenning and Sandel 2013; Scheffer et al. 2012):

1. The Paleocene-Eocene Thermal Maximum (Chapter 8), 56.1 million years ago, is a model system to study the transient and equilibrium responses of terrestrial ecosystems to a massive release of organic carbon into the atmosphere and ocean and consequent changes in temperature and ocean biogeochemistry.

2. The last deglaciation, 19,000 to 8,000 years ago, is a model system to study the ability of species to track temperature changes that were, at least regionally, comparable to those expected for this century.

3. The middle to late Holocene, 8,000 years ago to ~1800 AD, is a model system for the effects of hydrological variability and extreme events in a warmer world. Sometimes misperceived as stable, Holocene

climates were characterized by enhanced hydrological variability, with widespread ecological and societal impacts.

"Abrupt" has several definitions and application across many time scales (Williams et al. 2011, National Research Council 2013). We focus on climatic and ecological changes large and fast enough to challenge climate adaptation (i.e., over years to decades) or where an ecological tipping point is indicated by an ecological response much faster than the climatic forcing.

PALEOCENE-EOCENE THERMAL MAXIMUM (PETM)

During the Paleocene-Eocene Thermal Maximum, 56.1 million years ago (McInerney and Wing 2011), vast amounts of organic carbon from geobiological reservoirs were released rapidly into the atmosphere and ocean. This release is unequivocally signaled by a negative excursion in carbon isotopes (~4.6‰) in multiple marine and terrestrial records (McInerney and Wing 2011): a carbon pulse (ca. 5,000 to 10,000 gigatons carbon [GtC]) equivalent to or greater than all contemporary reserves of fossil fuels. Ocean acidification and carbonate undersaturation are indicated by decreased preservation of calcium carbonate in marine sediments (Zachos et al. 2005). Carbon release was on the order of thousands of years and less than 20,000 years (McInerney and Wing 2011). The perturbation and reequilibration of the carbon cycle and climate system lasted 150,000 to 200,000 years (McInerney and Wing 2011).

The favored explanation for the PETM until recently was degasing of methane from shallow-marine clathrates. However, the amount of carbon stored in Paleocene clathrates (~1000 GtC, Buffett and Archer 2004) is insufficient to explain the magnitude of carbon isotopic excursions and

Table 10.1. Abrupt climatic and ecological events in the geological record

Event	What happened?	When?	Rapidity of event	Ecological effects	Rapidity of response	References
Paleocene-Eocene Thermal Maximum (PETM), ETM2, ETM3	Release of 5–10 Gt organic carbon; ocean acidification; 4°C–5°C warming of tropics	PETM: 56.1 Ma ETM2: 53.7 Ma ETM3: 52.4 Ma	<20,000 years	Intercontinental species immigration; community turnover; dwarfing	Rapid response, lags unclear	McInerney and Wing 2011; DeConto et al. 2012
Glacial Terminations	Global temperature rise of 3°C–5°C, GHG rise, ice sheets melt	9 terminations over the past 800,000 years	5,000–10,000 years	Range shifts over 10^2 to 10^3 km, community turnover	Rapid response, lags unclear	Lüthi et al. 2008
Dansgaard-Oeschger Events	5°C warming in Greenland	20 events between 10,000 and 80,000 years ago	<30 years	Europe: afforestation from grasslands to wooded steppe	Rapid—lags likely but not yet measured	Bond et al. 1993; Allen et al. 1999
Bølling Warming	9°C–14°C warming in Greenland	14,700 years ago	1–3 years	Rapid afforestation of tundra, expansion of species from glacial refugia	Initial response: 8–16 years; Pinus expansion delayed 800 years	Ammann et al. 2013
Younger Dryas Initiation	5°C–9°C cooling in Greenland, 2°C cooling in Western Europe	12,700 years ago	1 year?	Rapid plant community turnover, declines of thermophilous species	<40 years, i.e. within sampling resolution	Brauer et al. 2008; Buizert et al. 2014; Rach et al. 2014
Younger Dryas Termination, Start of Holocene	8°C–12°C warming in Greenland, 3°C in Northern Europe	11,700 years ago	<60 years	Similar to Bølling Warming	Synchronous with warming; tree Betula arrival delayed 650 years	Birks 2015
8.2 ka Event	5°C–7°C cooling in Greenland	8,400 years ago	5 years	Similar to YD Initiation	0 to 40 years	Tinner and Lotter 2001; Alley and Ágústsdóttir 2005
Holocene Aridification	Waning of Northern Hemisphere monsoons and declines in subtropical precipitation	11,000 to 5,000 years ago, timing varies regionally	Thousands of years	Collapse of North African grasslands, retreat of forest ecotones, activation of dunes, C_3/C_4 shifts, altered fire regime	Locally abrupt (10^1–10^2 years) shifts embedded within slower aridification	Kuper and Kröpelin 2008
Holocene Megadroughts		Well documented by dendroclimatic records for last 2000 years; heightened variability 5,500 to 4,000 years ago	1–10 years	Slowed tree growth rates, mortality of mesic tree species, abandonment of early agricultural sites	Tree population collapses as fast as <10 years	Cook et al. 2010a; Cook et al. 2010c; Williams et al. 2011; Shuman 2012

carbonate undersaturation (Higgins and Schrag 2006). Consequently, other mechanisms for the PETM carbon release have been proposed (Higgins and Schrag 2006; McInerney and Wing 2011).

Of these, the recent Antarctic permafrost hypothesis (DeConto et al. 2012) elegantly unifies several lines of evidence. In this hypothesis, organic carbon was released by thawing of Antarctic reservoirs of permafrost and peat. Antarctica was unglaciated during the Paleocene, creating a large polar land area (~12 million km^2) available for carbon sequestration in peatland and permafrost. The PETM was followed by thermal maxima and releases of organic carbon into the atmosphere-ocean system (ETM2 and ETM3) at 53.7 million years ago and 52.4 million years ago. These carbon releases align with past configurations of the earth's slowly wobbling orbit that favor higher incoming sunlight at the poles (DeConto et al. 2012), suggesting that permafrost thawing was triggered by orbitally driven variations in incoming sunlight over the South Pole. The Antarctic peatland hypothesis can explain both the magnitude and rate of carbon release and the multiple thermal maxima. It also explains why thermal events are absent after Antarctica glaciated approximately 35 million years ago and why the amount of carbon released declined between subsequent events: each thermal maximum depleted the organic carbon reserves in Antarctic permafrost and peatlands, leaving less available for release in subsequent thermal events (DeConto et al. 2012). The Antarctic peatland hypothesis underscores the sensitivity of terrestrial carbon reserves to thawing and the potential of high-latitude soil carbon and vegetation feedbacks to accelerate warming.

The climatic, biogeochemical, and biological effects of the PETM were profound. Surface air temperatures rose by 4°C to 5°C in the tropics and 6°C to 8°C in the high latitudes, while the deep ocean warmed by 4°C to 5°C (Higgins and Schrag 2006). Biological responses varied strongly among taxonomic groups. The PETM event was most severe for benthic foraminifera, 30 percent to 50 percent of which went extinct. It also profoundly affected the ecology and evolution of terrestrial species (McInerney and Wing 2011). Mammal body sizes rapidly decreased, perhaps as a result of higher temperatures or reduced forage quality (Gingerich 2006; McInerney and Wing 2011). The duration of dwarfed body size is limited to the PETM interval, suggesting highly plastic responses of mammalian body size to environmental change. Connectivity and intercontinental dispersal increased among North America, Europe, and Asia, with long-lasting impacts on faunal distributions and evolution (Gingerich 2006).

Although there was no increase in global terrestrial extinctions during the PETM, turnover in plant communities was rapid (<10,000 years) and implies the local extirpation of some taxa and range expansions of others across distances up to 1,000 km (Wing et al. 2005; McInerney and Wing 2011). Interestingly, plants experienced few long-lasting impacts: pre-PETM and post-PETM floral assemblages are closely similar. Herbivory-damage analyses of PETM leaves suggest intensified and more specialized types of insect feeding under conditions of higher temperatures and atmospheric CO_2 (Currano et al. 2008). Several orders of modern mammals, including primates, first appear after the PETM (Gingerich 2006).

The magnitude of carbon cycle disruption during the PETM is similar to the high-end emission scenarios for the coming centuries, but the PETM onset is slower than Anthropocene onset by one to two orders of magnitude (i.e., a PETM carbon pulse and temperature rise over 10^3 to 10^4 years vs. an Anthropocene rise over 10^2 to 10^3 years). Hence, the ecological and evolutionary effects of the PETM should be conservative relative to those of the coming century.

THE LAST DEGLACIATION: ABRUPT TEMPERATURE CHANGE AND RAPID VEGETATION RESPONSE

How quickly can population, species, and ecosystems adjust to current rates of temperature rise? How long are the time lags between climatic forcing and ecological responses, and how long are ecological and evolutionary adaptive lags (Aitken et al. 2008)? Do slow responses enhance extinction risk, thus requiring adaptation interventions such as managed relocation? Are species distributions already in disequilibrium with twentieth- and early twenty-first-century temperature trends (Svenning and Sandel 2013; Goring and Williams 2017)? Concerns about adaptive lags and extinction risk are acute for species with long generation times, low fecundity, small ranges, low genetic diversity, or limited dispersal capabilities.

The abrupt temperature rises during past glacial terminations and Dansgaard-Oeschger events (Table 10.1; see also Chapter 8) offer a replicated series of model systems useful to study the response of biological systems to changes that were, at least regionally, as fast or faster than any expected for this century. Of these, the last deglaciation has received the most attention because it contains several abrupt temperature rises and reversals, occurred in the midst of—and likely contributed to—a global wave of extinctions of large-bodied vertebrates, and is the best-dated and -documented by proxies.

The last deglaciation (19,000 to 8,000 years ago) combines the Pleistocene-to-Holocene glacial-to-interglacial transition with several abrupt, decadal-scale flickers. Long-term trends include global temperature rise (3°C to 5°C), melting ice and rising sea level (120 m to 130 m), rising greenhouse gas concentrations (e.g., CO_2 from 190 ppm to 265 ppm; Clark et al. 2012), and plant distribution shifts of hundreds to thousands of kilometers (Williams et al. 2004). Superimposed on these longer-term trends is a series

of abrupt global climate reversals that manifested differently among regions. These reversals were triggered by rapid melting of ice sheets, meltwater pulses into the North Atlantic, and perturbations to global oceanic circulation, carbon cycle, and atmospheric transport of heat and moisture (Alley 2000; Alley and Ágústsdóttir 2005; Clark et al. 2012). The magnitude and rapidity of these events in Greenland and Western Europe (Figure 10.1) is remarkable:

- During the Bølling-Allerød onset, 14,700 years ago, temperatures rose in Greenland by 9°C to 14°C (Buizert et al. 2014) over 1 to 3 years (Steffensen et al. 2008). Monsoons strengthened in India and East Asia (Clark et al. 2012).

- During the Younger Dryas onset, 12,700 years ago, temperatures dropped 5°C to 9°C over Greenland perhaps within a single year (Steffensen et al. 2008; Buizert et al. 2014), marking a temporary return to near-glacial cold and dry conditions in the Northern Hemisphere (Alley 2000). Wind strength and atmospheric circulation in Western Europe apparently shifted within one year (Brauer et al. 2008). Monsoons weakened in India and East Asia (Clark et al. 2012). In Antarctica and circumpolar southern landmasses, temperatures rose as ocean circulation weakened and heat transport slowed from the South to North Atlantic.

- During the Younger Dryas termination (also the Holocene start), 11,700 years ago, temperatures rose by 8°C to 12°C (Buizert et al. 2014) within 60 years at Greenland (Steffensen et al. 2008).

- During the 8.2 ka event, from 8,400 to 8,200 years ago, Greenland temperatures dropped by 5°C to 7°C over five years (Alley and Ágústsdóttir 2005). In Western Europe, the cooling is weaker (~1°C to 2°C). The 8.2 ka event marks one final flicker of the glacial climate system as the Laurentide Ice Sheet collapsed, sending a pulse of cold meltwater into the North Atlantic. Monsoons were weaker in North Africa and southeastern Asia, producing regionally drier conditions (Alley and Ágústsdóttir 2005).

Multiproxy studies of lake sediments are critical for studying terrestrial vegeta-tion responses to abrupt climate change. First, multiple ecological and climatic prox-ies from the same sediment cores permit the precise establishment of the relative timing of events, thereby avoiding uncer-tainties in absolute age estimates from ra-diometric dating. Second, lake sediments include fossils from multiple terrestrial and aquatic taxonomic groups, differing in fe-cundity, dispersal capabilities, and genera-tion times, enabling study of these traits on ecological rates of response. Third, lake sediments include both fossil pollen and (sometimes) plant macrofossils. Fossil

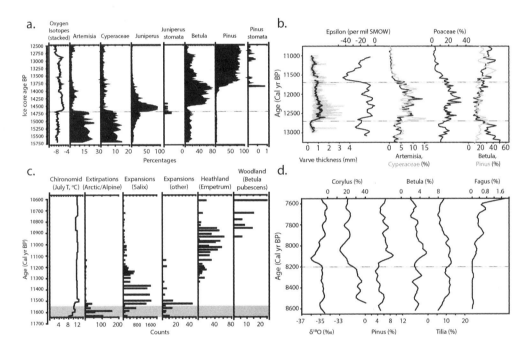

Figure 10.1. Fast and slow vegetation responses to the abrupt climate changes during the last deglaciation, based on four high-resolution multiproxy records from Western Europe. (A) Lake Gerzensee (Switzerland), where rapid warming at Bølling onset (14,700 years ago, dashed line) is captured by δ¹⁸O. Rapid woodland expansion and delayed *Pinus* migration are summarized by relative pollen abundances for *Artemisia*, Cyperaceae, *Juniperus*, *Betula*, and *Pinus* and stomatal counts for *Pinus* and *Juniperus* (Ammann et al. 2013). (B) Meerfelder Maar (Germany), where abrupt cooling and increased wind strength at the start of the Younger Dryas are indicated by hydrogen isotopes from terrestrially sourced or-ganic compounds, varve thickness, and varve mineral-ogy (Rach et al. 2014). Rapid conversion from forest to tundra is summarized by *Pinus, Betula, Artemisia*, Cypera-ceae, and Poaceae pollen abundances. (C) Lake Kråkenes (Norway), where abrupt warming at start of Holocene (highlighted by gray band) is inferred from chirono-mid assemblages (Birks 2015). This warming triggered immediate losses of some Arctic and alpine species and multiple stages of vegetation turnover, culminating 700 years later with the immigration of tree birch (*Betula pu-bescens*). (D) Soppensee Lake (Switzerland), where forest community composition responded within 0–40 years to the abrupt cooling at 8.2 ka (dashed line), in turn trig-gered by Laurentide Ice Sheet collapse (Tinner and Lotter 2001). Gains or losses were transitory for some tree taxa (*Pinus, Betula, Tilia*) and long lasting for others (e.g., *Corylus avellana, Fagus sylvatica*).

pollen is ubiquitous and a good proxy for plant abundance, although weighted toward wind-dispersed pollen types. Plant macrofossils are less common but definitive indicators of the presence of a source plant.

Here we present four examples (Figure 10.1), from different sites and time periods in Western Europe, where the climatic signal of abrupt climate change is particularly strong. These records demonstrate a mixture of fast (near instantaneous) and slow (century scale) response rates by terrestrial plant communities. Forest composition and many tree species abundances are sensitive to climate change, with rapid climatic tracking by some components. Often the lag between climatic forcing and initial vegetation response is statistically indistinguishable from zero. Other taxa have century-scale response times, presumably due to successional dynamics and migration rates.

At Lake Gerzensee in Switzerland (Figure 10.1A), initial biotic responses to rapid warming 14,685 years BP, at the start of the Bølling, occurred within 8 to 16 years (Ammann et al. 2013). Terrestrial and aquatic communities responded simultaneously, suggesting that intertaxonomic differences in generation time or climatic sensitivity had no effect. However, the complex sequence of vegetation changes triggered by the Bølling-Allerød warming lasted nearly 1,000 years. *Juniperus* and *Betula* were fast responders, with near-instantaneous expansion into shrub tundra. These woodlands persisted for 180 years, until replacement by *Betula* forests. *Pinus* arrived at Gerzensee by about 13,850 years ago, based on *Pinus* stomata and rapidly rising *Pinus* pollen abundances. The closest known glacial refugia for *Pinus* are in the southeastern Alps, 500 km from Gerzensee (Ammann et al. 2013) If we assume that *Pinus* was initially 500 km distant and that climates became locally suitable for *Pinus* at the start of the Bølling warming, then we can calculate an average migration rate for *Pinus* of 0.6 km/year. Actual migration rates could have been faster (if source populations were >500 km

away or migration occurred in a series of rapid jumps) or slower (if cryptic refugia were <500 km or if migration began prior to Bølling onset).

Meerfelder Maar in western Germany (Figure 10.1B) captures the effect of the rapid return to cold and dry conditions in Europe at the onset of the Younger Dryas (YD). At Meerfelder, the YD began with a gradual cooling at 12,850 years ago (based on hydrogen isotopes in plant biomarkers), followed by an abrupt shift (1 to 3 years) 170 years later to drier and windier climates (Rach et al. 2014). The initial cooling correlates closely to Greenland isotopic records but took several centuries to manifest, whereas the second shift was extraordinarily rapid (perhaps within a single year) and likely caused by expansion of sea ice over the North Atlantic, rapid atmospheric reorganization, and reduced moisture advection to Western Europe (Brauer et al. 2008; Rach et al. 2014). Changes in vegetation composition were synchronous with both events: declines in *Pinus* and *Betula* and rises in *Artemisia*, Cyperaceae, and Poaceae closely parallel the initial cooling and quickly accelerated with the second shift. The pollen sampling resolution at Meerfelder is 8 to 40 years, indicating short (at most, decadal-scale) time lags between regional cooling and aridification, mortality of *Pinus* and *Betula*, and replacement of woodlands with tundra. Meerfelder offers a cautionary note against assessing past rates of ecological response based on high-resolution but geographically dispersed records; a comparison of Meerfelder pollen to Greenland isotopic records would suggest an approximate 150-year vegetation lag after the Younger Dryas onset, when in fact any lag is at most a few decades.

Kråkenes Lake, in coastal Norway (Figure 10.1C), has a rich plant macrofossil record, which enables precise estimates of local turnover in plant community composition and the relative timing of immigration and extirpation events. July temperatures rose from 8°C to 11°C at Kråkenes at the start of the Holocene (Birks 2015). This rise began 11,630 years ago and lasted until 11,550 years BP, for an average of 3.75°C/century. Vegetation composition at Kråkenes abruptly changed between 11,600 and 11,550 years ago, synchronous with the temperature rise. Many Arctic and alpine plants were locally extirpated (e.g., *Koenigia islandica*) or reduced to trace presences (e.g., *Saxifraga rivularis, S. cespitosa*). This rapid response was followed by a succession of vegetation communities over the next 700 years. High abundances of *Salix herbacea* persisted for 150 years (until 11,400 years BP), followed by increases in grassland and fen taxa, then *Empetrum* heathlands. *Betula pubescens* (tree birch) apparently arrived at Kråkenes by 10,940 years BP, roughly 650 years after the initial event. The initial rise to 11°C should have been sufficient for B. *pubescens* survival, suggesting that dispersal limitation was the primary reason for the 600-year lag. During the Younger Dryas, the nearest known B. *pubescens* refugia are in Denmark and southern Sweden (Birks 2015), roughly 600 km away, suggesting a migration rate of ~1 km/year. In northern Norway, expansion of B. *pubescens* woodlands kept pace with temperature velocities (Birks 2015).

At Soppensee Lake, Switzerland, forest composition responded quickly to 8.2 ka cooling, with no time lags for several tree taxa (Figure 10.1D). The varved Soppensee sediments enable annual-scale analyses, and the pollen record at Soppensee was analyzed at decadal resolution (Tinner and Lotter 2001). *Corylus avellana* abundances collapsed, while *Pinus, Betula, Fagus sylvatica,* and *Tilia* all expanded. Time lags are short, ranging from 0 to 40 years, suggesting that forest composition turned over within a single tree (and human) generation. Most forest responses to the 8.2 ka event were temporary, but the event ended the early Holocene dominance of *Corylus avellana* and began the rise of *Fagus sylvatica*.

The large changes in abundances and geographic distributions often caused extirpations of plant populations at the trailing edge of species ranges and presumably caused species to move among microclimates within their established range. Despite these large shifts, extinctions of plant species were rare. One extinction, *Picea critchfieldii* (Jackson and Weng 1999), is well documented. This contrasts with the global wave of megafaunal extinctions that began 40,000 years ago and continued through the late Holocene (Koch and Barnosky 2006). However megafaunal populations on remote islands (Madagascar, New Zealand, Fiji, Mediterranean islands) did not suffer widespread extinctions during the last deglaciation, despite major environmental change and sea-level rise (but see Graham et al. 2016), suggesting that late-Holocene human arrival was the critical driver.

HYDROLOGICAL VARIABILITY AND ECOLOGICAL TIPPING POINTS DURING THE HOLOCENE

Hydrological variability and the frequency of extreme events is expected to increase over this century because a warmer atmosphere can hold more water vapor and because enhanced evaporation rates will deliver more water to the atmosphere while locally drying out land surfaces. Similarly, a defining characteristic of the Holocene is its heightened hydrological variability and the profound consequences of this variability for terrestrial ecosystems (Figure 10.2) and early agricultural societies (deMenocal 2001).

There is a persistent myth that the Holocene was climatically stable and benign, because of the muted temperature variability over the last 11,000 years. For example, global mean temperatures variations were on the order of <1°C to 2°C from the early to late Holocene (Marcott et al. 2013; Marsicek et al. 2018), which is small relative to the temperature swings discussed above. This myth of Holocene stability, however, overlooks profound variations in hydrological,

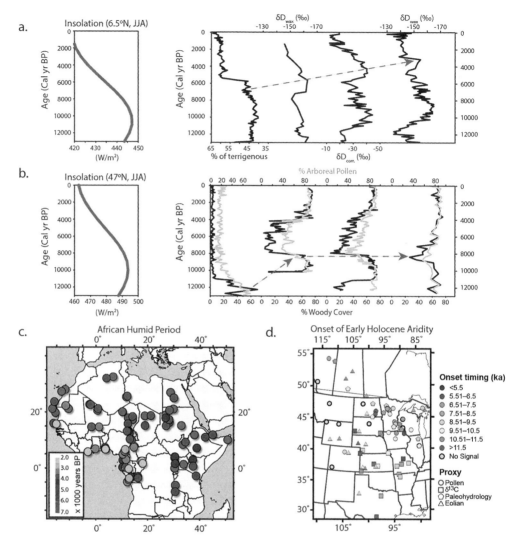

Figure 10.2. During the Holocene, slow and progressive regional aridification produced locally abrupt collapses of forest cover and precipitation. (A) Insolation trends at 6.5°N and a north-south transect of hydrogen isotopic records from North Africa, showing locally abrupt and regionally time-transgressive collapses in monsoonal rainfall, highlighted by dashed line (Shanahan et al. 2015). (B) Insolation trends at 47°N and an east-west transect of pollen records from the eastern Great Plains, showing locally abrupt and regionally time-transgressive collapses in forest cover (Williams et al. 2009). (C) Map showing early drying of northern Sahara and later drying in south (Shanahan et al. 2015). (D) Map showing earlier drying of sites in interior Great Plains and later drying of sites nearer the prairie-forest ecotone, based on a compilation of eolian, carbon isotopic, pollen, and paleohydrological records (Williams et al. 2010).

ecological, and societal systems during the Holocene. Hence, the Pleistocene-Holocene transition should be viewed instead as a fundamental shift in the type and drivers of climate variability, from a cold glacial world characterized by large temperature swings and positive feedbacks strongly governed by ice-sheet dynamics and cryosphere-ocean-atmosphere feedbacks to a warm interglacial world characterized by a wetter and more energetic atmosphere and governed by vegetation-atmosphere and ocean-atmosphere feedbacks.

High hydrological variability during the Holocene triggered major ecological and societal effects at regional to subcontinental scales, often abrupt, particularly in subtropical, semiarid, and temperate regions. Grasslands in North Africa collapsed during the middle Holocene and were replaced by the Sahara Desert (Figure 10.2), causing retreat of pastoral societies (Kuper and Kröpelin 2008). In the Great Plains, the Nebraska Sand Hills repeatedly flipped between active dunes and stabilized grasslands (Miao et al. 2007). Forests in the eastern Great Plains converted to prairie in the early Holocene, reverting to forest in the middle to late Holocene (Williams et al. 2010). Mesophilic trees in the temperate deciduous forests of the eastern United States (e.g., *Tsuga canadensis*, *Fagus grandifolia*) experienced repeated population crashes, likely tied to decadal- to centennial-scale hydrological variability (Foster et al. 2006; Booth et al. 2012; Shuman 2012). The Holocene expansion of *Pinus edulis* (piñon pine) and *Juniperus osteosperma* in the western United States was controlled by a hydrological ratchet (Jackson et al. 2009), with recruitment and range expansion during wet periods and persistence during dry periods. The strength of El Niño–Southern Oscillation has strongly varied during the Holocene (Cobb et al. 2013), with profound effects globally on precipitation patterns, flood risk, rates of tree growth, and fire regime. Severe megadroughts, lasting decades, were common in the western United States (Cook et al. 2010b) and eastern and southern Asia (Cook et al. 2010a) over the past several millennia, the latter tied to shifts in monsoonal precipitation (Zhang et al. 2008). Early agricultural societies in India, the Fertile Crescent, Egypt, Central America, and western and central United States were destabilized by drought and floods (deMenocal 2001; Munoz et al. 2015).

A major research question is to disentangle when these abrupt vegetation changes were caused by similarly abrupt extrinsic forcings (e.g., megadroughts) (Williams et al. 2011; Seddon et al. 2015), intrinsic tipping points within ecological systems that cause abrupt responses to hydrological variability (Liu et al. 2007), positive surface-atmosphere feedbacks (Claussen 2009), or simply internal dynamics with no climatic trigger (Williams et al. 2011; Booth et al. 2012; Seddon et al. 2015). Holocene hydrological variability and ecological response are spatially and temporally complex; rarely are events globally or even regionally synchronous. Rather, hydrological and vegetation variability occurred as a temporal mosaic (Williams et al. 2011), with episodes of heightened hydrological variability or reduced variability but few individual events that can be confidently correlated across sites. For example, the mid-Holocene collapse of *Tsuga canadensis* in eastern North America was once thought to have been a unique and synchronous event, caused by a single pest outbreak, but newly resolved records show multiple collapses, varying in timing across sites and regions, and the collapses have been linked to hydrological variability (Shuman 2012). Hydrological variability appears to have been particularly high between 5,500 and 4,000 years ago, for unclear reasons, with global effects on societies and ecosystems (deMenocal 2001; Shuman 2012).

Case studies from North Africa and the US Great Plains illustrate the abrupt responses of terrestrial vegetation to Holocene hydrological variability. Progressive aridification caused large-scale shifts in biome extent, decreases in tree density, and replacement of grasslands with savanna in semiarid regions (Figure 10.2). In North Africa, moisture availability peaked during the early Holocene during the African Humid Period, caused by orbitally driven shifts in summer solar radiation and enhanced monsoonal strength in the Northern Hemisphere subtropics (Figure 10.2A). North African aridification proceeded as the withdrawal of a monsoonal front, with northern locations first losing access to monsoonal precipitation and southern locations later (Shanahan et al. 2015). The southward withdrawal took thousands of years, but, at any given location, the decline in rainfall often was abrupt (decades to centuries) (Kuper and Kröpelin 2008; Shanahan et al. 2015). At Lake Yoa, tropical trees disappeared, followed by grasslands, and then deserts established (Kröpelin et al. 2008). Collapses of individual taxa or taxonomic groups at individual sites occurred quickly—on time scales of decades—but multiple collapses of multiple taxonomic groups played out over centuries for any given site (Kröpelin et al. 2008), and over a 3,000-year window across a transect of sites in North Africa (Shanahan et al. 2015).

The Great Plains also show gradual regional aridification combined with locally abrupt declines in tree densities (Figure 10.2B). The timing differs from North Africa, which was wet in the early Holocene, whereas the Great Plains were dry (Figure 10.2), likely because of regional atmospheric subsidence linked to intensified monsoons in the southwestern United States and enhanced evapotranspiration. Great Plains proxies consistently indicate aridification in the early Holocene and increasing moisture availability during the middle to late Holocene. For example, the prairie-forest ecotone shifted eastward during the early Holocene (prairie replacing forest) then westward during the middle to late Holocene (forest replacing prairie). At individual sites, rates of tree-grassland shifts are asymmetrical, with rapid deforestation (decades to centuries) and gradual reforestation (centuries to millennia). On the basis of close correspondences among proxies of vegetation, fire regime, and lake water balance, Nelson and Hu (2008) argued that these abrupt local vegetation changes were direct responses to locally rapid declines in water availability and vegetation-climate-fire feedbacks. Williams et al. (2010) argued that locally abrupt vegetation responses with a regionally time-transgressive pattern spanning multiple ecotones (grassland to desert, C_4 to C_3 grasses, forest to prairie) (Figure 10.2) indicate that site-specific ecological tipping points were the primary cause of locally abrupt responses.

Regardless of exact mechanism, the Holocene pattern is clear: semiarid systems are highly sensitive to moisture availability and can quickly shift to new states when moisture availability changes. The Holocene history of abrupt vegetation change in North Africa and Great Plains is consistent with contemporary observations of a forest-grassland bistability along precipitation gradients, with savannas a relatively uncommon intermediate state. It is also consistent with current concerns about globally enhanced rates of tree mortality due to warming and drought (Allen et al. 2010).

SUMMARY

The geological record is replete with examples with abrupt climatic and ecological change (Table 10.1). We can use the past to identify tipping points and tipping elements in the climate system and biological systems, identify governing processes, measure rates of ecological processes operating at decadal and longer timescales, and assess the adaptive capacity of organisms when confronted by abrupt environmental change. Just as some animals and ecosystems make better model systems for particular questions in medical and ecological research, some geological events offer particularly useful insights into contemporary questions in global change ecology.

Of the three model systems reviewed here, the PETM highlights the potential sensitivity of high-latitude peatlands and reservoirs of soil carbon to externally forced warming and the ability of soil-carbon feedbacks to rapidly accelerate global warming. Terrestrial ecosystems responded rapidly to the PETM, with some effects limited to the PETM itself (high rates of plant community turnover, mammalian body size dwarfing), others permanent (intercontinental immigration of species across land corridors made available through climate). Unlike the marine realm, there is no evidence for enhanced rates of terrestrial extinction during the PETM.

The last deglaciation shows that rapid climate change can have near-immediate effects on tree abundances and forest composition. Forest response times were consistently less than 20 to 40 years and often had no detectable time lag. There is no obvious difference in the initial response time between short-lived herbaceous and long-lived arboreal species, or between aquatic and terrestrial organisms. Nevertheless, at several sites, particularly Arctic and alpine sites, the initial temperature rise triggered a series of vegetation changes that continued for as long as 700 to 1,000 years after the initial warming. Likely causes for these longer-term responses are dispersal limitations for tree species, local succession, and soil development. Migration rates for two tree species are on the order of 0.6 to 1 km/year, albeit with substantial uncertainty.

During the Holocene, abrupt shifts in terrestrial ecosystems and early agricultural societies were common and often forced by a combination of hydroclimatic variability and intrinsic feedback loops and critical thresholds. These manifest in many ways, including shifts in dunes from active to vegetated states, rapid deforestation at prairie, savanna, and forest ecotones, population collapses of mesophilic tree species and abandonment of early agricultural sites. The Great Plains and North Africa show how slow regional aridification (over thousands of years) can produce locally rapid ecological shifts, driven by locally abrupt precipitation changes, critical thresholds, and/or positive feedbacks in semiarid systems between climate, fire regime, and fuel. We can expect similarly rapid and complex responses to twenty-first-century aridification and shifts in frequency of extreme events.

ACKNOWLEDGMENTS

We gratefully thank Matt McGlone for discussion and editing of this chapter and Brigitta Ammann, Hilary Birks, Achim Brauer, André Lotter, Oliver Rach, Tim Shanahan, Willy Tinner, and Willem van der Knaap for contributing data. This work benefited from discussions with University of Wisconsin–Madison graduate seminar participants in the spring of 2015, and was supported by the National Science Foundation (EF-1241868, DEB-1353896).

REFERENCES

Aitken, S. N., S. H. Yeaman, J. A. Holliday, T. Wang, and S. Curtis-McLane. 2008. "Adaptation, migration or extirpation: Climate change outcomes for tree populations." *Evolutionary Applications* 1: 95–111.

Allen, C. D., A. K. Macalady, H. Chenchouni, D. Bachelet, N. McDowell, M. Vennetier, T. Kitzberger, et al. 2010. "A global overview of drought and heat-induced tree mortality reveals emerging climate change risks for forests." *Forest Ecology and Management* 259: 660–684.

Alley, R. 2000. "The Younger Dryas cold interval as viewed from central Greenland." *Quaternary Science Reviews* 19: 213–226.

Alley, R. B., and A. M. Ágústsdóttir. 2005. "The 8k event: Cause and consequences of a major Holocene abrupt climate change." *Quaternary Science Reviews* 24: 1123–1149.

Ammann, B., U. J. van Raden, J. Schwander, U. Eicher, A. Gilli, S. M. Bernasconi, J. F. N. van Leeuwen, et al. 2013. "Responses to rapid warming at Termination 1a at Gerzensee (Central Europe): Primary succession, albedo, soils, lake development, and ecological interactions." *Palaeogeography, Palaeoclimatology, Palaeoecology* 391, pt. B: 111–131.

Birks, H. H. 2015. "South to north: Contrasting late-glacial and early-Holocene climate changes and vegetation responses between south and north Norway." *Holocene* 25: 37–52.

Booth, R. K., S. T. Jackson, V. A. Sousa, M. E. Sullivan, T. A. Minckley, and M. J. Clifford. 2012. "Multi-decadal drought and amplified moisture variability drove rapid forest community change in a humid region." *Ecology* 93: 219–226.

Brauer, A., G. H. Haug, P. Dulski, D. M. Sigman, and J. Negendank. 2008. "An abrupt wind shift in Western Europe at the onset of the Younger Dryas cold period." *Nature Geoscience* 1: 520–523.

Buffett, B., and D. Archer. 2004. "Global inventory of methane clathrate: Sensitivity to changes in the deep ocean." *Earth and Planetary Science Letters* 227: 185–199.

Buizert, C., V. Gkinis, J. P. Severinghaus, F. He, B. S. Lecavalier, P. Kindler, M. Leuenberger, et al. 2014. "Greenland temperature response to climate forcing during the last deglaciation." *Science* 345: 1177–1180.

Clark, P. U., J. D. Shakun, P. A. Baker, P. J. Bartlein, S. Brewer, E. J. Brook, A. E. Carlson, et al. 2012. "Global climate evolution during the last deglaciation." *Proceedings of the National Academy of Sciences* 109: E1134–E1142.

Claussen, M. 2009. "Late Quaternary vegetation-climate feedbacks." *Climate of the Past* 5: 203–216.

Cobb, K. M., N. Westphal, H. R. Sayani, J. T. Watson, E. Di Lorenzo, H. Cheng, R. L. Edwards, and C. D. Charles. 2013. "Highly variable El Niño–Southern Oscillation throughout the Holocene." *Science* 339: 67–70.

Cook, E. R., K. J. Anchukaitis, B. M. Buckley, R. D. D'Arrigo, G. C. Jacoby, and W. E. Wright. 2010a. "Asian monsoon failure and megadrought during the last millennium." *Science* 328: 486–489.

Cook, E. R., R. Seager, R. R. Heim, R. S. Vose, C. Herweijer, and C. Woodhouse. 2010b. "Megadroughts in North America: Placing IPCC projections of hydroclimatic change in a long-term palaeoclimate context." *Journal of Quaternary Science* 25: 48–61.

Currano, E. D., P. Wilf, S. L. Wing, C. C. Labandeira, E. C. Lovelock, and D. L. Royer. 2008. "Sharply increased insect herbivory during the Paleocene–Eocene Thermal Maximum." *Proceedings of the National Academy of Sciences* 105: 1960–1964.

DeConto, R. M., S. Galeotti, M. Pagani, D. Tracy, K. Schaefer, T. Zhang, D. Pollard, and D. J. Beerling. 2012. "Past extreme warming events linked to massive carbon release from thawing permafrost." *Nature* 484: 87–92.

deMenocal, P. B. 2001. "Cultural responses to climate change during the late Holocene." *Science* 292: 667–673.

Foster, D. R., W. W. Oswald, E. K. Faison, E. D. Doughty, and B. C. S. Hansen. 2006. "A climatic driver for abrupt mid-Holocene vegetation dynamics and the hemlock decline in New England." *Ecology* 87: 2959–2966.

Gingerich, P. D. 2006. "Environment and evolution through the Paleocene-Eocene thermal maximum." *Trends in Ecology & Evolution* 21: 246–253.

Goring, S. J., and J. W. Williams. 2017. "Effect of historic land-use and climate change on tree-climate relationships in the upper Midwestern United States." *Ecology Letters* 20: 461–470.

Graham, R. W., S. Belmecheri, K. Choy, B. Culleton, L. J. Davies, D. Froese, P. D. Heintzman, et al. 2016. "Timing and causes of a middle Holocene mammoth extinction on St. Paul Island, Alaska." *Proceedings of the National Academy of Sciences* 113: 9310–9314.

Higgins, J. A., and D. P. Schrag. 2006. "Beyond methane: Towards a theory for the Paleocene-Eocene Thermal Maximum." *Earth and Planetary Science Letters* 245: 523–537.

Jackson, S., and C. Weng. 1999. "Late Quaternary extinction of a tree species in eastern North America." *Proceedings of the National Academy of Sciences* 96: 13847–13852.

Jackson, S. T., J. L. Betancourt, R. K. Booth, and S. T. Gray. 2009. "Ecology and the ratchet of events: Climate variability, niche dimensions, and species distributions." *Proceedings of the National Academy of Sciences* 106: 19685–19692.

Koch, P. L., and A. D. Barnosky. 2006. "Late Quaternary extinctions: State of the debate." *Annual Review of Ecology, Evolution, and Systematics* 37: 215–250.

Kröpelin, S., D. Verschuren, A.-M. Lézine, H. Eggermont, C. Cocquyt, P. Francus, J.-P. Cazet, et al. 2008. "Climate-driven ecosystem succession in the Sahara: The past 6000 years." *Science* 320: 765–768.

Kuper, R., and S. Kröpelin. 2008. "Climate-controlled Holocene occupation in the Sahara: Motor of Africa's evolution." *Science* 313: 803–807.

Liu, Z., Y. Wang, R. Gallimore, F. Gasse, T. Johnson, P. deMenocal, J. Adkins, et al. 2007. "Simulating the transient evolution and abrupt change of Northern Africa atmosphere-ocean-terrestrial ecosystem in the Holocene." *Quaternary Science Reviews* 26: 1818–1837.

Marcott, S. A., J. D. Shakun, P. U. Clark, and A. C. Mix. 2013. "A reconstruction of regional and global temperature for the past 11,300 years." *Science* 339: 1198–1201.

Marsicek, J., B. N. Shuman, P. J. Bartlein, S. L. Shafer, and S. Brewer. 2018. "Reconciling divergent trends and millennial variations in Holocene temperatures." *Nature* 554: 92.

McInerney, F. A., and S. L. Wing. 2011. "The Paleocene-Eocene Thermal Maximum: A perturbation of carbon cycle, climate, and biosphere with implications for the future." *Annual Review of Earth and Planetary Sciences* 39: 489–516.

Miao, X., J. A. Mason, J. B. Swinehart, D. B. Loope, P. R. Hanson, R. J. Goble, and X. Liu. 2007. "A 10,000 year record of dune activity, dust storms, and severe drought in the central Great Plains." *Geology* 35: 119–122.

Munoz, S. E., K. E. Gruley, A. Massie, D. A. Fike, S. Schroeder, and J. W. Williams. 2015. "Cahokia's emergence and decline coincided with shifts of flood frequency on the Mississippi River." *Proceedings of the National Academy of Sciences* 112: 6319–6324.

National Research Council. 2013. "Abrupt impacts of climate change: Anticipating surprises." National Academy of Sciences, Washington, DC.

Nelson, D. M., and F. S. Hu. 2008. "Patterns and drivers of Holocene vegetational change near the prairie-forest ecotone in Minnesota: Revisiting McAndrews' transect." *New Phytologist* 179: 449–459.

Rach, O., A. Brauer, H. Wilkes, and D. Sachse. 2014. "Delayed hydrological response to Greenland cooling at the onset of the Younger Dryas in Western Europe." *Nature Geoscience* 7: 109–112.

Scheffer, M., M. Hirota, M. Holmgren, E. H. Van Nes, and F. S. Chapin. 2012. "Thresholds for boreal biome transitions." *Proceedings of the National Academy of Sciences* 109: 21384–21389.

Seddon, A. W., M. Macias-Fauria, and K. J. Willis. 2015. "Climate and abrupt vegetation change in Northern Europe since the last deglaciation." *The Holocene* 25: 25–36.

Shanahan, T. M., N. P. McKay, K. A. Hughen, J. T. Overpeck, B. Otto-Bliesner, C. W. Heil, J. King, C. A. Scholz, and J. Peck. 2015. "The time-transgressive termination of the African Humid Period." *Nature Geoscience* 8: 140–144.

Shuman, B. N. 2012. "Patterns, processes, and impacts of abrupt climate change in a warm world: The past 11,700 years." *WIREs Climate Change* https://doi.org/10.1002/wcc.152.

Steffensen, J. P., K. K. Andersen, M. Bigler, H. B. Clausen, D. Dahl-Jensen, H. Fischer, K. Goto-Azuma, et al. 2008. "High-resolution Greenland ice core data show abrupt climate change happens in few years." *Science* 321: 680–684.

Svenning, J.-C., and B. Sandel. 2013. "Disequilibrium vegetation dynamics under future climate change." *American Journal of Botany* 100: 1266–1286.

Tinner, W., and A. F. Lotter. 2001. "Central European vegetation response to abrupt climate change at 8.2 ka." *Geology* 29: 551–554.

Williams, J. W., J. L. Blois, J. L. Gill, L. M. Gonzales, E. C. Grimm, A. Ordonez, B. Shuman, and S. Veloz. 2013. "Model systems for a no-analog future: Species associations and climates during the last deglaciation." *Annals of the New York Academy of Sciences* 1297: 29–43.

Williams, J. W., J. L. Blois, and B. N. Shuman. 2011. "Extrinsic and intrinsic forcing of abrupt ecological change: Case studies from the late Quaternary." *Journal of Ecology* 99: 664–677.

Williams, J. W., B. Shuman, P. J. Bartlein, N. S. Diffenbaugh, and T. Webb III. 2010. "Rapid, time-transgressive, and variable responses to early-Holocene midcontinental drying in North America." *Geology* 38: 135–138.

Williams, J. W., B. N. Shuman, T. Webb III, P. J. Bartlein, and P. L. Leduc. 2004. "Late Quaternary vegetation dynamics in North America: Scaling from taxa to biomes." *Ecological Monographs* 74: 309–334.

Wing, S. L., G. J. Harrington, F. A. Smith, J. I. Bloch, D. M. Boyer, and K. H. Freeman. 2005. "Transient floral change and rapid global warming at the Paleocene-Eocene boundary." *Science* 310: 993–996.

Zachos, J. C., U. Röhl, S. A. Schellenberg, A. Sluijs, D. A. Hodell, D. C. Kelly, E. Thomas, M. Nicolo, I. Raffi, L. J. Lourens, H. McCarren, and D. Kroon. 2005. "Rapid acidification of the ocean during the Paleocene-Eocene Thermal Maximum." *Science* 308: 1611–1615.

Zhang, P., H. Cheng, R. L. Edwards, F. Chen, Y. Wang, X. Yang, J. Liu, et al. 2008. "A test of climate, sun, and culture relationships from an 1810-year Chinese cave record." *Science* 322: 940–942.

CHAPTER ELEVEN

A Neotropical Perspective on Past Human-Climate Interactions and Biodiversity

MARK B. BUSH

In this chapter we explore the effects of climate change on Amazonian and Andean ecosystems. Rather than documenting simple upslope and downslope range changes, we consider the heterogeneity of both modern and past events. Layers of complexity are added as biotic and abiotic factors are considered. Last, we add human influence into the mix, emphasizing that fire, which is almost uniquely associated with human activity in these forests, will accelerate and amplify the effects of climate change.

Tropical landscapes are hugely varied, ranging from ice-capped mountains to lowland tropical rainforest. Even lowland tropical rainforests are ecologically highly heterogeneous, often varying as a result of edaphic, hydrological, historical, or climatic factors. Heterogeneity of species composition and structure across many spatial scales is the natural state of tropical systems. At local scales, a single hectare of mature forest may support more than 350 species of tree, plus a wealth of treelets, vines, and epiphytes. This diversity has led to a widely held misconception is that all species are relatively rare in these forests. Certainly, the near monodominance seen in boreal forest trees is absent from neotropical systems, but tropical tree species are not all equally rare. Among the approximately 16,000 tree species in Amazonia, about 230 species routinely account for 50 percent of the stems (ter Steege et al. 2013). The reason for the abundance of these so-called hyperdominant species may be related to niche and competition success (Pitman et al. 2001), neutrality (Hubbell 2001), or past human activity (Roosevelt 2013). Few studies of the historical patterns leading to hyperdominance have been conducted in Amazonia, but the abundance of the palm

Mauritia may have been enhanced by human activity (Rull and Montoya 2014). Contrastingly, *Iriartea*, another hyperdominant palm, showed an overall increase in abundance in the past 3,000 years but probably as a result of wetter conditions rather than human activity (Bush and McMichael 2016). These data suggest that membership in the hyperdominant class of trees may be labile. Whatever the cause of the success of these most abundant species, their traits (canopy architecture, fruit production, phenology, leaf palatability, and flammability) play an important role in shaping current niche availability for other forest organisms.

Many factors contribute to fine-scale heterogeneity in forests, including microclimatic conditions, history, soil moisture gradients, flood regimes, time since last disturbance, soil type, and frugivore population size and type. In mountainous areas, the strongest predictor of species presence is temperature, which is negatively correlated with elevation (Richards 1996). Indeed, the most predictable temperature gradients in the tropics are associated not with latitude but with elevation. On the flank of the Andes, temperatures decrease approximately −5.5°C per 1000 m of vertical ascent. The steep slopes cause localities just 3 km apart to have temperatures differing by >5°C (Bush 2002).

Regional heterogeneity is also influenced by climatic differences, especially in total amount and seasonality of precipitation. Within lowland Amazonia there are evergreen forests receiving in excess of 10,000 mm of rain per year, whereas others may receive only 1,500 mm. If the rainfall is evenly spread through the year less is needed to maintain an evergreen forest than in one that experiences a 3-, 4-, or 5-month-long dry season (Silman 2011). Soils, too, play a critical role in determining the effects of rainfall. For example, plants growing on a slow-draining clay soil will be prone to waterlogging, while under the same rainfall regime plants growing on a free-draining white-sand soil may be vulnerable to drought.

Tropical climates have varied substantially over time. Within the past 2 million years there have been at least 22 ice ages, the last five of which were more extreme and lasted longer than previous ones. During ice ages, ice caps expanded in the mountains, with glaciers often extending about 1500 m–2000 m downslope, leaving terminal moraines at about 3200 m–3800 m elevation (wetter locations supported longer glaciers). Estimates of temperature change in the high Andes reveal a cooling of approximately 7°C–9°C during the last ice age (Hooghiemstra 1984; Van 't Veer and Hooghiemstra 2000). In the lowlands, the cooling was not as extreme, with many locations 2°C–4°C cooler than present for much of the ice age. Brief intervals, lasting a few thousand years, cooled the lowlands by about 5°C–6°C (Bush, de Oliveira, et al. 2004; Bush, Silman, et al. 2004. Changes in precipitation were highly variable in space and time but probably just as influential in reshaping communities as temperature.

Haffer (1969) proposed the refugial hypothesis as a means to account for modern biogeographic patterns in birds. Haffer's hypothesis that glacial-age aridity caused grasslands to replace most Amazonian rainforest is completely unsupported by paleoecological or molecular data (Colinvaux et al. 2001; Miller et al. 2008; Wang et al. 2017). The switching between wet and dry episodes was faster and not as profound as that implied by the refugial hypothesis. At the risk of oversimplifying the pattern, there was a tendency toward alternating periods of drier and wetter average conditions, each lasting approximately 10,000 years due to Earth's precessional variability in its orbit around the Sun (Bush et al. 2002; Cruz et al. 2005). This relationship was not invariant, and outside factors, such as the oceanic circulation of the Atlantic Ocean, periodically trumped the precessional signature (Mosblech et al. 2012; Stríkis et al.

2018). The most extreme drying may have occurred during interglacials rather than glacials.

A potential pitfall in thinking about climate is to use the current interglacial as a representative example of other interglacials. Rather than being warm and wet like the modern Holocene, analysis of long sedimentary records has shown that the last interglacial (Marine Isotope Stage 5e) was the driest time in the Bolivian Altiplano in the past half million years (Baker et al. 2001; Fritz et al. 2007). Multimillennial dry events also occurred within glacial periods, but the relatively low temperatures may have reduced evapotranspirative stress and thus had a relatively modest impact on vegetation in contrast to the refugial model (Sato and Cowling 2017).

During the last ice age, millennial-scale climate events, called Dansgaard-Oeschger cycles, had the capacity to cause approximate 5°C–10°C swings in average temperature in Greenland. These events often started with a very abrupt warming of ~5°C occurring in the space of 10–30 years (North Greenland Ice-Core Project [NGRIP] 2004), followed by 3,000–7,000 years of cooling. These same events affected the tropics but generally as changes in precipitation rather than temperature. The effect of these events was location specific, with eastern and western Amazonia responding in opposite ways (Cruz et al. 2009). In general, cold events in the Atlantic caused western Amazonia to become wetter while the eastern Amazon became drier. More locally, if the modern response to anomalously high temperatures in the Atlantic in 2010 is a guide, climatic responses would have been very heterogeneous (Aragão and Shimabukuro 2010). In 2010, anomalously warm temperatures in the northern tropical Atlantic caused profound drought in some areas of western Amazonia while other areas were largely unchanged. As climatic impacts of phenomena such as the El Niño–Southern Oscillation or the Atlantic Multidecadal Oscillation become more familiar, it becomes

apparent that it is very difficult to predict actual climatic outcomes for given locations (Marengo et al. 2008; Marengo et al. 2011). This uncertainty of outcome increases the greater the distance from the event. A common theme, however, is the importance of abrupt climate change in contributing to spatial and temporal climatic heterogeneity. It may well be that larger phenomena, such as Dansgaard-Oeschger cycles (Bond et al. 1993), were climatically far more variable within each event than is generally believed.

Although the Holocene has been portrayed as a period of climatic stability, some parts of the Andes, Amazonia, and southeastern Brazil experienced profound droughts between 9,000 and 5,000 years ago (Mayle et al. 2000; Abbott et al. 2003; Raczka et al. 2013). Lake Titicaca shrank to half its modern size (Seltzer et al. 1998), many shallower lakes dried out completely, and the proportion of grasses in the Carajas pollen record is higher than at any time in the past 100,000 years (Absy et al. 1991; Hermanowski et al. 2013). Humans who had only occupied the high Andes about 12,800 years ago (Rademaker et al. 2014) abandoned large areas, and instead of a signal of exponentially increasing human impact, there is a zone of archaeological "silence" (Núñez et al. 2001). More detailed investigations of this period show—probably like many previous dry events—a highly heterogeneous climate with intense wet events punctuating a time of overall lower precipitation (Hillyer et al. 2009).

Indeed, we should be cautious about simplifying complex climatic histories to infer millennial-scale wet or dry events. Biologically, wet interludes within an overall dry event make a huge difference, as water tables and root zones can be recharged with water, alleviating some of the effects of the drying. Similarly, as we think about biological impacts of dry events in the past or the future, the timing of those events is critically important. If wet-season rains are diminished but the length of the wet season is not affected, there would be a minimal biological response in that forest, as water is not limiting in the wet season. Such a change, however, does not preclude larger ecosystem effects as reduced flood-water could alter downstream hydrological regimes. If, by contrast, the dry season becomes longer, or the number of consecutive days without rain increases, the drought deficit and increased fire probability could reshape forests (Bush and Silman 2004).

Fire is a great game changer for biodiversity. In wet forests that do not burn naturally, a single fire will cause complete turnover in ground nesting bird species, and repeated fires quickly reduce forest biomass, simplify physical structure, and cause almost complete species turnover among trees (Barlow 2002; Barlow et al. 2003). In much of western lowland Amazonia, fire was not a natural component of Holocene ecosystems, but it did accompany human occupation. A 4,000-year sediment history from Lake Werth in southeastern Peru contained no sign of human occupation or fire in that catchment (Bush et al. 2007). By contrast, just a few kilometers away, Lakes Gentry, Parker, and Vargas all showed periods of human occupation and have complex fire histories (McMichael et al. 2013). Fires are difficult to ignite in these wet forests, so it is probable that humans took advantage of seasonal droughts to burn and clear forests.

Human impacts on Amazonian and high Andean (>3000 m elevation) systems were probably very localized until approximately 10,000 years ago (Dillehay 2009). The areal extent and temporal trajectory of human impacts in lowland Amazonia continues to be actively debated, but it is clear that, through fire, human action had the capacity to completely alter occupied landscapes.

BIOTIC MECHANISMS OF HETEROGENEITY

Past climate fluctuations caused re-assortments of species into novel assemblages,

under continuous but different adaptive or evolutionary pressures (Overpeck et al. 1985). Ecological disequilibrium was the norm and played a key role in maintaining species richness through time. Responses to climate change would have varied greatly according to landscape and the mobility of species; that is, birds can migrate much more freely than trees. Temperate system responses to ongoing warming cause upslope or poleward range shifts (Chapter 3). In the tropics, the species that live at high elevations have only an upslope route to escape warmth as there is little appreciable benefit in terms of cooling to migrate from 0° latitude to 10° latitude. On mountain flanks, tropical lowland taxa may be able to migrate upslope relatively short migrational distances to maintain a given temperature (Bush 2002). Upslope and downslope responses to past warming and cooling are readily documented in the Central American Isthmus and the flank of the Andes. An asymmetry exists, however, in the capacity of species to migrate in response to warming (upslope) versus cooling (downslope).

Downslope migration during cold events can be relatively rapid for fluvially dispersed species, with seeds that can withstand immersion in water for days or weeks. Seeds of many species are spread downslope by rivers and their ranges could expand quickly through a form of jump dispersal (Pielou 1979), but only as conditions cool sufficiently can cold-adapted species compete in the lowlands. In contrast, animal-dispersed species expand their ranges incrementally (Figure 11.1). During cooling events, many low-elevation species see no benefit, as they are not cold requiring, although some biotic interactions would be altered, creating a different competitive balance of "winners" and "losers." Consequently, past cooling enabled some montane species (e.g., *Hedyosmum*, *Podocarpus*, *Myrsine*) to expand their presence in the lowlands (Colinvaux et al. 2000), but the real "winners" during these times were probably lowland species favored by the new conditions. In other

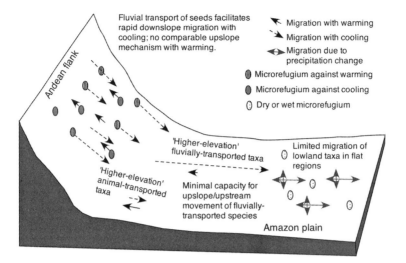

Figure 11.1. Selected factors influencing rates of plant migration following climate change. An elevational profile of the Andes and Amazon lowlands. Arrows represent the probable speed of seed transport. Microrefugia can increase the capacity of species or genotypes to expand rapidly. As temperature change is the strongest driver of migration in montane areas, the microrefugia would reflect unusual thermal microclimates. In the lowlands where soil moisture availability is more likely to influence the success of species, microrefugia would be primarily edaphically or hydrologically driven. Seeds of thermally inhibited (cold tolerant) species are transported by rivers downslope continuously, but establish only during cool periods. Fluvial upslope transport is not available, and therefore species cannot migrate upslope or must be moved by animals. Lowland taxa have the same dispersal constraints as the upslope counterparts but lack strong, directional abiotic gradients to facilitate migration. The net effect is that species with different dispersal traits have differential probabilities of dispersal, and downslope migration akin to jump dispersal in response to cooling is more likely than rapid upslope migration in response to warming.

words, the hyperdominants of the glacial forest were probably still lowland species, but not necessarily the same ones as today.

During warm events, upslope migration relies heavily on mammals and birds, and is probably much slower than downslope expansions. For the great majority of trees in the lowlands of Amazonia there is no escape from changing temperature. An insufficient gradient of temperature exists to favor survival at one edge of a range, and there is no escape using macrotopography, as the vast Amazon plain is so flat (Bush and Colinvaux 1990). As species ranges contracted upslope during warming, pockets of outlying individuals were left behind in climatic microrefugia (e.g., where a cool spring created a cool microclimate). Where greater ranges in topography existed, such as in the high Andes, steep slopes may have been a key predictor of microrefugial locations as they trapped moisture and inhibited the spread of fire (Valencia et al. 2016).

In lowland Amazonia, there is currently no evidence showing that trees migrated to warmer areas during glacial cooling. Certainly, forests were enriched by the cool-tolerant taxa, but there is no evidence of species loss due to cooling. Probably the temperature changes were within the range of evolutionary experience and hence not unduly threatening. Responses to moisture changes may have been more pressing. Amazonia almost certainly remained forested during Quaternary ice ages and interglacials, although the forest was compositionally, and perhaps structurally, different from those of today (Mayle et al. 2004). Species stressed by the glacial cooling, or by wet or dry events within the past, may simply have become rarer or more range restricted. For example, during times of drought, when low CO_2 concentrations would have increased drought stress, the most sensitive species may have been restricted to micro-refugia in low-lying wet areas or along watercourses (Rull 2009). Such microrefugia are increasingly being seen as vitally important to maintaining high biodiversity (McGlone and Clark 2005; Collins et al. 2013). As a consequence of local migration, microrefugia may have supported population fragmentation, local selection, and survival of particular genotypes. Population genetics shows that the legacies of microrefugia persist in genetically identifiable subpopulations for thousands of years after the amelioration of climatic conditions enabled the species to expand its range (McLachlan and Clark 2005). Estimates of range expansion for many tree species provide unrealistically high required rates of migration unless microrefugia are invoked to reduce migratory distances (Clark et al. 1998; Correa-Metrio et al. 2013). Under more extreme isolation and selection, fragmented populations could evolve in isolation to become a separate species. Thus, microrefugia contribute to patchiness of species abundances and to heterogeneity at both long and short temporal scales (Mosblech et al. 2011).

Colwell (2008) has suggested that because the temperature range in the past 2 million years was effectively $+1°C–2°C$ and $-5°C$ relative to modern in Amazonia, the upper temperature threshold of prior evolutionary experience is approaching. With more warming and limited options for migration, there may well be a temperature threshold beyond which there could be a wave of plant extinctions. The more immediate threat will come from climate destabilization as well as increased probability of drought and forest-changing fire, coupled with human land use.

THE PATH TO HOMOGENIZATION

People tend to simplify landscapes, introducing domesticated plants and animals, promoting fire, and altering soils and drainage to suit their needs (Kareiva et al. 2007).

Forests are emptied of wildlife by hunting, which reduces the dispersal ability of large-seeded plants and ultimately leads to altered forest composition and structure (Terborgh 2001; Peres et al. 2016).

A classic example of a synergistic climate-human mediated simplification of systems is the replacement of the Pleistocene megafauna. About 57 genera of megafauna went extinct at the end of the last ice age in South America (Barnosky and Lindsey 2010). Recent research in the Andes indicates that at least some of the extinctions were initiated by climate change (Rozas-Davila et al. 2016), although the final blows were by humans. The timing of apparent extinction differs between sites, ranging from ~16,000 years ago in the Peruvian Andes (Rozas-Davila et al. 2016) to 12,800 years ago in southeastern Brazil (Raczka et al. 2018) and about 12,500 years ago in Patagonia (Metcalf et al. 2016). The role of seed dispersers, grazers, and browsers—a role once performed by those megafauna—is now filled by people, horses, and cows. North America lost slightly fewer genera of megafauna than South America, but the species were for the most part quite different. Yet the replacement animals that today fulfill their roles are again humans, horses, and cows. The same story can be told of Australia, and even of Europe. Plants did not experience such a cataclysmic extinction at the end of the last ice age, but the homogenization of floras is evident worldwide as a result of human activity. On the once endemic-dominated Galapagos Islands, exotic plant species now outnumber native ones (Toral-Granda et al. 2017). In the Andes, unique montane forests have been replaced by planted stands of *Pinus* and *Eucalyptus*. In both the uplands and lowlands of South America, European and African grasses are replacing neotropical grasses. At continental and regional scales, humans continue to simplify and homogenize ecosystems.

A recent debate has focused on the role of pre-Columbian peoples in altering Amazonian landscapes. Some archaeologists and historical ecologists have argued that Amazonian ecosystems were so modified that they should be regarded as manufactured landscapes (e.g., Heckenberger et al. 2003; Erickson 2008; Roosevelt 2013; Levis et al. 2017). However, empirical data from >400 randomly distributed soil pits in western Amazonia did not produce a single example of a modified soil or a human artifact (McMichael et al. 2012). In light of the archaeological and paleoecological data it seems likely that while major river channels supported substantial but patchy human populations, the inferred disturbance should not be extrapolated to include areas much more than a day's walk from those centers of occupation (Piperno et al. 2015; McMichael et al. 2017).

The debate has generated two very different conclusions about the resilience of Amazonian forests (Bush and Silman 2007). On the one hand, if the forests are a manufactured landscape and are essentially a product of human disturbance, then they should be expected to be resilient to modern disturbance. Past human activities included forest clearance and burning, which would have opened the system to nutrient loss and extensive hunting. If humans occupied these forests in the past at the high densities advocated by some, then the forests would have been game poor. The empty-forest syndrome in which the trees remained but the animals were hunted out (Redford 1992) could easily have prevailed. According to this view, the forests will regrow to their present diversity even if their food webs are shattered and trees cut down. The more conservative view is that resilience in these systems comes from a closed nutrient system and relatively intact faunas that are sensitive to, and altered by, fire and human disturbance.

The dichotomy extends to the policy and conservation implications of these views. In the former, where resilience is assumed, it could lead to a policy advocating forest clearance in the expectation of regrowth. If this view is correct, all would be well; if it is incorrect, then the forest would be used and lost through misunderstanding. If the more conservative view of disturbance being inimical to forest persistence is correct, then the forest should be conserved. Should this view prove incorrect, the worst that could happen is that the forest be conserved for later use when an informed decision could be made regarding its loss (Bush and Silman 2007; Bush et al. 2015).

LINEAR AND NONLINEAR RESPONSES TO CLIMATE CHANGE

Cascading ecological change is a natural state following any disturbance, whether climatic or human-induced (the severity of which often determines the scale of the response). In general, recovery from a disturbance is broadly predictable, with species going through a recognizable succession. Similarly, response to climate change is highly predictable when the response is driven by abiotic factors, such as temperature gradients. Where responses become less predictable is when the primary control on the niche of the species is biotic rather than abiotic (Schemske et al. 2009). In natural tropical systems, the importance of biotic factors influencing local diversity often exceeds that of abiotic factors. Consequently, for tropical species, the realized niche of a species may be much smaller than its fundamental niche. If that is the case, then the potential for counterintuitive responses to climate change would be greatly increased; that is, the realized niche could be reshaped in any direction as a competitor or predator responds to climate forcing or to an alteration in its own biotic controls. The formation of no-analogue assemblages in which novel combinations of species co-occur has been shown to have occurred during past climate change and can be expected in the future.

Positive feedback mechanisms can accelerate change, inducing disproportionately large responses to a forcing. An example may be the Bolivian Altiplano where warming during the last interglacial caused abiotic factors to operate in a nonlinear manner through the positive feedback of lake shrinkage and changing microclimates (Bush et al. 2010). With the onset of the interglacial, warming initially caused a retreat of ice caps and an upslope expansion of tree species. Instead of trees continuing to move upslope as the warming progressed, however, a tipping point was reached, when evaporative loss progressively lowered the lake level. As the lake shrank, it no longer supported evaporation-driven convective rains and a moist microclimate. The drier conditions halted the upslope march of trees, and the landscape became saltbush dominated.

Another example of positive feedbacks fueled modeled projections of an "Amazonian dieback" in which 80 percent of Amazonian forests were replaced with savanna or bare ground in the coming 50–100 years (Betts et al. 2004). More recent papers have recognized a weakness in the precipitation estimates of the initial model, and the dire projections have been retracted (Huntingford et al. 2013). While the original model, sensu stricto, may have exaggerated impacts of reduced precipitation, the possibility that a combination of human-mediated land clearance and fire activity coupled with a destabilized climate could induce the predicted pattern is very real. The potential of deforestation to degrade the hydrological cycle could induce feedbacks that increase the likelihood of regional fires (Salati and Vose 1984; De Faria et al. 2017). Even if the result is not savannization of the landscape, the dispersal of humans ever deeper into Amazonian landscapes produces a different modification to landscape than any produced by pre-Columbian societies. Grazing animals, African grasses, and weeds from all over the world, paired with defaunation of forests, will homogenize systems and reduce biodiversity (Solar et al. 2015). Here the positive feedback mechanism is one of new means of access begetting settlement, which promotes more road building and yet more change (Nepstad et al. 2001; Barber et al. 2014). People introduce fire either accidentally or deliberately, and during times of drought, those fires transform forest ecology. Thus, the biodiversity risk to Amazonia comes from both government policies that encourage occupation of forest areas and periodic drought rather than from a simple protracted reduction in precipitation.

CONCLUSIONS

For almost half a century a discussion has centered on whether temperature, atmospheric CO_2 fluctuations, or precipitation shapes Amazonian forests during climatic cycles. More recently, a focus on microrefugia has emphasized the individualistic nature of species' responses. While the species driven upslope and downslope by glacial-interglacial temperature changes would have thermal microrefugia, lowland taxa would probably have been challenged to survive changes in soil moisture availability rather than temperature. Not all taxa are equally able to migrate, with waterborne tree taxa able to disperse faster than animal-dispersed taxa. The expansion of species during periods of cool (downslope) migration would be expected to be faster than during warm (upslope) migrations because rivers flow downslope.

The standard response of migrations poleward or upslope in response to future warming may have limited applicability to the tropics. Migration is certainly possible in mountainous areas, where maintaining vertical corridors to facilitate that movement should be a conservation priority. Microrefugia are probably important to the long-term survival of many species in both mountainous and lowland landscapes, and to the extent that these settings are identifiable, they, too, should be priorities for conservation. In the great flatlands of the Amazon Basin, microrefugia are probably more important than migration in species survival. Here the clash between the probable location of microrefugia along watercourses that are also preferred sites for human habitation may result in a direct conservation conflict.

Although Amazonian forests have withstood millennia of climate change, the new threat posed is not from climate change alone but from the synergy of future megadroughts coupled with human activity. Unprecedented numbers of people, more widely distributed than ever before, will increase forest flammability through land-use changes and climate volatility, and will thus provide new ignition sources. Human activity homogenizes landscapes and ultimately leads to the loss of biodiversity. While the scale of pre-Columbian human occupation is still debated, all would agree that continued modern ecosystem replacement with exotic crops such as soybeans and African grasses will lead to an unparalleled loss of biodiversity and a completely unsustainable use of the landscape.

REFERENCES

Abbott, M. B., B. B. Wolfe, A. P. Wolfe, G. O. Seltzer, R. Aravena, B. G. Mark, P. J. Polissar, D. T. Rodbell, H. D. Rowe, and M. Vuille. 2003. "Holocene paleohydrology and glacial history of the central Andes using multiproxy lake sediment studies." *Palaeogeography, Palaeoclimatology, Palaeoecology* 194: 123–138.

Absy, M. L., A. Clief, M. Fournier, L. Martin, M. Servant, A. Sifeddine, F. d. Silva, F. Soubiès, K. T. Suguio, and T. van der Hammen. 1991. "Mise en évidence de quatre phases d'ouverture de la forêt dense dans le sud-est de L'Amazonie au cours des 60,000 dernières années: Première comparaison avec d'autres régions tropicales." *Comptes rendus Academie des Sciences Paris*, ser. II, 312: 673–678.

Aragão, L. E. O. C., and Y. E. Shimabukuro. 2010. "The incidence of fire in Amazonian forests with implications for REDD." *Science* 328: 1275–1278.

Baker, P. A., C. A. Rigsby, G. O. Seltzer, S. C. Fritz, T. K. Lowenstein, N. P. Bacher, and C. Veliz. 2001. "Tropical climate changes at millennial and orbital timescales on the Bolivian Altiplano." *Nature* 409: 698–701.

Barber, C. P., M. A. Cochrane, C. M. Souza Jr., and W. F. Laurance. 2014. "Roads, deforestation, and the miti-

gating effect of protected areas in the Amazon." *Biological Conservation* 177: 203–209.

Barlow, J., et al. 2002. "Effects of ground fires on understorey bird assemblages in Amazonian forests." *Biological Conservation* 105: 157–169.

Barlow, J., C. A. Peres, B. O. Lagan, and T. Haugaasen. 2003. "Large tree mortality and the decline of forest biomass following Amazonian wildfires." *Ecology Letters* 6: 6–8.

Barnosky, A. D., and E. L. Lindsey. 2010. "Timing of Quaternary megafaunal extinction in South America in relation to human arrival and climate change." *Quaternary International* 217: 10–29.

Betts, R. A., P. M. Cox, M. Collins, P. P. Harris, C. Huntingford, and C. D. Jones. 2004. "The role of ecosystem-atmosphere interactions in simulated Amazonian precipitation decrease and forest dieback under global climate warming." *Theoretical and Applied Climatology* 78: 157–175.

Bond, G., W. Broecker, S. Johnsen, J. McManus, L. Labeyrie, J. Jouzel, and G. Bonani. 1993. "Correlations between climate records from the North Atlantic sediments and Greenland Ice." *Nature* 365: 143–147.

Bush, M. B. 2002. "Distributional change and conservation on the Andean flank: A palaeoecological perspective." *Global Ecology and Biogeography* 11: 463–467.

Bush, M. B., and P. A. Colinvaux. 1990. "A long record of climatic and vegetation change in lowland Panama." *Journal of Vegetation Science* 1: 105–119.

Bush, M. B., P. E. De Oliveira, P. A. Colinvaux, M. C. Miller, and E. Moreno. 2004. "Amazonian paleoecological histories: One hill, three watersheds." *Palaeogeography, Palaeoclimatology, Palaeoecology* 214: 359–393.

Bush, M. B., J. A. Hanselman, and W. D. Gosling. 2010. "Non-linear climate change and Andean feedbacks: An imminent turning point?" *Global Change Biology* 16: 3223–3232.

Bush, M. B., C. N. H. McMichael, D. R. Piperno, M. R. Silman, J. Barlow, C. A. Peres, M. Power, and M. W. Palace. 2015. "Anthropogenic influence on Amazonian forests in pre-history: An ecological perspective." *Journal of Biogeography* 42: 2277–2288.

Bush, M. B., and C. N. H. McMichael. 2016. "Holocene variability of an Amazonian hyperdominant." *Journal of Ecology* 104: 1370–1378.

Bush, M. B., M. C. Miller, P. E. de Oliveira, and P. A. Colinvaux. 2002. "Orbital forcing signal in sediments of two Amazonian lakes." *Journal of Paleolimnology* 27: 341–352.

Bush, M. B., and M. R. Silman. 2004. "Observations on Late Pleistocene cooling and precipitation in the lowland neotropics." *Journal of Quaternary Science* 19: 677–684.

Bush, M. B., and M. R. Silman. 2007. "Amazonian exploitation revisited: Ecological asymmetry and the policy pendulum." *Frontiers in Ecology and the Environment* 5: 457–465.

Bush, M. B., M. R. Silman, M. B. de Toledo, C. R. S. Listopad, W. D. Gosling, C. Williams, P. E. de Oliveira, and C. Krisel. 2007. "Holocene fire and occupation in Amazonia: Records from two lake districts." *Philosophical Transactions of the Royal Society of London B* 362: 209–218.

Bush, M. B., M. R. Silman, and D. H. Urrego. 2004. "48,000 years of climate and forest change from a biodiversity hotspot." *Science* 303: 827–829.

Clark, J. S., C. Fastie, G. Hurtt, S. T. Jackson, C. Johnson, G. A. King, M. Lewis, J. Lynch, S. Pacala, C. Prentice, E. W. Schupp, T. I. Webb, and P. Wyckoff. 1998. "Reid's paradox of rapid plant migration." *BioScience* 48: 13–24.

Colinvaux, P., P. De Oliveira, and M. Bush. 2000. "Amazonian and neotropical plant communities on glacial time-scales: The failure of the aridity and refuge hypotheses." *Quaternary Science Reviews* 19: 141–169.

Colinvaux, P., G. Irion, M. Räsänen, M. Bush, and J. N. De Mello. 2001. "A paradigm to be discarded: Geological and paleoecological data falsify the Haffer & Prance refuge hypothesis of Amazonian speciation." *Amazoniana* 16: 609–646.

Collins, A. F., M. B. Bush, and J. P. Sachs. 2013. "Microrefugia and species persistence in the Galápagos highlands: A 26,000-year paleoecological perspective." *Frontiers in Genetics* 4. https://doi.org/10.3389/fgene.2013.00269.

Colwell, R. K., G. Brehm, C. L. Cardelus, A. C. Gilman, and J. T. Longino. 2008. "Global warming, elevational range shifts, and lowland biotic attrition in the wet tropics." *Science* 322: 258–261.

Correa-Metrio, A., M. Bush, S. Lozano-García, and S. Sosa-Nájera. 2013. "Millennial-scale temperature change velocity in the continental northern neotropics." *PLOS One* 8: e81958.

Cowling, S. A. 2004. "Tropical forest structure: A missing dimension to Pleistocene landscapes." *Journal of Quaternary Science* 19: 733–743.

Cruz, F., M. Vuille, S. J. Burns, X. Wang, H. Cheng, M. Werner, R. L. Edwards, I. Karmann, A. S. Auler, and H. Nguyen. 2009. "Orbitally driven east-west antiphasing of South American precipitation." *Nature Geoscience* 2: 210–214.

Cruz Jr., F. W., S. J. Burns, I. Karmann, W. D. Sharp, M. Vuille, A. O. Cardoso, J. A. Ferrari, P. L. Silva Dias, and O. Viana Jr. 2005. "Insolation-driven changes in atmospheric circulation over the past 116,000 years in subtropical Brazil." *Nature* 434: 63–66.

De Faria, B. L., P. M. Brando, M. N. Macedo, P. K. Panday, B. S. Soares-Filho, and M. T. Coe. 2017. "Current and future patterns of fire-induced forest degradation in Amazonia." *Environmental Research Letters* 12: 095005.

Dillehay, T. D. 2009. "Probing deeper into first American studies." *Proceedings of the National Academy of Sciences* 106: 971–978.

Erickson, C. L. 2008. "Amazonia: The historical ecology of a domesticated landscape." In *The Handbook of South American Archaeology*, ed. H. Silverman and W. H. Isbell, 157–183. Springer.

Fritz, S. C., P. A. Baker, G. O. Seltzer, A. Ballantyne, P. M. Tapia, H. Cheng, and R. L. Edwards. 2007. "Quaternary glaciation and hydrologic variation in the South American tropics as reconstructed from the Lake Titicaca drilling project." *Quaternary Research* 68: 410–420.

Haffer, J. 1969. "Speciation in Amazonian forest birds." *Science* 165: 131–137.

Heckenberger, M. J., A. Kuikuro, U. T. Kuikuro, J. C. Russell, M. Schmidt, C. Fausto, and B. Franchetto. 2003. "Amazonia 1492: Pristine forest or cultural parkland?" *Science* 301: 1710–1714.

Hillyer, R., B. G. Valencia, M. B. Bush, M. R. Silman, and M. Steinitz-Kannan. 2009. "A 24,700-year paleolimnological history from the Peruvian Andes." *Quaternary Research* 71: 71–82.

Hooghiemstra, H. 1984. "Vegetational and climatic history of the high plain of Bogota, Colombia." *Dissertaciones Botanicae* 79. J. Cramer.

Hubbell, S. P. 2001. *The Unified Neutral Theory of Biodiversity and Biogeography*. Princeton University Press.

Huntingford, C., P. Zelazowski, D. Galbraith, L. M. Mercado, S. Sitch, R. Fisher, M. Lomas, et al. 2013. "Simulated resilience of tropical rainforests to CO_2-induced climate change." *Nature Geoscience* 6: 268–273.

Kareiva, P., S. Watts, R. McDonald, and T. Boucher. 2007. "Domesticated nature: Shaping landscapes and ecosystems for human welfare." *Science* 316: 1866–1869.

Levis, C., F. R. Costa, F. Bongers, M. Peña-Claros, C. R. Clement, A. B. Junqueira, E. G. Neves, E. K. Tamanaha, F. O. Figueiredo, and R. P. Salomão. 2017. "Persistent effects of pre-Columbian plant domestication on Amazonian forest composition." *Science* 355: 925–931.

Marengo, J. A., C. A. Nobre, J. Tomasella, M. F. Cardoso, and M. D. Oyama. 2008. "Hydro-climatic and ecological behaviour of the drought of Amazonia in 2005." *Philosophical Transactions of the Royal Society B: Biological Sciences* 363: 1773–1778.

Marengo, J. A., J. Tomasella, L. M. Alves, W. R. Soares, and D. A. Rodriguez. 2011. "The drought of 2010 in the context of historical droughts in the Amazon region." *Geophysical Research Letters* 38: L12703.

Mayle, F., R. Burbridge, and T. Killeen. 2000. "Millennial-scale dynamics of southern Amazonian rain forests." *Science* 290: 2291.

Mayle, F. E., D. J. Beerling, W. D. Gosling, and M. B. Bush. 2004. "Responses of Amazonian ecosystems to climatic and atmospheric CO_2 changes since the Last Glacial Maximum." *Philosophical Transactions of the Royal Society of London B* 359: 499–514.

McGlone, M. S., and J. S. Clark. 2005. "Microrefugia and Macroecology." In *Climate Change and Biodiversity*, ed. T. Lovejoy and L. Hannah, 157–159. Yale University Press.

McLachlan, J. S., and J. S. Clark. 2005. "Molecular indicators of tree migration capacity under rapid climate change." *Ecology* 86: 2088–2098.

McMichael, C. H., M. B. Bush, M. R. Silman, D. R. Piperno, M. Raczka, L. C. Lobato, M. Zimmerman, S. Hagen, and M. Palace. 2013. "Historical fire and bamboo dynamics in western Amazonia." *Journal of Biogeography* 40: 299–309.

McMichael, C. H., D. R. Piperno, and M. B. Bush. 2017. "Response to Levis et al.: Persistent effects of pre-Columbian plant domestication on Amazonian forest composition." *Science* 358. https://doi.org/10.1126/science.aan8347.

McMichael, C., D. R. Piperno, M. B. Bush, M. R. Silman, A. R. Zimmerman, M. F. Raczka, and L. C. Lobato. 2012. "Sparse pre-Columbian human habitation in western Amazonia." *Science* 336: 1429–1431.

Metcalf, J. L., C. Turney, R. Barnett, F. Martin, S. C. Bray, J. T. Vilstrup, L. Orlando, R. Salas-Gismondi, D. Loponte, M. Medina, et al. 2016. "Synergistic roles of climate warming and human occupation in Patagonian megafaunal extinctions during the Last Deglaciation." *Science Advances* 2: e1501682.

Mosblech, N. A., M. B. Bush, W. D. Gosling, D. Hodell, L. Thomas, P. van Calsteren, A. Correa-Metrio, B. G. Valencia, J. Curtis, and R. van Woesik. 2012. "North Atlantic forcing of Amazonian precipitation during the last ice age." *Nature Geoscience* 5: 817–820.

Mosblech, N. A., M. B. Bush, and R. Van Woesik. 2011. "On metapopulations and microrefugia: Palaeoecological insights." *Journal of Biogeography* 38: 419–429.

Nepstad, D., G. Carvalho, A. C. Barros, A. Alencar, J. P. Capobianco, J. Bishop, P. Moutinho, P. Lefebvre, U. L. Silva, and E. Prins. 2001. "Road paving, fire regime feedbacks, and the future of Amazon forests." *Forest Ecology and Management* 154: 395–407.

NGRIP. 2004. "High-resolution record of Northern Hemisphere climate extending into the last interglacial period." *Nature* 431: 147–151.

Núñez, L., M. Grosjean, and I. Cartajena. 2001. "Human dimensions of late Pleistocene/Holocene arid events in southern South America." In *Interhemispheric Climate Linkages*, ed. M. Vera, 105–117. Academic Press.

Overpeck, J. T., T. I. Webb, and I. C. Prentice. 1985. "Quantitative interpretation of fossil pollen spectra: Dissimilarity coefficients and the method of modern analogs." *Quaternary Research* 23: 87–708.

Peres, C. A., T. Emilio, J. Schietti, S. J. Desmoulière, and T. Levi. 2016. "Dispersal limitation induces long-term biomass collapse in overhunted Amazonian forests." *Proceedings of the National Academy of Sciences* 113: 892–897.

Pielou, E. C. 1979. *Biogeography*. Wiley.

Piperno, D. R., C. McMichael, and M. B. Bush. 2015. "Amazonia and the Anthropocene: What was the spatial extent and intensity of human landscape modification in the Amazon Basin at the end of prehistory?" *Holocene* 25: 1588–1597.

Pitman, N. C. A., J. W. Terborgh, M. R. Silman, P. V. Nunez, D. A. Neill, C. E. Cerón, W. E. Palacios, and M. Aulestia. 2001. "Dominance and distribution of tree species in upper Amazonian terra firme forests." *Ecology* 82: 2101–2117.

Raczka, M. F., P. E. De Oliveira, M. B. Bush, and C. H. McMichael. 2013. "Two paleoecological histories spanning the period of human settlement in southeastern Brazil." *Journal of Quaternary Science* 28: 144–151.

Raczka, M. F., P. D. De Oliveira, and M. B. Bush. 2018. "Megafaunal extinction in south-eastern Brazil." *Quaternary Research* 89: 103–118.

Redford, K. 1992. "The empty forest." *BioScience* 42: 412–422.

Richards, P. W. 1996. *The Tropical Rain Forest*. 2nd ed. Cambridge University Press.

Roosevelt, A. C. 2013. "The Amazon and the Anthropocene: 13,000 years of human influence in a tropical rainforest." *Anthropocene* 4: 69–87.

Rozas-Davila, A., B. G. Valencia, and M. B. Bush. 2016. "The functional extinction of Andean megafauna." *Ecology* 97: 2533–2539.

Rull, V. 2009. "Microrefugia." *Journal of Biogeography* 36: 481–484.

Rull, V., and E. Montoya. 2014. "*Mauritia flexuosa* palm swamp communities: Natural or human-made? A palynological study of the Gran Sabana region (northern South America) within a neotropical context." *Quaternary Science Reviews* 99: 17–33.

Salati, E., and P. B. Vose. 1984. "Amazon basin: A system in equilibrium." *Science* 225: 129–138.

Sato, H., and S. A. Cowling. 2017. "Glacial Amazonia at the canopy-scale: Using a biophysical model to understand forest robustness." *Quaternary Science Reviews* 171: 38–47.

Schemske, D. W., G. G. Mittelbach, H. V. Cornell, J. M. Sobel, and K. Roy. 2009. "Is there a latitudinal gradient in the importance of biotic interactions?" *Annual Review of Ecology, Evolution, and Systematics* 40: 245–269.

Seltzer, G. O., S. Cross, P. Baker, R. Dunbar, and S. Fritz. 1998. "High-resolution seismic reflection profiles from Lake Titicaca, Peru/Bolivia: Evidence for Holocene aridity in the tropical Andes." *Geology* 26: 167–170.

Silman, M. R. 2011. "Plant species diversity in Amazonia." In *Tropical Rainforests and Climate Change*, 2nd ed., ed. M. B. Bush, J. R. Flenley, and W. D. Gosling, 269–288. Praxis.

Solar, R. R. d. C., J. Barlow, J. Ferreira, E. Berenguer, A. C. Lees, J. R. Thomson, J. Louzada, M. Maués, N. G. Moura, and V. H. Oliveira. 2015. "How pervasive is biotic homogenization in human-modified tropical forest landscapes?" *Ecology Letters* 18: 1108–1118.

Stríkis, N. M., F. W. Cruz, E. A. Barreto, F. Naughton, M. Vuille, H. Cheng, A. H. Voelker, H. Zhang, I. Karmann, and R. L. Edwards. 2018. "South American monsoon response to iceberg discharge in the North Atlantic." *Proceedings of the National Academy of Sciences*: 201717784.

ter Steege, H., N. C. A. Pitman, D. Sabatier, C. Baraloto, R. P. Salomão, J. E. Guevara, O. L. Phillips, et al. 2013. "Hyperdominance in the Amazonian tree flora." *Science* 342: 1243092.

Terborgh, J., L. Lopez, P. Nuñez, M. Rao, G. Shahabuddin, G. Orihuela, M. Riveros, R. Ascanio, G. H. Adler, T. D. Lambert, and L. Balbas. 2001. "Ecological meltdown in predator-free forest fragments." *Science* 5548: 1923–1925.

Toral-Granda, M. V., C. E. Causton, H. Jäger, M. True-
man, J. C. Izurieta, E. Araújo, M. Cruz, K. K. Zander, A.
Izurieta, and S. T. Garnett. 2017. "Alien species pathways to the Galapagos Islands, Ecuador." *PLOS One* 12:
e0184379.

Valencia, B. G., F. Matthews-Bird, D. H. Urrego, J. J. Williams, W. D. Gosling, and M. Bush. 2016. "Andean
microrefugia: Testing the Holocene to predict the Anthropocene." *New Phytologist* 212: 510–522.

Van 't Veer, R., and H. Hooghiemstra. 2000. "Montane
forest evolution during the last 650,000 yr in Colombia: A multivariate approach based on pollen record
Funza-I." *Journal of Quaternary Science* 15: 329–346.

Wang, X., R. L. Edwards, A. S. Auler, H. Cheng, X. Kong,
Y. Wang, F. da Cruz, J. A. Dorale, and H.-W. Chiang.
2017. "Hydroclimate changes across the Amazon lowlands over the past 45,000 years." *Nature* 541: 204–207.

PART IV

What Does the Future Hold?

CHAPTER TWELVE

Modeling Species and Vegetation Distribution under Climate Change

PABLO IMBACH, PEP SERRA-DIAZ,
LEE HANNAH, EMILY FUNG, AND
ELIZABETH H. T. HIROYASU

Biodiversity responses to climate change include range changes, phenological changes, and genetic selection of traits to match new conditions (Bellard et al. 2012). Biogeographic approaches, usually based on modeling the biophysical envelope suitable for a species, are widely used for exploring impacts on range distributions. Dynamic vegetation modeling (DVM) provides insights into biogeochemical-physical responses, including carbon, water, nutrient cycling, and vegetation changes (Bellard et al. 2012). Earth-system modeling couples atmospheric, oceanic, and ecological processes at global scales. More recently, approaches based on functional traits have been proposed that transcend species-based biogeography (Violle et al. 2014).

WHY MODEL SPECIES AND VEGETATION DISTRIBUTIONS?

Species and vegetation are key components of the biosphere, with climate-driven changes in their distributions affecting biodiversity, water provision, carbon sequestration, and many other ecosystems services. Species are expected to move independently with climate change (Parmesan 1996), so to fully understand ecosystem change, the range changes of individual species need to be understood. Movement of species' ranges will alter plant and animal assemblages—the species that compose ecosystems will disassemble and reassemble in novel combinations in response to future climate change. Modeling of species is therefore key to understanding both species and ecosystem responses to climate change.

The general structure of ecosystems will change as result of aggregated species responses. Plant functional types—for example, evergreen needleleaf forest, deciduous shrub land, C_3 grassland—are determined by availability of water, species tolerances, temperature, and other environmental factors. These structural and functional attributes can be simulated without resolving individual species. Many ecosystem service changes can be approximated by understanding vegetation and plant functional type responses to climate change, thus providing further insights into climate-driven change.

Therefore, species and vegetation modeling together can provide a more complete view of biotic impacts that can help address specific climate change questions. Both types of models are useful in impact studies to understand the sensitivity of biodiversity to changes in climate, for example, by providing insights into changes in species location and into changes in ecosystems and their functioning. Modeling results can be incorporated into land use and conservation planning tools, such as Marxan, to aid in local or regional adaptation planning.

SPECIES MODELS

There are a wide variety of species modeling tools available to researchers and impact practitioners. These models are known as species distribution models (SDMs) or ecological niche models, but they are also referred to as bioclimatic envelope models. All of them simulate species distributions under current or changed climates (hence "species distribution models"), and species distributions are determined by the fundamental climatic tolerances, or niche of species (hence "ecological niche models" [ENM]). The term *SDM* is used in this chapter, but ENM is generally synonymous.

All SDMs are based on the assumption that current distributions of species provide information about the climatic conditions under which the species' populations can survive and reproduce (the "fundamental niche"), and few directly address competitive interactions or other limitations that create the realized niche of a species (Pearman et al. 2008).

Among the earliest of such models is BioClim (Nix 1986), a bioclimatic envelope approach that uses simple correlative methods to assess the range of conditions suitable for a species. Other, more recent additions to the species modeling quiver include tools with more sophisticated statistical treatment of species-climate relationships. These include modeling tools such as GARP (genetic algorithm for rule set production) and MaxEnt (a machine learning model based on maximum entropy theory), and statistical tools such as GLM (generalized linear modeling), GDM (generalized dissimilarity modeling), GAM (generalized additive modeling), and others. MaxEnt is the most widely used SDM and has performed relatively well in comparative tests (Elith et al. 2011).

ENSEMBLE MODELING TO DEAL WITH UNCERTAINTY

Dealing with uncertainty is a major issue in SDM applications. Tests have shown that SDM uncertainty may be greater than general circulation model (GCM) uncertainty for future climate scenarios. Sources of uncertainty include incomplete or biased species observation data and variation in number or type of climate variables used in model parameterization.

One technique that has been shown to reduce uncertainty in biodiversity impact studies is ensemble modeling. Ensemble approaches use a large number of combinations of GCMs and SDM algorithms to assess the impacts of climate change (Araújo and New 2007; Figure 12.1). To achieve a sound result, it is important to drive each SDM with each individual GCM, rather than averaging the GCM outputs and using that

Box 12.1   Modeling Species Distribution Ranges: Practical Issues

Data availability and sampling size: A large proportion of the world's biodiversity has not been correctly identified, increasing uncertainty in model outputs (Cayuela et al. 2009). Limited availability of species occurrence records also affects model performance and output applications (Cayuela et al. 2009). In some cases data paucity occurs for rare or endemic species as a result of sampling difficulties; in others, data are not yet electronically available (Graham et al. 2004). In general, a small sample size (<30 occurrence points) may affect model performance due to poor representation of the environmental complexity of the species range distribution and a limited capacity to detect outlier values (Wisz et al. 2008). There is no agreement on minimum sampling size. Some authors suggest that a sampling size of fewer than 70 observations decreases model performance (Kadmon et al. 2003), whereas others propose using no fewer than 30–40 species occurrences (Wisz et al. 2008).

Data quality (time and space bias): Many studies rely on herbaria collections and natural history data sets to obtain observation records; nonetheless, in many cases species occurrence points are time and space biased with reference to the real species range distribution (Bean et al. 2012). For example, occurrence points cover a particular area of the total range or species, or sampling covers particular periods of their life cycle (i.e., reproductive or dispersal) with smaller range distributions (Cayuela et al. 2009). Ideally, sampling efforts should include systematic surveys along the entire species range distributions to capture the full environmental gradient. Absence data are not always available, but they are highly desirable for better describing the range of conditions suitable for a species. However, algorithms exist that can fit models by generating pseudoabsences or comparing the presence distribution to the background environment (Jarnevich et al. 2015).

Threshold selection: Many SDM studies provide outputs in binary formats (0 or 1 values) depicting suitable (potentially present) or unsuitable (potentially absent) areas for a species. Binary outputs are estimated from suitability maps of continuous values using a threshold value to define presence and absence (Bean et al. 2012). Many threshold selection approaches exist and have been evaluated under biased and limited occurrence data (Liu et al. 2005; Bean et al. 2012); however, the selection of an appropriate threshold is finally case specific and depends on the objective of the study (Liu et al. 2005).

Finally, some SDM methods require considerable data preparation, model parameterization, and evaluation to obtain accurate results. Tools have been developed to simplify their implementation; for example, the SDM toolbox (Brown 2014) and ensemble modeling packages (Warren et al. 2010; Ranjitkar et al. 2014) facilitate data processing and intermodel comparisons (Thuiller et al. 2014).

---

to drive the SDMs. This is because change in climate parameters such as temperature and precipitation are not independent in a specific GCM simulation. The combined effects of multiple climate variables must be maintained when driving the SDMs, and then subsequently the SDM outputs are combined to create an ensemble.

Ensemble approaches are most suitable when there is no evidence that any one model performs better than others. However, model intercomparison has shown

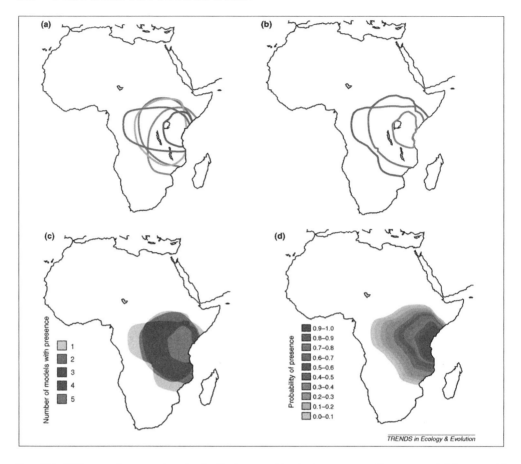

Figure 12.1. Different options to analyze the distribution range of a species under future climate as a result of an ensemble modeling approach (based on artificial data for Africa and hypothetical models). (A) results from each bioclimatic model output are represented by each line, (B) mapping areas suitable for at least one model (larger polygon), all models (smaller polygon) and areas where at least half of the models indicate suitable areas (medium polygon), (C) map indicating the number of models simulating suitable areas for a species, and (D) the likelihood of species presence in the future based on a probability density function. (From Araújo and New 2007.)

that MaxEnt and GDM perform better than other models in standard species modeling tasks (Elith et al. 2006). Where data suggest that some models have stronger performance, it is preferable to limit the ensemble to those stronger-performing models. This is particularly true if using all available models means that default settings and in-discriminate sets of input variables would be used.

WHAT HAVE WE LEARNED FROM SDMS IN THE CONTEXT OF GLOBAL CHANGE?

Species distribution models have provided a wide range of projections of future range shifts under global change. Maps of potential species distributions have been produced for plants, vertebrates, invertebrates, and freshwater and marine species. SDM projections in the terrestrial realm have generally emphasized poleward and upward range shifts—the predicted fingerprint of climate change on species distributions in a warming world. Insights from SDMs have also highlighted counterintuitive range dynamics. In Australia, an analysis of 464 bird species showed equatorward and multidirectional shifts in addition to poleward shifts (VanDerWal et al. 2013). Wisz et al. (2015) described a potential interconnection of marine fish species through the Arctic. Their simulation indicated that 41 and 44 new species might enter the Pacific and the Atlantic, respectively, predominantly east-west range expansions due to the breakdown of a northern (cold) thermal barrier. Thus, SDMs have helped identify potential novel species assemblages.

Extinction risk from climate change has been estimated using suites of SDM simulations representing multiple taxa and multiple geographic regions (Thomas et al. 2004; see Midgley and Hannah in this volume). These estimates show major extinction risk vulnerability associated with climate change and have been influential in national and international climate change policy discussions. Initial estimates that 18 percent–34 percent of all species (midrange) might be at risk of extinction due to climate change have been criticized on methodological grounds. However, subsequent analyses incorporating the improved methods have largely confirmed these ballpark figures, both using SDMs (Malcolm et al. 2006) and non-SDM modeling (Sinervo et al. 2010).

Conservation outcomes have been explored using SDMs, including the effectiveness of protected areas and protected areas networks. Climate-driven range shifts have been found to cause some species to move out of reserves, effectively reducing species representation in protected areas (Araújo 2011). Additions of new protected areas have been found to be useful in counteracting this effect, with area increases of as little as 5 percent–10 percent in national protected areas networks restoring the losses due to climate change (Hannah et al. 2007).

Large-scale analyses of biodiversity have multiplied in response to improving species occurrence data in museum and global data repositories, especially the Global Biodiversity Information Facility (GBIF). As the data on species, global circulation models, and new algorithms for SDMs are growing, so are the multiple potential projections that can be derived.

SDM climate change projections face limitations. Dispersal capacity is not accounted for, yet varies widely among species and is critical in determining species' ability to occupy newly suitable range. To account for dispersal effects, studies using SDM projections have often used no-dispersal and full-dispersal scenarios. That is, scenarios assuming either that the species is not able to disperse to new suitable climate (no dispersal) or that the species is able to colonize any new suitable climate regardless of the distance to current suitable habitat (full dispersal). These two contrasting scenarios can provide a bounding box to gauge the importance of species dispersal on range shifts. More sophisticated methods have been proposed in which SDM suitability maps are produced at different time steps, and are combined with a dispersal algorithms (Engler and Guisan 2009; Midgley et al. 2010). Approaches incorporating dispersal have shown the importance of assembling climatic requirements to other species features to derive realistic projections of species distributions under global change. Recent approaches also explicitly consider the effects of land-use change on the output of SDMs (Beltrán et al. 2014).

New SDM approaches derive metrics related to species exposure to climate change. Several metrics can be used to characterize different features of the effects of climate change to species distribution and conservation (Figure 12.2): area metrics, population metrics, and velocity metrics (see Serra-Diaz et al. 2014).

Area Metrics

Area metrics are calculated using the relationship between current and future areas

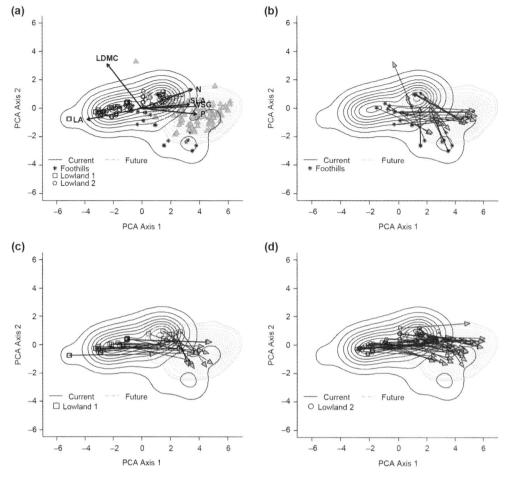

Figure 12.2. The community weighted mean (CWM) trait values, of 127 old-growth forest plots in northern Costa Rica, are ordered across two dimensions resulting from a principal component analysis (PCA). PCA dimensions reflect climatic features related to precipitation: (i) higher/lower values in PCA Axis 1 indicate higher precipitation seasonality/higher annual precipitation, and (ii) higher or lower values in PCA Axis 2 reflect higher seasonal precipitation and precipitation during the wettest month/higher annual temperature and temperature seasonality. CWM trait values were calculated for leaf area (LA), specific leaf area (SLA), leaf dry-matter content (LDMC), leaf nitrogen (N) and phosphorus (P) content, and wood basic specific gravity (WSG). No-filled and gray symbols represent forest plots under current and future climate, respectively. Black and gray contour lines represent the probability density function of forest plots across the PCA dimensions under current and future climate respectively. Arrows in (A) show the relationship between each trait across each PCA dimension. Arrows in (B, C, D) indicate shifts in CWM values induced by climate change for each plot by forest type: foothill forests (B) and two different lowland forests (C, D). (From Chain-Guadarrama et al. 2017.)

of projected suitable climate. Net area change (also called species range change) is by far the most commonly used, in which the amount of future suitable habitat is compared to the present suitable habitat. This metric reflects potential shrinkage, maintenance, or expansion of a given species under climate change. Less commonly applied but useful metrics from landscape ecology can be used to understand potential shifts in the fragmentation of suitable areas.

Population-Level Metrics

Population-level metrics assess the exposure to climate change of currently known populations of a species. Indexes include the number of populations with increasing or decreasing suitability or the number of populations for which suitability is lost.

Velocity Metrics

Velocity metrics are a set of metrics that aim at quantifying the pace of suitability changes with reference to species sensitivity. These metrics are reported in units of space per units of time (e.g., km/decade, m/year). Bioclimatic velocity is a local index that informs, for each grid cell, the velocity needed to encounter the same probability of species occurrence within a local neighborhood distance (Serra-Diaz et al. 2014; Figure 12.3). This metric is similar to velocity of climate change, except that probability of occurrence is substituted for climatic similarity. Thus, bioclimatic velocity integrates both climate aspects and species-specific sensitivities to climate. This metric has been shown to improve the estimates of range shift (Comte and Grenouillet 2015). The biotic velocity (Ordoñez and Williams 2013) is an analogue-climate based metric that measures the distance needed by a population to reach its nearest suitable site.

The use of SDM to provide estimates of species range change under climate change

has attracted critiques. The relative ease of use of the algorithms and the increasing data availability on species occurrence have boosted its use, often without full consideration of the array of initial necessary quality control in data or with poor understanding of model assumptions (Araújo and Peterson 2012). One of the most discussed assumptions of this modeling approach is that the current distribution of the species is in equilibrium with climate. We know, however, that species are still responding to climatic rebound from the last glacial period and that the long-term effects of land-use change may not yet be fully manifested in species' current distri-

Figure 12.3. Maps show the averaged bioclimatic velocity of climate change for endemic tree species in California. The bioclimatic velocity indicates a measure of the changes in climatic suitability (toward increased or decreased suitability) for a species within its potential suitable range and was estimated as the ratio of temporal to spatial gradients (magnitude of change over time and space, respectively) under climate change. The velocity was averaged for two GCMs. The inset figure shows that different velocities can occur within short distances. PIBA = Pinus balfouriana, PICO = Pinus coulteri, PIMU = Pinus muricata, PISA = Pinus sabiniana, QUDO = Quercus douglasii, QUEN = Quercus engelmannii, QULO = Quercus lobata, QUWI = Quercus wislizeni. (From Serra-Diaz et al. 2014.)

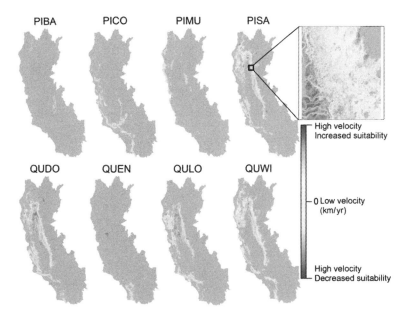

butions (Svenning and Sandel 2013). These factors limit the reliability of response curves derived from species distributions under current climate. Another critique is that SDMs are correlative rather than mechanistic, and therefore cannot resolve the effects of competition and other biotic interactions. Further, CO_2 fertilization effect may reduce water stress, making SDM projections for plants overly pessimistic (Keenan et al. 2011), or thermoregulatory behavior in ectotherms may potentially buffer species from a warmer climate (Kearney, Shine, and Porter 2009).

INCREASING COMPLEXITY: FROM SPECIES TO ECOSYSTEMS

There are an increasing number of modeling tools capable of moving beyond species modeling, to simulate biological dynamics at landscape or larger scales. These tools allow for consideration of competition, disturbance, dispersal and population dynamics. Below we briefly highlight some of the modeling approaches available to climate change biologists.

Gap Models

Gap models are used at a very fine spatial scale (0.1 ha) to examine species interactions and vegetation change (Shugart 2002). The gap model projects growth, death, and regeneration of individual trees, allowing for the investigation of how individual tree changes can impact whole-forest composition. Gap models have been used in a variety of ecosystems, though first and most notably in forest systems, where they can simulate compostion and change in forest landscapes in response to climate step function or glacial-interglacial changes or simulate dynamic mosaic patterns (Shugart 2002). Gap models may be tiled together to simulate change at landscape or larger scales.

Landscape Models

Landscape models simulate different processes—typically succession and disturbance—and the interaction between them. A widely used model is LANDIS-II, a model that integrates natural spatial processes, such as seed dispersal, and disturbances including fire, harvest, and wind (Mladenoff 2004), adapting the simulation time steps to the ecological processes, especially ecophysiological processes related to growth (Scheller et al. 2007). Climate change may be simulated in LANDIS and other landscape models by using SDM outputs as habitat suitability values. BIOMOVE is a landscape model developed specifically to address dispersal and disturbance processes associated with climate change (Midgley et al. 2010).

Adding Information on Population Processes

SDM outputs may be combined with population models for studies of population dynamics under climate change. In this approach, SDM outputs are used to define habitat patches in a population model that simulates effects of habitat disturbance, life history, and distributions (Keith et al. 2008). Integral projection models (IPMs) analyze individual population states over time to project population dynamics (Merow et al. 2014).

Species to Functional Traits

Buckley et al. (2011) have found that using functional and physiological traits of a species may be more powerful for predicting change under global warming than SDMs alone. Coupling species-specific trait physiology with SDMs, using fully mechanistic approaches or using outputs from mechanistic models as predictors in correlative models (Buckley and Kingsolver 2012), may provide more robust insights into species' response to climate change (Buckley et al. 2011). Identifying functional and phylogenetic traits that affect dispersal and response to climate change will be key to simulating more complex processes (Buckley and Kingsolver 2012). An alternative method for incorporating life-history traits is to couple a suitability model with an age–life stage model, to assess the relative impact of climate change and disturbance on the persistence of a population (Swab et al. 2012). In species that are facing significant impacts from both climate change and altered disturbance regimes, this can be a powerful tool for managers to compare threats and how they interact to produce novel outcomes for species of interest (Swab et al. 2012). BIOMOVE is one platform for such analyses (Midgley 2010).

MACROSCALE MODELING: FROM BIOGEOGRAPHY APPROACHES TO GLOBAL DYNAMIC VEGETATION MODELING

Earth-system models address effects of the land surface on climate, including interactions between atmosphere, ocean, land, ice, and biosphere, through land surface models (Levis 2010). Increasingly, dynamic global vegetation models (DGVMs) are being incorporated into Earth-system models, effectively coupling DGVMs and GCMs, replacing the older practice of driving DGVMs with GCM simulations.

Climate is affected by land surface characteristics influenced by vegetation. Albedo and roughness are affected by vegetation type, and these surface characteristics in turn have effects on plant transpiration and biosphere interactions with greenhouse gases (Levis 2010). Thus, the understanding generated by coupling DGVMs and GCMs in Earth-system models is important to our evolving understanding of climate dynamics. Earth-system models allow for continued inquiries begun with DGVMs, including efforts to understand climate change, CO_2 fertilization, fires, and land-use change

and their feedbacks on regional and global scale biogeochemical-physical processes.

Plant functional types are now commonly used to represent groups of species with similar characteristics (morphological, phenological, and physiological), an improvement over predetermined biome-based approaches (Peng 2000), that can be used to represent vegetation of a site under a range of historical, present, or future climates (Levis 2010). PFTs are usually represented by leaf and stem area, height and vegetation cover that allows simulation of photosynthesis, and respiration and transpiration given physiological plant traits. Simulating these processes helps estimate primary productivity and distribution of carbon pools in leaves, stems, roots, litter, and the soil (Levis 2010).

However, Earth-system models and DGVMs are often limited by oversimplification of vegetation types (PFTs and their parameterizations) and competition of individual plants (Yang et al. 2015). PFTs are usually parameterized with traits with constant values and assuming constant relationship between form and function under current and change climates (Wullschleger et al. 2015). Traits, however, have been found to have great variation within PFTs (van Bodegom et al. 2012). These approaches are also limited in their simulation of competition, as competition is simulated between PFTs rather than among individuals (Yang et al. 2015). Representation of processes is being improved by developing individual-based (i.e., tree) frameworks (Scheiter and Higgins 2009) and by more sophisticated simulation of seed dispersal and plant migration processes (Higgins and Harte 2006). Improving representation of vegetation structure and functions could help reduce current uncertainties in simulations of surface processes at regional and global scales related to water, carbon, and nutrient cycling, as well as litter decomposition, tree-line advancement at high latitudes, and others (Wullschleger et al. 2014).

Trait-based biogeography, or the functional component of biodiversity, is expected to bridge the gap between changes in species distribution and biogeochemical cycles simulated by DGVMs that drive ecosystem functions and services (Violle et al. 2014).

CONCLUSIONS

SDM tools have been extensively used to assess potential impacts of land use, climate change, and other disturbances on biodiversity to support conservation planning for adaptation to global change. Many efforts have been based on correlative approaches modeling the equilibrium distribution of individual species under climate. Ongoing efforts for improved modeling aim at developing mechanistic approaches under dynamic environmental conditions that account for populations, multiple disturbances, and species functional-trait responses to environmental gradients. Eventually, methods and tools will evolve into a continuum of modeling approaches from species to macroscale vegetation patterns that improve our understanding of global change threats to biodiversity and potential impacts to society in order to support planning and implementation of coordinated conservation efforts across scales.

REFERENCES

Araújo, Miguel B., and Mark New. 2007. "Ensemble forecasting of species distributions." Trends in Ecology and Evolution 22 (1): 42–47. https://doi.org/10.1016/j.tree.2006.09.010.

Araújo, Miguel B., Diogo Alagador, Mar Cabeza, David Nogués-Bravo, and Wilfried Thuiller. 2011. "Climate change threatens european conservation areas." Ecology Letters 14 (5): 484–492. https://doi.org/10.1111/j.1461-0248.2011.01610.x.

Araújo, Miguel B., and A. Townsend Peterson. 2012. "Uses and misuses of bioclimatic envelope modeling." Ecology 93 (7): 1527–1539. https://doi.org/10.1890/11-1930.1.

Bellard, Céline, Cleo Bertelsmeier, Paul Leadley, Wilfried Thuiller, and Franck Courchamp. 2012. "Impacts of climate change on the future of biodiversity." Ecology Letters 15 (4): 365–377. https://doi.org/10.1111/j.1461-0248.2011.01736.x.

Beltrán, Bray J., Janet Franklin, Alexandra D. Syphard, Helen M. Regan, Lorraine E. Flint, and Alan L. Flint. 2014. "Effects of climate change and urban development on the distribution and conservation of vegetation in a Mediterranean type ecosystem." International Journal of Geographical Information Science 28 (8): 1561–1589. https://doi.org/10.1080/13658816.2013.846472.

Buckley, Lauren B., and Joel G. Kingsolver. 2012. "Functional and phylogenetic approaches to forecasting species' responses to climate change." Annual Review of Ecology, Evolution, and Systematics 43 (1): 205–226. https://doi.org/10.1146/annurev-ecolsys-110411-160516.

Buckley, Lauren B., Stephanie A. Waaser, Heidi J. MacLean, and Richard Fox. 2011. "Does including physiology improve species distribution model predictions of responses to recent climate change?" Ecology 92 (12): 2214–2221.

Chain-Guadarrama, A., P. Imbach, S. Vilchez-Mendoza, L. A. Vierling, and B. Finegan. 2017. "Potential trajectories of old-growth neotropical forest functional composition under climate change." Ecography 41. https://doi.org/10.1111/ecog.02637.

Comte, Lise, and Gaël Grenouillet. 2015. "Distribution shifts of freshwater fish under a variable climate: Comparing climatic, bioclimatic and biotic velocities." Diversity and Distributions 21 (9): 1014–1026. https://doi.org/10.1111/ddi.12346.

Elith, Jane, Steven J. Phillips, Trevor Hastie, Miroslav Dudík, Yung En Chee, and Colin J. Yates. 2011. "A statistical explanation of MaxEnt for ecologists." Diversity and Distributions 17 (1): 43–57. https://doi.org/10.1111/j.1472-4642.2010.00725.x.

Elith, J., C. H. Graham, R. P. Anderson, M. Dudík, S. Ferrier, A. Guisan, et al. 2006. "Novel methods improve prediction of species' distributions from occurrence data." Ecography: 129–151.

Engler, Robin, and Antoine Guisan. 2009. "MigClim: Predicting plant distribution and dispersal in a changing climate." Diversity and Distributions 15 (4): 590–601. https://doi.org/10.1111/j.1472-4642.2009.00566.x.

Higgins, Paul A. T., and John Harte. 2006. "Biophysical and biogeochemical responses to climate change depend on dispersal and migration." BioScience 56 (5): 407–417. https://doi.org/10.1641/0006-3568(2006)056[0407:BABRTC]2.0.CO;2.

Kearney, Michael, Richard Shine, and Warren P. Porter. 2009. "The potential for behavioral thermoregulation to buffer 'cold-blooded' animals against climate warming." Proceedings of the National Academy of Sciences 106 (10): 3835–3840.

Keenan, T., J. Maria Serra, F. Lloret, M. Ninyerola, and S. Sabate. 2011. "Predicting the future of forests in the Mediterranean under climate change, with niche- and process-based models: CO_2 matters!" Global Change Biology 17 (1): 565–579. https://doi.org/10.1111/j.1365-2486.2010.02254.x.

Keith, D. A, H. R. Akcakaya, W. Thuiller, G. F. Midgley, R. G. Pearson, S. J. Phillips, H. M. Regan, M. B. Araújo, and T. G. Rebelo. 2008. "Predicting extinction risks under climate change: Coupling stochastic population models with dynamic bioclimatic habitat models." *Biology Letters* 4 (5): 560–563. https://doi.org/10.1098/rsbl.2008.0049.

Levis, Samuel. 2010. "Modeling vegetation and land use in models of the Earth system." *Wiley Interdisciplinary Reviews: Climate Change* 1 (6): 840–856. https://doi.org/10.1002/wcc.83.

Merow, Cory, Johan P. Dahlgren, C. Jessica E. Metcalf, Dylan Z. Childs, Margaret E. K. Evans, Eelke Jongejans, Sydne Record, Mark Rees, Roberto Salguero-Gómez, and Sean M. McMahon. 2014. "Advancing population ecology with integral projection models: A practical guide." *Methods in Ecology and Evolution* 5 (2): 99–110. https://doi.org/10.1111/2041-210X.12146.

Midgley, G. F., I. D. Davies, C. H. Albert, R. Altwegg, L. Hannah, G. O. Hughes, and W. Thuiller. 2010. "BioMove: An integrated platform simulating the dynamic response of species to environmental change." *Ecography* 33 (3): 612–616.

Mladenoff, David J. 2004. "LANDIS and forest landscape models." *Ecological Modelling* 180 (January): 7–19. https://doi.org/10.1016/j.ecolmodel.2004.03.016.

Nix, H. A. 1986. "A biogeographic analysis of Australian elapid snakes." In *Atlas of Elapid Snakes of Australia*, ed. R. Longmore, 4–15. Australian Government Publishing Service.

Ordoñez, Alejandro, and John W. Williams. 2013. "Climatic and biotic velocities for woody taxa distributions over the last 16,000 years in eastern North America." *Ecology Letters* 16 (6): 773–781. https://doi.org/10.1111/ele.12110.

Parmesan, Camille. 1996. "Climate and species' range." *Nature*. https://doi.org/10.1038/382765a0.

Pearman, Peter B., Antoine Guisan, Olivier Broennimann, and Christophe F. Randin. 2008. "Niche dynamics in space and time." *Trends in Ecology and Evolution* 23 (3): 149–158. https://doi.org/10.1016/j.tree.2007.11.005.

Peng, Changhui. 2000. "From static biogeographical model to dynamic global vegetation model: A global perspective on modelling vegetation dynamics." *Ecological Modelling* 135 (1): 33–54.

Scheiter, Simon, and Steven Higgins. 2009. "Impacts of climate change on the vegetation of Africa: An adaptive dynamic vegetation modelling approach." *Global Change Biology* 15 (9): 2224–2246. https://doi.org/10.1111/j.1365-2486.2008.01838.x.

Scheller, Robert M., James B. Domingo, Brian R. Sturtevant, Jeremy S. Williams, Arnold Rudy, Eric J. Gustafson, and David J. Mladenoff. 2007. "Design, development, and application of LANDIS-II, a spatial landscape simulation model with flexible temporal and spatial resolution." *Ecological Modelling* 201 (3–4): 409–419. https://doi.org/10.1016/j.ecolmodel.2006.10.009.

Serra-Diaz, Josep M., Janet Franklin, Miquel Ninyerola, Frank W. Davis, Alexandra D. Syphard, Helen M. Regan, and Makihiko Ikegami. 2014. "Bioclimatic velocity: The pace of species exposure to climate change." *Diversity and Distributions* 20 (2): 169–180. https://doi.org/10.1111/ddi.12131.

Shugart, H. H. 2002. "Forest gap models." In *Encyclopedia of Global Environmental Change*, ed. Harold A. Mooney and Josep G. Canadell, 2: 316–323. John Wiley & Sons.

Svenning, Jens-Christian, and Brody Sandel. 2013. "Disequilibrium vegetation dynamics under future climate change." *American Journal of Botany* 100 (7): 1266–1286.

Swab, Rebecca M., Helen M. Regan, David A. Keith, Tracey J. Regan, and Mark K. J. Ooi. 2012. "Niche models tell half the story: Spatial context and life-history traits influence species responses to global change." *Journal of Biogeography* 39: 1266–1277. https://doi.org/10.1111/j.1365-2699.2012.02690.x.

VanDerWal, Jeremy, Helen T. Murphy, Alex S. Kutt, Genevieve C. Perkins, Brooke L. Bateman, Justin J. Perry, and April E. Reside. 2013. "Focus on poleward shifts in species' distribution underestimates the fingerprint of climate change." *Nature Climate Change* 3 (3): 239–243.

Violle, Cyrille, Peter B. Reich, Stephen W. Pacala, Brian J. Enquist, and Jens Kattge. 2014. "The emergence and promise of functional biogeography." *Proceedings of the National Academy of Sciences* 111 (38): 13690–13696. https://doi.org/10.1073/pnas.1415442111.

Wisz, M. S., O. Broennimann, P. Gronkjaer, P. R. Moller, S. M. Olsen, D. Swingedouw, R. B. Hedeholm, E. E. Nielsen, A. Guisan, and L. Pellissier. 2015. "Arctic warming will promote Atlantic-Pacific fish interchange." *Nature Climate Change* 5 (3): 261–265.

Wullschleger, Stan D., Howard E. Epstein, Elgene O. Box, Eugénie S. Euskirchen, Santonu Goswami, Colleen M. Iversen, Jens Kattge, Richard J. Norby, Peter M. Van Bodegom, and Xiaofeng Xu. 2014. "Plant functional types in earth system models: Past experiences and future directions for application of dynamic vegetation models in high-latitude ecosystems." *Annals of Botany* 114 (1): 1–16.

Yang, Yanzheng, Qiuan Zhu, Changhui Peng, Han Wang, and Huai Chen. 2015. "From plant functional types to plant functional traits: A new paradigm in modelling global vegetation dynamics." *Progress in Physical Geography* 39 (4): 514–535. https://doi.org/10.1177/0309133315582018.

CHAPTER THIRTEEN

Climate Change and Marine Biodiversity

WILLIAM W. L. CHEUNG AND
MIRANDA C. JONES

INTRODUCTION

The ocean covers 71 percent of Earth's surface and is home to 226,000 described eukaryotic species, including 35 animal phyla, 14 of which are exclusively marine. Marine biodiversity is important for a range of ecosystem services, such as fisheries yield, carbon sequestration, and recreation, and it further supports traditional culture and values (Millennium Ecosystem Assessment 2005). Despite this importance, marine biodiversity is threatened by human activities such as overfishing, habitat destruction, and pollution (Dulvy, Sadovy, and Reynolds 2003; Pitcher and Cheung 2013; United Nations 2017). There is also growing concern over the impact of climate change on marine systems, with longer-term shifts in the conditions caused by increasing CO_2 emissions, such as altered temperature, oxygen levels, and acidity, moving outside the bounds of previous climatic variability with which changes and adaptations in marine communities have been associated (Pörtner et al. 2014; Gattuso et al. 2015). Climate change is likely to further interact with other human stressors, such as the removal by fisheries of predators or prey, the release of substances that disrupt organisms' physiology into the ecosystem, and the degradation of essential marine habitats.

The effects of atmospheric climate change manifest themselves through factors such as temperature, ocean circulation, acidity, salinity, and the density structure of the water column (IPCC 2014). Globally averaged ocean temperature in the upper 75 m increased at a rate of over 0.1°C per decade between 1971 and 2010 over the mean of this period (IPCC 2014). Simultaneously, surface ocean acidity has increased by approximately 30 percent since preindustrial levels (Doney et al. 2009), while oxygen concentration in the open ocean decreased by 3–5 µmol kg^{-1} per decade (IPCC 2014). Global oxygen content has decreased by more than 2 percent since 1960 (Schmidtko et al. 2017) and is projected to continue to decrease in the twenty-first century (Bopp et al. 2013) as a result of increase in stratification and decrease in mix-layer depth.

Marine biodiversity responds to changing temperature and other ocean conditions manifested through physiology, population dynamics, distributions, and phenology (Poloczanska et al. 2013; Pörtner et al. 2014). These responses to ocean and atmospheric changes have been projected to lead to altered patterns of species richness (Cheung et al. 2009), changes in community structure (MacNeil et al. 2010), ecosystem functions (Petchey et al. 1999), and consequential changes in marine goods and services (Cheung and Sumaila 2008; Sumaila et al. 2011; Madin et al. 2012).

This chapter aims to provide an overview of marine biodiversity under climate change and ocean acidification. First, we describe overarching theories and hypotheses of how changing ocean conditions affect marine organisms. Second, we discuss observed and projected effects of climate change and ocean acidification on major marine taxa. We then examine the potential interactions of climate impacts on marine biodiversity with other non-CO_2-related human stressors. Finally, we discuss the key challenges and opportunities for marine biodiversity conservation under climate change. Topics related to ocean acidification and coral reefs under climate change are covered in separate chapters (Chapters 5 and 14).

MECHANISMS OF CLIMATE CHANGE EFFECTS ON MARINE BIODIVERSITY

Marine organisms are adapted to and dependent on the maintenance of a characteristic window of environmental conditions within their natural environment (Pörtner 2001). Organism physiology performs optimally at a certain temperature range; performance decreases when environmental temperature increases or decreases from that range with upper and lower temperature limits at which body functions that are necessary for survival are halted. Such a relationship between temperature and physiological performance of organisms applies to a wide range of organismal groups, from plants to vertebrates. For marine water-breathing ectotherms, theories suggest that the characteristic window of optimal physiological performance is largely determined by the relationship between oxygen availability for metabolism and growth, and the level of available oxygen for such purposes is referred to as aerobic scope (Pauly 2010a; Pörtner et al. 2017).

An organism's temperature tolerance limits are dependent on its organization complexity, the temperature stability that the species is adapted to, its life stage, and the existence of other physiological stressors (Figure 13.1). For example, some polar species have adapted to stable cold polar environments, leading to fewer red blood cells, oxygen-binding proteins (Nikinmaa 2002), and enzymes that are especially sensitive to temperature. Such physiological adaptations render these polar species particularly sensitive to ocean warming. Larval and spawning fishes also have higher temperature sensitivity. Moreover, aerobic scope (and temperature tolerance) decreases with limitation in food supply, increase in acidity, or decrease in oxygen level (Pörtner et al. 2017). Evaluating the effect of rapidly changing climate on species' survival and distributions therefore requires linking the geographic pattern of climatic change to the physiological sensitivity of study organisms (Deutsch et al. 2008).

Organisms can sustain physiological processes in low-oxygen areas only under certain environmental and biological conditions. Low temperatures reduce energetic demands and the need for oxygen.

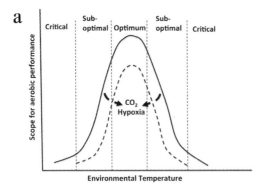

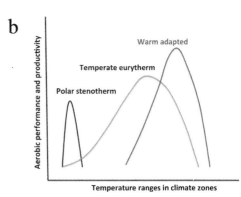

Figure 13.1. Relationship between temperature and physiological performance of marine water-breathing ectotherms. (A) Temperature tolerance decreases with multiple environmental stressors, and (B) temperature sensitivity is dependent on the characteristics of the environmental the organisms are adapted to. (Redrawn from Pörtner and Knust 2007.)

Organisms with small body size are also better adapted to low-oxygen areas because of their morphology (e.g., surface area to volume ratio) and reduced energy demands (Childress and Seibel 1998; Levin et al. 2009; Pauly 2010b). As only species with specialized adaptations can survive, biodiversity is therefore frequently low in an area with low oxygen levels, which may further become a "dead zone" in extreme cases (Levin 2003; Breitburg et al. 2018).

Understanding organisms' thermal tolerances and the geographic pattern of climate change helps explain and predict the survival and distributions of marine organisms under changing climate (Cheung and Pauly 2017). The theory of optimal foraging predicts that organisms tend to be distributed in environments that maximize their growth and reproduction, thereby providing a pathway through which the thermal window of an organism may affect its biogeography. Thus, as the ocean warms, marine organisms respond by shifting their distributions and/or seasonality to maintain themselves in habitats that lie within their preferred temperature limits. Shifted species distribution ranges follow temperature clines from high to low, reflecting a lateral gradient at the basin scale (Pinsky

et al. 2013; Poloczanska et al. 2013) or a vertical temperature gradient to deeper waters (Dulvy et al. 2008; Pauly 2010b).

Temperature may also act indirectly on a species' survival and distribution by influencing phenology, dispersal, predation pressure, and available food supply. Change in the distribution and abundance of prey and predators affects growth and mortality of other species. As temperature affects the rate of egg and larval development, warmer temperatures will also decrease the chance of predation at this phase in the life cycle. In addition, as the duration of the larval stage will determine the length of time they are subjected to movement by ocean currents, increased temperatures will indirectly affect population connectivity, community structure, and regional to global patterns of biodiversity (O'Connor et al. 2007).

REPONSES OF MARINE BIODIVERSITY TO CLIMATE CHANGE

Global patterns of species richness of marine animals are related to habitats (e.g., coastal vs. open ocean), temperature, and historical geographic factors (Tittensor et al. 2010). Marine animals include zooplankton, invertebrates, fish, reptiles, birds, and mammals, with species richness being highest in coastal and shelf seas. Although the richness of coastal species is highest in the western Pacific, that for oceanic species is highest in regions along mid-latitudinal zones. As temperature is one of the most important environmental predictors for marine animals' species richness, ocean warming is expected to have large implications for the global pattern of oceanic biodiversity. Thus, biological responses to ocean-atmospheric changes have been projected to lead to altered future patterns of species richness (Cheung et al. 2009; Jones and Cheung 2015), resulting in changes in community structure (MacNeil et al. 2010), ecosystem function (Petchey et al. 1999), and marine goods and services (Cheung and Sumaila 2008; Sumaila et al. 2011; Madin et al. 2012).

Phytoplanktons, Macroalgae, and Seagrasses

Ocean warming affects the growth, abundance, distribution, phenology, and community structure of phytoplankton. Increases in temperature initially increase the metabolic rates, and thus productivity, of many phytoplankton species. However, when temperature at a particular location exceeds a species' thermal window, its abundance will decrease, and it will eventually be replaced by species with a higher thermal tolerance (Thomas et al. 2012). Thus, on the basis of the thermal windows of phytoplankton, their distributions are projected to shift poleward, with a large decline in tropical diversity (Thomas et al. 2012). Also, ocean warming favors smaller species, resulting in a decrease in size structure of phytoplankton community (Flombaum et al. 2013). The species-specific thermal tolerances of macroalgae and seagrasses lead to similar types of responses to ocean warming as phytoplankton, such as distribution shifts (Lima et al. 2007). Increases in CO_2 levels appear to have positive effects on primary producers as a result of increased rate of photosynthesis and productivity, although some calcifying algae will be vulnerable to ocean acidification, and the indirect trophic impacts from changes in primary production are uncertain (Martin and Gattuso 2009; Arnold et al. 2012).

Macroinvertebrates and Fishes

The extant number of marine fishes and marcoinvertebrates that have been described amount to approximately 15,000 species and 200,000 species, respectively (Cheung, Pitcher, and Pauly 2005). As described above, marine invertebrates and fishes, as water-breathing ectotherms, are physiologically and ecologically sensitive to changes in ocean properties. Their responses to changes in ocean temperature, oxygen level, and acidity can therefore generally be predicted from their physiology of thermal tolerance, oxygen capacity, and acid-base regulation, as well as changes in ocean productivity.

Overall, climate change and ocean acidification lead to alteration of population dynamics, shifts in biogeography and seasonality, and changes in species' morphologies, such as body size and behavior. These responses are potentially linked to altered genetic frequencies (Thomas et al. 2001; Parmesan and Yohe 2003), although evidence from Pleistocene glaciations has shown that species are more likely to exhibit ecological responses to climate change, such as shifts in range distributions, than evolutionary responses, through local adaptation (Bradshaw and Holzapfel 2006).

Observations and theory have indicated that marine fish and invertebrates frequently undergo shifts in distribution in response to changing environmental factors, to areas within their physiological limits. Shifts are therefore most commonly toward higher latitudes (Perry et al. 2005; Hiddink and ter Hofstede 2008; Doney et al. 2012; Poloczanska et al. 2013), deeper waters (Dulvy et al. 2008), and, in general, following temperature velocity (Pinsky et al. 2013). Overall, marine examples of shifting distributions due to climate change are more striking than their terrestrial counterparts because of their greater rapidity (Edwards and Richardson 2004; Parmesan and Yohe 2003; Cheung et al. 2009). Meta-analyses of observed range shift in the past decades have shown that zooplanktons and fishes are moving poleward, on average, at rates of tens to hundreds of kilometers per decade (Poloczanska et al. 2013) (see Chapter 5, Figure 5.2A). For example, Beaugrand et al. (2009) described northward movement in calanoid copepod zooplankton assemblages in the North Atlantic at a mean rate of up to 23.16 km per year over 48 years. These changes in planktonic communities were paralleled by a northward migration of both commercial and noncommercial fish species (Brander 2003; Beare et al. 2004; Perry et al. 2005). In the European continental shelf, a response to warming has been demonstrated in the abundances of 72 percent of the 50 most common species inhabiting UK waters (Simpson et al. 2011). These shifts reflect the influx of warmer-water-adapted marine species to regions with colder waters that have been observed elsewhere (Arvedlund 2009; Fodrie et al. 2010). Analysis of survey data further shows that range shifts for

invertebrates and fishes in North American shelf seas are in directions consistent with gradients of temperature changes (Pinsky et al. 2013).

Observed changes in the species composition of catches from 1970 to 2006 that are partly attributed to long-term ocean warming suggest increasing dominance of warmer-water species in subtropical and higher-latitude regions, concurrent with a reduction in the abundance of subtropical species in equatorial waters (Cheung et al. 2013a), with implications for fisheries. Using global fisheries catch data, an index called the mean temperature of catch (MTC), computed from the average preferred temperature of each species reported

Figure 13.2. Average mean temperature of catch (MTC) and sea-surface temperature (SST) from (upper) nontropical and (lower) tropical large marine ecosystems (*left panels*). The right panels are schematic representation of hypotheses explaining the change in MTC over time. (From Cheung et al. 2013.)

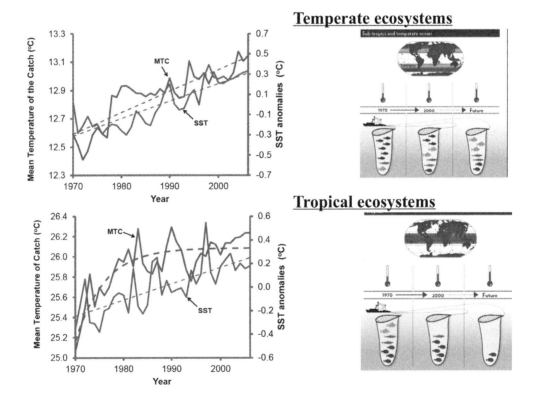

Temperate ecosystems

Tropical ecosystems

in fisheries data weighted by their annual catches, was calculated for all the large marine ecosystems (LMEs) of the world from 1970 to 2006. After accounting for the effects of fishing and large-scale oceanographic variability, global MTC increased at a rate of 0.19°C per decade between 1970 and 2006, whereas MTC in nontropical areas increased at a rate of 0.23°C per decade (Figure 13.2). In tropical areas, MTC increased initially because of the reduction in the proportion of subtropical species catches but subsequently stabilized as scope for further tropicalization of communities became limited (Figure 13.2).

Range shifts for fishes and invertebrates are projected to continue in the twenty-first century under climate change (Jones and Cheung 2015; Cheung et al. 2009). Although in temperate climates, local extinctions may be compensated by local invasions as species move into newly suitable habitat, thereby leading to an overall change in community structure, tropical regions may see declines in species richness as the scope for community tropicalization is reached. For example, projections of distributions for over 800 exploited fishes and invertebrates using multiple species distribution models result in a predicted average poleward latitudinal range shift of 15.5–25.6 km per decade under low and high emission scenarios (Representative Concentration Pathways 2.6 and 8.0, respectively) (Jones and Cheung 2015). This rate of shift is consistent with observed rates of shift in the twentieth century (Dulvy et al. 2008; Perry et al. 2005). Predicted distribution shifts resulted in large-scale changes in patterns of species richness through species invasions (occurring in new areas) and local extinctions (disappearing from previously occurring areas) (Figure 13.3). Hotspots of high local invasion are common in high-latitude regions, whereas local extinctions are concentrated near the equator.

Climate change modifies phenology, periodic biological phenomena, of marine fishes and invertebrates so that critical

phases remain synchronized with climatic alterations. For example, phenophases of the spawning season have been shown to be negatively correlated with mean sea-surface temperature (SST) the preceding winter for 27 species in the North Sea (Greve et al. 2005), and earlier spring migrations have also been noted (Sims et al. 2001; Clarke et al. 2003). Phenological responses are highly taxon or species specific, resulting from sensitivity to climatic fluctuations as well as factors such as temperature, light, or food availability (Edwards and Richardson 2004). Meta-analysis of observed phonological shifts suggests that seasonal events of marine species have advanced by an average of 4.4 days per decade (Poloczanska et al. 2013). Altered phenology and timing of development may also lead to altered dispersal. For species whose offspring develop in the water column, for example, the duration of the larval stage will determine the length of time that larvae are subject to movement by ocean currents (O'Connor et al. 2007).

Both theory and empirical observations further support the hypothesis that warming and reduced oxygen will reduce the body size of marine fishes and invertebrates (Pauly and Cheung 2018). The preferred minimum oxygen tolerance threshold of an organism varies across species, body size, and life stage, and is highest for large organisms. As fish increase in size (weight), mass-specific oxygen demand increases more rapidly than oxygen supply (Pauly 1997). Thus, fish reach a maximum body size when oxygen supply is balanced by oxygen demand. Moreover, the scope for aerobic respiration and growth decreases when size increases, with oxygen supply per unit body weight therefore decreasing. The decrease in food conversion efficiency that this implies, all else being equal, decreases the biomass production of fish and invertebrate populations.

Simulation model projections suggest decreases in the maximum body size of fishes under scenarios of ocean warming,

a.

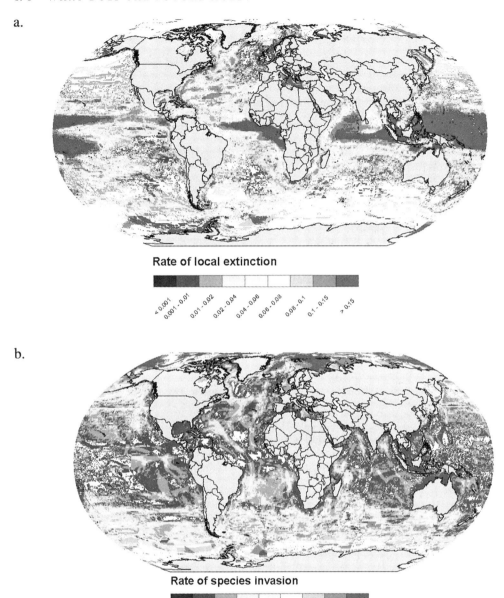

Rate of local extinction

<0.001 0.001 - 0.01 0.01 - 0.02 0.02 - 0.04 0.04 - 0.06 0.06 - 0.08 0.08 - 0.1 0.1 - 0.15 >0.15

b.

Rate of species invasion

<0.01 0.01 - 0.03 0.03 - 0.06 0.06 - 0.1 0.1 - 0.15 0.15 - 0.2 0.2 - 0.3 0.3 - 0.5 >0.5

Figure 13.3. Projected intensity of (A) local extinction and (B) species invasion between 2000 and 2050 under the RCP 8.5 scenario. Shading shows areas of high (three models) and moderate (two models) agreement. (From Jones and Cheung 2014.)

as well as decreases in oxygen levels. Specifically, the integrated biological responses of over 600 species of marine fishes due to changes in distribution, abundance, and body size were examined, on the basis of explicit representations of ecophysiology, dispersal, distribution, and population dynamics (Cheung et al. 2013b). The result was that assemblage-averaged maximum body weight is expected to shrink by 14 percent–24 percent globally from 2000 to 2050 under a high-emission scenario. The pro-

Observed change in surface temperature 1901–2012

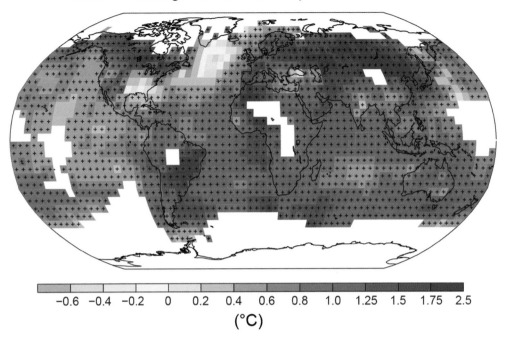

-0.6 -0.4 -0.2 0 0.2 0.4 0.6 0.8 1.0 1.25 1.5 1.75 2.5
(°C)

RCP 2.6   RCP 8.5
Change in average surface temperature (1986–2005 to 2081–2100)

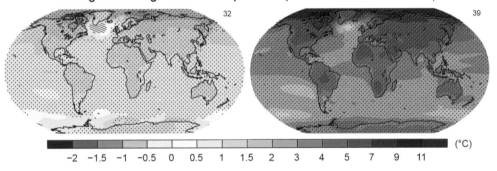

-2 -1.5 -1 -0.5 0 0.5 1 1.5 2 3 4 5 7 9 11   (°C)

Change in average precipitation (1986–2005 to 2081–2100)

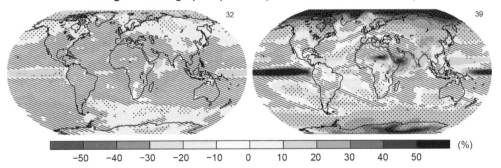

-50 -40 -30 -20 -10 0 10 20 30 40 50   (%)

Plate 1. Observed and simulated future climate change. Observed change in surface temperature 1901–2012 (*top*). Simulated future change in temperature for two different emissions (radiative forcing) scenarios—RCP 2.6 and RCP 8.5 (*middle*). Simulated future change in precipitation and model agreement for our RCP 2.6 and RCP 8.5 (*bottom*). RCP 8.5 approximates a business-as-usual global emissions trajectory, whereas RCP 2.6 approximates a world in which there is moderate to strong global policy to restrict climate change. (Figure SPM.7; Figure SPM.1 (b) from *Climate Change 2014: Synthesis Report. Contribution of Working Groups I, II and III to the Fifth Assessment Report of the Intergovernmental Panel on Climate Change*.)

Plate 2. Coral bleaching. Corals have bleached in every major reef system on the planet due to climate change. When regional surface water temperatures exceed a fixed threshold, corals expel their symbiotic zooxanothellae, resulting in coral bleaching over large areas (*top*). Bleached corals (*bottom*) suffer high mortality, but some corals may recover following bleaching. (Photographs by Ove Hoegh-Guldberg, Global Change Institute, University of Queensland.)

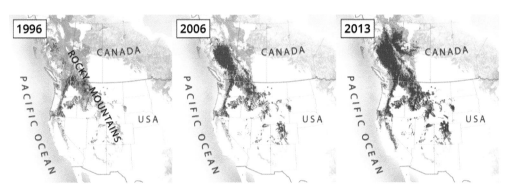

Plate 3. Bark beetle-killed trees (*top*). Warming temperatures have moved bark beetles into temperature bands in which synchronous outbreaks occur, destroying millions of trees across huge areas of North America (*bottom*; red indicates area affected by bark beetle outbreak). (Used with permission of National Geographic. Martin Gamache, NGM Staff; Shelley Sperry. Sources: Canadian Forest Service; Barbara Bentz and Jeanne Paschke, U.S. Forest Service; Aaron McGill, Alberta Environment and Sustainable Resource Development; Tim Ebata, British Columbia MFLNRO, University of British Columbia; Brian Aukema, University of Minnesota.)

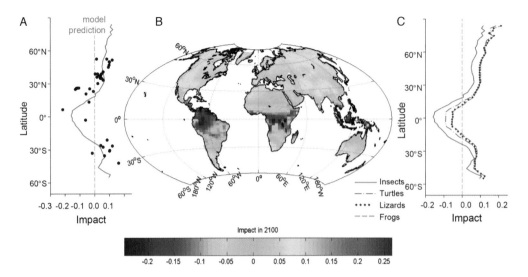

Plate 4. Predicted impact of warming on the thermal performance of ectotherms in 2100. (A) Impact versus latitude for insects using thermal performance curves fit to intrinsic population growth rates measured for each species and for a global model (red line) in which performance curves at each location are interpolated from empirical linear relationships between seasonality and both warming tolerance and thermal safety margin. (B and C) Results from the simplified conceptual model are shown globally for insects (B) for which performance data are most complete, and versus latitude for three additional taxa of terrestrial ectotherms: frogs and toads, lizards, and turtles (C), for which only warming tolerance was available. On the basis of patterns in warming tolerance, climate change is predicted to be most deleterious for tropical representatives of all four taxonomic groups. Performance is predicted to increase in mid- and high latitudes because of the thermal safety margins observed there for insects, and provisionally attributed to other taxa. Note that blue denotes negative impact on performance, while performance is enhanced by temperature change. (Redrawn from Deutsch et al. 2008.)

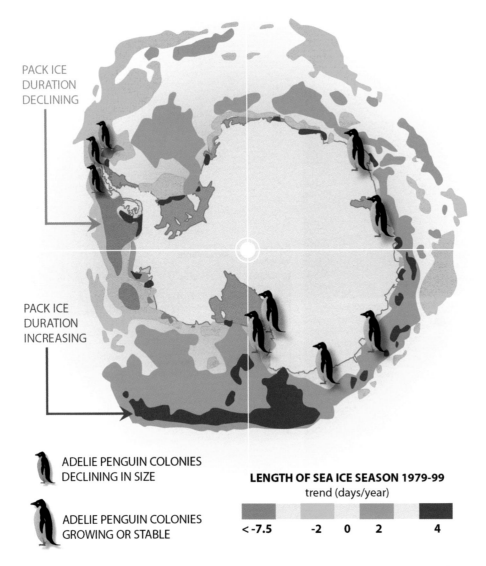

PACK ICE
DURATION
DECLINING

PACK ICE
DURATION
INCREASING

ADELIE PENGUIN COLONIES
DECLINING IN SIZE

ADELIE PENGUIN COLONIES
GROWING OR STABLE

LENGTH OF SEA ICE SEASON 1979-99
trend (days/year)

< -7.5   -2   0   2   4

Plate 5. Penguin population responses to climate change. Some penguin populations in Antarctica are declining due to warming temperatures and declining sea ice. Other penguin populations are increasing as climate change-driven winds bring more cold water to the surface, resulting in increases in ice cover. (Courtesy of I. Gaffney.)

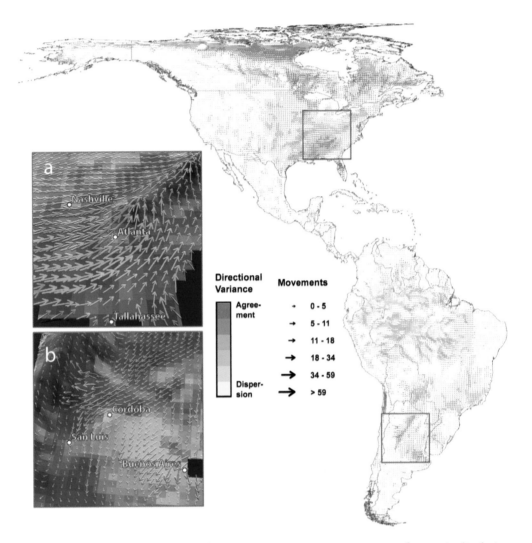

Plate 6. Projected climate-driven movements for 2,903 vertebrate species. Range movements from species distribution models are averaged across 10 future climate projections. Arrows represent the direction of modeled movements from unsuitable climates to suitable climates via routes that avoid human land uses. The sizes of the arrows represent the number of projected species movements as a proportion of current species richness. The colors of the arrows reflect the level of agreement in the direction of movement across species and routes. The insets are maps of (A) a high concentration of movements through southeastern North America and into the Appalachian Mountains and (B) areas of movement through the Sierras de Córdoba and into the Andes and the southern Pampas. Lighter blue shading in these topographical overlays indicates more intensive human activity. (From Lawler et al. 2013.)

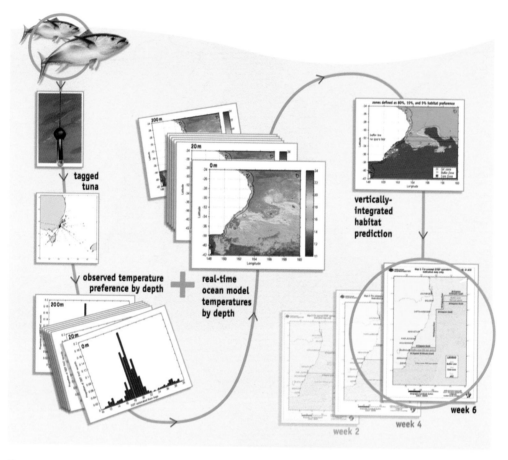

Plate 7. Dynamic ocean management for Australian tuna fishery. Southern bluefin tuna (*Thunnus maccoyii*) (*upper left*) are the target of a high-value fishery in southeast Australia. The fishery is managed using a model of tuna response to ocean temperature that is updated every 2 weeks. This management system provides adaptation to climate change because it responds to evolving ocean conditions. The modeling process (green arrows) begins with tagging tuna with pop-up tags that record water temperature and depth (*upper right*). Based on tuna water temperature preferences, a model is constructed of tuna habitat. Model simulations are run every two weeks using sea surface temperature information from satellites and near-real-time temperature recording at depth (*middle*). Based on the model simulations, a management zone is defined that is accessible only to fishing vessels that hold permits for southern bluefin tuna. Fishing vessels use GPS to ensure that they are respecting the management boundaries. Boundaries could be any shape, but for simplicity are usually a straight line (*lower right*). (Figure supplied by CSIRO.)

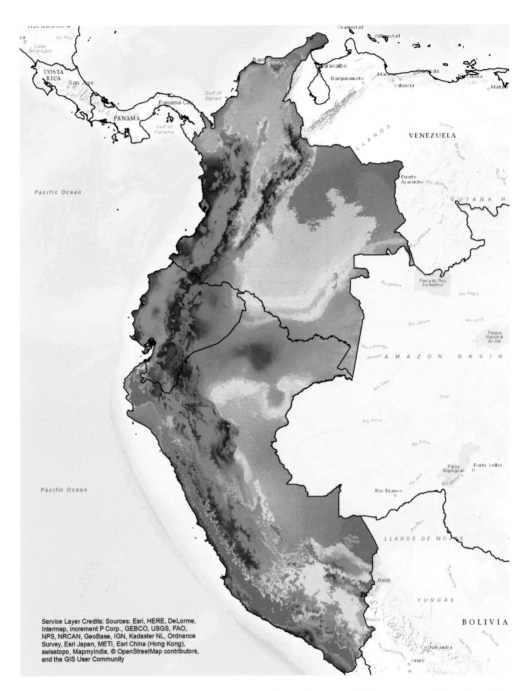

Plate 8. Conservation planning for climate change. Models of species' current and future ranges can be used to select priority areas for new protection that conserve species both in their present ranges and in their likely future range. Conservation planning software can suggest optimal solutions across hundreds or thousands of species' present and future ranges. In this example, 4,000 species in the northern tropical Andes are used to suggest new areas for protection that will improve representation of species' future ranges in protected area systems of the region. The results for the Zonation conservation planning software and Network Flow algorithm (purple areas) are compared. Note the high level of agreement on priority areas for additional protection. (Courtesy of Patrick R. Roehrdanz.)

jected magnitude of decrease in body size is consistent with experimental (Forster et al. 2012; Cheung et al. 2013b) and field observations (Baudron et al. 2014). About half of this shrinkage is due to change in distribution and abundance, and the remainder to changes in physiological performance. The tropical and intermediate latitudinal areas will be heavily affected, with an average reduction of more than 20 percent. Decreases in growth and body size are likely to reduce the biomass production of fish populations, and hence fishery catches, and potentially alter trophic interactions.

Ocean acidification is also expected to affect marine animals, although the sensitivity varies largely between taxonomic groups. Available studies suggest that sensitivity is high for organisms that form calcium carbonate exoskeletons or shells (Kroeker et al. 2013; Wittman and Pörtner 2013). Also, elevated CO_2 levels lead to reduced tolerance to warming (Wittman and Pörtner 2013). Sensitivity to ocean acidification appears to be highest for tropical coral reefs and mollusks, followed by crustacean and cold-water corals. Their ecological responses may become more unpredictable when trophic interactions are considered (Busch et al. 2013) (see Chapter 14).

Reptiles, Mammals, and Birds

Some species of seabirds, marine mammals, and sea turtles are affected by changes in temperature and other ocean conditions. Marine reptiles, including sea turtles, sea snakes, saltwater crocodile, and marine iguanas (from the Galapagos Islands), are ectotherms and thus inherently sensitive to ocean warming. All sea turtles have demonstrated poleward distribution shifts that are correlated with ocean warming (Pörtner et al. 2014). As the gender of sea turtles is also temperature dependent, warming alters their sex ratio toward an increased number of females. In addition, warming increases eggs and hatchling mortality and influences the phenology

of nesting (Weishampel, Bagley, and Ehrhart 2004; Mazaris et al. 2008), whereas increases in extreme weather events, such as flooding, increase nest damage and further negatively affect reproductive success (Van Houtan and Bass 2007). Using SST as a predictor, the distribution of loggerhead turtles is projected to expand poleward in the Atlantic Ocean by the end of the twenty-first century relative to 1970–1989 (Witt et al. 2010). In contrast, leatherback turtles are projected to decrease in abundance because of reduced hatching success with warming (Saba et al. 2012).

Climate change affects seabird populations indirectly through changes in the distribution and productivity of prey and directly through impacts on vulnerable life stages. Seabirds' foraging grounds and their food supply are strongly affected by the distribution and availability of mid-trophic-level prey that are in turn affected by climate change and resulting changes in ocean productivity (Hoegh-Guldberg and Bruno 2010; Trathan et al. 2014). For example, Cassin's auklet in the California Current ecosystem is projected to decline by 11 percent–45 percent by the end of the twenty-first century due to ocean warming (~2°C), with consideration of changes in upwelling intensity (Wolf et al. 2010). Distributions of some seabirds are also affected by changing habitat structure, such as sea ice in the case of the emperor penguin (Jenouvrier et al. 2012). Increased frequency of extreme weather events also affects penguin populations. For example, storms are a major source of mortality of Magellanic penguin (*Spheniscus magellanicus*) chicks at Punta Tombo, Argentina, with storm intensity being positively related to chick mortality. As storm frequency and intensity are projected to increase in the region under climate change, populations of Magellanic penguin are therefore expected to be negatively affected (Boersma and Rebstock 2014). Warming also affects the phenology of seabirds, frequently leading to earlier breeding and the extension of breeding seasons in high-latitude species (Sydeman and Bograd 2009).

Mechanisms driving the responses of marine mammal population to climate change are similar to those for seabirds, with changes in marine mammal distributions further being observed to influence shifts in the distributions of their prey species (Salvadeo et al. 2010; Moore and Barlow 2011). Changes in sea-ice structure also affect the foraging and breeding grounds of polar marine mammals such as walrus (*Odobenus rosmarus*). Overall, in a study that includes all extant (115 species) of cetacean and pinnipeds, species richness of cetaceans is projected to increase in mid-latitude regions, while that of pinnipeds and cetaceans is projected to decrease at low latitudes by the mid-twenty-first century relative to the late 1990s under the SRES A1B scenario (Kaschner et al. 2011).

INTERACTIONS WITH OTHER HUMAN DRIVERS

Climate change and ocean acidification add to the list of anthropogenic stressors affecting marine biodiversity. The main nonclimate anthropogenic stressors include overfishing, pollution, and habitat destruction. These drives will interact with and confound marine biodiversity changes that are related to climate change and ocean acidification (Planque et al. 2010), increasing the sensitivity of marine organisms to climate stressors. For example, intensive fishing leads to the depletion of large predatory species and the truncated age structure in targeted populations, with an increased dominance of juveniles and small-bodied, fast-turnover species. Such communities tend to track changes in ocean conditions more closely (Perry, Barange, and Ommer 2010). Also, in Tasmania, biological communities in exploited areas are shown to be more sensitive to ocean changes relative to areas protected from fishing (Bates et al. 2013). One of the most important pathways

of pollution impact on marine organisms is also through nutrient enrichment from the discharge of sewage, and agricultural and industrial waste into the ocean, ultimately leading to oxygen depletion (Diaz and Rosenberg 2008), thereby rendering marine organisms more vulnerable to ocean warming. The combination of multiple human stressors thus reduces the resilience of marine biodiversity to impacts from climate change and ocean acidification, as well as reducing their predictability, making detection, mitigation, and adaptation to climate-related effects on marine biodiversity more difficult.

IMPLICATIONS FOR MARINE BIODIVERSITY CONSERVATION

Effective management of ecosystems under climate change increases the resilience of ecosystems and the adaptive capacity of management systems, for example, by reducing other human perturbations. Impacts of climate change on marine biodiversity can be moderated by reducing stresses from overfishing, habitat degradation, pollution runoff, land-use transformation, and invasive species. As such, effective implementation of ecosystem-based management that considers a much wider range of environmental and human stressors as well as objectives is important to increase the adaptive capacity of marine social-ecological systems and biodiversity to climate change and ocean acidification.

Adaptive marine conservation and management are also important in uncertain future ocean ecosystems (Walters and Martell 2004). The reduced predictability of marine ecosystems due to climate change will make it more difficult to provide accurate assessments of the current and future status of marine biodiversity. Also, changing baseline oceanographic and ecological conditions may affect the effectiveness of existing conservation and management measures such as marine protected areas. The application of adaptive management approach through the incorporation of monitoring programs that are designed for a changing ocean, and the subsequent usage of the data to improve monitoring is thus important. Monitoring will include data for indicators at the pressure, state, and response levels, thereby promoting fast decision responses to changing and uncertain conditions and allowing a suite of possible responses to be maintained.

Marine protected areas (MPAs), for example, are a major tool to conserve marine biodiversity and have been shown to enhance population resilience to climate-driven disturbance (Micheli et al. 2012). However, climate change–induced changes in environmental suitability and consequential species' distribution shifts may lead to both emigration and immigration of species from or into an MPA. This will alter the specific species assemblage being conserved, potentially losing species of conservation value and reducing the efficacy of the MPA. There is therefore a need to increase the robustness and enhance the resilience of protected areas themselves to climate change. For example, by assessing the degree of future environmental change in proposed protected areas, conservation planning may be used to protect against biodiversity loss (Levy and Ban 2013). Implementing networks of MPAs may also increase the likelihood of effectively conserving species following climate change–induced range shifts (McLeod et al. 2009; Gaines et al. 2010).

Climate change may also affect the effectiveness of conservation and management, thus increasing the risk to marine biodiversity. In particular, the increased likelihood of abrupt and unpredictable changes in the productive potential and migratory behavior of exploited fish stocks may threaten to disrupt cooperative management arrangements. For example, the distribution of Atlantic mackerel (*Scomber scombrus*) in the northeastern Atlantic Ocean recently shifted northward, believed to be driven by changing ocean conditions. The Atlantic mackerel

fisheries were believed to be sustainable as evidenced by its certification by the Marine Stewardship Council. However, following the species' northward shift to waters around Iceland and the Faroe Islands, these countries unilaterally increased their quota, leading to international dispute in quota allocation with countries sharing the straddling mackerel stock. This results in destabilization of management of the mackerel fisheries and the suspension of its Marine Stewardship Council certification (Miller et al. 2013; Sumaila et al. 2011). Such disputes are projected to increase as ocean warming increases with climate change.

KEY CHALLENGES

The observed and projected impacts of climate change on marine biodiversity and socio-ecological systems discussed here highlight the fact that changing ocean conditions associated with climate change will bring increased challenges to efforts to conserve biodiversity and manage fisheries already struggling to cope with the impacts of overexploitation and economic underperformance. Although these challenges may be reduced by methodologies to assess likely responses and the possible ways of mitigating or adapting to them, limitations in prediction methodologies contribute uncertainties and present additional challenges in their application and use.

Long-term observation data and monitoring programs, essential to detection and attribution of the responses of marine biodiversity to climate change and ocean acidification, are limited globally. It is suggested that time series that span at least multiple decades are needed to detect long-term trends in the ocean from natural variability (e.g., for net primary production; Henson et al. 2010). Also, analysis explicitly linking biological responses to environmental change between levels of organization (from individual to ecosystem) is also needed to provide integrated multiscale understanding of climate change effects on marine bio-diversity. The role of evolutionary and phe-notypic responses to determining climate change impacts on marine biodiversity is still uncertain. The interaction of multiple anthropogenic threats and predator-prey interactions further contributes uncertainty to predicting the likely impact of climate change on specific populations and species. For example, both fisheries and warming waters are thought to have caused the de-cline in sandeel and consequential decline in breeding success of black-legged kitti-wakes and common guillemot (Pinnegar, Watt, and Kennedy 2012). Disentangling the impact of these threats and projecting possible scenarios of change into the future therefore remains a challenge despite recent advances in this area (Ainsworth et al. 2011).

The resolution of climate model projec-tions may not match with scales that are relevant to regional or local marine biodi-versity. For example, coastal processes of coupled ocean-atmosphere global climate models included in the Fifth Assessment Report of the IPCC are not well resolved for many coastal oceanographic processes, such as eastern boundary coastal upwell-ings. This limits the application of these projections to study local-scale marine bio-diversity. Furthermore, climate model er-ror is not evenly distributed over different depths and also varies geographically.

The inherent mistrust of uncertainty and skepticism toward necessarily uncertain fore-casts also needs to be overcome to promote adaptive capacity in marine conservation and management. A range of options from dif-ferent climatic and socioeconomic scenarios could therefore be viewed as a suite of pos-sible scenarios (Jones et al. 2014; Haward et al. 2013), consideration of which will mini-mize the surprises and risks climate change may impose on the marine ecosystem and the people who depend on it. Methodologies have further been proposed that incorporate a system's capacity to adapt into a framework for climate change scenario analysis (van Vuuren et al. 2012). This may promote the systematic integration of environmental, biophysical, and socioeconomic scenarios to explore opportunities for adaptation and re-silience to climate change impacts.

SUMMARY

Marine biodiversity is being affected by cli-mate change and ocean acidification. The main biological responses to ocean warm-ing and deoxygenation can be predicted from physiological principles of tempera-ture tolerance that agree with observations and simulation modeling. Poleward shifts in species distribution are generally con-sistent across organism groups, from phy-toplanktons and invertebrates to fishes and marine mammals. Hotspots of change in species richness are projected to occur in high latitudinal regions (species gains) and in the tropics (species loss) under climate change. Projected ocean acidification nega-tively affects biological processes across taxonomic groups, with mollusks and trop-ical corals being particularly sensitive. The combination of these multiple CO_2-related stressors and other human drivers will ex-acerbate the impacts directly on the biology of the organisms or indirectly through the food web. Climate change further impacts the effectiveness of marine conservation and management measures. This highlights the need for ecosystem-based management that is adaptive to climate change and ocean acidification. However, improved manage-ment is not expected to fully compensate for impacts from climate change, particu-larly for sensitive ecosystems. Thus, mitiga-tion of greenhouse gas emission, in addi-tion to adaptation measures, is necessary for effective marine biodiversity conservation.

REFERENCES

Ainsworth, C. H., J. F. Samhouri, D. S. Busch, and W. W. L. Cheung. 2011. "Potential impacts of climate change on northeast Pacific marine foodwebs and fisheries." *ICES Journal of Marine Science* 68: 1217–1229.

Arvedlund, Michael. 2009. "First records of unusual marine fish distributions—Can they predict climate changes?" *Journal of the Marine Biological Association of the United Kingdom* 89 (4): 863–866. https://doi.org/10.1017/S002531540900037X.

Beare, D., B. Finlay, E. Jones, K. Peach, E. Portilla, T. Greig, E. McKenzie, and D. Reid. 2004. "An increase in the abundance of anchovies and sardines in the north-western North Sea since 1995." *Global Change Biology* 10: 1209–1213. https://doi.org/10.1111/j.1365-2486.2004.00790.x.

Beaugrand, G. 2009. "Decadal changes in climate and ecosystems in the North Atlantic Ocean and adjacent seas." *Deep Sea Research Part II: Topical Studies in Oceanography* 56: 656–673. https://doi.org/10.1016/j.dsr2.2008.12.022.

Boersma, P. D., and G. A. Rebstock. 2014. "Climate change increases reproductive failure in Magellanic penguins." *PLOS One* 9 (1): e85602.

Bradshaw, W. E., and C. M. Holzapfel. 2006. "Evolutionary response to rapid climate change." *Science* 312: 1477–1478.

Brander, K. M. 2003. "Fisheries and climate." In *Marine Science Frontiers for Europe*, ed. G. Wefer, F. Lamy, and F. Mantoura, 29–38. Springer-Verlag.

Breitburg, D., L. A. Levin, A. Oschlies, M. Grégoire, F. P. Chavez, D. J. Conley, V. Garçon, D. Gilbert, D. Gutiérrez, K. Isensee, et al. 2018. "Declining oxygen in the global ocean and coastal waters." *Science* 359: eaam7240.

Cheung, W. W. L., and U. Sumaila. 2008. "Trade-offs between conservation and socio-economic objectives in managing a tropical marine ecosystem." *Ecological Economics* 66 (1): 193–210. https://doi.org/10.1016/j.ecolecon.2007.09.001.

Cheung, W., T. Pitcher, and D. Pauly. 2005. "A fuzzy logic expert system to estimate intrinsic extinction vulnerabilities of marine fishes to fishing." *Biological Conservation* 124 (July): 97–111. https://doi.org/10.1016/j.biocon.2005.01.017.

Cheung, William W. L., Vicky W. Y. Lam, Jorge L. Sarmiento, Kelly Kearney, Reg Watson, and Daniel Pauly. 2009. "Projecting global marine biodiversity impacts under climate change scenarios." *Fish and Fisheries* 10: 235–251. https://doi.org/10.1111/j.1467-2979.2008.00315.x.

Cheung, W. W. L., and D. Pauly. 2016. "Impacts and effects of ocean warming on marine fishes." In *Explaining Ocean Warming: Causes, Scale, Effects and Consequences*, ed. D. Laffoley and J. M. Baxter, 239–254. Gland, Switzerland: IUCN.

Childress, J. J., and B. A. Seibel. 1998. "Life at stable low oxygen levels: Adaptations of animals to oceanic oxygen minimum layers." *Journal of Experimental Biology* 201 (8): 1223–1232.

Clarke, R. A., C. J. Fox, D. Viner, and M. Livermore. 2003. "North Sea cod and climate change—Modelling the effects of temperature on population dynamics." *Global Change Biology*: 1669–1680. https://doi.org/10.1046/j.1529-8817.2003.00685.x.

Deutsch, Curtis A., Joshua J. Tewksbury, Raymond B. Huey, Kimberly S. Sheldon, Cameron K. Ghalambor, David C. Haak, and Paul R. Martin. 2008. "Impacts of climate warming on terrestrial ectotherms across latitude." *Proceedings of the National Academy of Sciences* 105 (18): 6668–6672. https://doi.org/10.1073/pnas.0709472105.

Diaz, R. J., and R. Rosenberg. 2008. "Spreading dead zones and consequences for marine ecosystems." *Science* 321: 926–929.

Doney, S. C., V. J. Fabry, R. A. Feely, and J. A. Kleypas. 2009. "Ocean acidification: The other CO_2 problem." *Annual Review of Marine Science* 1: 169–192.

Doney, Scott C., Mary Ruckelshaus, J. Emmett Duffy, James P. Barry, Francis Chan, Chad A. English, Heather M. Galindo, et al. 2012. "Climate change impacts on marine ecosystems." *Annual Review of Marine Science* 4 (1): 11–37. https://doi.org/10.1146/annurev-marine-041911-111611.

Dulvy, Nicholas K., Stuart I. Rogers, Simon Jennings, Vanessa Stelzenmller, Stephen R. Dye, and Hein R. Skjoldal. 2008. "Climate change and deepening of the North Sea fish assemblage: A biotic indicator of warming seas." *Journal of Applied Ecology* 45. https://doi.org/10.1111/j.1365-2664.2008.01488.x.

Dulvy, Nicholas K., Yvonne Sadovy, and John D. Reynolds. 2003. "Extinction vulnerability in marine populations." *Fish and Fisheries* 4: 25–64.

Edwards, Martin, and Anthony J. Richardson. 2004. "Impact of climate change on marine pelagic phenology and trophic mismatch." *Nature* 430 (7002): 881–884. https://doi.org/10.1038/nature02808.

Flombaum, P., J. L. Gallegos, R. A. Gordillo, J. Rincon, L. L. Zabala, N. Jilao, D. M. Karl, et al. 2013. "Present and future global distributions of the marine cyanobacteria Prochlorococcus and Synechococcus." *Proceedings of the National Academy of Sciences* 110 (24): 9824–9829.

Fodrie, F. Joel, Kenneth L. Heck, Sean P. Powers, William M. Graham, and Kelly L. Robinson. 2010. "Climate-related, decadal-scale assemblage changes of seagrass-associated fishes in the northern Gulf of Mexico." *Global Change Biology* 16 (1): 48–59. https://doi.org/10.1111/j.1365-2486.2009.01889.x.

Gaines, S. D., C. White, M. H. Carr, and S. R. Palumbi. 2010. "Designing marine reserve networks for both conservation and fisheries management." *Proceedings of the National Academy of Sciences* 107: 18286–18293.

Gattuso, J.-P., A. Magnan, R. Bille, W. W. L. Cheung, E. L. Howes, F. Joos, D. Allemand, et al. 2015. "Contrasting futures for ocean and society from different anthropogenic CO_2 emissions scenarios." *Science* 349 (6243): aac4722. https://doi.org/10.1126/science.aac4722.

Greve, W., S. Prinage, H. Zidowitz, J. Nast, and F. Reiners. 2005. "On the phenology of North Sea ichthyoplankton." *ICES Journal of Marine Science* 62 (7): 1216–1223. https://doi.org/10.1016/j.icesjms.2005.03.011.

Haward, Marcus, Julie Davidson, Michael Lockwood, Marc Hockings, Lorne Kriwoken, and Robyn Allchin. 2013. "Climate change, scenarios and marine biodiversity conservation." *Marine Policy* 38 (March): 438–446. https://doi.org/10.1016/j.marpol.2012.07.004.

Henson, S. A., J. L. Sarmiento, J. P. Dunne, L. Bopp, I. D. Lima, S. C. Doney, J. John, and C. Beaulieu. 2010. "Detection of anthropogenic climate change in satellite records of ocean chlorophyll and productivity." *Biogeosciences* 7: 621–640.

Hiddink, J. G., and R. ter Hofstede. 2008. "Climate induced increases in species richness of marine fishes." *Global Change Biology* 14 (3): 453–460. https://doi.org/10.1111/j.1365-2486.2007.01518.x.

Hoegh-Guldberg, O., and J. F. Bruno. 2010. "The impact of climate change on the world's marine ecosystems." *Science* 328 (5985): 1523–1528.

Houtan, K. S. Van, and O. L. Bass. 2007. "Stormy oceans are associated with declines in sea turtle hatching." *Current Biology* 17 (15): R590–R591.

Jenouvrier, S., M. Holland, J. Stroeve, C. Barbraud, H. Weimerskirch, M. Serreze, and H. Caswell. 2012. "Effects of climate change on an emperor penguin population: Analysis of coupled demographic and climate models." *Global Change Biology* 18 (9): 2756–2770.

Jones, Miranda C., and William W. L. Cheung. 2015. "Multi-model ensemble projections of climate change effects on global marine biodiversity." *ICES Journal of Marine Science* 72: 741–752.

Jones, Miranda C., Stephen R. Dye, John K. Pinnegar, Rachel Warren, and William W. L. Cheung. 2014. "Using scenarios to project the changing profitability of fisheries under climate change." *Fish and Fisheries* (May). https://doi.org/10.1111/faf.12081.

Kaschner, Kristin, Derek P. Tittensor, Jonathan Ready, Tim Gerrodette, and Boris Worm. 2011. "Current and future patterns of global marine mammal biodiversity." *PLOS One* 6 (5): e19653. https://doi.org/10.1371/journal.pone.0019653.

Kroeker, K. J., R. L. Kordas, R. Crim, I. E. Hendriks, L. Ramajo, G. S. Singh, C. M. Duarte, and J.-P. Gattuso. 2013. "Impacts of ocean acidification on marine organisms: Quantifying sensitivities and interaction with warming." *Global Change Biology* 19: 1884–1896.

Levin, L. A. 2003. "Oxygen minimum zone benthos: Adaptation and community response to hypoxia." *Oceanography and Marine Biology: An Annual Review* 41: 1–45.

Levin, L. A., W. Ekau, A. J. Gooday, F. Jorissen, J. J. Middelburg, S. W. A. Naqvi, C. Neira, N. N. Rabalais, and J. Zhang. 2009. "Effects of natural and human-induced hypoxia on coastal benthos." *Biogeosciences* 6 (10): 2063–2098.

Levy, Jessica S., and Natalie C. Ban. 2013. "A method for incorporating climate change modelling into marine conservation planning: An Indo–West Pacific example." *Marine Policy* 38 (March): 16–24. https://doi.org/10.1016/j.marpol.2012.05.015.

Lima, F. P., P. A. Ribeiro, N. Queiroz, and S. J. Hawkins. 2007. "Do distributional shifts of northern and southern species of algae match the warming pattern?" *Global Change Biology* 13: 2592–2604.

MacNeil, M. Aaron, Nicholas A. J. Graham, Joshua E. Cinner, Nicholas K. Dulvy, Philip A. Loring, Simon Jennings, Nicholas V. C. Polunin, Aaron T. Fisk, and Tim R. McClanahan. 2010. "Transitional states in marine fisheries: Adapting to predicted global change." *Philosophical Transactions of the Royal Society of London B: Biological Sciences* 365 (1558): 3753–3763. https://doi.org/10.1098/rstb.2010.0289.

Madin, Elizabeth M. P., Natalie C. Ban, Zoë A. Doubleday, Thomas H. Holmes, Gretta T. Pecl, and Franz Smith. 2012. "Socio-economic and management implications of range-shifting species in marine systems." *Global Environmental Change* 22 (February): 137–146. https://doi.org/10.1016/j.gloenvcha.2011.10.008.

Martin, S., and J. P. Gattuso. 2009. "Response of Mediterranean coralline algae to ocean acidification and elevated temperature." *Global Change Biology* 15: 2089–2100.

Mazaris, A. D., A. S. Kallimanis, S. P. Sgardelis, and J. D. Pantis. 2008. "Do long-term changes in sea surface temperature at the breeding areas affect the breeding dates and reproduction performance of Mediterranean loggerhead turtles? Implications for climate change." *Journal of Experimental Biology* 367 (2): 219–226.

McLeod, Elizabeth, Rodney Salm, Alison Green, and Jeanine Almany. 2009. "Designing marine protected area networks to address the impacts of climate change." *Frontiers in Ecology and the Environment* 7 (7): 362–370. https://doi.org/10.1890/070211.

Micheli, Fiorenza, Andrea Saenz-Arroyo, Ashley Greenley, Leonardo Vazquez, Jose Antonio Espinoza Montes, Marisa Rossetto, and Giulio A. De Leo. 2012. "Evidence that marine reserves enhance resilience to climatic impacts." *PLOS One* 7 (7): e40832. https://doi.org/10.1371/journal.pone.0040832.

Millennium Ecosystem Assessment. 2005. *Ecosystems and Human Well-being: Opportunities and Challenges for Business and Industry.* World Resources Institute.

Miller, Kathleen A., Gordon R. Munro, U. Rashid Sumaila, and William W. L. Cheung. 2013. "Governing marine fisheries in a changing climate: A game-theoretic perspective." *Canadian Journal of Agricultural Economics/Revue Canadienne d'agroeconomie* 61 (2): 309–334. https://doi.org/10.1111/cjag.12011.

Moore, Jeffrey E., and Jay Barlow. 2011. "Bayesian state-space model of fin whale abundance trends from a 1991–2008 time series of line-transect surveys in the California Current." *Journal of Applied Ecology* 48 (5): 1195–1205. https://doi.org/10.1111/j.1365-2664.2011.02018.x.

Nikinmaa, M. 2002. "Oxygen-dependent cellular functions—Why fishes and their aquatic environment are a prime choice of study." *Comparative Biochemistry and Physiology Part A: Molecular & Integrative Physiology* 133: 1–16.

O'Connor, Mary I., John F. Bruno, Steven D. Gaines, Benjamin S. Halpern, Sarah E. Lester, Brian P. Kinlan, and Jack M. Weiss. 2007. "Temperature control of larval dispersal and the implications for marine ecology, evolution, and conservation." *Proceedings of the National Academy of Sciences* 104 (4): 1266–1271. https://doi.org/10.1073/pnas.0603422104.

Parmesan, Camille, and Gary Yohe. 2003. "A globally coherent fingerprint of climate change impacts across natural systems." *Nature* 421 (6918): 37–42.

Pauly, D. 1997. "Geometric constraints on body size." *Trends in Ecology & Evolution* 12 (11): 442–443.

Pauly, D. 2010a. "The state of fisheries." In *Conservation Biology for All*, ed. S. N. S. Sodhi and P. R. Ehrlich, 118–120. Oxford University Press.

Pauly, D. 2010b. *Gasping Fish and Panting Squids: Oxygen, Temperature and the Growth of Water-Breathing Animals.* International Ecology Institute.

Pauly, D., and W. W. L. Cheung. 2017. "Sound physiological knowledge and principles in Odelling shrinking of fishes under climate change." *Global Change Biology* 25: e15–e26.

Perry, Allison L., Paula J. Low, Jim R. Ellis, and John D. Reynolds. 2005. "Climate change and distribution shifts in marine fishes." *Science* 308: 1912–1915. https://doi.org/10.1126/science.1111322.

Perry, R. Ian, Manuel Barange, and Rosemary E. Ommer. 2010. "Global changes in marine systems: A social-ecological approach." *Progress in Oceanography* 87 (1–4): 331–337.

Petchey, Owen L., P. Timon Mcphearson, Timothy M. Casey, and Peter J. Morin. 1999. "Environmental warming alters food-web structure and ecosystem function." *Nature* 402 (November): 69–72.

Pinnegar, J., T. Watt, and K. Kennedy. 2012. "Climate change risk assessment for the marine and fisheries sector." In *UK Climate Change Risk Assessment*. Project Code GA0204. Defra.

Pinsky, Malin L., Boris Worm, Michael J. Fogarty, Jorge L. Sarmiento, and Simon A. Levin. 2013. "Marine taxa track local climate velocities." *Science* 341 (6151): 1239–1242. https://doi.org/10.1126/science.1239352.

Pitcher, T. J., and W. W. L. Cheung. 2013. "Fisheries: Hope or despair?" *Marine Pollution Bulletin* 74: 506–516.

Planque, Benjamin, Jean-Marc Fromentin, Philippe Cury, Kenneth F. Drinkwater, Simon Jennings, R. Ian Perry, and Souad Kifani. 2010. "How does fishing alter marine populations and ecosystems sensitivity to climate?" *Journal of Marine Systems* 79 (3–4): 403–417. https://doi.org/10.1016/j.jmarsys.2008.12.018.

Poloczanska, Elvira S., Christopher J. Brown, William J. Sydeman, Wolfgang Kiessling, David S. Schoeman, Pippa J. Moore, Keith Brander, et al. 2013. "Global imprint of climate change on marine life." *Nature Climate Change* 3 (10): 919–925. https://doi.org/10.1038/nclimate1958.

Pörtner, H. 2001. "Climate induced temperature effects on growth performance, fecundity and recruitment in marine fish: Developing a hypothesis for cause and effect relationships in Atlantic cod (*Gadus morhua*) and common eelpout (*Zoarces viviparus*)." *Continental Shelf Research* 21 (18–19): 1975–1997. https://doi.org/10.1016/S0278-4343(01)00038-3.

Pörtner, H. O., D. Karl, P. W. Boyd, W. Cheung, S. E. Lluch-Cota, Y. Nojiri, D. N. Schmidt, and P. O. Zavialov. 2014. "Ocean systems." In *Climate Change 2014: Impacts, Adaptation, and Vulnerability. Part A: Global and Sectoral Aspects. Contribution of Working Group II to the Fifth Assessment Report of the Intergovernmental Panel on Climate Change*, ed. C. B. Field, V. R. Barros, D. J. Dokken, K. J. Mach, M. D. Mastrandrea, T. E. Bilir, M. Chatterjee, et al., 1–138. Cambridge University Press.

Pörtner, H.-O., C. Bock, and F. C. Mark. 2017. "Oxygen- and capacity-limited thermal tolerance: Bridging ecology and physiology." *Journal of Experimental Biology* 220: 2685–2696.

Saba, V. S., C. A. Stock, J. R. Spotila, F. V. Paladino, and P. S. Tomillo. 2012. "Projected response of an endangered marine turtle population to climate change." *Nature Climate Change* 2 (11): 814–820.

Salvadeo, C. J., D. Lluch-Belda, A. Gómez-Gallardo, J. Urbán-Ramírez, and C. D. MacLeod. 2010. "Climate change and a poleward shift in the distribution of the Pacific white-sided dolphin in the northeastern Pacific." *Endangered Species Research* 11 (March): 13–19. https://doi.org/10.3354/esr00252.

Schmidtko, S., L. Stramma, and M. Visbeck. 2017. "Decline in global oceanic oxygen content during the past five decades." *Nature* 542: 335–339.

Simpson, Stephen D., Simon Jennings, Mark P. Johnson, Julia L. Blanchard, Pieter-Jan Schön, David W. Sims, and Martin J. Genner. 2011. "Continental shelf-wide response of a fish assemblage to rapid warming of the sea." *Current Biology* 21 (September): 1565–1570. https://doi.org/10.1016/j.cub.2011.08.016.

Sims, D. W., M. J. Genner, A. J. Southward, and S. J. Hawkins. 2001. "Timing of squid migration reflects North Atlantic climate variability." *Proceedings of the Royal Society of London B: Biological Sciences* 268 (1485): 2607–2611. https://doi.org/10.1098/rspb.2001.1847.

Sumaila, U. Rashid, William W. L. Cheung, Vicky W. Y. Lam, Daniel Pauly, and Samuel Herrick. 2011. "Climate change impacts on the biophysics and economics of world fisheries." *Nature Climate Change* (November): 1–8. https://doi.org/10.1038/nclimate1301.

Sydeman, W. J., and S. J. Bograd. 2009. "Marine ecosystems, climate and phenology: Introduction." *Marine Ecology Progress Series* 393: 185–188.

Thomas, C. D., E. J. Bodsworth, R. J. Wilson, A. D. Simmons, Z. G. Davies, M. Musche, and L. Conradt. 2001. "Ecological and evolutionary processes at expanding range margins." *Nature* 411: 577–581.

Thomas, M. K., C. T. Kremer, C. A. Klausmeier, and E. Litchman. 2012. "A global pattern of thermal adaptation in marine phytoplankton." *Science* 338: 1085–1089. https://doi.org/10.1126/science.1224836.

Tittensor, Derek P., Camilo Mora, Walter Jetz, Heike K. Lotze, Daniel Ricard, Edward Vanden Berghe, and Boris Worm. 2010. "Global patterns and predictors of marine biodiversity across taxa." *Nature* 466 (7310): 1098–1101. https://doi.org/10.1038/nature09329.

Trathan, Phil N., Pablo García-Borboroglu, Dee Boersma, Charles-André Bost, Robert J. M. Crawford, Glenn T. Crossin, Richard J. Cuthbert, et al. 2014. "Pollution, habitat loss, fishing, and climate change as critical threats to penguins." *Conservation Biology: The Journal of the Society for Conservation Biology* 29 (1): 31–34. https://doi.org/10.1111/cobi.12349.

United Nations. 2017. *The First Global Integrated Marine Assessment: World Ocean Assessment I.* Cambridge University Press. https://doi.org/10.1017/9781108186148.

Vuuren, Detlef P. van, Keywan Riahi, Richard Moss, Jae Edmonds, Allison Thomson, Nebojsa Nakicenovic, Tom Kram, et al. 2012. "A proposal for a new scenario framework to support research and assessment in different climate research communities." *Global Environ-mental Change* 22 (February): 21–35. https://doi.org/10.1016/j.gloenvcha.2011.08.002.

Weishampel, J. F., D. A. Bagley, and L. M. Ehrhart. 2004. "Earlier nesting by loggerhead sea turtles following sea surface warming." *Global Change Biology* 10 (8): 1424–1427.

Witt, M. J., L. A. Hawkes, M. H. Godfrey, B. J. Godley, and A. C. Broderick. 2010. "Predicting the impacts of climate change on a globally distributed species: The case of the loggerhead turtle." *Journal of Experimental Biology* 213 (6): 901–911.

Wittman, A. C., and H. O. Pörtner. 2013. "Sensitivities of extant animal taxa to ocean acidification." *Nature Climate Change* 3. https://doi.org/10.1038/nclimate1982.

Wolf, S. G., M. A. Snyder, W. J. Sydeman, D. F. Doak, and D. A. Croll. 2010. "Predicting population consequences of ocean climate change for an ecosystem sentinel, the seabird Cassin's auklet." *Global Change Biology* 16 (7): 1923–1935.

Anticipating Climate-Driven Movement Routes

Joshua J. Lawler

There is clear evidence that the distributions of many species have been changing in ways that correspond with recent changes in climate (Chen et al. 2011; Pinsky et al. 2013). As climates continue to change, the ranges of many species will very likely continue to shift to track suitable climates. Some species will be able to keep pace with changing climatic conditions and others will not (Nathan et al. 2011; Schloss, Nuñez, and Lawler 2012). A species' ability to track suitable climates will depend not only on the nature of the climatic changes but also on the species' ability to move across landscapes—many of which are dominated by human activity.

A recent study explored how species might move across the landscapes of the Western Hemisphere to track projected changes in climate (Lawler et al. 2013). The study drew on the projections of climatic niche models built for 2,903 species of birds, mammals, and amphibians (Lawler et al. 2009). The authors used the models to identify areas that might become newly climatically suitable and those that might become climatically unsuitable for each species by the end of the twenty-first century under 10 different future climate projections. They then mapped potential movement routes that individuals of each species might follow to either move into areas projected to become newly climatically suitable or out of areas projected to become climatically unsuitable.

To model the potential effects of patterns of human activity on movement routes, the authors used the Human Influence Index (HII; Sanderson et al. 2002). The HII maps patterns of human impact using estimates of human population densities, distance to roads, land cover, and nighttime lights. Movement routes were mapped using Circuitscape, a modeling tool that maps landscape connectivity on the basis of electrical circuit theory (McRae et al. 2008). Circuitscape treats a landscape as a circuit board in which each grid cell in a digital map is a resistor with an assigned resistance value. Resistance values for the 2,903 spe-

cies were assigned according to the values of the HII, with areas of higher human influence receiving higher resistance values. Movement was then modeled from each grid cell in a digital map that represented the area projected to become climatically unsuitable to the area of the map that was projected to be climatically suitable in the future. Likewise, movement was modeled from the area that is currently suitable into each of the grid cells in areas projected to become newly climatically suitable. When the results from the individual species analyses were combined and averaged across the 10 climate change projections, clear patterns of potential climate-driven movements emerged.

There are likely to be several places in North and South America that will act as critical conduits for species movements in a changing climate (see Plate 6). Two areas with pronounced, consistent, projected movements of species are in southeastern North America where species are projected to move northeast, up into the Appalachian Mountains and along their length to the north, and in central Argentina, where species are projected to move southward into the southern Pampas and up in the Sierras de Córdoba and Andes. These are places

183

where the projected changes in climate are likely to force species movements poleward and upward in elevation. They are also places where the human presence on the landscape is likely to channel movements in a specific way.

In general, the tundra, boreal forest and taiga, north temperate broadleaf and mixed forests, and the tropical and subtropical moist broadleaf forest ecoregions are projected to experience the largest number of movements (controlling for current species richness). The far northern latitudes, the Amazon Basin, and northern Paraguay and southeastern Bolivia are projected to experience many movements and are generally less affected by human activity. Thus, these areas, as well as the areas of intense anticipated movement in central Argentina and the southern Appalachians, may be ones where conservation efforts to protect connectivity may serve moving species well.

Protecting and enhancing connectivity to allow species to track changes in climate are often-cited adaptation strategies for conserving biodiversity in the face of climate change (Heller and Zavaleta 2009). Targeting connectivity efforts at areas identified to be critical for facilitating species movements is a first step in planning for climate-driven species movements. However, the hemispheric-scale study described here was conducted at a coarse spatial resolution and thus does not provide the fine-scale analyses needed to target specific sites for on-the-ground conservation efforts. Several studies have begun to explore how to identify specific locations that, if protected or restored, have the potential to facilitate climate-driven species movements. These studies have focused on linking snapshots of potentially climatically suitable habitat through time, mapping routes that follow climatic gradients (Nuñez et al. 2013), and connecting similar (and diverse sets of) topographically defined zones (e.g., Brost and Beier 2012). By combining projections from coarser resolution studies with fine-scale

connectivity mapping efforts, it may be possible to identify specific areas in which focused conservation actions will facilitate climate-driven species movements.

REFERENCES

Brost, B., and P. Beier. 2012. "Use of land facets to design linkages for climate change." *Ecological Applications* 22 (1): 87–103.

Chen, I. C., J. K. Hill, R. Ohlemueller, D. B. Roy, and C. D. Thomas. 2011. "Rapid range shifts of species associated with high levels of climate warming." *Science* 333 (6045): 1024–1026.

Heller, N. E., and E. S. Zavaleta. 2009. "Biodiversity management in the face of climate change: A review of 22 years of recommendations." *Biological Conservation* 142 (1): 14–32.

Lawler, J. J., S. L. Shafer, D. White, P. Kareiva, E. P. Maurer, A. R. Blaustein, and P. J. Bartlein. 2009. "Projected climate-induced faunal change in the Western Hemisphere." *Ecology* 90 (3): 588–597.

Lawler, J. J., Aaron S. Ruesch, J. D. Olden, and B. H. McRae. 2013. "Projected climate-driven faunal movement routes." *Ecology Letters* 16 (8): 1014–1022.

McRae, B. H., B. G. Dickson, T. H. Keitt, and V. B. Shah. 2008. "Using circuit theory to model connectivity in ecology, evolution, and conservation." *Ecology* 89: 2712–2724. https://doi.org/10.1890/07-1861.1.

Nathan, R., N. Horvitz, Y. P. He, A. Kuparinen, F. M. Schurr, and G. G. Katul. 2011. "Spread of North American wind-dispersed trees in future environments." *Ecology Letters* 14 (3): 211–219.

Nuñez, T. A., J. J. Lawler, B. H. McRae, D. J. Pierce, M. B. Krosby, D. M. Kavanagh, P. H. Singleton, and J. J. Tewksbury. 2013. "Connectivity planning to address climate change." *Conservation Biology* 27 (2): 407–416. https://doi.org/10.1111/cobi.12014.

Pinsky, M. L., B. Worm, M. J. Fogarty, J. L. Sarmiento, S. A. Levin. 2013. "Marine taxa track local climate velocities." *Science* 341 (6151): 1239–1242.

Sanderson, E. W., M. Jaiteh, M. A. Levy, K. H. Redford, A. V. Wannebo, and G. Woolmer. 2002. "The human footprint and the last of the wild the human footprint is a global map of human influence on the land surface, which suggests that human beings are stewards of nature, whether we like it or not." *BioScience* 52 (10): 891–904. https://doi.org/10.1641/0006-3568(2002)052[0891:THFATL]2.0.CO;2.

Schloss, C. A., T. A. Nuñez, and J. J. Lawler. 2012. "Dispersal will limit ability of mammals to track climate change in the Western Hemisphere." *Proceedings of the National Academy of Science* 109: 8606–8611.

CHAPTER FOURTEEN

Impacts of Ocean Acidification on Marine Biodiversity

JOAN A. KLEYPAS

INTRODUCTION

Rising concentrations of CO_2 in the atmosphere cause ocean pH to decline. This fact has been known for some time, but the impacts of declining pH on marine organisms and ecosystems became widely recognized only about 10–15 years ago (Kleypas et al. 2006). Ocean acidification has the same root cause as climate change—that is, increases in atmospheric CO_2 concentration—but climate change and ocean acidification present different challenges to life on Earth. Acidification is primarily a marine phenomenon (although freshwater systems may also be affected). Wherever the ocean is in contact with the atmosphere, Henry's law dictates that the concentration of CO_2 in seawater come to equilibrium with that of the atmosphere. Thus, as atmospheric CO_2 increases, so does the concentration of CO_2 in seawater. Inorganic carbon exists in multiple forms in seawater: CO_{2aq}, H_2CO_3 (carbonic acid), HCO_3^- (bicarbonate) and CO_3^{2-} (carbonate). An increase in CO_2 in seawater causes a shift in the equilibrium CO_2 chemistry so that the first three compounds (CO_{2aq}, carbonic acid, and bicarbonate) increase while carbonate decreases (Figure 14.1). Ocean pH also decreases, and collectively these are referred to as "changes in the CO_2 system in seawater." Ocean acidification technically refers to a reduction in ocean pH regardless of the process causing it (e.g., respiration, addition of acids), but it most commonly refers to the global aspect of how rising atmospheric CO_2 impacts the suite of ocean carbon chemistry properties. The chemistry of ocean acidification is well established, and the decrease in ocean pH in response to rising atmospheric CO_2 is

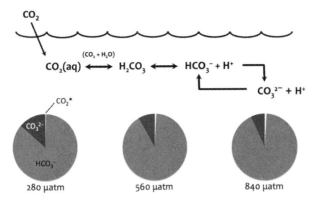

Figure 14.1. CO_2 equilibrium in seawater, showing how the increase in hydrogen ion concentration (due to the dissociation of carbonic acid, H_2CO_3) leads to the conversion of carbonate (CO_3^{2-}) to bicarbonate (HCO_3^-). This results in a shift in relative concentrations of the inorganic carbon compounds as atmospheric forcing increases from a CO_2 concentration of 280 μatm (the preindustrial level) through a doubling (560 μatm) and tripling (840 μatm).

well documented. But at local scales, ocean acidification occurs against a backdrop of many processes that can affect the CO_2 system in seawater and that operate from hourly to seasonal and longer time scales.

Ocean acidification was originally described in terms of geochemical effects, that is, how changes in the CO_2 system in seawater affect the formation and dissolution of the biomineral calcium carbonate ($CaCO_3$). Initially it was thought that ocean acidification was primarily a problem for marine calcifiers, particularly those organisms with only weak control over their skeletal-building process, such as algae, corals, and foraminifera. It is now widely recognized that the full suite of chemical changes in the CO_2 system impact not only $CaCO_3$ formation and dissolution, but also biogeochemistry, physiology, and behavior (Figure 14.2), and across a wide group of organisms, from protists to fish. Thus, the challenge of understanding how ocean acidification will affect marine biodiversity is perhaps as complicated as understanding how temperature increases will affect biodiversity.

Even though ocean acidification and its impacts constitute a relatively new field of study, the exponential growth in research provides a large body of evidence from which the question "How will ocean acidification affect marine biodiversity?" can be addressed. This chapter first summarizes the effects of ocean acidification from a geochemical, physiological and behavioral, and biogeochemical perspective, and then addresses how those effects ultimately have an impact on marine biodiversity.

HOW ACIDIFICATION IMPACTS ORGANISMS

Geochemical

The most widely recognized consequence of ocean acidification is its effects on the precipitation and dissolution of $CaCO_3$. Previous studies illustrate that the rates of precipitation and dissolution of abiotic $CaCO_3$ are a function of the ion concentrations of calcium and carbonate. This relationship is demonstrated by the distribution of $CaCO_3$ on the seafloor; $CaCO_3$ sediments are absent where the overlying waters are undersaturated with respect to aragonite and calcite. In fact, the most recent analogue for a global ocean acidification event—the Paleocene-Eocene Thermal Maximum 55 million years ago—was identified in the geologic record as the sudden disappearance of $CaCO_3$ sediments

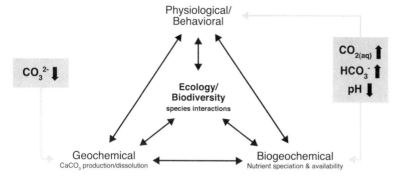

Figure 14.2. Simplified representation of how the various components of the CO$_2$ system in seawater (gray boxes) impact marine organisms, which collectively affect biodiversity.

from much of the ocean floor (Zachos et al. 2005). Most CaCO$_3$ production in the ocean is not abiotic, however, but biologically mediated. Almost all calcifying organisms exert some control over their calcification by isolating seawater and elevating its ionic concentrations of calcium and carbonate. Nonetheless, the rates of biotic precipitation in many (but not all) organisms reflect the saturation state of the surrounding seawater. Thus, biotic calcification has both geochemical and physiological components. The physiological aspects of biotic calcification are further addressed below.

The geochemical impacts of ocean acidification are considered a particular threat to marine biodiversity within ecosystems where calcium carbonate production provides the substrate that supports biodiversity. Coral reefs, cold-water reefs, and oyster reefs form where CaCO$_3$ production exceeds its dissolution (Plate 2). Dissolution is often overlooked in its importance, but when dissolution rates increase, the ability to grow or maintain reef structures is compromised even if the reef-building organisms are able to maintain calcification rates. The loss of reef substrate is exacerbated by the fact that ocean acidification enhances the biological breakdown and dissolution of CaCO$_3$ (Andersson and Gledhill 2013).

Physiological and Behavioral

Ocean acidification is known to have both favorable and unfavorable impacts on the physiology of marine organisms. The most common theme in any summary of ocean acidification impacts on physiology is the high degree of variability in the responses. Physiologically, organisms respond to acidification with alterations in photosynthesis, calcification, growth, and reproduction. The wide range of responses to ocean acidification largely reflects both how an organism obtains energy and how it allocates it to each one of these processes. If ocean acidification leads to greater energy uptake, then the organism may benefit physiologically. If ocean acidification presents an energetic demand, then the organism must respond with a reallocation of energy from other processes. To complicate matters, the responses vary within single species with genotypic plasticity and genetic diversity, with physiological state, and in the presence of other environmental stressors.

PRIMARY PRODUCTION

Photosynthetic rates of marine primary producers that are carbon limited should benefit, for example, from an increase in concentrations of CO$_2$ and/or HCO$_3^-$, at least when other nutrients are not limiting. Carbon fixation rates in marine phytoplankton do tend to increase with ocean acidification, but only slightly. The genetic diversity of many phytoplankton groups seems to explain the wide range of results in experiments testing the responses of phytoplankton species to ocean acidification. A comprehensive study of 16 ecotypes of the picoplankton *Ostreococcus tauri* revealed that although all ecotypes experienced increases in photosynthesis and growth, the responses ranged by a factor of two (Schaum et al. 2013). Such studies are important because they reveal the potential of genetic variation in maintaining phytoplankton populations into the future. For marine macroalgae and seagrasses, Koch et al. (2013) surmised that because most are C_3 photosynthesizers and carbon limited, their photosynthesis and growth rates are also likely to benefit from increases in CO_2 and HCO_3^- concentrations. Increases in production can alter food-web dynamics, and in some species it appears that increases in primary production are accompanied by changes in carbon to nutrient stoichiometry, which reduces their nutritional value for herbivores (e.g., Verspagen et al. 2014). For at least three species of seagrasses (*Cymodocea nodosa*, *Rupia maritima*, and *Potamogeton perfoliatus*), the benefits of increased growth appear to be compromised by a loss in phenolic acid concentrations that act to deter grazing (Arnold et al. 2012).

CALCIFICATION

Calcification rates of many species are likely to decline. Some marine calcifiers have the capacity to internally buffer the chemistry of their internal $CaCO_3$-precipitating fluids, but within many taxa, such as corals, coralline algae, foraminifera, and echinoderms, that capacity is weak or has energetic costs. Sometimes those costs can be supplemented through increases in photosynthesis or feeding, but there is considerable variation in calcification responses, even within a taxonomic group. Massive species of the reef-building coral genus *Porites*, for example, appear to be more tolerant of low saturation states than are branching coral species

within the genus *Acropora*. Within mollusks, calcification rates of most pteropod species are consistently reduced, but in the adults of many marine gastropods and bivalves the calcification responses vary strongly. Even within the coccolithophore *Emiliania huxleyi*, the calcification response to ocean acidification varies widely among genotypes (Raven and Crawfurd 2012).

REPRODUCTION

Many marine organisms invest a large amount of energy in reproduction, either through asexual means or in the production of sexual gametes. Ocean acidification has been found to affect the fecundity and quality of gametes that are produced, fertilization success, larval survival, and recruitment, but these impacts are certainly not universal. Some species, such as copepods, allocate more energy toward reproduction with ocean acidification (Fitzer et al. 2012). Reproduction in most fish species appears to remain robust to the projected decrease in ocean pH, but the responses are extremely varied (Secretariat of the Convention on Biological Diversity 2014).

ANIMAL BEHAVIOR

Changes in animal behavior are a rather surprising effect of ocean acidification. Ocean acidification impairs the chemosensory response in a number of fish species, with consequences for their ability to learn (Chivers et al. 2014), locate suitable habitat or food, and avoid predation (Munday et al. 2014). When ocean acidification has an impact on an organism's fitness, it can lead to a compensatory behavioral change. In the gastropod *Littorina littorea* for example, ocean acidification caused reduced shell thickness, which did not affect the organism physiologically but did result in enhanced escape activity when a predator was present (Bibby et al. 2007). Foraging rate of a deep-sea urchin increased under lowered pH conditions, presumably to compensate for an altered ability to detect food (Barry et al. 2014).

MULTIPLE LIFE STAGES

Ocean acidification is well known for its differential impacts on marine life stages. Many studies show, for example, that the nonadult stages of marine species are more vulnerable to ocean acidification than are the adults. This ranges from decreased survivorship to delayed development to reduced post-settlement survival, often creating bottlenecks in the population dynamics. In some cases, this could prevent the overpopulation of nuisance species, such as the corallivorous crown-of-thorns starfish, *Acanthaster planci* (Uthicke et al. 2013), but for other groups an increased vulnerability of early life stages could result in reduced recruitment rates unless those rates remain oversaturated at recruitment sites (Russell et al. 2012). Successful recruitment of some calcifying species appears to be hindered by inadequate calcification during shell formation or early growth.

ACCLIMATION AND ADAPTATION

The capacity for species to cope with ocean acidification builds on a trove of mechanisms from acclimation, genetic adaptation, mobility, and behavioral changes. Two excellent reviews of this topic present evidence that both acclimation and genetic adaptation to acidification are ongoing (Kelly and Hofmann 2013; Sunday et al. 2014). This is a large, emerging frontier that offers many tools for predicting the ability of species to cope with the changing environment, but several generalizations are worth mentioning here. Genetic evolution in response to acidification is more likely within species with large population sizes and rapid generation times (Lohbeck, Riebesell, and Reusch 2012). For species with longer generation times, adaptation is more likely to result from selection in the standing vari-

ation within the genetic pool, and it seems logical that species that thrive in a variable environment, such as coastal regions, may have a greater standing variation than those from more stable environments (Sunday et al. 2014).

Biogeochemical

The direct biogeochemical impacts of ocean acidification typically refer to the effects of pH on the speciation of metals such as iron and copper, which can be limiting to carbon and nitrogen fixation in phytoplankton communities. Far more is known, however, about the indirect effects of ocean acidification on ocean biogeochemistry. One rather robust biogeochemical response is that nitrogen fixation in some diazotrophs species (e.g., *Trichodesmium*) increases under elevated pCO_2 conditions (Hutchins, Mulholland, and Fu 2009), with potential impacts on the availability of nitrogen and on marine phytoplankton community composition, although this certainly requires more research before extrapolating to the global scale. The responses of other functional groups within microbial systems have been highly varied but could have strong biogeochemical feedbacks that could in turn affect biodiversity. Most heterotrophic bacterial studies have found very little effect on respiration, carbon demand, or growth efficiency (Motegi et al. 2013), although the benthic bacterial community near CO_2 vent sites in the Mediterranean is more diverse than those in nearby lower-CO_2 conditions (Kerfahi et al. 2014).

Another interesting biogeochemical aspect of ocean acidification is how it affects the production of organic compounds that are used as behavioral cues in other species. Coral larvae, for example, detect organic compounds produced by coralline algae as a cue for settlement, but coralline algae produce fewer of those compounds under low-pH conditions and coral larval settlement is reduced (Doropoulos and Diaz-Pulido 2013).

THE CONSEQUENCES FOR MARINE BIODIVERSITY

These geochemical, physiological, and biogeochemical effects of ocean acidification illustrate that the title of the Kroeker et al. (2010) review holds true today: "Meta-analysis reveals negative yet variable effects of ocean acidification on marine organisms." As described in the preceding section, the sheer number of ways that ocean acidification affects species leads to the compelling conclusion that ocean acidification will result in changes in marine biodiversity (Figure 14.2), but predicting the nature of those changes is a big challenge. Ocean acidification will affect marine biodiversity when it leads to a net change in the number of species and/or their relative dominance within a community. This can happen either directly or indirectly. The fitness of some species may simply be so compromised by ocean acidification that they cannot survive within a particular community. The indirect impacts of acidification are many and entail the complex interactions of symbiosis, competition, predator-prey relationships, availability of substrate, and changes in species functionality. How these interactions will play out in terms of biodiversity are perhaps best demonstrated by observations.

PRESENT-DAY OBSERVATIONS OF OCEAN ACIDIFICATION EFFECTS ON BIODIVERSITY

We need to understand how the diversity of marine ecosystems will respond to ocean acidification, as diversity is known to enhance ecosystem stability and complementarity and provide resistance to invasive species. As reviewed by Duffy and Stachowicz (2006), the diversity of both producers and consumers can strongly affect ecosystem processes, while genetic and species diversity impart a stabilizing effect to communities and increase their resilience to

disturbance. Few studies, however, have directly measured changes in biodiversity under ocean acidification conditions. Most of the information about future biodiversity within planktonic systems is derived from large mesocosm experiments, in which the CO_2 system chemistry is manipulated to simulate future conditions. Within benthic systems, most of the information on community response to ocean acidification has been derived from natural CO_2 vents that create gradients in the CO_2 system in seawater.

Microbial and Planktonic Systems

Studies of the long-term effects of ocean acidification on planktonic biodiversity are few, and the results have been inconsistent. Mesocosm experiments on the responses of heterotrophic microbial communities have so far revealed no consistent patterns (Weinbauer, Mari, and Gattuso 2011), although when combined with the effects of future temperature, acidification often seems to amplify those effects (Lindh et al. 2013).

The responses of phytoplankton communities may reflect the local adaptation of the communities to natural variability. The community compositions of phytoplankton communities in polar regions—Southern Ocean (Tortell et al. 2008) and Arctic Ocean (Schulz et al. 2013)—showed shifts in species compositions at elevated CO_2; but the compositions of two coastal plankton communities were largely unaffected by acidification (Nielsen, Jakobsen, and Hansen 2010). For calcifying plankton (including foraminifers, coccolithophores, mollusks, and echinoderms), an analysis of North Atlantic assemblages for the period 1960–2009 revealed that the largest changes were due to temperature rather than acidification (Beaugrand et al. 2013).

Benthic Systems

Some of the most compelling evidence of marine biodiversity changes comes from benthic systems located near shallow submarine CO_2 vents. Although caution should be used when interpreting the results from these sites, the responses of various taxa reflect the expected geochemical, physiological, and biogeochemical impacts of elevated CO_2, providing confidence that the observed changes in biodiversity are representative of a future state. Studies of the communities near CO_2 vents in the Mediterranean Sea were the first to document significant changes in communities along pH gradients from ambient (8.2) to as low as 7.4 (Hall-Spencer et al. 2008). The density and diversity of calcifying organisms declined; sea anemones, seagrasses, and macroalgae increased (including invasive species); and larval settlement of invertebrates declined. The loss of calcifying species reflects the direct impact of low pH on their ability to produce their $CaCO_3$ skeletons. The increase in seagrasses and algae is likely to reflect both an enhancement of growth and an increased ability to compete with slower-growing calcifiers for space.

Similar shifts occur at CO_2 vents within coral reef habitats in Japan (Inoue et al. 2013) and Papua New Guinea (Fabricius et al. 2011). In Japan, the ecosystem shifted from domination by hard corals to soft corals. In Papua New Guinea, the biodiversity of the coral community declined as pH decreased from 8.2 to 7.8, at which point a single coral type dominated. Below 7.8, carbonate accretion ceased, and seagrasses became dominant (Fabricius et al. 2011). The associated decline in biodiversity has been attributed to the loss of habitat complexity (Fabricius et al. 2014).

MECHANISMS OF BIODIVERSITY CHANGE

These field studies often confirm results from benthic mesocosm studies. Mesocosm and field manipulation experiments provide insights into the mechanisms of how ocean acidification is acting to alter biodiversity. A common theme in this chapter is the variability in the impacts of ocean acidification on species. This in itself is a driver of biodiversity loss because of how the variable impacts translate to changes in predation and competition (Hale et al. 2011). Several experiments have highlighted how changes in substrate, competition, predator-prey interactions, and functional roles have led to changes in biodiversity.

Habitat Complexity

Habitat complexity, particularly where there is a heterogeneous mix of large and small spaces (St. Pierre and Kovalenko 2014), is an important factor determining species richness across a variety of habitats, such as reefs, coralline turfs, and mangroves. The architectural complexity of reef $CaCO_3$ buildups, for example, provides habitats for a wide variety of organisms. As ocean acidification acts to break down that complexity, biodiversity declines (Fabricius et al. 2014). The biodiversity of deepwater coral banks also appears to be correlated with proximity to the aragonite saturation horizon, where the $CaCO_3$ sediments remains stable (Cairns 2007).

Species Competition

Changes in competition have been documented in several marine systems, particularly for competition between benthic algal species. Some taxa that respond positively to ocean acidification, such as matt-forming algae, can gain the competitive edge in kelp and reef environments (Connell et al. 2013), eliminating the availability of bare space for other species and leading to homogenization of the benthic community (Kroeker, Micheli, and Gambi 2013). Despite the increased ability of noncalcifying benthic algae to compete with calcifying algae, however, increased growth of benthic algae (which can elevate pH locally via photosynthetic removal of CO_2) can at the same time enhance increased calcification rates in calcifying algae. Ocean acidification's effect on competition between animals has been rarely investigated, but one study illustrated a reversal in dominance between two damselfish species (McCormick, Watson, and Munday 2013). The increased competitiveness of invasive species, particularly marine algae, has been a major concern with ocean acidification, as many invasive species share traits such as rapid dispersal and ability to tolerate a wide range of conditions that enable them to outcompete native species in a changing environment (Dukes and Mooney 1999).

Predator-Prey Interactions

The outcomes of predator-prey relationships depend on the relative responses of both prey and predator (Allan et al. 2013). Many important functional groups, such as grazers, play key roles in maintaining ecosystem structure, such as algae-grazing sea urchins in reef, rocky intertidal, and kelp habitats. Many studies in the lab and field suggest that sea urchins will be affected by ocean acidification and, along with that, their functional role as grazers. Thus, the expected ecological shift would be toward macroalgae-dominated systems. Yet indirect impacts can override the direct impacts of ocean acidification. Asnaghi et al. (2013) found that the urchins in their ocean acidification experiments were more strongly affected by the elimination of coralline algae in their diet, which led to weaker tests and increased vulnerability to predation, than by a direct impact of ocean acidification on urchin calcification. And even in simple systems, the impacts of ocean acidification on a key species such as a sea urchin can lead to an increase or a decrease in biodiversity. In macroalgae-dominated systems, reduced grazing would lead to a reduction in biodiversity, but in rocky barrens, reduced grazing would lead to an increase in biodiversity by allowing macroalgae to become reestablished (Asnaghi et al. 2013).

Multiple Stressors

Decoupling the competition and predator-prey linkages from field data is challenging, particularly because even weak interactions among species have stabilizing effects on food webs and ecosystems (O'Gorman and Emmerson 2009). This becomes ever more challenging given that multiple stressors are acting simultaneously. Rising temperature, decreasing oxygen, and a host of local to regional impacts can interact either antagonistically or synergistically with ocean acidification, and the ability of an organism to maintain normal functioning within a suite of changing environmental variables is extremely difficult to predict. Following the concept of oxygen and capacity-limited thermal tolerance (OCLTT), ocean acidification may reduce an organism's tolerance to other variables (Pörtner 2012). OCLTT has been supported by experimental studies with invertebrates, but less so in studies of temperature tolerance in fish. At the community level, the concept of co-tolerance described by Vinebrooke (2004) suggests that biodiversity (and ecosystem functioning) will be less impacted when the tolerance to two stressors (e.g., temperature and ocean acidification) is positively correlated than if they are negatively correlated or not correlated. If acidification leads to smaller calcifying plankton, for example, and that shift imparts a higher degree of temperature tolerance, then natural selection will lead to long-term co-tolerance to the two stressors. There are likely to be complications, however, due to ecological feedbacks; the function of a smaller calcifying phytoplankton within the community, for example, may also change.

SUMMARY

The past decade of research has moved us far from the simple assumption that ocean acidification is a problem primarily for marine calcifiers. Ocean acidification causes a suite of changes in the CO_2 system in seawater, and there is ample evidence that each of these changes can affect marine organisms in different ways. Increases in CO_2 and bicarbonate ion concentrations can stimulate primary production, decreases in carbonate ion concentration disrupt the balance of production and dissolution of calcium carbonate, and decreasing pH presents physiological challenges. We now know that not all calcifiers will be directly affected by ocean acidification. We also know that the impacts of ocean acidification on organisms vary widely between and within taxonomic groups, within species, and within the life-history stages of species. Marine biodiversity is most directly impacted by the simple loss of species that can no longer function within a high-CO_2 ocean. Marine biodiversity is likely to be just as strongly affected by the indirect effects of ocean acidification, through changes in competition, predator-prey interactions, and loss of complex substrates. With so much variation in responses, and so much uncertainty in predicting how marine communities will respond, it is understandable to be overwhelmed and conclude that predicting the outcomes, much less managing our ecosystems, is out of reach. However, some encouraging patterns are emerging. First, many projections of how ocean acidification will affect marine biodiversity appear to be validated by observations from regions with naturally high CO_2 levels, such as submarine volcanic vents. These sites uniformly illustrate changes in community composition due to a loss of calcifiers, a loss of habitat complexity, and an increase in benthic algae, with an associated loss in biodiversity. Second, the increase in our understanding of genetic adaptation is explaining much of the variation in the responses of species and communities. The combination of natural laboratories and genetic tools thus provides hope for understanding the adaptive capacities of species to ocean acidification as well as the marine communities in which they live.

ACKNOWLEDGMENTS

The author greatly acknowledges the many researchers whose work could not be properly cited here. This chapter originally included nearly 120 cited references, which had to be pared down to a mere 40. The author welcomes requests to provide an amended list of references that support the statements provided herein. The author's institution, the National Center for Atmospheric Research, is supported by the National Science Foundation.

REFERENCES

Allan, B. J. M., P. Domenici, M. I. McCormick, S. A. Watson, and P. L. Munday. 2013. "Elevated CO_2 affects predator-prey interactions through altered performance." *PLOS One* 8 (3). https://doi.org/10.1371/journal.pone.0058520.

Andersson, A. J., and D. Gledhill. 2013. "Ocean acidification and coral reefs: Effects on breakdown, dissolution, and net ecosystem calcification." *Annual Review of Marine Science* 5: 321–348. https://doi.org/10.1146/annurev-marine-121211-172241.

Arnold, T., C. Mealey, H. Leahey, A. W. Miller, J. M. Hall-Spencer, M. Milazzo, and K. Maers. 2012. "Ocean acidification and the loss of phenolic substances in marine plants." *PLOS One* 7 (4). https://doi.org/10.1371/journal.pone.0035107.

Asnaghi, V., M. Chiantore, L. Mangialajo, F. Gazeau, P. Francour, S. Alliouane, and J.-P. Gattuso. 2013. "Cascading effects of ocean acidification in a rocky subtidal community." *PLOS One* 8 (4). https://doi.org/10.1371/journal.pone.0061978.

Barry, J. P., C. Lovera, K. R. Buck, E. T. Peltzer, J. R. Taylor, P. Walz, P. J. Whaling, and P. G. Brewer. 2014. "Use of a free ocean CO_2 enrichment (FOCE) system to evaluate the effects of ocean acidification on the foraging behavior of a deep-sea urchin." *Environmental Science & Technology* 48 (16): 9890–9897. https://doi.org/10.1021/es501603r.

Beaugrand, G., A. McQuatters-Gollop, M. Edwards, and E. Goberville. 2013. "Long-term responses of North Atlantic calcifying plankton to climate change." *Nature Climate Change* 3 (3): 263–267. https://doi.org/10.1038/nclimate1753.

Bibby, R., P. Cleall-Harding, S. Rundle, S. Widdicombe, and J. Spicer. 2007. "Ocean acidification disrupts induced defences in the intertidal gastropod *Littorina littorea*." *Biology Letters* 3 (6): 699–701.

Cairns, S. D. 2007. "Deep-water corals: An overview with special reference to diversity and distribution of deep-water Scleractinian corals." *Bulletin of Marine Science* 81 (3): 311–322.

Chivers, D. P., M. I. McCormick, G. E. Nilsson, P. L. Munday, S. A. Watson, M. G. Meekan, M. D. Mitchell, K. C. Corkill, and M. C. O. Ferrari. 2014. "Impaired learning of predators and lower prey survival under elevated CO_2: A consequence of neurotransmitter interference." *Global Change Biology* 20 (2): 515–522. https://doi.org/10.1111/gcb.12291.

Connell, S. D., K. J. Kroeker, K. E. Fabricius, D. I. Kline, and B. D. Russell. 2013. "The other ocean acidification problem: CO_2 as a resource among competitors for ecosystem dominance." *Philosophical Transactions of the Royal Society B: Biological Sciences* 368 (1627). https://doi.org/10.1098/rstb.2012.0442.

Doropoulos, C., and G. Diaz-Pulido. 2013. "High CO_2 reduces the settlement of a spawning coral on three common species of crustose coralline algae." *Marine Ecology Progress Series* 475: 93–99. https://doi.org/10.3354/meps10096.

Duffy, J. E., and J. J. Stachowicz. 2006. "Why biodiversity is important to oceanography: Potential roles of genetic, species, and trophic diversity in pelagic ecosystem processes." *Marine Ecology Progress Series* 311: 179–189. https://doi.org/10.3354/meps311179.

Dukes, J. S., and H. A. Mooney. 1999. "Does global change increase the success of biological invaders?" *Trends in Ecology & Evolution* 14 (4): 135–139. https://doi.org/10.1016/s0169-5347(98)01554-7.

Fabricius, K. E., G. De'ath, S. Noonan, and S. Uthicke. 2014. "Ecological effects of ocean acidification and habitat complexity on reef-associated macroinvertebrate communities." *Proceedings of the Royal Society B: Biological Sciences* 281 (1775). https://doi.org/10.1098/rspb.2013.2479.

Fabricius, K. E., C. Langdon, S. Uthicke, C. Humphrey, S. Noonan, G. De'ath, R. Okazaki, N. Muehllehner, M. S. Glas, and J. M. Lough. 2011. "Losers and winners in coral reefs acclimatized to elevated carbon dioxide concentrations." *Nature Climate Change* 1 (3): 165–169. https://doi.org/10.1038/NCLIMATE1122.

Fitzer, S. C., G. S. Caldwell, A. J. Close, A. S. Clare, R. C. Upstill-Goddard, and M. G. Bentley. 2012. "Ocean acidification induces multi-generational decline in copepod naupliar production with possible conflict for reproductive resource allocation." *Journal of Experimental Marine Biology and Ecology* 418: 30–36. https://doi.org/10.1016/j.jembe.2012.03.009.

Hale, R., P. Calosi, L. McNeill, N. Mieszkowska, and S. Widdicombe. 2011. "Predicted levels of future ocean acidification and temperature rise could alter community structure and biodiversity in marine benthic communities." *Oikos* 120 (5): 661–674. https://doi.org/10.1111/j.1600-0706.2010.19469.x.

Hall-Spencer, J. M., R. Rodolfo-Metalpa, S. Martin, E. Ransome, M. Fine, S. M. Turner, S. J. Rowley, D. Tedesco, and M. C. Buia. 2008. "Volcanic carbon dioxide vents show ecosystem effects of ocean acidification."

Nature 454 (7200): 96–99. https://doi.org/10.1038/nature07051.

Hutchins, D. A., M. R. Mulholland, and F. X. Fu. 2009. "Nutrient cycles and marine microbes in a CO_2-enriched ocean." *Oceanography* 22 (4): 128–145.

Inoue, S., H. Kayanne, S. Yamamoto, and H. Kurihara. 2013. "Spatial community shift from hard to soft corals in acidified water." *Nature Climate Change* 3 (7): 683–687. https://doi.org/10.1038/nclimate1855.

Kelly, Morgan W., and Gretchen E. Hofmann. 2013. "Adaptation and the physiology of ocean acidification." *Functional Ecology* 27 (4): 980–990. https://doi.org/10.1111/j.1365-2435.2012.02061.x.

Kerfahi, D., J. M. Hall-Spencer, B. M. Tripathi, M. Milazzo, J. Lee, and J. M. Adams. 2014. "Shallow water marine sediment bacterial community shifts along a natural CO_2 gradient in the Mediterranean Sea off Vulcano, Italy." *Microbial Ecology* 67 (4): 819–828. https://doi.org/10.1007/s00248-014-0368-7.

Kleypas, J. A., R. A. Feely, V. J. Fabry, C. Langdon, C. L. Sabine, and L. L. Robbins. 2006. "Impacts of ocean acidification on coral reefs and other marine calcifiers: A guide for future research." Report of a workshop held 18–20 April 2005, St. Petersburg, FL, sponsored by NSF, NOAA, and the U.S. Geological Survey.

Koch, M., G. Bowes, C. Ross, and X. H. Zhang. 2013. "Climate change and ocean acidification effects on seagrasses and marine macroalgae." *Global Change Biology* 19 (1): 103–132. https://doi.org/10.1111/j.1365-2486.2012.02791.x.

Kroeker, K. J., R. L. Kordas, R. N. Crim, and G. G. Singh. 2010. "Meta-analysis reveals negative yet variable effects of ocean acidification on marine organisms." *Ecology Letters* 13 (11): 1419–1434. https://doi.org/10.1111/j.1461-0248.2010.01518.x.

Kroeker, K. J., F. Micheli, and M. C. Gambi. 2013. "Ocean acidification causes ecosystem shifts via altered competitive interactions." *Nature Climate Change* 3 (2): 156–159. https://doi.org/10.1038/nclimate1680.

Lindh, M. V., L. Riemann, F. Baltar, C. Romero-Oliva, P. S. Salomon, E. Graneli, and J. Pinhassi. 2013. "Consequences of increased temperature and acidification on bacterioplankton community composition during a mesocosm spring bloom in the Baltic Sea." *Environmental Microbiology Reports* 5 (2): 252–262. https://doi.org/10.1111/1758-2229.12009.

Lohbeck, Kai T., Ulf Riebesell, and Thorsten B. H. Reusch. 2012. "Adaptive evolution of a key phytoplankton species to ocean acidification." *Nature Geoscience* 5 (5): 346–351. https://doi.org/10.1038/ngeo1441.

McCormick, M. I., S. A. Watson, and P. L. Munday. 2013. "Ocean acidification reverses competition for space as habitats degrade." *Scientific Reports* 3. https://doi.org/10.1038/srep03280.

Motegi, C., T. Tanaka, J. Piontek, C. P. D. Brussaard, J. P. Gattuso, and M. G. Weinbauer. 2013. "Effect of CO_2 enrichment on bacterial metabolism in an Arc-

tic fjord." *Biogeosciences* 10 (5): 3285–3296. https://doi
.org/10.5194/bg-10-3285-2013.

Munday, P. L., A. J. Cheal, D. L. Dixson, J. L. Rummer,
and K. E. Fabricius. 2014. "Behavioural impairment
in reef fishes caused by ocean acidification at CO_2
seeps." *Nature Climate Change* 4 (6): 487–492. https://doi
.org/10.1038/nclimate2195.

Nielsen, L. T., H. H. Jakobsen, and P. J. Hansen. 2010.
"High resilience of two coastal plankton communities to twenty-first century seawater acidification: Evidence from microcosm studies." *Marine Biology Research* 6 (6): 542–555. https://doi
.org/10.1080/17451000903476941.

O'Gorman, E. J., and M. C. Emmerson. 2009. "Perturbations to trophic interactions and the stability of complex food webs." *Proceedings of the National Academy of Sciences of the United States of America* 106 (32): 13393–13398.
https://doi.org/10.1073/pnas.0903682106.

Pörtner, H. O. 2012. "Integrating climate-related stressor
effects on marine organisms: Unifying principles
linking molecule to ecosystem-level changes." *Marine Ecology Progress Series* 470: 273–290. https://doi.org/
10.3354/meps10123.

Raven, J. A., and K. Crawfurd. 2012. "Environmental
controls on coccolithophore calcification." *Marine Ecology Progress Series* 470: 137–166. https://doi.org/10.3354/
meps09993.

Russell, B. D., C. D. G. Harley, T. Wernberg, N. Mieszkowska, S. Widdicombe, J. M. Hall-Spencer, and S. D.
Connell. 2012. "Predicting ecosystem shifts requires
new approaches that integrate the effects of climate
change across entire systems." *Biology Letters* 8 (2): 164–
166. https://doi.org/10.1098/rsbl.2011.0779.

Schaum, E., B. Rost, A. J. Millar, and S. Collins. 2013.
"Variation in plastic responses of a globally distributed picoplankton species to ocean acidification."
Nature Climate Change 3 (3): 298–302. https://doi.org/
10.1038/nclimate1774.

Schulz, K. G., R. G. J. Bellerby, C. P. D. Brussaard, J.
Budenbender, J. Czerny, A. Engel, M. Fischer, S. Koch-
Klavsen, S. A. Krug, S. Lischka, A. Ludwig, M. Mey-
erhofer, G. Nondal, A. Silyakova, A. Stuhr, and U.
Riebesell. 2013. "Temporal biomass dynamics of an
Arctic plankton bloom in response to increasing lev-
els of atmospheric carbon dioxide." *Biogeosciences* 10 (1):
161–180. https://doi.org/10.5194/bg-10-161-2013.

Secretariat of the Convention on Biological Diversity.
2014. "An updated synthesis of the impacts of ocean
acidification on marine biodiversity." In *Technical Series
No. 75*, ed. S. Hennige, J. M. Roberts, and P. Williamson. Montreal: Secretariat of the Convention on Biological Diversity.

St. Pierre, J. I., and K. E. Kovalenko. 2014. "Effect of habitat complexity attributes on species richness." *Ecosphere*
5 (2). https://doi.org/10.1890/es13-00323.1.

Sunday, J. M., P. Calosi, S. Dupont, P. L. Munday, J. H.
Stillman, and T. B. H. Reusch. 2014. "Evolution in an
acidifying ocean." *Trends in Ecology & Evolution* 29 (2):
117–125. https://doi.org/10.1016/j.tree.2013.11.001.

Tortell, P. D., C. D. Payne, Y. Y. Li, S. Trimborn, B. Rost,
W. O. Smith, C. Riesselman, R. B. Dunbar, P. Sedwick,
and G. R. DiTullio. 2008. "CO_2 sensitivity of Southern
Ocean phytoplankton." *Geophysical Research Letters* 35 (4).
https://doi.org/10.1029/2007gl032583.

Uthicke, S., D. Pecorino, R. Albright, A. P. Negri, N.
Cantin, M. Liddy, S. Dworjanyn, P. Kamya, M. Byrne,
and M. Lamare. 2013. "Impacts of ocean acidification on early life-history stages and settlement of the
coral-eating sea star *Acanthaster planci*." *PLOS One* 8 (12).
https://doi.org/10.1371/journal.pone.0082938.

Verspagen, J. M. H., D. B. Van de Waal, J. F. Finke, P. M.
Visser, and J. Huisman. 2014. "Contrasting effects of
rising CO_2 on primary production and ecological stoichiometry at different nutrient levels." *Ecology Letters* 17
(8): 951–960. https://doi.org/10.1111/ele.12298.

Vinebrooke, R. D., K. L. Cottingham, J. Norberg, M.
Scheffer, S. I. Dodson, S. C. Maberly, and U. Sommer.
2004. "Impacts of multiple stressors on biodiversity
and ecosystem functioning: The role of species co-
tolerance." *Oikos* 104 (3): 451–457.

Weinbauer, M. G., X. Mari, and J.-P. Gattuso. 2011. "Effects of ocean acidification on the diversity and activity of heterotrophic marine microorganisms." In *Ocean
Acidification*, ed. J. P. Gattuso and L. Hansson, 83–98.
Oxford University Press.

Zachos, J. C., U. Rohl, S. A. Schellenberg, A. Sluijs, D.
A. Hodell, D. C. Kelly, E. Thomas, M. Nicolo, I. Raffi,
L. J. Lourens, H. McCarren, and D. Kroon. 2005.
"Rapid acidification of the ocean during the Paleo-
cene-Eocene Thermal Maximum." *Science* 308 (5728):
1611–1615.

CHAPTER FIFTEEN

Tropical Forests in a Changing Climate

JAMES E. M. WATSON, DANIEL B. SEGAN, AND JOSHUA TEWKSBURY

OVERVIEW

Tropical forest landscapes are the most bio-diverse terrestrial biomes on the planet, supporting 50 percent of described species and an even larger number of undescribed species (Dirzo and Raven 2003). Thermophilic forests similar to today's tropical forests first appeared in the early Cenozoic (50–65 Mya), and today they span five distinct biogeographical regions: tropical Americas, Africa, Southeast Asia, Madagascar, and New Guinea, with small outliers in Australia, South Asia, and many tropical islands (Corlett and Primack 2006) (Box 15.1). The exceptional species richness at a global scale is produced by two commonly shared complementary patterns: exceptionally high local diversity, with species richness in many tropical forests sites 20 times greater than richness in comparable temperate forest sites, and greater levels of species turnover along environmental gradients (Dirzo and Raven 2003). In addition, genetic divergence between populations is much higher in the tropics, which makes them globally important ecosystems when considering long-term evolutionary processes.

Tropical forests play a critical role in regulating global climate, accounting for one-third of land-surface productivity and evapotranspiration (Malhi 2012). Although debates still rage about how much carbon tropical forests contain relative to other ecosystems, there is no doubt that tropical forest ecosystems play an important role in safeguarding the world against human-forced climate change and that efforts to Reduce Emissions from Deforestation and Forest Degradation (REDD+) need to be financed (IPCC 2013; Watson et al. 2018).

Box 15.1 **Are Tropical Forests Climatically Stable?**

One way of broadly ascertaining how vulnerable tropical forests are going to be with future climate change is to assess their "stability" at the ecoregional scale. Climate stability has been defined as the similarity between current and future climate (in the 2050s) using a six-variable envelope-based gauge to represent general climate patterns and seasonality (Watson et al. 2013). This gauge combines patterns of annual mean temperature, mean diurnal temperature range, mean annual temperature range, annual precipitation, precipitation seasonality, and precipitation of the driest quarter, and it assesses the degree to which regions are likely to shift outside of their current climate envelopes. By this measure, climates in tropical forest ecoregions in South America and Southeast Asia are less stable than climates in tropical forest ecoregions in Africa and Oceania (Figures 15.1 and 15.2B). When tropical forest climate stability is integrated with the ecoregion's "natural integrity" (defined as the proportion of intact natural vegetation found in each ecoregion, and thus a function of past land use), a measure of global ecosystem vulnerability emerges that comprises elements of exposure to climate change and an ecoregions adaptive capacity (Figures 15.1 and 15.2C). It is clear that some tropical ecoregions are very vulnerable to a changing climate, with East African, southeastern South American, and northern Southeast Asian forests standing out as areas of particular concern.

If a formal carbon market for REDD+ does eventuate, it will have significant financial implications. A recent broad global analysis also showed that between 2000 and 2005, reduced carbon emissions from the 17.2 million ha (19.6 percent) of protected humid tropical forests were worth approximately US$6.2–7.4 billion (Scharlemann et al. 2010). Tropical forests bolster national economies through tourism and timber revenues and provide essential ecosystem services to many thousands of local communities (de Groot et al. 2012).

The level of recent tropical forest loss has been enormous. In 2000, tropical forests covered 10 percent–13 percent of the terrestrial surface of the earth; today, intact tropical forests make up less than 5 percent of the terrestrial surface of the earth (Mackey et al. 2014; Watson et al. 2018). Absolute loss rates for dense forests between 2000 and 2012 were highest in the tropical Americas (39,900 km² per year), followed by tropical Asia (22,300 km² per year), and Africa (11,000 km² per year). Relative rates were highest in Asia (0.62 percent per year),

followed by the Americas (0.45 percent per year) and Africa (0.28 percent per year) (Malhi et al. 2014). Remaining tropical forests not severely affected by land clearance have been subject to pervasive disturbances from logging, defaunation, wildfire, and fragmentation. Their separate and combined effects have resulted in marked forest degradation, loss of biodiversity, and impairment of ecological processes (Figure 15.1). Many human-modified tropical forest landscapes exist as complex mosaics of primary- and second-growth forest, with a patchwork of regenerating areas on fallow or abandoned agricultural land. Importantly, though often ignored, tropical forested ecosystems around the world are already experiencing the impacts of human-forced climate change.

In this chapter, we briefly outline the magnitude of impact human activities have had on tropical forests to date. We then describe likely responses of species and ecosystems across the tropical forest biome under climate change forecasts, focusing on vulnerabilities of individuals and

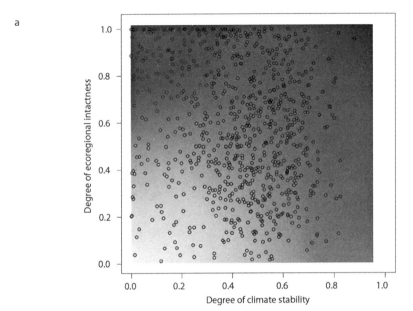

a

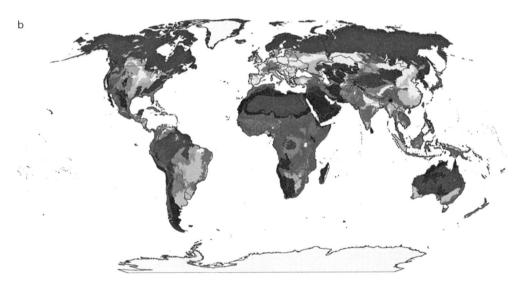

b

Figure 15.1. The relationship between climate stability and mean intactness in tropical forested ecoregions by different region (n = 308). Tropical forested ecoregions are defined broadly as those containing a tropical or subtropical moist broadleaf forests, dry broadleaf forests, coniferous forests, and mangroves. Climate stability was defined as a measure of similarity between forecasted climatic conditions in 2050 and the current climate of the ecoregion. The climate stability shown here is the average over the results from the seven GCMs (see Iwamura et al. 2013 for additional details). Vegetation intactness was calculated using the GlobCover V2 data set, as 1 minus the proportion of the ecoregional area where native vegetation has been transformed through agricultural development and urbanization. The intactness axis has been transformed to a normal distribution for presentation purposes by taking the square root values. (Figure and data adopted from Watson et al. 2013.)

population, communities, and ecosystem. Although these direct impacts will be important, we argue that the human response to climate change in tropical forest landscapes are going to significantly increase the vulnerability of already highly modified tropical forest systems, and we describe some of the likely major consequences. We conclude with some suggested strategies for conservation of tropical forests that take into account both the biological and human response to current and future climate change.

IMPACTS OF ANTHROPOGENIC CLIMATE CHANGE TO DATE

Since the mid-1970s land-surface temperatures in tropical rainforest regions have been increasing by ~0.25°C per decade as a result of increases in the concentration of atmospheric greenhouse gases (Malhi et al. 2009). In regions that have been highly deforested, this warming has been exacerbated by the direct loss of tree cover, which decreases land-surface evapotranspiration and alters cloud formation patterns (Davin and de Noblet-Ducoudré 2010). Observational studies in tropical forest mountain transects suggest that there has been an upward shift in mean species distributions of species (Colwell et al. 2008). There is already direct, observational evidence of changes in the hydrological cycle; over the past 70 years, many tropical regions have seen a steady increase in the seasonality of rainfall and a steady decrease in rainfall predictability (Feng, Porporato, and Rodriguez-Iturbe 2013). Two large-scale rainfall exclusion experiments in Brazil have demonstrated that large trees are particularly vulnerable to drought (da Costa et al. 2010), and field surveys across Amazonia following an intense drought in 2005 demonstrated a widespread increase in tree mortality and a decrease in rates of carbon sequestration (Phillips et al. 2009). Because the tropics play an important role in regulating global atmospheric circulation, changes in tropical forest cover almost certainly have global consequences for temperature and precipitation patterns (Figure 15.2).

What Does the Future Look Like for Tropical Forests?

Atmospheric CO_2 has risen from preindustrial concentrations of 280 parts per million (ppm) to 400 ppm in 2014, and it could to rise to 600–1,000 ppm by 2100 under some emissions scenarios (IPCC 2013). Anthropogenic greenhouse gas emissions have committed tropical forests to increased temperatures, altered rainfall patterns, and resulted in increases in the frequency and severity of extreme events (Collins 2013). The speed and magnitude of these changes will have large-scale impacts on tropical forest biodiversity, as the climates throughout the tropics will fall outside of the historic range of variability for tropical forest regions up to a decade faster than any other major terrestrial ecosystem and depending on emission scenarios, possibly within the next 25–45 years (Mora et al. 2013). This extremely rapid departure from nineteenth- and twentieth-century climates will affect tropical forest communities in a wide range of ways, and our capacity to predict those impacts is uneven.

Predicted changes in the physical climate can be broken into three broad categories—changes in temperature, precipitation, and CO_2. Mean land-surface temperatures in tropical forest regions are expected to increase by 3°C–6°C this century (Zelazowski et al. 2011). The warming predicted is similar in extent to the warming of the Paleocene-Eocene Thermal Maximum (PETM), but the speed of the warming is unprecedented, perhaps two orders of magnitude faster than previous periods of rapid tropical warming (Jaramillo et al. 2010). Global precipitation is positively linked to global temperature increase, but in tropical regions the most robust signal is for changes to, and increases in, rainfall seasonal-

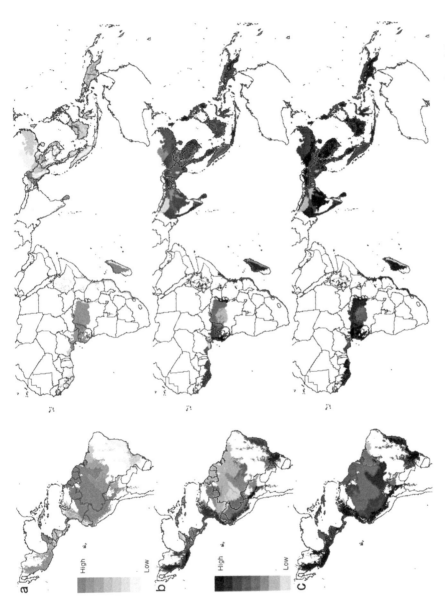

Figure 15.2. Global distribution of tropical forest ecoregional vegetation intactness (A), climate stability (B), and an overlay of climate stability and vegetation intactness (C). Ecoregions that have high relative climate stability and high vegetation intactness are depicted in black. Ecoregions that have low relative climate stability but high vegetation intactness are depicted in dark gray. Ecoregions that have both low relative climate stability and low levels of vegetation intactness are depicted in light gray. The tropical forested ecoregions of Asia are more climatically stable but less intact, relative to the tropical forests of the Amazon and Congo Basins. The darker colors on panel A indicate higher ecoregional intactness, while darker colors on panel B represent more relative stable climates. (Adapted from Watson et al. 2013.)

ity. This is characterized by more intense dry seasons, stronger and more frequent droughts, and stronger, longer-lasting heat waves (Malhi et al. 2014). This pattern is particularly clear in eastern Amazonia (Collins 2013). The latest IPCC report predicted that the only tropical forest regions that will experience significant declines in total rainfall are Central America and northern and eastern Amazonia.

DIRECT IMPACTS

Tolerance, Adaptation, and Acclimation

Species at all latitudes will respond to warming through a combination of physiological tolerance and acclimation, evolutionary adaptation, migration, and dispersal. Our understanding of individual species' resilience to rapid climate change, created through these different response mechanisms, is far from complete, and it is now clear that it varies considerably across taxa (Corlett 2011; Malhi et al. 2014). In plants, substantial evidence is emerging that tropical trees have significantly narrower thermal niches than temperate trees (Araújo et al. 2013), and there is some evidence that tropical trees have lower temperature acclimation capacities compared to temperate trees (Corlett 2011). There is also broad-based evidence for warming-induced slowdowns in tropical tree growth rates (Way and Oren 2010). Projecting these impacts into the future is complicated by uncertainties surrounding the degree to which CO_2 fertilization may increase tropical forest productivity and potentially make up for heat-induced growth reductions (Pan et al. 2011). In animals, narrower thermal tolerances are well established in the scientific literature, with increasingly strong links between thermal tolerance and elevational range sizes (Sheldon and Tewksbury 2014). Evidence on acclimation is less strong for animals, but there is some suggestion that tropical animal species may have lower ac-

climation capacities than their temperate counterparts (Somero 2010).

A fundamental point of concern for tropical biodiversity focuses on the relatively narrow thermal niches of tropical organisms that have evolved in climates exhibiting low levels of variation across seasonal to millennial time scales (Ghalambor et al. 2006). This narrow thermal tolerance, coupled with the fact that current temperatures in the tropics are already near or at thermally optimal conditions for many organisms (Deutsch et al. 2008), creates a situation in which a moderate increase in temperature could lead to a marked decline in fitness for many tropical organisms. The debate around this issue is a major source of uncertainty in predicting how tropical forest biodiversity will respond to different scenarios of warming (Corlett 2011), and sources of uncertainty come from the degree to which shifts in mean temperature versus extreme temperatures will drive changes in population fitness, and the importance of microclimatic refugia in buffering populations from climate change (see Chapter 7 in this volume).

Movement

Although many temperate species have access to both latitudinal and elevational temperature gradients, it is thought that more than 50 percent of animal and plant species in tropical regions have ranges so restricted that they lack contact with a latitudinal temperature gradient (Wright, Muller-Landau, and Schipper 2009). The narrow thermal tolerance of tropical species, combined with the lack of latitudinal temperature gradients, suggests that a primary response for tropical species should be elevational migration (Colwell et al. 2008). Some empirical evidence confirms this, with upslope migration in response to climate change already more common in tropical than in temperate communities (Freeman and Class Freeman 2014). This pattern is likely to strengthen through time as climate change intensifies, and yet there are multiple factors that may limit the positive effects of mountains as buffers against the loss of species under climate change.

First, high-elevation habitat is not extensive across the tropics. Even in mountainous areas, the amount of area inevitably declines as altitude increases. As a result, extrapolations of species-area relationships suggest a limited capacity for mountains to act as refugia for the many lower-elevation species seeking to take advantage of elevational gradients to escape the negative effects of rising temperatures. Greater than 70 percent of the terrestrial surface of the earth is below 1,000 m, and thus the amount of land available at higher elevations is relatively small.

Second, in tropical regions, there is a greater potential for biotic attrition (Colwell et. al. 2008). In temperate zones, as species migrate to higher elevations, lower elevations presumably become available for colonization by species from lower latitudes, but in the tropics, there are no floral and faunal assemblage preadapted to landscapes warmer than the current temperatures in lowland tropical areas (Feeley and Silman 2010; Colwell et al. 2008). This has led some researchers to conclude that large-scale migration upslope will lead to a significant reduction in species in lower-elevation landscapes (Colwell et al. 2008).

Finally, in many lowland tropical areas, upslope migration may simply not be feasible because of the distance that tropical fauna would need to travel. The pace of climate change suggests that many tropical species—particularly those with limited dispersal capacities or difficulties crossing gaps—will be unable to migrate rapidly enough to keep pace with the change (Schloss et al. 2012). Plant populations appear to track changes in climate less well than animal populations, as they frequently display longer time lags and lack of response compared to animals (Corlett and Westcott 2013). Even in more mobile animal groups, such as mammals, limited dispersal capaci-

ties, low tolerance to changing temperatures, and the rapid pace of warming are predicted to lead to large tropical areas in which greater than 30 percent of mammals are unable to keep pace with climate change and will find themselves in very different climate envelopes over the next 75 years from what they experience today (Schloss, Nuñez, and Lawler 2012).

The degree to which microrefugia, behavior modification, adaptation, acclimation, and tolerance can collectively ward off attrition or outright loss of lowland tropical species is still a matter of debate. Past records of high plant diversity during previous warming periods suggest considerable resilience, but the fact that the current warming is orders of magnitude faster than previous warming events (Jaramillo et al. 2010), and is coming after a prolonged period of cooler tropical temperatures with little evolutionary selection for extreme warm tolerance (Corlett 2011), suggests significant risk of large scale losses in biodiversity.

Interactions

Multiple lines of evidence highlight the importance of biotic interactions in creating and sustaining biodiversity across the tropics (Schemske et al. 2009). Understanding these interactions and how they are affected by the different components of climate change will be critical for accurate predictions of the impacts on biodiversity as whole (Gilman et al. 2010). Direct changes in biotic interactions will occur as a result of changes in direct temperature, precipitation, and CO_2-dependent dominance relationships, shifts in phenology, and changes in rainfall and temperature-dependent mutualisms and in community-level interactions. Movement in the face of climate change as described above will also create changes in biotic interactions. Because elevational migration in response to climate change will be highly variable among species, and because tropical species have both narrower thermal tolerances and narrower elevational ranges than their temperate counterparts, it is believed that upslope movement will result in significantly higher rates of community disruption, loss of species interactions, and species extinction in tropical communities than in subtropical and temperate communities (Urban, Tewksbury, and Sheldon 2012).

Ecosystems

It is well established that vegetation cover change in tropical forests affects both regional and global precipitation patterns by altering the amount of water returned to the atmosphere via evapotranspiration. Therefore, tropical forest loss to date is thought to have already affected precipitation patterns, with severe drought a significant concern in many tropical areas. This feedback, coupled with increases in drought length and severity, particularly in the eastern Amazon, can lead to increased fire risk and reduced forest cover. Because few tropical plant species are adapted to fire, and most tropical rainforests are in fact fire sensitive (Barlow and Peres 2008), the synergy between droughts and more frequent fires significantly increases the risk of a tropical forest dieback scenario, characterized by large-scale replacement of forest with savanna or shrub (Brando et al. 2014). The probability of such large-scale change, especially in the Amazon forests, has been the subject of considerable debate in the scientific literature over the past decade. High levels of uncertainty surround the potential role of CO_2 fertilization, which could buffer tropical systems from droughts through increased productivity, but there is increasing evidence for tipping points brought on by heat, drought, and fire (Wright, Muller-Landau, and Schipper 2009).

INDIRECT THREATS

As the dominant conditions that drive species presence and abundance are reshaped by climate change across tropical forest ecosystems in the ways outlined above, human populations that occupy and are in many cases dependent on them, are also adapting to the changing climatic conditions (Maxwell et al. 2016). Whether planned or unplanned, these indirect impacts of climate change are thought to be increasingly affecting many species and ecosystems (Segan et al. 2015). Although there are many possible human responses to climate change that could lead to indirect impacts, two stand out as very likely to occur and have significant ramifications for tropical forest ecosystems across the world.

Increased Agriculture Movement into Tropical Forests

The conversion of natural ecosystems is the greatest driver of biodiversity decline globally, largely because tropical forest ecosystems are being rapidly replaced with croplands that support few species (Laurance, Sayer, and Cassman 2014). Human population growth and increased global demand for tropical agricultural commodities are trends that are likely to continue in many regions and the dynamics of each are being affected by climate change. For example, increasingly severe dry seasons in tropical forest ecosystems are likely to exacerbate the impacts of changes in land use on tropical forest biodiversity, as prolonged or more intense dry periods may allow for increased accessibility, thus removing a current barrier to resource development (Brodie, Post, and Laurance 2012). As a consequence of this, the economic feasibility of forest colonization and logging could increase, and many of the last remaining "remote" forests could become accessible to large-scale exploitation. There is evidence that severity of the dry season is already a strong, positive predictor of deforestation pressure in the Amazon (Laurance et al. 2002), implying that drying trends in certain tropical forests could increase their vulnerability. While increased drought frequency in

tropical forests may create economic opportunities for some, past droughts have
also had adverse impacts on local communities through impaired water quality,
fish die-offs, and hindered riverine transport that local communities rely on to stay
connected to markets. All these lead to additional changes as local communities attempt to cope with their new environment
(Marengo et al. 2008).

Societal responses to climate change are
also magnifying agricultural pressures on
tropical forests. Driven by the need to mitigate greenhouse gas emissions, crop-based
biofuel production has increased rapidly in
recent years, especially in Southeast Asia
(Fargione et al. 2008). Accompanied by increased demand for food, this has led to a
substantial expansion of agricultural lands
in all tropical forested regions to create new
areas for biofuel production or to provide
replacement sites for food production when
existing croplands are switched to biofuel
production. Recent analysis of drivers of
deforestation globally suggest that at least
half of deforestation in the past decade was
the result of agricultural pressure (Lawson
et al. 2014). Rising demands for land to
grow biofuels is likely to increase opportunity costs for conservation, reducing the
competitiveness of carbon offsets and other
payment for ecosystem service programs
designed to slow forest destruction.

Increasing Threats from Changing
Fire Regimes

Synergies between climate change and human-lit fires represent another severe threat
to tropical forest ecosystems that is likely to
increase in importance with future warming. As substantial expanses of the tropics
could become both warmer and drier in the
future, there is likely to be an increase in
the incidence, magnitude, and duration of
human-lit fires (Nepstad et al. 2008). Importantly, human changes to land use also
increases forest vulnerability to fire, magnifying the impacts of climatic drying. For

example, as human populations increase
and land use changes, drought cycles in Indonesia become increasingly coupled with
fire cycles (Field, van der Werf, and Shen
2009). Moreover, logged and fragmented
tropical forests are far more vulnerable to
fire than intact forests, because canopy loss
leads to desiccation, which allows fire to
penetrate more deeply into forest remnants.
The synergies between human activities
and their impact on tropical forests can also
take surprising and unexpected forms. For
instance, research from the Amazon suggests that as humans vacate rural areas, the
ability to control fires decreases, resulting
in overall increased fire risk (Schwartz et
al. 2015). The complex feedback loops between human activity and physical climate
cnanges necessitate that planners directly
account for their interactions when assessing risk or designing interventions.

CONCLUSIONS: CONSEQUENCES
FOR CONSERVATION

Climate change will directly influence
biodiversity in tropical regions in myriad
ways. The narrow thermal niches of many
tropical species, combined with a general
lack of latitudinal temperature gradients,
means that uphill movement is likely to
be a common adaptive response in tropical forests. For those species that lack the
evolutionary adaptive response enabling
them to track a climate gradient, rapid climate change will clearly be a serious threat.
Even for those species that are able to track
their climate via movement, their response
over time will lead to novel ecological interactions that are likely to reshape tropical
forest ecosystems. However, it is important
that we recognize that the direct impacts of
climate change are particularly challenging
to predict and come with huge uncertainty
around how individual species will adapt
to a changing climate, what this is likely
to mean for community composition and
ecological interactions in different tropi-

cal regions, and the synergy between these changes and other important ecological processes that drive tropical forest function (Scheffers et al. 2016). This lack of knowledge significantly affects our ability to assess species risk to climate change and the most effective actions to avert this risk, and at the same time may lead to both overestimations and underestimations of the degree of risk species face when it comes to climate change.

Our lack of ability to predict the direct consequences of climate change should not be leveraged as an excuse for inaction. We are far more certain in our ability to understand and predict how human communities will respond in tropical systems—and it is human activities that are driving the current extinction crisis we face. We can use this information to assess species risk to potential indirect affects to climate change and identify suitable conservation strategies. Importantly, many conservation strategies are already known to be effective in protecting biodiversity from human activities, including well-managed protected areas, fire regulation, and some payment-for-ecosystem-services schemes like REDD+. All of these can be used in ways that benefit climate and help species overcome many of the short-term challenges we have outlined in this chapter. With the large amount of uncertainty around species response to climate change, there are two clear overarching conservation priorities for tropical biodiversity. First, global mitigation efforts that limit the extent of climate change will be critical for species with few adaptation options in the short term and all species over the long term. Second, it is necessary to prevent large-scale land clearing, wherever possible, in intact ecosystems, especially those with latitudinal and elevational gradients (Watson et al. 2018). By protecting large, functional tropical forest ecosystems, we will ensure that tropical species can maximize their adaptive capacity to overcome the challenges posed by future climate change.

REFERENCES

Araújo, Miguel B., Francisco Ferri-Yáñez, Francisco Bozinovic, Pablo A. Marquet, Fernando Valladares, and Steven L. Chown. 2013. "Heat freezes niche evolution." *Ecology Letters* 16 (9): 1206–1219. https://doi.org/10.1111/ele.12155.

Barlow, Jos, and Carlos A. Peres. 2008. "Fire-mediated dieback and compositional cascade in an Amazonian forest." *Philosophical Transactions of the Royal Society B: Biological Sciences* 363 (1498): 1787–1794. https://doi.org/10.1098/rstb.2007.0013.

Brando, Paulo Monteiro, Jennifer K. Balch, Daniel C. Nepstad, Douglas C. Morton, Francis E. Putz, Michael T. Coe, Divino Silvério, et al. 2014. "Abrupt increases in Amazonian tree mortality due to drought-fire interactions." *Proceedings of the National Academy of Sciences* 111 (17): 6347–6352. https://doi.org/10.1073/pnas.1305499111.

Brodie, Jedediah, Eric Post, and William F. Laurance. 2012. "Climate change and tropical biodiversity: A new focus." *Trends in Ecology & Evolution* 27 (3): 145–150. https://doi.org/10.1016/j.tree.2011.09.008.

Collins, M., R. Knutti, J. Arblaster, J.-L. Dufresne, T. Fichefet, P. Friedlingstein, X. Gao, W. J. Gutowski, T. Johns, G. Krinner, M. Shongwe, C. Tebaldi, A. J. Weaver, and M. Wehner. 2013. "Long-term climate change: Projections, commitments and irreversibility." In *Climate Change 2013: The Physical Science Basis. Contribution of Working Group I to the Fifth Assessment Report of the Intergovernmental Panel on Climate Change*, ed. T. F. Stoker, D. Qin, G.-K. Plattner, M. Tignor, S. K. Allen, J. Boschung, A. Nauels, Y. Xia, V. Bex, and P. M. Midgley, 1029–1136. Cambridge University Press.

Colwell, Robert K., Gunnar Brehm, Catherine L. Cardelus, Alex C. Gilman, and John T. Longino. 2008. "Global warming, elevational range shifts, and lowland biotic attrition in the wet tropics." *Science* 322: 258–261.

Corlett, Richard T. 2011. "Impacts of warming on tropical lowland rainforests." *Trends in Ecology & Evolution* 26 (11): 606–613. https://doi.org/10.1016/j.tree.2011.06.015.

Corlett, Richard T., and Richard B. Primack. 2006. "Tropical rainforests and the need for cross-continental comparisons." *Trends in Ecology & Evolution* 21 (2): 104–110. https://doi.org/10.1016/j.tree.2005.12.002.

Corlett, Richard T., and David A. Westcott. 2013. "Will plant movements keep up with climate change?" *Trends in Ecology and Evolution* 28 (8): 482–488. https://doi.org/10.1016/j.tree.2013.04.003.

Costa, Antonio Carlos Lola da, David Galbraith, Samuel Almeida, Bruno Takeshi Tanaka Portela, Mauricio da Costa, João de Athaydes Silva Junior, Alan P. Braga, et al. 2010. "Effect of 7 yr of experimental drought on vegetation dynamics and biomass storage of an eastern Amazonian rainforest." *New Phytologist* 187 (3): 579–591. https://doi.org/10.1111/j.1469-8137.2010.03309.x.

Davin, Edouard L., and Nathalie de Noblet-Ducoudré. 2010. "Climatic impact of global-scale deforestation:

Radiative versus nonradiative processes." *Journal of Climate* 23 (1): 97–112. https://doi.org/10.1175/2009JCLI3102.1.

Deutsch, Curtis A., Joshua J. Tewksbury, Raymond B. Huey, Kimberly S. Sheldon, Cameron K. Ghalambor, David C. Haak, and Paul R. Martin. 2008. "Impacts of climate warming on terrestrial ectotherms across latitude." *Proceedings of the National Academy of Sciences* 105 (18): 6668–6672. https://doi.org/10.1073/pnas.0709472105.

Dirzo, Rodolfo, and Peter H. Raven. 2003. "Global state of biodiversity and loss." *Annual Review of Environment and Resources* 28 (1): 137–167. https://doi.org/10.1146/annurev.energy.28.050302.105532.

Fargione, Joseph, Jason Hill, David Tilman, Stephen Polasky, and Peter Hawthorne. 2008. "Land clearing and the biofuel carbon debt." *Science* 319 (5867): 1235–1238. https://doi.org/10.1126/science.1152747.

Feeley, Kenneth J., and Miles R. Silman. 2010. "Biotic attrition from tropical forests correcting for truncated temperature niches." *Global Change Biology* 16: 1830–1836.

Feng, Xue, Amilcare Porporato, and Ignacio Rodriguez-Iturbe. 2013. "Changes in rainfall seasonality in the tropics." *Nature Climate Change* 3 (9): 811–815. https://doi.org/10.1038/nclimate1907.

Field, Robert D., Guido R. van der Werf, and Samuel S. P. Shen. 2009. "Human amplification of drought-induced biomass burning in Indonesia since 1960." *Nature Geoscience* 2 (3): 185–188. https://doi.org/10.1038/ngeo443.

Freeman, Benjamin G, and Alexandra M. Class Freeman. 2014. "Rapid upslope shifts in New Guinean birds illustrate strong distributional responses of tropical montane species to global warming." *Proceedings of the National Academy of Sciences* 111 (12): 4490–4494. https://doi.org/10.1073/pnas.1318190111.

Ghalambor, Cameron K., Raymond B. Huey, Paul R. Martin, Joshua J. Tewksbury, and George Wang. 2006. "Are mountain passes higher in the tropics? Janzen's hypothesis revisited." *Integrative and Comparative Biology* 46 (1): 5–17. https://doi.org/10.1093/icb/icj003.

Gilman, S. E., M. C. Urban, J. Tewksbury, G. W. Gilchrist, and R. D. Holt. 2010. "A framework for community interactions under climate change." *Trends in Ecology and Evolution* 25: 325–331. https://doi.org/10.1016/j.tree.2010.03.002.

Groot, Rudolf de, Luke Brander, Sander van der Ploeg, Robert Costanza, Florence Bernard, Leon Braat, Mike Christie, et al. 2012. "Global estimates of the value of ecosystems and their services in monetary units." *Ecosystem Services* 1 (1): 50–61. https://doi.org/10.1016/j.ecoser.2012.07.005.

IPCC. 2013. *Climate Change 2013: The Physical Science Basis: Working Group I Contribution to the Fifth Assessment Report of the Intergovernmental Panel on Climate Change.* International Panel on Climate Change.

Jaramillo, Carlos, Diana Ochoa, Lineth Contreras, Mark Pagani, Humberto Carvajal-Ortiz, Lisa M. Pratt, Srinath Krishnan, et al. 2010. "Effects of rapid global warming at the Paleocene-Eocene boundary on neotropical vegetation." *Science* 330 (6006): 957–961. https://doi.org/10.1126/science.1193833.

Laurance, William F., Thomas E. Lovejoy, Heraldo L. Vasconcelos, Emilio M. Bruna, Raphael K. Didham, Philip C. Stouffer, Claude Gascon, Richard O. Bierregaard, Susan G. Laurance, and Erica Sampaio. 2002. "Ecosystem decay of Amazonian forest fragments: A 22-year investigation." *Conservation Biology* 16 (3): 605–618. https://doi.org/10.1046/j.1523-1739.2002.01025.x.

Laurance, William F., Jeffrey Sayer, and Kenneth G. Cassman. 2014. "Agricultural expansion and its impacts on tropical nature." *Trends in Ecology & Evolution* 29 (2): 107–116. https://doi.org/10.1016/j.tree.2013.12.001.

Lawson, Sam, A. Blundell, Bruce Cabarle, Naomi Basik, Michael Jenkins, and Kerstin Canby. 2014. *Consumer Goods and Deforestation: An Analysis of the Extent and Nature of Illegality in Forest Conversion for Agriculture and Timber Plantations.* Forest Trends.

Mackey, Brendan, Dominick A. DellaSala, Cyril Kormos, David Lindenmayer, Noelle Kumpel, Barbara Zimmerman, Sonia Hugh, et al. 2014. "Policy options for the world's primary forests in multilateral environmental agreements." *Conservation Letters* 8 (2): 139–147. https://doi.org/10.1111/conl.12120.

Malhi, Yadvinder. 2012. "The productivity, metabolism and carbon cycle of tropical forest vegetation." *Journal of Ecology* 100 (1): 65–75. https://doi.org/10.1111/j.1365-2745.2011.01916.x.

Malhi, Yadvinder, Luiz E. O. C. Aragão, David Galbraith, Chris Huntingford, Rosie Fisher, Przemyslaw Zelazowski, Stephen Sitch, Carol McSweeney, and Patrick Meir. 2009. "Exploring the likelihood and mechanism of a climate-change-induced dieback of the Amazon rainforest." *Proceedings of the National Academy of Sciences* (February). https://doi.org/10.1073/pnas.0804619106.

Malhi, Yadvinder, Toby A. Gardner, Gregory R. Goldsmith, Miles R. Silman, and Przemyslaw Zelazowski. 2014. "Tropical forests in the Anthropocene." *Annual Review of Environment and Resources* 39 (1): 125–159. https://doi.org/10.1146/annurev-environ-030713-155141.

Maxwell, S. L., O. Venter, K. R. Jones, and J. E. M. Watson. 2015. "Integrating human responses to climate change into conservation vulnerability assessments and adaptation planning." *Annals of the New York Academy of Sciences* 1355 (1): 98–116. https://doi.org/10.1111/nyas.12952.

Marengo, J. A., C. A. Nobre, J. Tomasella, M. F. Cardoso, and M. D. Oyama. 2008. "Hydro-climatic and ecological behaviour of the drought of Amazonia in 2005." *Philosophical Transactions of the Royal Society B: Biological Sciences* 363 (1498): 1773–1778. https://doi.org/10.1098/rstb.2007.0015.

Mora, Camilo, Abby G. Frazier, Ryan J. Longman, Rachel S. Dacks, Maya M. Walton, Eric J. Tong, Joseph J. Sanchez, et al. 2013. "The projected timing of climate departure from recent variability." *Nature* 502 (7470): 183–187. https://doi.org/10.1038/nature12540.

Nepstad, Daniel C., Claudia M. Stickler, Britaldo Soares-Filho, and Frank Merry. 2008. "Interactions among Amazon land use, forests and climate: Prospects for a near-term forest tipping point." *Philosophical Transactions of the Royal Society B: Biological Sciences* 363 (1498): 1737–1746. https://doi.org/10.1098/rstb.2007.0036.

Pan, Yude, Richard A. Birdsey, Jingyun Fang, Richard Houghton, Pekka E. Kauppi, Werner A. Kurz, Oliver L. Phillips, et al. 2011. "A large and persistent carbon sink in the world's forests." *Science* 333 (6045): 988–993. https://doi.org/10.1126/science.1201609.

Phillips, Oliver L., Luiz E. O. C. Aragao, Simon L. Lewis, Joshua B. Fisher, Jon Lloyd, Gabriela Lopez-Gonzalez, Yadvinder Malhi, et al. 2009. "Drought sensitivity of the Amazon rainforest." *Science* 323 (5919): 1344–1347. https://doi.org/10.1126/science.1164033.

Scharlemann, Joern P. W., Valerie Kapos, Alison Campbell, Igor Lysenko, Neil D. Burgess, Matthew C. Hansen, Holly K. Gibbs, Barney Dickson, and Lera Miles. 2010. "Securing tropical forest carbon: The contribution of protected areas to REDD." *Oryx* 44 (3): 352–357.

Scheffers, B. R., L. De Meester, T. C. L. Bridge, A. A. Hoffmann, J. M. Pandolfi, R. T. Corlett, S. H. M. Butchart, et al. 2016. "The broad footprint of climate change from genes to biomes to people." *Science* 354 (6313). https://doi.org/10.1126/science.aaf7671.

Schemske, Douglas W., Gary G. Mittelbach, Howard V. Cornell, James M. Sobel, and Kaustuv Roy. 2009. "Is there a latitudinal gradient in the importance of biotic interactions?" *Annual Review of Ecology, Evolution and Systematics* 40 (1): 245–269. https://doi.org/10.1146/annurev.ecolsys.39.110707.173430.

Schloss, Carrie A., Tristan A. Nuñez, and Joshua J. Lawler. 2012. "Dispersal will limit ability of mammals to track climate change in the Western Hemisphere." *Proceedings of the National Academy of Sciences* 109 (22): 8606–8611.

Schwartz, Naomi B., Maria Uriarte, Victor H. Gutiérrez-Vélez, Walter Baethgen, Ruth DeFries, Katia Fernandes, and Miguel A. Pinedo-Vasquez. 2015. "Climate, landowner residency, and land cover predict local scale fire activity in the western Amazon." *Global Environmental Change* 31 (March): 144–153. https://doi.org/10.1016/j.gloenvcha.2015.01.009.

Segan, D. B., D. G. Hole, C. I. Donatti, C. Zganjar, S. Martin, S. H. M. Butchart, and J. E. M. Watson. 2015. "Considering the impact of climate change on human communities significantly alters the outcome of species and site-based vulnerability assessments." *Diversity and Distributions* 21 (9): 1101–1111. https://doi.org/10.1111/ddi.12355.

Sheldon, Kimberly S., and Joshua J. Tewksbury. 2014. "The impact of seasonality in temperature on thermal tolerance and elevational range size." *Ecology* 95 (8): 2134–2143.

Somero, G. N. 2010. "The physiology of climate change: How potentials for acclimatization and genetic adaptation will determine 'winners' and 'losers.'" *Journal of Experimental Biology* 213 (6): 912–920. https://doi.org/10.1242/jeb.037473.

Urban, Mark C., Josh J. Tewksbury, and Kimberly S. Sheldon. 2012. "On a collision course: Competition and dispersal differences create no-analogue communities and cause extinctions during climate change." *Proceedings of the Royal Society B: Biological Sciences* 279 (1735): 2072–2080.

Watson, J. E. M., T. Iwamura, and N. Butt. 2013. "Mapping vulnerability and conservation adaptation strategies under climate change." *Nature Climate Change* 3 (11): 989–994. https://doi.org/10.1038/nclimate2007.

Watson, James E. M., Tom Evans, Oscar Venter, Brooke Williams, Ayesha Tulloch, Claire Stewart, Ian Thompson, et al. 2018. "The exceptional value of intact forest ecosystems." *Nature Ecology & Evolution* 2: 599–610.

Way, Danielle A., and Ram Oren. 2010. "Differential responses to changes in growth temperature between trees from different functional groups and biomes: A review and synthesis of data." *Tree Physiology* 30 (6): 669–688. https://doi.org/10.1093/treephys/tpq015.

Wright, S. Joseph, Helene C. Muller-Landau, and Jan Schipper. 2009. "The future of tropical species on a warmer planet." *Conservervation Biology* 23 (6): 1418–1426. https://doi.org/10.1111/j.1523-1739.2009.01337.x.

Zelazowski, Przemyslaw, Yadvinder Malhi, Chris Huntingford, Stephen Sitch, and Joshua B. Fisher. 2011. "Changes in the potential distribution of humid tropical forests on a warmer planet." *Philosophical Transactions of the Royal Society A: Mathematical, Physical and Engineering Sciences* 369 (1934): 137–160. https://doi.org/10.1098/rsta.2010.0238.

Postponing the Amazon Tipping Point

Daniel Nepstad

The Amazon forest is a giant from any perspective. It is the planet's largest tropical forest drained by the largest river—the source of one-fifth of all freshwater that reaches the oceans. It is the greatest cornucopia of biological and indigenous cultural diversity.

The Amazon forest is also a giant in shaping the Earth's climate. Like an enormous planetary cooling system, it influences the global circulation of air and vapor by evaporating vast amounts of water into the atmosphere—converting the equatorial Sun's intense radiant energy into latent heat. Even where long dry seasons are the norm, the deep root systems of Amazon trees allow the forest to absorb soil water stored many meters beneath the soil surface to supply the leaf canopy high above the ground (Nepstad et al. 1994). When patches of Amazon forest are replaced by cattle pasture, the amount of vapor pumped into the atmosphere declines.

The wood of Amazon forest trees is another reason this ecosystem shapes global climate. An amount of carbon equivalent to nearly a decade's worth of global emissions from human activities is stored in Amazon wood—that is, outside of the atmosphere, where it would contribute to global warming. This biological carbon leaks into the atmosphere during and after droughts that are severe enough to kill big Amazon trees, such as the droughts of 2005, 2007, 2010, and 2016. It also leaks out when humans deliberately fell the forest to grow crops or livestock, when they degrade the forest to harvest timber, or when the fires they set to manage the land escape into neighboring forests, killing trees (Nepstad et al. 2008).

Positive feedbacks in the Amazon forest-rainfall-fire system could be important in determining the future of this ecosystem in the face of climate change and further expansion of land-use activities. First, the Amazon rainfall system itself depends on the forest. The same year-round forest evapotranspiration that shapes global circulation patterns sustains rainfall patterns in the Amazon region (Salati and Vose 1984). Simply stated, the large-scale conversion of forest to pasture and cropland along the eastern flank of the Amazon forest means that there is less vapor to supply rainfall systems to the west and southwest. A potential "tipping point" could be reached if forest clearing becomes extensive enough to suppress rainfall below the minimum amount that is necessary to sustain closed-canopy forests (Lovejoy and Nobre 2018).

A second positive feedback is between forest degradation and forest fire: once a forest burns, the chances of subsequent fires increase. In pre-Columbian times, Amazon forests appear to have caught fire a few times every millennium, when severe droughts occurred. Then, as now, during years of normal rainfall, most of the forests of the Amazon extended like giant firebreaks across the landscape; the dense shade of the lofty tree canopies supplied with water by deep root systems kept the forest floor too damp to carry a fire. Today, forest fire is a frequent occurrence along the agricultural and livestock frontiers partly because forests have become more susceptible to fire, either because they were already subjected to an earlier fire or through logging operations that punch holes in the forest canopy, allowing light to penetrate to the forest floor. They have also increased in frequency because there are more ignition sources—more people using fire to clear the land and kill shrubs and trees that invade cattle pastures. If Amazon forests burn repeatedly, they can eventually be invaded by flammable grasses and herbs, further increasing the likelihood of recurring fire—in a second tipping point (Balch et al. 2015).

With our knowledge of Amazon forest feedbacks and tipping points we can assemble two plausible scenarios for the future of this globally significant ecosystem—one in which the Amazon forest becomes a major source of added carbon emissions to the atmosphere as the forest-rainfall-fire system drives regional degradation. In a second, the Amazon forest becomes a major sink of atmospheric carbon, figuring prominently in our success in keeping the planet below a 1.5°C average temperature increase, as expressed in the 2015 Paris Agreement.

In the first "business as usual" future scenario, deforestation accelerates, reversing the recent slowdown in deforestation of the Brazilian Amazon (Nepstad et al. 2014), as forest degradation through logging and fire increases. Major droughts continue to become more frequent and severe, increasing the occurrence of forest fires in vast swaths of primary and degraded forests. The expansion of forage grasses, crop fields, and scrub vegetation further inhibits rainfall. The "scrubification" of Amazon forests that have been degraded by drought, logging, repeated burning, and invasion by highly flammable grasses and fires begins to appear across the eastern and southern Amazon regions (Figure CS6.1)—this is accelerated when deforestation frontiers expand along the newly paved highways from Porto

Figure CS6.1. The forests of the Amazon Basin are being altered through severe droughts, land use deforestation, logging, and increased frequencies of forest fire. Some of these processes are self-reinforcing through positive feedbacks and create the potential for a large-scale tipping point. For example, forest fire kills trees, increasing the likelihood of subsequent burning. This effect is magnified when tree death allows forests to be invaded by flammable grasses. Deforestation provides ignition sources to flammable forests, contributing to this dieback. Climate change contributes to this tipping point by increasing drought severity, reducing rainfall and raising air temperatures, particularly in the eastern Amazon Basin. (Figure 4-8 from Settele et al. 2014.)

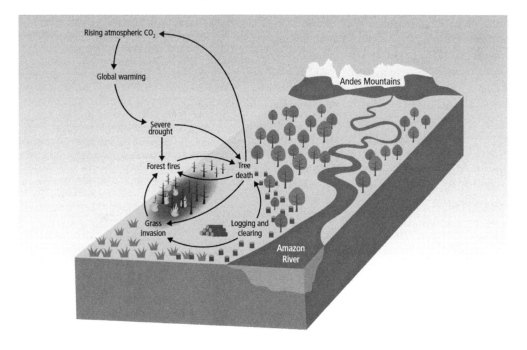

Velho to Manaus, Brazil; from Santarem to Cachimbo, Brazil; and from Pucallpa, Peru, to Cruzeiro do Sul, Brazil.

In a scenario of "managed resilience," the slowdown in deforestation that has been taking place in the Brazilian Amazon continues to deepen and spreads to the other Amazon forest nations. Logging continues but through reduced-impact practices; forest fire–control programs established across the Amazon region successfully extinguish forest fires soon after they ignite, and the number of fires needing extinguishing declines through better fire management. This fire prevention program has also allowed forest recovery to take place on nearly half of the forestlands that had been cleared for livestock and crops but were then abandoned because of inadequate soils or infrastructure. This regional forest recovery removes carbon dioxide from the atmosphere as it reestablishes the year-round supply of water vapor to the atmosphere that sustains the rainfall system of the Amazon and that shapes global circulation patterns.

The Amazon forest can be part of the problem of climate change—exacerbating global warming—or it can be part of the solution. Tropical forests could provide a quarter or more of the emissions reductions needed by 2030 to avoid a 2°C increase in global temperature (Griscom et al. 2017). What is remarkable is that the managed resilience scenario is within reach—it is a viable choice that can be achieved with political will, the right market signals, and the right types of financial investments and incentives for landholders. The ecological and climatic integrity of the Amazon can be maintained for decades if Amazon so-

cieties secure and deepen the innovations in public policies, law enforcement, and agricultural innovation that already avoided more than 6 billion tons of carbon dioxide emissions from deforestation in Brazil. In the long term, however, the ecological integrity of this giant ecosystem will depend on humanity's success in slowing climate change.

REFERENCES

Balch, J., P. Brando, D. Nepstad, et al. 2015. "Susceptibility of southeastern Amazon forests to fire: Insights from a large-scale burn experiment." *Bioscience* 65 (9): 893–905.

Griscom, B. W., J. Adams, P. Ellis, R. A. Houghton, G. Lomax, D. A. Miteva, W. H. Schlesinger, D. Shoch, J. V. Siikamäki, P. Smith, and P. Woodbury. 2017. "Natural climate solutions." *Proceedings of the National Academy of Sciences* 114 (44): 11645–11650.

Lovejoy, T. E., and C. Nobre. 2018. "Amazon tipping point." *Science Advances* 4 (2): eaat2340.

Nepstad, D., G. Carvalho, A. C. Barros, A. Alencar, J. P. Capobianco, J. Bishop, P. Moutinho, P. Lefebvre, and U. L. Silva Jr. 2001. "Road paving, fire regime feedbacks, and the future of Amazon forests." *Forest Ecology and Management* 154: 395–407.

Nepstad, D., C. Reis de Carvalho, E. Davidson, P. Jipp, P. Lefebvre, G. Hees Negreiros, E. Silva, T. Stone, S. Trumbore, and S. Vieira. 1994. "The role of deep roots in the hydrologic and carbon cycles of Amazonian forests and pastures." *Nature* 372: 666–669.

Nepstad, D, C. M. Stickler, B. Soares-Filho, and F. Merry. 2008. "Interactions among Amazon land use, forests and climate: Prospects for a near-term forest tipping point." *Philosophical Transactions of the Royal Society B—Biological Sciences.* https://doi.org/10.1098/rstb.2007.0036.

Nepstad, D., D. McGrath, C. Stickler, et al. 2014. "Slowing Amazon deforestation through public policies and interventions in beef and soy supply chains." *Science* 344: 1118–1123.

Salati, E., and P. Vose. 1984. "The Amazon: A system in equilibrium." *Science* 225 (4648): 129–138.

CHAPTER SIXTEEN

Temperate and Boreal Responses to Climate Change

LAUREN B. BUCKLEY AND
JANNEKE HILLERISLAMBERS

OVERVIEW

Climates are seasonally and otherwise variable in temperate and boreal regions, which shapes the environmental sensitivities of plants and animals and determines how they will respond to climate change (see Figure 16.1). For example, the pronounced latitudinal temperature gradient in these regions will result in widespread poleward range shifts, and the cool temperatures and limited growing seasons that characterize the region will generally result in increased productivity with warming. We briefly outline the magnitude of climate change expected in temperate and boreal regions, and review the sensitivity of organisms to the climatic elements projected to change. We then describe ongoing and potential future responses of organisms and ecosystems to these climate changes. Finally, we examine ecological disturbances associated with climate change and how these stressors may interact to affect biodiversity in temperate and boreal regions.

Temperate and boreal ecosystems have warmed more than tropical systems over the history of temperature records (e.g., 0.7°C–1.1°C in the United States since 1895; Melillo, Richmond, and Yohe 2014), with greater warming toward the poles expected to continue in the future. There have not been consistent changes to precipitation regimes, although the proportion of precipitation falling as snow has declined. Thus, growing seasons have and will continue to extend, as long as increasing temperatures or declining snowpack do not result in increased summer drought (in summer dry and snow-dependent areas). Winter extreme events (e.g., frosts) have generally declined, whereas summer extreme heat events have increased—these trends are expected to continue (Melillo, Richmond, and Yohe 2014).

SENSITIVITY TO CLIMATE CHANGE AND VARIABILITY

Physiological Sensitivities to Climate Change

Organisms in more poleward ecosystems have generally been thought to be more vulnerable to climate change because they are expected to experience a greater magnitude of temperature change (Parmesan 2006). However, recent studies highlight the importance of considering how seasonality influences physiological performance. These studies also illustrate the complexities involved in predicting vulnerability. For example, insects tend to evolve more specialized thermal physiology in constant tropical environments relative to seasonally variable temperate (and boreal) environments (Deutsch et al. 2008). Compilations of thermal tolerance data for both ectotherms (Sunday et al. 2014) and endotherms (Khaliq et al. 2014) confirm that the breadth of thermal tolerances increases with latitude. Declines in lower thermal limits with latitude and invariant upper thermal limits tend to produce this pattern, which yields increases in thermal safety margins (the difference between heat tolerance and habitat air temperatures) with latitude. An analysis based on this relationship between temperature and fitness suggested that, as a result of these physiological differences, climate change will decrease the fitness of tropical ectotherms while increasing the fitness of temperate and boreal ectotherms (Deutsch et al. 2008).

However, studies using similar approaches have cautioned against concluding that temperate and boreal organisms are less sensitive to climate change. For one, the latitudinal gradients in the breadth of

temperature tolerances found in the studies are likely overestimates, because they are based on thermal safety margins derived from air temperatures, which do not reflect the body temperatures actually experienced by organisms. Indeed, an analysis based on estimated ectotherm body temperatures reveals that thermal stress events can occur at high latitudes largely as a result of radiation spikes, which provides an explanation for the latitudinal invariance of heat tolerance (Sunday et al. 2014). Studies subsequent to Deutsch et al. (2008) have also pointed out that the initial analysis based on mean thermal conditions may have underestimated the fitness detriments associated with thermal stress events in temperate areas. Kingsolver et al. (2013) predicted that climate change impacts would be most severe for insects at mid-latitudes (20°–40°) due to an increased frequency of heat stress events, which reduce fitness. However, capitalizing on an extended growing season may enable organisms to counter some of the projected fitness loss. Similarly, Vasseur et al. (2014) found that fitness detriments associated with anticipated increases in the incidence of extreme events will outweigh fitness increases associated with shifts in mean temperatures and make temperate and boreal organisms most vulnerable to climate change. These results suggest that behavioral buffering will be essential to ectothermic organisms across latitudes to alleviate the impacts of climate change (Sunday et al. 2014).

Studies on plants show some similar trends and complexities. A recent meta-analysis (Way and Oren 2010), for example, showed that warmer temperatures are expected to increase plant performance (measured as growth or physiological responses) to a greater extent in temperate and boreal ecosystems than in tropical forests, where growth declines with increased warming was often seen. Moreover, growth responses to increased temperatures were less variable for tropical than temperate trees, suggesting that tropical species have narrower thermal tolerances. This meta-analysis also suggested that deciduous trees were more stimulated by increased temperatures than are evergreen trees. Finally, other studies demonstrate that water availability (influenced by temperature) will play a large role in determining the sensitivity of tree species to increased temperatures, with increases in growth primarily expected in locations where water availability does not decline.

Seasonality

The impacts of winters with warming temperatures and shorter durations will be substantial for organisms but have been largely underappreciated by analyses focused on activity seasons (reviewed in Williams, Henry, and Sinclair 2014). Responses to warming winters will be shaped by the contrasting life cycles of temperate and boreal organisms (Figure 16.1).

Organisms that remain active through winter may be able to capitalize on prolonged opportunities for activity and energy acquisition associated with warmer temperatures, although negative impacts may also result from the increased activity and duration of interacting species such as

pathogens (see the section "Interacting and Indirect Stressors" below). A key response among organisms that are dormant during the winter will be shifts in energy use. For endothermic animals below their thermal neutral zone, warm temperatures will reduce the energy required to maintain stable body temperatures. Ectothermic animals will experience increased metabolic rates, and the higher energy use may threaten depletion of overwintering energy reserves,

Figure 16.1. Warming has occurred across seasons in the continental United States between the past (dashed line) and current (solid lines) time period. (Data derive from US weather stations and global historical climatological networks accessed via the National Climate Data Center, http://www.ncdc.noaa.gov/cag/.) We depict potential shifts (+ = increase, − = decrease, Δ = shift in time) in biological processes associated with warming across the season (with arrows indicating the timing of relevant temperature shifts) for an organism that uses temperature as a phenological cue. We assume that the winter temperatures are below the thermal neutral zone of an endotherm. Responses to increased winter temperatures will depend on whether the organism is active in winter and will include shifts in rates of energy use and acquisition. The growing season is expected to extend for an organism using temperature as a phenological cue, resulting in increased exposure to the thermally variable spring season. Although increased temperatures may accelerate growth and development, the organism may experience an increased incidence of thermal stress events.

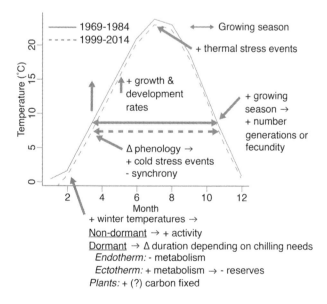

ultimately decreasing overwintering survival or future fecundity (Williams, Henry, and Sinclair 2014). Some animals and plants will experience prolonged dormant periods as warm winters fail to meet their chilling or vernalization requirements. The duration of dormancy will be reduced for other organisms without such requirements.

Shifts in seasonality can have complex effects, due to both temperature variability and to differences among organisms in responses to seasonality. For example, although a longer activity season can provide greater opportunities for development, growth, and reproduction, earlier emergence (e.g., tree budding) in the variable spring season may also risk cold injury (Aitken et al. 2008). Those temperate organisms that rely on photoperiod cues (more reliable than temperature cues in thermally variable environments) can avoid cold injury, but at the same time, those organisms could be prevented from capitalizing on an extended season. A failure to shift phenology can be particularly problematic for species that closely interact with other species; differential phenological shifts may result in mismatches that reduce resource availability or otherwise limit essential interactions (Visser and Both 2005; Buitenwerf et al. 2015).

RESPONSES TO CLIMATE CHANGE

Organismal Responses

Temperate and boreal areas are characterized by steep latitudinal clines in temperature and regions of high topographic variability. These properties are reflected in the velocity of climate change, or the speed at which an organism would need to move to keep pace with climate change. Values are low relative to other ecosystems for temperate coniferous (0.11 km/yr over the twenty-first century) and broadleaf (0.35 km/yr) forests and boreal forests and taiga (0.43 km/yr) (Loarie et al. 2009), and

potentially even lower on mountain slopes. Movement in response to climate change may thus be a viable option for many temperate organisms, which implies that biodiversity may be maintained by species replacement—although montane organisms may lose habitat and face declines in viability as their climatic niche moves upslope through climate change. Even mobile organisms such as mammals will face dispersal limitations (Schloss, Nuñez, and Lawler 2012). Temperate regions, with topographic variability and pronounced seasonality, offer the greatest opportunity for organisms to make a favorable move but also the greatest potential detriments of a wrong move (Buckley, Tewksbury, and Deutsch 2013).

Many species have responded and will respond to climate change in temperate and boreal regions by shifting their distributions as expected (Parmesan 2006)—poleward and upward. However, both historical and recent distribution shifts have been more variable than the baseline expectation. For example, a resurvey of montane mammals after a century found that only half shifted upslope as expected in response to a 3°C increase in mean temperature (Moritz et al. 2008). Unexpected distribution shifts have been particularly pronounced for plants because their distributions tend to be constrained by multiple environmental factors as well as dispersal limitations (particularly following glacial retreat; Aitken et al. 2008). For example, a recent study in California documented downward range shifts, which were surprising but linked to changes in water availability (Crimmins et al. 2011; Dobrowski et al. 2011). Unexpected plant range shifts may drive disruptions of plant-animal interactions and additionally reassemble communities into novel configurations.

Other common responses temperate and boreal organisms are likely to exhibit (due to pronounced seasonality) are shifts in phenology (Parmesan 2006; Buitenwerf et al. 2015). For example, phenological shifts along with enhanced rates of growth and development have increased the number of generations that some insects with seasonal life cycles are able to complete (Altermatt 2009), and many plant species are flowering earlier as a result of warming (Parmesan 2006). However, much like observed range shifts, there is large species-to-species variation in the extent and magnitude of phenological shifts (Parmesan 2007).

Of course, organisms unable spatially or temporally to track shifts in suitable climate space with warming (i.e., climate velocity) can also persist through evolutionary adaption or acclimation (Aitken et al. 2008). Such responses may be particularly viable in temperate and boreal systems where consistent seasonality may enable selection for directional change, but the broad organismal thermal tolerances associated with seasonal environments may slow rates of physiological adaptation. Existing within-species adaptation to local thermal regimes may either constrain or facilitate species responses to climate change, in terms of both phenological and range shifts (Aitken et al. 2008).

Ecosystem Responses

How the diverse responses of individual organisms will aggregate into biodiversity changes is difficult to forecast in temperate and boreal systems. Thus, the biodiversity consequences of ecosystem changes may be easier to predict and more pronounced. Shifts in tree line provide a clear illustration of ecosystem and biome shifts (Figure 16.2). For example, conifer forests have extended their upper elevational limit at multiple sites across the Sierra Nevada of California (Millar et al. 2004). In fact, several recent studies have suggested that tree growth is most sensitive to climate at biome boundaries (Ettinger, Ford, and HilleRisLambers 2011), implying even more rapid rates of biome shifts with future warming.

Figure 16.2. Historical photos provide a key tool for assessing shifts in vegetation and tree line. For example, comparing this 2014 image (bottom; copyright 2014 Ian Breckheimer) of Spray Park on Washington's Mt. Rainier taken from the slopes of Mt. Pleasant (46.927667 N, −121.838111 W) to historical photos (top; by W. P. Romans, courtesy of the Washington State Historical Society) suggests vegetation changes.

Several factors complicate this simple prediction and are likely to lead to complex biome shifts at tree line. First, the primary constraint on tree growth at some tree lines (both altitudinal and latitudinal) is likely to be water availability, not growing season length, which implies that biome shifts will depend on changes in both temperature and water availability (Littell, Peterson, and Tjoelker 2008). Indeed, a recent meta-analysis found only 52 percent of monitored tree lines showed a significant advance, despite recent warming (Harsch et al. 2009). A second factor constraining tree-line shifts is dispersal limitation, especially latitudinally (Aitken et al. 2008). Tree cover is likely to increase in grasslands and shrublands where moisture is not limiting (Grimm et al. 2013). Other, moisture-limited temperate woodlands may become grasslands (Breshears et al. 2005).

Temperate tree species are also expected to shift their ranges poleward into boreal forests in response to warming. Increases in the recruitment of temperate seedlings and saplings and decreases in recruitment of boreal species have been observed as boreal forests warm (Fisichelli, Frelich, and Reich 2013), but altered trophic interactions may constrain the expansion of temperate forests (Frelich et al. 2012). Shifts in the latitude and elevation of temperate and boreal forests are widely anticipated (Grimm et al. 2013; Melillo, Richmond, and Yohe 2014), and the ecotone between temperate and boreal forests has already shifted upward in some regions (Beckage et al. 2008). Experimental warming conducted over three growing seasons found that photosynthesis and growth were reduced for boreal tree species near their warm-edge range limit but enhanced for temperate species near their cold range limit (Reich et al. 2015). This study suggests that leaf-level responses of photosynthesis and respiration will be indicative of whole-plant responses.

Ecosystem changes associated with vegetation shifts will also likely affect biodiversity. Productivity increases are expected and have been broadly observed as warming extends the growing season (Boisvenue and Running 2006). However, productivity increases associated with the longer growing season may be countered by thermal and drought stress in the summer and increased respiration in the fall. Similarly, climate change impacts on disturbance (e.g., fire regimes, pests) may also limit the increased productivity we would expect from warming-induced increases in growing season length (Kurz et al. 2008), although the movement of less flammable temperate species into boreal forests may offset potential wildfire increases in response to warming (Terrier et al. 2013).

The effects of climate change on temperate and boreal biomes and disturbance regimes may also result in positive feedbacks to the climate system. Shifts in vegetation types induced by climate change could amplify local warming through albedo and transpiration effects, as is predicted for boreal regions (Bonan and Pollard 1992). Climate change–induced changes to disturbance regimes (e.g., fire, pests) may also have consequences for carbon dynamics (Running 2008). For example, warming-induced changes to the life cycle of the mountain pine beetle *Dendroctonus ponderosae* caused a region in British Columbia to go from being a carbon sink to a carbon source (Kurz et al. 2008).

Interacting and Indirect Stressors and Feedbacks

The responses to climate change discussed here will interact, often synergistically, with changing natural disturbance regimes. Climate change is expected to shift the frequency, intensity, duration, and timing of natural disturbances, which shape temperate and boreal forests. These natural disturbances include both abiotic (e.g., fire, drought, storms) and biotic (e.g., introduced species, insect and pathogen outbreaks) factors and may be further exacerbated by human stressors such as changes in land use (Dale et al. 2001). For example, wildfire is central to the dynamics of temperate and boreal forests because it drives tree mortality, successional dynamics, nutrient cycling, and hydrology (Dale et al. 2001). Warmer spring and summer temperatures and earlier spring snowmelt are expected to increase fire frequency and intensity along with extending wildfire seasons (Westerling et al. 2006). Climate change is also likely to increase the severity and duration of drought, which can combine, often synergistically, with other physiological stresses such as increased temperature to increase tree mortality above baseline levels. This tree mortality can in turn facilitate insect and pathogen outbreaks (Allen et al. 2010). Some widespread tree die-offs, such as pine mortality spanning over a million hectares in the southwestern United States, have already been linked to anomalous droughts associated with climate change (Breshears et al. 2005).

Climate change can also have direct impacts on diseases through the temperature dependence of host immunity and pest and parasite growth and reproduction (reviewed in Altizer et al. 2013). These two processes may either increase or decrease the incidence or severity of a disease depending on how the thermal optimum of the parasites and their hosts compares to baseline environmental temperatures and those following climate change. The relative shifts in host immunity and parasite growth and reproduction will determine disease dynamics. Rates will be influenced both by shifts in mean environmental conditions and increases in climate variability, which will tend to decrease host immunity as a result of climate warming (Altizer et al. 2013).

Climate change will also alter the seasonality of immunity and transmission (Figure 16.3). Temperature seasonality in temperate and boreal systems drives evolution to synchronize the population growth and virulence of parasites with the susceptibility of hosts. Climate change will disrupt

this synchrony in many systems with varied outcomes, as described below.

SHIFTS IN SEASONALITY

Increases in parasite growth and reproduction may accelerate their transmission cycles such that outbreaks occurring every several years may become an annual occurrence (Figure 16.3A). Mild winters may enable overwinter survival of parasites and extend the transmission season (Figure 16.3B). However, warming may lead summer temperatures to exceed the upper thermal tolerance of parasites and depress transmission in summer (Figure 16.3C). Shifts in seasonality may also result from shifts in host behavior. Milder winters may drive organisms that previously underwent long-distance migrations to remain resident, potentially elongating the transmission season and leading to parasite accumulation. For example, monarch but-

Figure 16.3. The seasonal dynamics of disease transmission may shift in numerous ways following warming depending on thermal tolerance and phenological cues. (A) Warming may increase rates of growth and development and accelerate semi-annual cycles to annual cycles. (B) Warmed summer temperatures may exceed the thermal tolerance of pathogens, shifting the peak transmission season to the spring and fall. (C) Warming may enable overwinter survival, leading to year-round transmission. (D) Warming may drive phenological shifts, which could disrupt synchrony with hosts and expose more sensitive life stages to transmission.

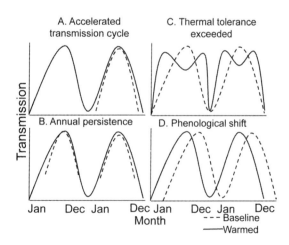

terfly (*Danaus plexippus*) populations that have become year-round residents of the United States as a result of milder winters exhibit higher rates of parasite infection than migrating populations (reviewed in Altizer et al. 2013). Conversely, augmented host migration or range shifts may enable release from parasites.

SHIFTS IN SYNCHRONY

Hosts and parasites may differ in their phenological responses to temperature and photoperiod, shifting synchrony following climate change (Figure 16.3D). For example, mild winters can increase the synchrony of Lyme disease vectors (ticks) and their hosts (mice) and increase disease incidence (Ostfeld 2010). Another mechanism by which climate change can increase disease transmission is exposing a more sensitive life stage to parasites.

An increased prevalence of insect pests will be another important form of ecological disturbance. Recent temperature increases have enabled bark beetles in North American temperate and boreal forests to survive winter, complete additional generations in a season, and expand their distributions; these shifts have expanded the areas of recent infestations beyond those observed over the past 125 years (Raffa et al. 2008). Climate change will continue to redistribute insect pests and enable them to invade new habitats and forest types (Logan, Regniere, and Powell 2003).

CONCLUSIONS: BIODIVERSITY CONSEQUENCES

Warming will directly influence biodiversity in temperate and boreal regions through altering energetics, rates of growth and development (including primary productivity), seasonal timing, and thermal stress. Effects will likely be complex, but we outline several generalizations here that we believe are likely to occur:

• Broad thermal tolerances associated with seasonal climates are likely to allow many temperate and boreal species to benefit from warming, especially if they are able to avoid thermal or drought stress events (for animals and plants, respectively). In all, we expect fewer negative direct responses to climate change for temperate and boreal species than for tropical species.

• Impacts on biodiversity in temperate and boreal regions are likely to be mediated by changes to seasonality and temperature variability. Additionally, negative biodiversity consequences resulting from direct responses of species to climate change may stem primarily from the loss of synchrony with interacting species and the disruption of communities rather than a decline in population viability.

• Many climate change impacts driven by seasonality and temperature variability will occur in winter rather than the more thoroughly studied growing season.

• Impacts on diversity and ecosystem processes will likely be driven by ecological disturbances associated with climate change and their interaction with warming rather than warming directly. Specifically, increases in the incidence of disease, insect pests, and drought may lead to biodiversity declines and decreased productivity. More prominent wildfires also have the potential to diminish biodiversity, but fire may alternately increase biodiversity by reverting forests to earlier successional stages.

• Biodiversity changes in temperate and boreal systems will be particularly challenging to predict because they will depend on the incidence of extreme and rare events. Seasonality and thermal variability have led temperate and boreal species to evolve broad environmental tolerances, which may lessen the impacts of warming. However, climate change may push rare events beyond the environmental tolerance of species and result in thermal stress.

REFERENCES

Aitken, S. N., S. Yeaman, J. A. Holliday, T. Wang, and S. Curtis-McLane. 2008. "Adaptation, migration or extirpation: Climate change outcomes for tree populations." *Evolutionary Applications* 1 (1): 95–111.

Allen, Craig D., Alison K. Macalady, Haroun Chenchouni, Dominique Bachelet, Nate McDowell, Michel Vennetier, Thomas Kitzberger, et al. 2010. "A global overview of drought and heat-induced tree mortality reveals emerging climate change risks for forests." *Forest Ecology and Management* 259 (4): 660–684.

Altermatt, F. 2009. "Climatic warming increases voltinism in European butterflies and moths." *Proceedings of the Royal Society B* 277: 1281–1287.

Altizer, Sonia, Richard S. Ostfeld, Pieter T. J. Johnson, Susan Kutz, and C. Drew Harvell. 2013. "Climate change and infectious diseases: From evidence to a predictive framework." *Science* 341 (6145): 514–519.

Beckage, Brian, Ben Osborne, Daniel G. Gavin, Carolyn Pucko, Thomas Siccama, and Timothy Perkins. 2008. "A rapid upward shift of a forest ecotone during 40 years of warming in the Green Mountains of Vermont." *Proceedings of the National Academy of Sciences* 105 (11): 4197–4202.

Boisvenue, Céline, and Steven W. Running. 2006. "Impacts of climate change on natural forest productivity: Evidence since the middle of the 20th century." *Global Change Biology* 12 (5): 862–882.

Bonan, Gordon B., and David Pollard. 1992. "Effect of boreal forest vegetation on global climate." *Nature* 359: 716–718.

Breshears, David D., Neil S. Cobb, Paul M. Rich, Kevin P. Price, Craig D. Allen, Randy G. Balice, William H. Romme, et al. 2005. "Regional vegetation die-off in response to global-change-type drought." *Proceedings of the National Academy of Sciences* 102 (42): 15144–15148.

Buckley, Lauren B., Joshua J. Tewksbury, and Curtis A. Deutsch. 2013. "Can terrestrial ectotherms escape the heat of climate change by moving?" *Proceedings of the Royal Society B: Biological Sciences* 280 (1765): 20131149.

Buitenwerf, Robert, Laura Rose, and Steven I. Higgins. 2015. "Three decades of multi-dimensional change in global leaf phenology." *Nature Climate Change* 5: 364–368.

Crimmins, S. M., S. Z. Dobrowski, J. A. Greenberg, J. T. Abatzoglou, and A. R. Mynsberge. 2011. "Changes in climatic water balance drive downhill shifts in plant species' optimum elevations." *Science* 331 (6015): 324–327.

Dale, Virginia H., Linda A. Joyce, Steve McNulty, Ronald P. Neilson, Matthew P. Ayres, Michael D. Flannigan, Paul J. Hanson, et al. 2001. "Climate change and forest disturbances: Climate change can affect forests by altering the frequency, intensity, duration, and timing of fire, drought, introduced species, insect and pathogen outbreaks, hurricanes, windstorms, ice storms, or landslides." *BioScience* 51 (9): 723–734.

Deutsch, C. A., J. J. Tewksbury, R. B. Huey, K. S. Sheldon, C. K. Ghalambor, D. C. Haak, and P. R. Martin. 2008. "Impacts of climate warming on terrestrial ectotherms across latitude." *Proceedings of the National Academy of Sciences* 105 (18): 6668–6672.

Dobrowski, S. Z., S. M. Crimmins, J. A Greenberg, J. T Abatzoglou, and A. R. Mynsberge. 2011. "Response to comments on 'Changes in climatic water balance drive downhill shifts in plant species' optimum elevations.'" *Science* 334 (6053): 177.

Ettinger, Ailene K., Kevin R. Ford, and Janneke HilleRisLambers. 2011. "Climate determines upper, but not lower, altitudinal range limits of Pacific Northwest conifers." *Ecology* 92 (6): 1323–1331.

Fisichelli, Nicholas A., Lee E. Frelich, and Peter B. Reich. 2013. "Climate and interrelated tree regeneration drivers in mixed temperate-boreal forests." *Landscape Ecology* 28 (1): 149–159.

Frelich, Lee E., Rolf O. Peterson, Martin Dovčiak, Peter B. Reich, John A. Vucetich, and Nico Eisenhauer. 2012. "Trophic cascades, invasive species and body-size hierarchies interactively modulate climate change responses of ecotonal temperate-boreal forest." *Philosophical Transactions of the Royal Society B: Biological Sciences* 367 (1605): 2955–2961.

Grimm, Nancy B., F. Stuart Chapin III, Britta Bierwagen, Patrick Gonzalez, Peter M. Groffman, Yiqi Luo, Forrest Melton, et al. 2013. "The impacts of climate change on ecosystem structure and function." *Frontiers in Ecology and the Environment* 11 (9): 474–482.

Harsch, Melanie A., Philip E. Hulme, Matt S. McGlone, and Richard P. Duncan. 2009. "Are treelines advancing? A global meta-analysis of treeline response to climate warming." *Ecology Letters* 12 (10): 1040–1049.

Khaliq, Imran, Christian Hof, Roland Prinzinger, Katrin Böhning-Gaese, and Markus Pfenninger. 2014. "Global variation in thermal tolerances and vulnerability of endotherms to climate change." *Proceedings of the Royal Society B: Biological Sciences* 281 (1789): 20141097.

Kingsolver, Joel G., Sarah E. Diamond, and Lauren B. Buckley. 2013. "Heat stress and the fitness consequences of climate change for terrestrial ectotherms." *Functional Ecology* 27 (6): 1415–1423.

Kurz, Werner A., C. C. Dymond, G. Stinson, G. J. Rampley, E. T. Neilson, A. L. Carroll, T. Ebata, and L. Safranyik. 2008. "Mountain pine beetle and forest carbon feedback to climate change." *Nature* 452 (7190): 987–990.

Littell, Jeremy S., David L. Peterson, and Michael Tjoelker. 2008. "Douglas-fir growth in mountain ecosystems: Water limits tree growth from stand to region." *Ecological Monographs* 78 (3): 349–368.

Loarie, S. R., P. B. Duffy, H. Hamilton, G. P. Asner, C. B. Field, and D. D. Ackerly. 2009. "The velocity of climate change." *Nature* 462 (7276): 1052–1055.

Logan, Jesse A., Jacques Regniere, and James A. Powell. 2003. "Assessing the impacts of global warming on forest pest dynamics." *Frontiers in Ecology and the Environment* 1 (3): 130–137.

Melillo, Jerry M., T. C. Richmond, and G. W. Yohe. 2014. *Climate Change Impacts in the United States: The Third National Climate Assessment.* U.S. Global Change Research Program.

Millar, Constance I., Robert D. Westfall, Diane L. Delany, John C. King, and Lisa J. Graumlich. 2004. "Response of subalpine conifers in the Sierra Nevada, California, USA, to 20th-century warming and decadal climate variability." *Arctic, Antarctic, and Alpine Research* 36 (2): 181–200.

Moritz, C., J. L. Patton, C. J. Conroy, J. L. Parra, G. C. White, and S. R. Beissinger. 2008. "Impact of a century of climate change on small-mammal communities in Yosemite National Park, USA." *Science* 322 (5899): 261–264.

Ostfeld, Richard. 2010. *Lyme Disease: The Ecology of a Complex System.* Oxford University Press.

Parmesan, C. 2006. "Ecological and evolutionary responses to recent climate change." *Annual Review of Ecology, Evolution and Systematics* 37: 637–669.

Parmesan, C. 2007. "Influences of species, latitudes and methodologies on estimates of phenological response to global warming." *Global Change Biology* 13 (9): 1860–1872.

Raffa, Kenneth F., Brian H. Aukema, Barbara J. Bentz, Allan L. Carroll, Jeffrey A. Hicke, Monica G. Turner, and William H. Romme. 2008. "Cross-scale drivers of natural disturbances prone to anthropogenic amplification: The dynamics of bark beetle eruptions." *Bioscience* 58 (6): 501–517.

Reich, Peter B., Kerrie M. Sendall, Karen Rice, Roy L. Rich, Artur Stefanski, Sarah E. Hobbie, and Rebecca A. Montgomery. 2015. "Geographic range predicts photosynthetic and growth response to warming in co-occurring tree species." *Nature Climate Change.* http://www.nature.com/nclimate/journal/vaop/ncurrent/full/nclimate2497.html.

Running, Steven W. 2008. "Ecosystem disturbance, carbon, and climate." *Science* 321 (5889): 652–653.

Schloss, C. A., T. A. Nuñez, and J. J. Lawler. 2012. "Dispersal will limit ability of mammals to track climate change in the Western Hemisphere." *Proceedings of the National Academy of Sciences* 109 (22): 8606–8611.

Sunday, Jennifer M., Amanda E. Bates, Michael R. Kearney, Robert K. Colwell, Nicholas K. Dulvy, John T. Longino, and Raymond B. Huey. 2014. "Thermal-safety margins and the necessity of thermoregulatory behavior across latitude and elevation." *Proceedings of the National Academy of Sciences* 111: 5610–5615.

Terrier, Aurélie, Martin P. Girardin, Catherine Périé, Pierre Legendre, and Yves Bergeron. 2013. "Potential changes in forest composition could reduce impacts of climate change on boreal wildfires." *Ecological Applications* 23 (1): 21–35.

Vasseur, David A., John P. DeLong, Benjamin Gilbert, Hamish S. Greig, Christopher D. G. Harley, Kevin S. McCann, Van Savage, Tyler D. Tunney, and Mary I. O'Connor. 2014. "Increased temperature variation poses a greater risk to species than climate warming." *Proceedings of the Royal Society B: Biological Sciences* 281 (1779): 20132612.

Visser, M. E., and C. Both. 2005. "Shifts in phenology due to global climate change: The need for a yardstick." *Proceedings of the Royal Society B* 272 (1581): 2561–2569.

Way, Danielle A., and Ram Oren. 2010. "Differential responses to changes in growth temperature between trees from different functional groups and biomes: A review and synthesis of data." *Tree Physiology* 30 (6): 669–688.

Westerling, Anthony L., Hugo G. Hidalgo, Daniel R. Cayan, and Thomas W. Swetnam. 2006. "Warming and earlier spring increase western US forest wildfire activity." *Science* 313 (5789): 940–943.

Williams, Caroline M., Hugh A. L. Henry, and Brent J. Sinclair. 2014. "Cold truths: How winter drives responses of terrestrial organisms to climate change." *Biological Reviews* 90: 214–235.

CHAPTER SEVENTEEN

Climate Change Impacts on Mountain Biodiversity

ANTOINE A. GUISAN, OLIVIER
BROENNIMANN, ALINE BURI,
CARMEN CIANFRANI, MANUELA D'AMEN,
VALERIA DI COLA, RUI FERNANDES,
SARAH M. GRAY, RUBÉN G. MATEO,
ERIC PINTO, JEAN-NICOLAS PRADERVAND,
DANIEL SCHERRER, PASCAL VITTOZ,
ISALINE VON DÄNIKEN, AND
ERIKA YASHIRO

SETTING THE SCENE: CLIMATE CHANGE IN MOUNTAIN AREAS

Mountains cover 12.3 percent of the terrestrial area outside Antarctica (Table 17.1; Körner et al. 2011) and harbor a proportionally higher amount of biodiversity than lowlands, including several biodiversity hotspots, such as tropical cloud forests (Dimitrov et al. 2012) or endemic-rich alpine grasslands (Engler et al. 2011). High-elevation habitats present particular ecological challenges for life (e.g., low temperatures, short windows for reproduction, fluctuation in food availability, high solar radiation, hypoxia), restricting colonization to adapted life forms (Scridel 2014). Mountain regions seem also to be warming at a higher rate than other regions (Rangwala et al. 2013), which makes these ecosystems quite sensitive to natural and anthropogenic impacts (e.g., climate, fires, land-use changes; Beniston 1994; Beniston et al. 1997).

Mountains were identified early as sensitive to climate change (e.g., Beniston 1994; Guisan et al. 1995) for being (1) climatically sensitive areas (Nogués-Bravo et al. 2007); (2) "islands" separated by lowland areas (Pauchard et al. 2009); (3) sensitive high-elevation ecosystems, with no escape for species toward higher elevations (Theurillat and Guisan 2001); and (4) prone to conflict between nature conservation and use of mountains to provide services to humans (Gret-Regamey et al. 2012).

According to the IPCC (Beniston et al. 1996 and following reports; see http://www.ipcc.ch), mountains are already affected by climate change, as shown in the past decades by the following:

Table 17.1.   Global area of bioclimatic mountain belts

Thermal belts	Area (Mio km²)	M (%)	T (%)
Nival (<3.5°C, season <10d)	0.53	3.24	0.40
Upper alpine (<3.5°C, 10d< season <54d)	0.75	4.53	0.56
Lower alpine (<6.4°C, 54d< season <94d)	2.27	13.74	1.68
Tree line			
Upper montane (>6.4°C ≤10°C)	3.29	20.53	2.51
Lower montane (>10°C ≤15°C)	3.74	22.64	2.78
Remaining mountain area with frost (<15°C)	1.34	8.11	0.99
Remaining mountain area without frost (>15°C)	4.49	27.22	3.34
Total	**16.51**	**100.00**	**12.26**

Note: See http://www.mountainbiodiversity.org. Temperatures refer to growth season mean air temperatures: M (%) = percentage of total mountain area (100% = 16.5 Mio km²); T (%) = percentage of total terrestrial area outside Antarctica (100% = 134.6 Mio km²). From Körner, Paulsen, and Spehn (2011).

• Shrinkage of most mountain glaciers, with greatest ice losses in Patagonia, Alaska, northwestern United States, southwestern Canada, and the European Alps (Arendt et al. 2012), and glaciers in mountains of tropical areas also particularly affected (Thompson et al. 2006).

• Significant change in snow cover, mainly toward overall reduction and greater seasonal variations (Brown and Robinson 2011), resulting in greater snow-free growing seasons (~5 days per decade; Choi et al. 2010).

Predictions of future climate change for mountain ranges worldwide involve an average temperature change of 2°C–3°C by 2070 and 3°C–5°C by the end of the century (Nogués-Bravo et al. 2007), with greater increases for mountains in northern latitudes than in temperate and tropical climates. An additional threat is represented by invasive species (e.g., plants), which are predicted to increasingly invade mountains under climate change (Pauchard et al. 2009; Petitpierre et al. 2016).

Many mountain species are already responding to the effects of ongoing climate warming (Pounds et al. 1999; Lenoir et al.

2008), and these trends are expected to intensify in the future. Here, we review the changes already observed and future model-based projections in mountain regions worldwide for many different groups and ecosystems (Figure 17.1).

MOUNTAIN FORESTS AND TREE LINES

About 78 percent of the world's mountain areas are below the natural tree line and therefore potentially forested (Paulsen and Körner 2014). In the past 100 years the vast majority of tree lines around the globe advanced upward (Harsch et al. 2009), correlated with an increasing growth rate of established trees (Salzer et al. 2009). However, the advance of the tree line is not necessarily tracking the current warming rate, likely due to the slow tree growth and very sensitive recruitment to climatic conditions. Hence, it may take a long time before a new equilibrium is reached between climate and tree line (Körner 2012). Despite such lag in tree-line advance, both mechanistic and correlative modeling approaches predict an advance of the tree line and a consequent reduction of the alpine and nival (i.e., higher than alpine, where vegetation becomes patchy and scarce) areas (Gehrig-Fasel et al. 2007; Körner 2012).

BIRDS

O: Rapid latitudinal or altitudinal range shifts; upslope shifts already happened, but some are toward suboptimal habitats due to human avoidance in the lowlands; truly climate-driven shifts also observed.

P: Severe range reduction and extinctions predicted.

FORESTS AND TREELINES

O: Upslope treeline range shifts ongoing, but delayed; most observed species/community range and composition shifts due to land-use changes.

P: Continued treeline advance will reduce available alpine grassland habitat, leading to changes in species diversity.

SOIL FAUNA

O: Still poorly known, but seem driven by soil, climatic factors and food resources.

P: Change in food availability expected to drive species composition shifts; range shifts expected to follow thermal isocline, with slow species' dispersal rates possibly causing some extinctions. No model predictions.

AMPHIBIANS

O: Threatened in tropical mountains (e.g. cloud forests), with strong declines observed in South America and elsewhere; species also affected in non-tropical mountains.

P: Increased risks predicted in the next decades for narrowly-distributed mountain species.

MAMMALS

O: Phenological changes, local extinctions, reduced genetic diversity and upslope species shifts at median rate of 11 m per decade.

P: Differential elevation changes among mammal species expected to modify food web structures; severe range reduction and connectivity alteration.

FRESHWATER ECOSYSTEMS

O: Glacial melting and changes in precipitation affecting volume, water temperature, and nutrient inputs.

P: Predicted species range shifts could alter community composition; temperature-generalist species abundance predicted to increase, while temperature specialists and ectotherms likely to decrease.

SOIL MICROORGANISMS

O: Follow climatic gradients similar to macroorganisms; strong dependence on soil factors.

P: Soil responses might mitigate pure climatic effects, but strong link with plants could make them indirectly sensitive; no model predictions available, but expected to partly respond as macroorganisms do.

REPTILES

O: Upslope range shifts by lower elevation species, greater extinction risk for high-elevation viviparous species constrained by thermal physiology, dispersal, and biotic interactions.

P: Cold-adapted species to lose habitats, whereas warm-adapted species to gain suitable surfaces.

INSECTS

O: Rapid changes in phenology, community composition, and species elevational range shifts (e.g. from 90-300 m in only a few decades for butterflies).

P: Models predict community homogenization and spatial mismatches in insect-host plant relationships (e.g. trophic shifts).

GRASSLAND

O: Upslope species range shifts (e.g. summit flora enrichment); greater changes in the alpine than subalpine grasslands.

P: High rates of alpine/nival species extinctions, but not before e.g. 40-80 years.

O: *Observed trends*
P: *Predicted trends*

Figure 17.1. Summaries of the main observed (O) and predicted (P) trends for the responses of the different biodiversity groups to climate change in mountain areas.

Additionally, most models project strong changes in composition and structure of temperate and Mediterranean mountain forests, affecting biodiversity and ecosystem services, such as protection against rockfalls and avalanches (Elkin et al. 2013). Some of these changes are already visible: there is an elevational discrepancy between adult trees and seedling distributions in mountain forests (Lenoir et al. 2009), as well as elevational shifts of species (Lenoir et al. 2008) and plant communities (Peñuelas and Boada 2003).

In tropical latitudes the highly diverse montane cloud forests, which shelter an exceptionally high biodiversity of plants and bryophytes, are severely threatened by climate and land-use changes. These fragmented tropical ecosystems depend on high amounts of rainfall, humidity, and condensation from cloud formations. They are therefore projected to be highly affected even by moderate changes in temperature and rainfall regimes (Ponce-Reyes et al. 2013).

ALPINE AND SUBALPINE GRASSLANDS

There is much evidence of upward shifts of alpine plants toward higher elevations (Pauli et al. 2012). These shifts are usually associated with a shrinking of high-elevation habitats (Gottfried et al. 2012). The replacement of alpine grasslands by lower elevation (e.g., forest) species has led to a loss of biodiversity and ecosystem functions (Elumeeva et al. 2013). Changes in water availability in dry regions influence plant species richness as well, with a decrease already observed in Mediterranean mountains (Pauli et al. 2012). Twenty-first-century climate change scenarios predict a massive reduction of high-elevation grassland plant diversity and high community turnover, which possibly will change the structures of current natural ecosystems (Engler et al. 2011). In the European Alps, Dullinger et al. (2012) predicted a range re-

duction of around 44 percent–50 percent for 150 high-mountain species, including several endemics, with possible lags in extinctions (extinction debt). Species that already occur near mountaintops with no possibility to escape upward have a greater risk of extinction (e.g., Engler et al. 2011; Dullinger et al. 2012). In some mountains, an increasing frequency of drought events and fires is also expected to dramatically homogenize species composition (Bendix et al. 2009). Nonetheless, mountain systems have pronounced microclimatic variation, which may allow species to persist locally (Scherrer and Korner 2011). The melting of permanent snow and ice may also provide new potential habitats at higher elevations, although the formation of soils may take several hundred years (Engler et al. 2011).

MOUNTAIN SOIL MICROORGANISMS

Mountain soil microorganisms follow environmental gradients in a similar way to macroorganisms (Pellissier et al. 2014). They play an essential role in ecosystem functioning (Averill et al. 2014) and response to climate change (Singh et al. 2010), and influence plant distributions (Pellissier et al. 2013). Several studies examined the effects of simulated climate change on alpine microbial communities (e.g., Streit et al. 2014). Also experimental warming of 4°C over four consecutive years had a smaller effect on soil bacterial community composition than did seasonal variations, pH, and salinity (Kuffner et al. 2012).

Despite a general increase in average temperatures over 17 years in the Italian Alps, no significant change in bacterial community could be detected (Margesin et al. 2014). Other soil warming experiments have shown an increase of certain fungal taxa (Fujimura et al. 2008), and precipitation treatments showed high resistance of fungi to drought (Barnard et al. 2015). Temperature warming may also lead to increased emergence of fungal diseases and associated biodiversity losses in mountain ecosystems (Fisher et al. 2012), causing amphibian extinctions in tropical mountain ecosystems (see "Mountain Amphibians" section). Few projections of microbial species responses to climate change exist for mountain areas, which illustrates the pressing need for predictive models of fungi and bacteria under diverse climate change scenarios in these systems.

MOUNTAIN SOIL FAUNA

Belowground animals seem to be distributed along elevational gradients similar to aboveground animals, although details of their distribution and ecology remain poorly known (Decaëns 2010). Yet this distribution seems more driven by physical factors, such as climate, soil structure, and food resources, than by competition (Sylvain and Wall 2011). There is little reported about observed changes caused by climatic change.

Climatic change is predicted to affect soils through changing snow-cover distribution and increased soil temperature (Zinger et al. 2009). Poorly developed soils of mountain areas and their faunal communities are predicted to be more affected than others by climate change (Hagvar and Klanderud 2009). Field experiments in Norway testing the effect of soil fertilization and warming on mountain soil biodiversity reported an increased biomass of microarthropods and a higher dominance of fast-growing species (Hagvar and Klanderud 2009). Wu et al. (2014) highlighted shifts in microarthropod community composition due to climate change and identified food availability to be more important than abiotic factors (i.e., temperature and water availability) in regulating microarthropod response to climate changes. Soil organisms are predicted to disperse following thermal isocline changes (Berg et al. 2010) and extinctions are mainly expected to occur for small soil organisms that may

not have a sufficient dispersal rate to follow the temperature isocline modification. Another limiting factor for migration of species is the depth of soil organic matter. High-elevation soils have thin soil organic matter layers, which may restrict the possibility of vertical movement (Swift et al. 1998). Finally, invasion by exotic species is predicted to occur more rapidly with climate change, as already has been observed for European earthworms invading American forests (Parkinson et al. 2004).

MOUNTAIN INSECTS

Climate impacts on mountain insect populations have already been documented—for example, extinctions along the lower-elevation boundaries for the Apollo butterfly in France (Descimon et al. 2005). Loss of open grassland habitats due to forest expansion is expected to produce a significant decrease in mountain insect diversity (Menéndez 2007). Butterflies species display a particularly fast response to climate change by elevation shifts, between 90 m and 300 m in a few decades (Wilson et al. 2007; Merrill et al. 2008). This elevational shift, also reported for other insect taxa (e.g., Sheldon 2012), can cause alteration of community diversity and biotic interactions (see Pradervand et al. 2014 on bumblebees). Morphological and physiological adaptations to the cold in mountain environments, such as abundant hairs or muscles generating heat in bumblebees (Hegland et al. 2009), could make some species more sensitive to climate change. Range contractions are expected in some slow-growing species that require low temperatures to induce diapause (Menéndez 2007). Phenological changes were also recorded in relation to earlier spring events, which can cause declines for larvae (more vulnerable to weather fluctuations) and females (reduced availability of trophic resource; Boggs and Inouye 2012).

Studies forecasting the effect of climate change on insect species in mountain environments are still scarce. Future projections of butterfly species suggest a potential altitudinal shift of 650 m by the year 2100 (Merrill et al. 2008). Pradervand et al. (2014) predicted community homogenization and species-specific responses of bumblebees to climate change along an elevation gradient, with the most affected species being those currently restricted to high elevation. For the same mountain area, Descombes et al. (2015) projected a potential reduction between 37 percent and 50 percent of the suitable areas for butterflies, depending on their faculty to broaden or not their diet to new host plants.

MOUNTAIN AMPHIBIANS

Climate change is expected to be a major threat to amphibian biodiversity, because the timing of the seasonal activities of hibernation, aestivation, and reproduction of amphibian species is tightly related to temperature conditions (Di Rosa et al. 2007). Amphibians on tropical mountains are particularly at risk (Pounds et al. 2006). In these ecosystems, species have experienced little temperature variation through geological time, so they may have little acclimation ability to rapid changes in the thermal regimes (Wake and Vredenburg 2008). Most species of montane, diurnal frogs from South and Central America have declined, often in relation to chytrid fungi disease outbreaks (e.g., Pounds et al. 2006). Local extinctions and declines have also been reported for nontropical mountains (e.g., Di Rosa et al. 2007).

Amphibian extinctions are projected to accumulate with temperature warming in the coming decades, especially narrowly distributed mountain species, such as cloud forest and mountaintop salamanders and frogs in Central and South America (Wake and Vredenburg 2008). For instance, Milanovich et al. (2010) projected short-term shifts in climatic distributions for plethodontid salamanders in the southern Appala-

chian Mountains, with significant declines in suitable habitat, especially for those species with southern and/or smaller ranges. Taking a broader perspective, Lawler et al. (2010) projected a very high species turnover (exceeding 60 percent) in the Andes under the 2071–2100 future scenarios. Other studies predicting climate change risk in European countries projected a potential high range contraction for many mountain amphibians and mountain areas (e.g., D'Amen et al. 2011; Popescu et al. 2013).

MOUNTAIN REPTILES

Declines of mountain reptiles have also been attributed to climate change (Popescu et al. 2013). The latitude-dependent impacts of climate change reported by Cadby et al. (2010) are also relevant along elevation. Many viviparous species of *Sceloporus* lizards in Mexico, confined to high-elevation "islands" by thermal physiology and/or ecological interactions, face high extinction risk as a result of their restricted migration potential, combined with upward migrations by low-elevation taxa (Sinervo et al. 2010). In the European Alps, an observed upward shift of a lizard species (*Podarcis muralis*) has already been observed (Vittoz et al. 2013). Global warming has also an impact on morphological and phenological traits. For instance, the observed increase in body size of the lizard species *Lacerta vivipara* in France appears related to temperature rising in the past 20 years (Chamaille-Jammes et al. 2006), and the birth date of viviparous lizard (*Niveoscincus ocellatus*) offspring has been strongly influenced by the temperature increase in Australian mountains (Cadby et al. 2010).

Future global warming is predicted to have a strong impact on mountain reptiles, with upward shifts or local reductions depending on the species (Guizado-Rodriguez et al. 2012). Climate change is predicted to shift the lower distribution limit of Patagonian lizards *Liolaemus*, but not the upper limit due to topographic constraint, with resulting range shrinkages (Bonino et al. 2014). Most isolated populations of reptile mountain species with low dispersal abilities will typically be at higher risk of local extinction (Chamaille-Jammes et al. 2006).

MOUNTAIN BIRDS

Mountain ecosystems maintain a wide diversity of birds during breeding and post-breeding seasons (Martin 2013). Birds are mobile species that are likely to respond more easily than other organisms to climate change, by rapid latitudinal or elevational shifts (Chen et al. 2011). Yet land use has a particularly high importance for birds and the impact of climate change may largely be modulated by changes in land-use management. In Europe, several species have retreated from high human-populated lowland areas toward suboptimal but still suitable mountain areas (Martin 2013). Maggini et al. (2011) found an upward shift in at least a third of the studied breeding species in Switzerland, estimated at a rate of 102 m/decade (Vittoz et al. 2013). Birds are also shifting their distributions upslope in other areas, as reported for New Guinea (Freeman and Freeman 2014).

A further drastic reduction of breeding bird distribution, potentially leading in a number of cases to extinction, is also predicted for Switzerland (Maggini et al. 2014). Where less suitable areas are available at higher elevation, species face the disappearance of their specific habitats, as in the mountains of the Australian Wet Tropics (Williams et al. 2003). As mountain tropical and cloud forests get increasingly threatened, the hundreds of bird species that they shelter are of particularly great concern (Sekercioglu et al. 2008).

MOUNTAIN MAMMALS

Surprisingly few studies exist on observed climate change impacts on mountain mammals, yet those studies do report impacts (Vittoz et al. 2013). The distributions of several species have shifted toward higher elevations, at an estimated median rate of 11 m/decade, a rate almost two times faster than previously reported (Chen et al. 2011). Phenological changes have also been observed. Yellow-bellied marmots (*Marmota flaviventris*) in the Rocky Mountains emerge from hibernation 38 days earlier than 23 years ago, due to warming spring air temperature (Inouye et al. 2000). Some local extinctions have already been observed, for instance, the American pika (*Ochotona princeps*) lost 7 of its 25 populations in the United States (Beever et al. 2011). Habitat loss and fragmentation caused by climate change can also lead to decreased gene flow and reduced genetic diversity in alpine mammals (Rubidge et al. 2012). Differential range changes among mammal species along elevation can further change food web structures. Lurgi et al. (2012) observed a decrease in predator-prey mass ratios at low and intermediate elevations in the Pyrenees following migration of prey with different body mass or ecological requirements. The population dynamics of big mammals, such as ungulates, seem also indirectly influenced by climate change–induced modifications in plant phenology causing earlier spring onset (Pettorelli et al. 2007). A consequence could be that current protected areas may not be able to protect mountain mammals effectively. The current extent of suitable areas for the giant panda in the Qinling Mountains of China may shrink by up to 62 percent and shift farther north, outside the current network of nature reserves (Fan et al. 2014). The alteration of connectivity is predicted by Wasserman et al. (2013) for marten (*Martes americana*) populations in the US northern Rocky Mountains following an increase of the fragmentation under climate change. Also, 19 of the 31 American pika subspecies are predicted to lose more than 98 percent of their suitable habitat under a 7°C increase of the warmest months of the year (Calkins et al. 2012).

MOUNTAIN FRESHWATER BIODIVERSITY

Species in high-elevation lakes and ponds are considered sentinels of climate change because they respond rapidly to environmental alteration (Adrian et al. 2009) and are less likely to be influenced by confounding factors (e.g., increased nutrient input) than their lower-elevation counterparts (Füreder et al. 2006). Hobbs et al. (2010) found that climate change is directly and indirectly altering food-web composition and structure in these systems, with community dynamics being affected by nutrient enrichment (Thompson et al. 2008), reduction in thermocline mixing (Winder et al. 2009), and glacial melting (Slemmons et al. 2013). Remote sensing studies show alpine ponds reduced in size or disappearing from low elevations but increased toward higher elevations (Salerno et al. 2014). With glacial melting and increased vegetation growth at high elevation, increased disruptions of thermal stratification and nutrient inputs are expected to lead to earlier spring phytoplankton blooms and turnover in algae and invertebrate composition (Smol et al. 2005; Weidman et al. 2014). The distribution of temperature specialists and generalists is also predicted to change: 33 percent of Odonata cold-water specialists may be at risk of extinction, whereas 63 percent of warm-water specialists and generalists are predicted to increase in abundance (Rosset and Oertli 2011).

In rivers and streams, an increase in water flow, volume, and temperature was observed as a result of increased glacial melt and decreased number of snow days, altering river and stream biodiversity (Brown et al. 2007; e.g., on fish species, see Scheurer 2009; on bacterial composition, see Wilhelm et al. 2013; on invertebrates, see Jacobsen et al. 2012). Ectotherms appear to be prone to local extinctions when water warms beyond a species' maximal thermal limit (Eby et al. 2014), when the timing of developmental stages and predator-prey

relationships are altered (MacDonald et al. 2014), and when the likelihood of diseases increases (Zimmerli et al. 2007). Range shifts are also occurring, for example, in the cutthroat trout (*Oncorhynchus clarkii lewisii*) in the Upper Colorado River Basin (Roberts et al. 2013). Change in precipitation is also affecting species assemblages by causing drought-sensitive species to become locally extinct with decreased precipitation (e.g., salamanders; Currinder et al. 2014) or increased sediment transportation with increased precipitation (e.g., fishes; Chen et al. 2015). Continued changes in climate will further threaten these highly sensitive ectotherm species, with, for instance, 63 percent of the cutthroat trout populations being predicted to disappear in the next 70 years (Isaak et al. 2012). However, in some areas, like France, climate change may cause mountain fish species' richness and trait diversity to increase (Buisson and Grenouillet 2009).

CONCLUSION

We have reviewed current evidences and projections of climate change impact on biodiversity in mountain ecosystems, spanning a large range of organisms and habitats, from soil microbes (and fauna) to plants, insects, and several vertebrate groups. Climate change effects, observed in nature or based on experimental evidences, were reported for nearly all taxonomic groups but with various amplitudes. Model projections were also reported for many groups but are still missing for several key organisms, such as communities inhabiting soil and freshwater (except fish) systems. As a result, a proper overall synthesis and comparison across taxa and ecosystems are still difficult for mountain regions. Large, multisite assessment of climate change impact on most biodiversity components is lacking. Model projections that combine predictions for the different biodiversity groups and their interactions within a given ecosystem type (e.g., terrestrial, aquatic) are still rare. For instance, predicted change in soil biodiversity is likely to affect plant communities, and thus the structure and dynamics of terrestrial ecosystems and their services, but these are rarely found in the same mountain study. Similarly, trophic relations in freshwater systems are likely to modify the predicted responses of individual groups of organisms in these highly dynamic systems.

REFERENCES

Adrian, R., C. M. O'Reilly, H. Zagarese, S. B. Baines, D. O. Hessen, W. Keller, D. M. Livingstone, R. Sommaruga, D. Straile, and E. Van Donk. 2009. "Lakes as sentinels of climate change." *Limnology and Oceanography* 54: 2283–2297.

Averill, C., B. L. Turner, and A. C. Finzi. 2014. "Mycorrhiza-mediated competition between plants and decomposers drives soil carbon storage." *Nature* 505: 543–545.

Barnard, R. L., C. A. Osborne, and M. K. Firestone. 2015. "Changing precipitation pattern alters soil microbial community response to wet-up under a Mediterranean-type climate." *ISME Journal* 9: 946–957. https://doi.org/10.1038/ismej.2014.192.

Beever, E. A., C. Ray, J. L. Wilkening, P. F. Brussard, and P. W. Mote. 2011. "Contemporary climate change alters the pace and drivers of extinction." *Global Change Biology* 17: 2054–2070.

Bendix, J., H. Behling, T. Peters, M. Richter, and E. Beck. 2009. "Functional biodiversity and climate change along an altitudinal gradient in a tropical mountain rainforest." In *Tropical Rainforests and Agroforests under Global Change*, ed. T. Tscharntke, C. Leuschner, E. Veldkamp, H. Faust, E. Guhardja, and A. Bidin, 239–268. Springer.

Beniston, M., ed. 1994. *Mountain Environments in Changing Climates*. Routledge.

Beniston, M., H. F. Diaz, and R. S. Bradley. 1997. "Climatic change at high elevation sites: An overview." *Climatic Change* 36: 233–251.

Beniston, M., D. G. Fox, S. Adhikary, R. Andressen, A. Guisan, J. I. Holten, J. Innes, J. Maitima, M. F. Price, L. Tessier, et al. 1996. "Impacts of climate change on mountain regions." In *Climate Change 1995: Impacts, Adaptations and Mitigation of Climate Change: Scientific-Technical Analyses*, ed. Robert T. Watson, Marufu C. Zinyowera, Richard H. Moss, and David J. Dokken, 191–213. Cambridge University Press.

Berg, M. P., E. T. Kiers, G. Driessen, M. van der Heijden, B. W. Kooi, F. Kuenen, M. Liefting, H. A. Verhoef, and J. Ellers. 2010. "Adapt or disperse: Understanding species persistence in a changing world." *Global Change Biology* 16: 587–598.

Boggs, C. L., and D. W. Inouye. 2012. "A single climate driver has direct and indirect effects on insect population dynamics." *Ecology Letters* 15: 502–508.

Bonino, M. F., D. L. M. Azócar, J. A. Schulte II, and F. B. Cruz. 2014. "Climate change and lizards: Changing species' geographic ranges in Patagonia." *Regional Environmental Change* 15 (6): 1121–1132.

Brown, L. E., D. M. Hannah, and A. M. Milner. 2007. "Vulnerability of alpine stream biodiversity to shrinking glaciers and snowpacks." *Global Change Biology* 13: 958–966.

Brown, R. D., and D. A. Robinson. 2011. "Northern Hemisphere spring snow cover variability and change over 1922–2010 including an assessment of uncertainty." *Cryosphere* 5: 219–229.

Buisson, L., and G. Grenouillet. 2009. "Contrasted impacts of climate change on stream fish assemblages along an environmental gradient." *Diversity and Distributions* 15: 613–626.

Cadby, C. D., G. M. While, A. J. Hobday, T. Uller, and E. Wapstra. 2010. "Multi-scale approach to understanding climate effects on offspring size at birth and date of birth in a reptile." *Integrative Zoology* 5: 164–175.

Calkins, M. T., E. A. Beever, K. G. Boykin, J. K. Frey, and M. C. Andersen. 2012. "Not-so-splendid isolation: Modeling climate-mediated range collapse of a montane mammal *Ochotona princeps* across numerous ecoregions." *Ecography* 35: 780–791.

Chamaille-Jammes, S., M. Massot, P. Aragon, and J. Clobert. 2006. "Global warming and positive fitness response in mountain populations of common lizards *Lacerta vivipara*." *Global Change Biology* 12: 392–402.

Chen, I. C., J. K. Hill, R. Ohlemuller, D. B. Roy, and C. D. Thomas. 2011. "Rapid range shifts of species associated with high levels of climate warming." *Science* 333: 1024–1026.

Chen, J.-P., C. K. C. Wen, P.-J. Meng, K. L. Cherh, and K.-T. Shao. 2015. "Ain't no mountain high enough: The impact of severe typhoon on montane stream fishes." *Environmental Biology of Fishes* 98: 35–44.

Choi, G., D. A. Robinson, and S. Kang. 2010. "Changing Northern Hemisphere snow seasons." *Journal of Climate* 23: 5305–5310.

Currinder, B., K. K. Cecala, R. M. Northington, and M. E. Dorcas. 2014. "Response of stream salamanders to experimental drought in the southern Appalachian Mountains, USA." *Journal of Freshwater Ecology* 29: 579–587.

D'Amen, M., P. Bombi, P. B. Pearman, D. R. Schmatz, N. E. Zimmermann, and M. A. Bologna. 2011. "Will climate change reduce the efficacy of protected areas for amphibian conservation in Italy?" *Biological Conservation* 144: 989–997.

Decaëns, T. 2010. "Macroecological patterns in soil communities." *Global Ecology and Biogeography* 19: 287–302.

Descimon, H., P. Bachelard, E. Boitier, and V. Pierrat. 2005. "Decline and extinction of *Parnassius apollo* populations in France, continued." *Studies on the Ecology and Conservation of Butterflies in Europe* 1: 114–115.

Descombes, P., J.-N. Pradervand, J. Golay, A. Guisan, and L. Pellissier. 2015. "Simulated shifts in trophic niche breadth modulate range loss of alpine butterflies under climate change." *Ecography* 39: 796–804.

Di Rosa, I., F. Simoncelli, A. Fagotti, and R. Pascolini. 2007. "The proximate cause of frog decline?" *Nature* 447 (E4–E5): 44.

Dimitrov, D., D. Nogues-Bravo, and N. Scharff. 2012. "Why do tropical mountains support exceptionally high biodiversity? The eastern arc mountains and the drivers of Saintpaulia diversity." *PLOS One* 7 (11): e48908. http://doi.org/10.1371/journal.pone.0048908.

Dullinger, S., A. Gattringer, W. Thuiller, D. Moser, N. E. Zimmermann, A. Guisan, W. Willner, C. Plutzar, M. Leitner, and T. Mang. 2012. "Extinction debt of high-mountain plants under twenty-first-century climate change." *Nature Climate Change* 2: 619–622.

Eby, L. A., O. Helmy, L. M. Holsinger, and M. K. Young. 2014. "Evidence of climate-induced range contractions in bull trout *Salvelinus confluentus* in a Rocky Mountain watershed, USA." *PLOS One* 9: e98812.

Elkin, C., A. G. Gutierrez, S. Leuzinger, C. Manusch, C. Temperli, L. Rasche, and H. Bugmann. 2013. "A 2 degrees C warmer world is not safe for ecosystem services in the European Alps." *Global Change Biology* 19: 1827–1840.

Elumeeva, T. G., V. G. Onipchenko, A. V. Egorov, A. B. Khubiev, D. K. Tekeev, N. A. Soudzilovskaia, and J. H. C. Cornelissen. 2013. "Long-term vegetation dynamic in the Northwestern Caucasus: Which communities are more affected by upward shifts of plant species?" *Alpine Botany* 123: 77–85.

Engler, R., C. F. Randin, W. Thuiller, S. Dullinger, N. E. Zimmermann, M. B. Araújo, P. B. Pearman, et al. 2011. "21st century climate change threatens mountain flora unequally across Europe." *Global Change Biology* 17: 2330–2341.

Fan, J., J. Li, R. Xia, L. Hu, X. Wu, and G. Li. 2014. "Assessing the impact of climate change on the habitat distribution of the giant panda in the Qinling Mountains of China." *Ecological Modelling* 274: 12–20.

Fisher, M. C., D. A. Henk, C. J. Briggs, J. S. Brownstein, L. C. Madoff, S. L. McCraw, and S. J. Gurr. 2012. "Emerging fungal threats to animal, plant and ecosystem health." *Nature* 484: 186–194.

Freeman, B. G., and A. M. C. Freeman. 2014. "Rapid upslope shifts in New Guinean birds illustrate strong distributional responses of tropical montane species to global warming." *Proceedings of the National Academy of Sciences* 111: 4490–4494.

Fujimura, K. E., K. N. Egger, and G. H. Henry. 2008. "The effect of experimental warming on the root-associated fungal community of *Salix arctica*." *ISME Journal* 2: 105–114.

Füreder, L., R. Ettinger, A. Boggero, B. Thaler, and H. Thies. 2006. "Macroinvertebrate diversity in Alpine lakes: Effects of altitude and catchment properties." *Hydrobiologia* 562: 123–144.

Gehrig-Fasel, J., A. Guisan, and N. E. Zimmermann. 2007. "Tree line shifts in the Swiss Alps: Climate change or land abandonment?" *Journal of Vegetation Science* 18: 571–582.

Gottfried, M., H. Pauli, A. Futschik, M. Akhalkatsi, P. Barančok, J. L. B. Alonso, G. Coldea, J. Dick, B. Erschbamer, and G. Kazakis. 2012. "Continent-wide response of mountain vegetation to climate change." *Nature Climate Change* 2: 111–115.

Gret-Regamey, A., S. H. Brunner, and F. Kienast. 2012. "Mountain ecosystem services: Who cares?" *Mountain Research and Development* 32: S23–S34.

Guisan, A., J. I. Holten, R. Spichiger, and L. Tessier, eds. 1995. *Potential Ecological Impacts of Climate Change in the Alps and Fennoscandian Mountains.* Conservatoire et Jardin Botanique de Genève.

Guizado-Rodriguez, M. A., C. Ballesteros-Barrera, G. Casas-Andreu, V. L. Barradas-Miranda, O. Tellez-Valdes, and I. H. Salgado-Ugarte. 2012. "The impact of global warming on the range distribution of different climatic groups of *Aspidoscelis costata costata*." *Zoological Science* 29: 834–843.

Hagvar, S., and K. Klanderud. 2009. "Effect of simulated environmental change on alpine soil arthropods." *Global Change Biology* 15: 2972–2980.

Harsch, M. A., P. E. Hulme, M. S. McGlone, and R. P. Duncan. 2009. "Are treelines advancing? A global meta analysis of treeline response to climate warming." *Ecology letters* 12: 1040–1049.

Hegland, S. J., A. Nielsen, A. Lazaro, A. L. Bjerknes, and O. Totland. 2009. "How does climate warming affect plant-pollinator interactions?" *Ecology Letters* 12: 184–195.

Hobbs, W. O., R. J. Telford, H. J. B. Birks, J. E. Saros, R. R. Hazewinkel, B. B. Perren, É. Saulnier-Talbot, and A. P. Wolfe. 2010. "Quantifying recent ecological changes in remote lakes of North America and Greenland using sediment diatom assemblages." *PLOS One* 5: e10026.

Inouye, D. W., B. Barr, K. B. Armitage, and B. D. Inouye. 2000. "Climate change is affecting altitudinal migrants and hibernating species." *Proceedings of the National Academy of Sciences* 97: 1630–1633.

Isaak, D. J., C. C. Muhlfeld, A. S. Todd, R. Al-Chokhachy, J. Roberts, J. L. Kershner, and S. W. Hostetler. 2012. "The past as prelude to the future for understanding 21st-century climate effects on Rocky Mountain trout." *Fisheries* 37: 542–556.

Jacobsen, D., A. M. Milner, L. E. Brown, and O. Dangles. 2012. "Biodiversity under threat in glacier-fed river systems." *Nature Climate Change* 2: 361–364.

Körner, C. 2012. *Alpine Treelines: Functional Ecology of the Global High Elevation Tree Limits.* Springer Science and Business Media.

Körner, C., J. Paulsen, and E. M. Spehn. 2011. "A definition of mountains and their bioclimatic belts for global comparisons of biodiversity data." *Alpine Botany* 121: 73–78.

Kuffner, M., B. Hai, T. Rattei, C. Melodelima, M. Schloter, S. Zechmeister-Boltenstern, R. Jandl, A. Schindlbacher, and A. Sessitsch. 2012. "Effects of season and experimental warming on the bacterial community in a temperate mountain forest soil assessed by 16S rRNA gene pyrosequencing." FEMS Microbiology Ecology 82: 551–562.

Lawler, J. J., S. L. Shafer, B. A. Bancroft, and A. R. Blaustein. 2010. "Projected climate impacts for the amphibians of the Western Hemisphere." Conservation Biology 24: 38–50.

Lenoir, J., J. C. Gégout, P. A. Marquet, P. De Ruffray, and H. Brisse. 2008. "A significant upward shift in plant species optimum elevation during the 20th century." Science 320: 1768–1771.

Lenoir, J., J. C. Gegout, J. C. Pierrat, J. D. Bontemps, and J. F. Dhote. 2009. "Differences between tree species seedling and adult altitudinal distribution in mountain forests during the recent warm period (1986–2006)." Ecography 32: 765–777.

Lurgi, M., B. C. Lopez, and J. M. Montoya. 2012. "Climate change impacts on body size and food web structure on mountain ecosystems." Philosophical Transactions of the Royal Society B: Biological Sciences 367: 3050–3057.

MacDonald, R. J., S. Boon, J. M. Byrne, M. D. Robinson, and J. B. Rasmussen. 2014. "Potential future climate effects on mountain hydrology, stream temperature, and native salmonid life history." Canadian Journal of Fisheries and Aquatic Sciences 71: 189–202.

Maggini, R., A. Lehmann, M. Kery, H. Schmid, M. Beniston, L. Jenni, and N. Zbinden. 2011. "Are Swiss birds tracking climate change? Detecting elevational shifts using response curve shapes." Ecological Modelling 222: 21–32.

Maggini, R., A. Lehmann, N. Zbinden, N. E. Zimmermann, J. Bolliger, B. Schroder, R. Foppen, H. Schmid, M. Beniston, and L. Jenni. 2014. "Assessing species vulnerability to climate and land use change: The case of the Swiss breeding birds." Diversity and Distributions 20: 708–719.

Margesin, R., S. Minerbi, and F. Schinner. 2014. "Long-term monitoring of soil microbiological activities in two forest sites in South Tyrol in the Italian alps." Microbes and Environments 29: 277–285.

Martin, K. 2013. "The ecological values of mountain environments and wildlife: The impacts of skiing and related winter recreational activities on mountain environments." In The Impacts of Skiing and Related Winter Recreational Activities on Mountain Environments, ed. C. Rixen and A. Rolando, 3–29. Bentham E-Books.

Menéndez, R. 2007. "How are insects responding to global warming?" Tijdschrift voor Entomologie 150: 355.

Merrill, R. M., D. Gutierrez, O. T. Lewis, J. Gutierrez, S. B. Diez, and R. J. Wilson. 2008. "Combined effects of climate and biotic interactions on the elevational range of a phytophagous insect." Journal of Animal Ecology 77: 145–155.

Milanovich, J. R., W. E. Peterman, N. P. Nibbelink, and J. C. Maerz. 2010. "Projected loss of a salamander diversity hotspot as a consequence of projected global climate change." PLOS One 5: e12189.

Nogués-Bravo, D., M. B. Araújo, M. P. Errea, and J. P. Martínez-Rica. 2007. "Exposure of global mountain systems to climate warming during the 21st Century." Global Environmental Change 17: 420–428.

Parkinson, D., M. McLean, and S. Scheu. 2004. "Impact of earthworms on other biota in forests soils, with some emphasis on cool temperate montane forests." In Earthworm Ecology, 2nd ed., ed. C. A. Edwards, 241–259. CRC Press.

Pauchard, A., C. Kueffer, H. Dietz, C. C. Daehler, J. Alexander, P. J. Edwards, J. R. Arevalo, et al. 2009. "Ain't no mountain high enough: Plant invasions reaching new elevations." Frontiers in Ecology and the Environment 7: 479–486.

Pauli, H., M. Gottfried, S. Dullinger, O. Abdaladze, M. Akhalkatsi, J. L. B. Alonso, G. Coldea, J. Dick, B. Erschbamer, and R. F. Calzado. 2012. "Recent plant diversity changes on Europe's mountain summits." Science 336: 353–355.

Paulsen, J., and C. Körner. 2014. "A climate-based model to predict potential treeline position around the globe." Alpine Botany 124: 1–12.

Pellissier, L., E. Pinto-Figueroa, H. Niculita-Hirzel, M. Moora, L. Villard, J. Goudet, N. Guex, M. Pagni, I. Xenarios, I. Sanders, and A. Guisan. 2013. "Plant species distributions along environmental gradients: Do belowground interactions with fungi matter?" Frontiers in Plant Science 4. https://doi.org/10.3389/fpls.2013.00500.

Pellissier, L., H. Niculita-Hirzel, A. Dubuis, M. Pagni, N. Guex, C. Ndiribe, N. Salamin, et al. 2014. "Soil fungal communities of grasslands are environmentally structured at a regional scale in the Alps." Molecular Ecology 23: 4274–4290.

Peñuelas, J., and M. Boada. 2003. "A global change induced biome shift in the Montseny Mountains (NE Spain)." Global Change Biology 9: 131–140.

Petitpierre, B., K. McDougall, T. Seipel, O. Broennimann, A. Guisan, and C. Kueffer. 2016. "Will climate change increase the risk of plant invasions into mountains?" Ecological Applications 26: 530–544.

Pettorelli, N., F. Pelletier, A. von Hardenberg, M. Festa-Bianchet, and S. D. Cote. 2007. "Early onset of vegetation growth vs. rapid green-up: Impacts on juvenile mountain ungulates." Ecology 88: 381–390.

Ponce-Reyes, R., E. Nicholson, P. W. J. Baxter, R. A. Fuller, and H. Possingham. 2013. "Extinction risk in cloud forest fragments under climate change and habitat loss." Diversity and Distributions 19: 518–529.

Popescu, V. D., L. Rozylowicz, D. Cogălniceanu, I. M. Niculae, and A. L. Cucu. 2013. "Moving into protected areas? Setting conservation priorities for Romanian reptiles and amphibians at risk from climate change." PLOS One 8: e79330.

Pounds, J. A., M. P. L. Fogden, and J. H. Campbell. 1999. "Biological response to climate change on a tropical mountain." Nature 398: 611–615.

Pounds, J. A., M. R. Bustamante, L. A. Coloma, J. A. Consuegra, M. P. L. Fogden, P. N. Foster, E. La Marca, et al. 2006. "Widespread amphibian extinctions from epidemic disease driven by global warming." *Nature* 439: 161–167.

Pradervand, J.-N., L. Pellissier, C. F. Randin, and A. Guisan. 2014. "Functional homogenization of bumblebee communities in alpine landscapes under projected climate change." *Climate Change Responses* 1 (1). https://doi.org/10.1186/s40665-014-0001-5.

Rangwala, I., E. Sinsky, and J. R. Miller. 2013. "Amplified warming projections for high altitude regions of the northern hemisphere mid-latitudes from CMIP5 models." *Environmental Research Letters* 8: 024040.

Roberts, J. J., K. D. Faush, D. P. Peterson, and M. B. Hooten. 2013. "Fragmentation and thermal risks from climate change interact to affect persistence of native trout in the Colorado River basin." *Global Change Biology* 19: 1383–1398.

Rosset, V., and B. Oertli. 2011. "Freshwater biodiversity under climate warming pressure: Identifying the winners and losers in temperate standing waterbodies." *Biological Conservation* 144: 2311–2319.

Rubidge, E. M., J. L. Patton, M. Lim, A. C. Burton, J. S. Brashares, and C. Moritz. 2012. "Climate-induced range contraction drives genetic erosion in an alpine mammal." *Nature Climate Change* 2: 285–288.

Salerno, F., S. Gambelli, G. Viviano, S. Thakuri, N. Guyennon, C. D'Agata, C. Diolaiuti, C. Smiraglia, F. Stefani, and D. Bocchiola. 2014. "High alpine ponds shift upwards as average temperatures increase: A case study of the Ortles-Cevedale mountain group (Southern Alps, Italy) over the last 50 years." *Global and Planetary Change* 120: 81–91.

Scherrer, D., and C. Korner. 2011. "Topographically controlled thermal-habitat differentiation buffers alpine plant diversity against climate warming." *Journal of Biogeography* 38: 406–416.

Scheurer, K., C. Alewell, D. Baenninger, and P. Burkhardt-Holm. 2009. "Climate and land-use changes affecting river sediment and brown trout in alpine countries: A review." *Environmental Science and Pollution Research* 16: 232–242.

Scridel, D. 2014. "Ecology and conservation of birds in upland and alpine habitats: A report on the BOU's Annual Conference held at the University of Leicester, 1–3 April 2014." *Ibis* 156: 896–900.

Sekercioglu, C. H., S. H. Schneider, J. P. Fay, and S. R. Loarie. 2008. "Climate change, elevational range shifts, and bird extinctions." *Conservation Biology* 22: 140–150.

Sheldon, A. L. 2012. "Possible climate-induced shift of stoneflies in a southern Appalachian catchment." *Freshwater Science* 31: 765–774.

Sinervo, B., F. Mendez-de-la-Cruz, D. B. Miles, B. Heulin, E. Bastiaans, M. V. S. Cruz, R. Lara-Resendiz, et al. 2010. "Erosion of lizard diversity by climate change and altered thermal niches." *Science* 328: 894–899.

Singh, B. K., R. D. Bardgett, P. Smith, and D. S. Reay. 2010. "Microorganisms and climate change: Terrestrial feedbacks and mitigation options." *Nature Reviews Microbiology* 8: 779–790.

Slemmons, K. E. H., J. E. Saros, and K. Simon. 2013. "The influence of glacial meltwater on alpine aquatic ecosystems: A review." *Environmental Science: Processes and Impacts* 15: 1794–1806.

Smol, J. P., A. P. Wolfe, H. J. B. Birks, M. S. V. Douglas, V. J. Jones, A. Korhola, R. Pienitz, K. Rühland, S. Sorvari, and D. Antoniades. 2005. "Climate-driven regime shifts in the biological communities of arctic lakes." *Proceedings of the National Academy of Sciences* 102: 4397–4402.

Swift, M. J., O. Andrén, L. Brusaard, M. Briones, M.-M. Couteaux, K. Ekschmitt, A. Kjoller, P. Loiseau, and P. Smith. 1998. "Global change, soil biodiversity, and nitrogen cycling in terrestrial ecosystems: three case studies." *Global Change Biology* 4: 729–743.

Sylvain, Z. A., and D. H. Wall. 2011. "Linking soil biodiversity and vegetation: Implications for a changing planet." *American Journal of Botany* 98: 517–527.

Theurillat, J. P., and A. Guisan. 2001. "Potential impact of climate change on vegetation in the European Alps: A review." *Climatic Change* 50: 77–109.

Thompson, L. G., E. Mosley-Thompson, H. Brecher, M. Davis, B. Leon, D. Les, P.-N. Lin, T. Mashiotta, and K. Mountain. 2006. "Abrupt tropical climate change: Past and present." *Proceedings of the National Academy of Sciences* 103: 10536–10543.

Thompson, P. L., M.-C. St.-Jacques, and R. D. Vinebrooke. 2008. "Impacts of climate warming and nitrogen deposition on alpine plankton in lake and pond habitats: An in vitro experiment." *Arctic, Antarctic, and Alpine Research* 40: 192–198.

Vittoz, P., D. Cherix, Y. Gonseth, V. Lubini, R. Maggini, N. Zbinden, and S. Zumbach. 2013. "Climate change impacts on biodiversity in Switzerland: A review." *Journal for Nature Conservation* 21: 154–162.

Wake, D. B., and V. T. Vredenburg. 2008. "Are we in the midst of the sixth mass extinction? A view from the world of amphibians." *Proceedings of the National Academy of Sciences* 105: 11466–11473.

Wasserman, T. N., S. A. Cushman, J. S. Littell, A. J. Shirk, and E. L. Landguth. 2013. "Population connectivity and genetic diversity of American marten (Martes americana) in the United States northern Rocky Mountains in a climate change context." *Conservation Genetics* 14: 529–541.

Weidman, P. R., D. W. Schindler, P. L. Thompson, and R. D. Vinebrooke. 2014. "Interactive effects of higher temperature and dissolved organic carbon on planktonic communities in fishless mountain lakes." *Freshwater Biology* 59: 889–904.

Wilhelm, L., G. A. Singer, C. Fasching, T. J. Battin, and K. Besemer. 2013. "Microbial biodiversity in glacier-fed streams." *ISME Journal* 7: 1651–1660.

Williams, S. E., E. E. Bolitho, and S. Fox. 2003. "Climate change in Australian tropical rainforests: an impending environmental catastrophe." *Proceedings of the Royal Society B: Biological Sciences* 270: 1887–1892.

Wilson, R. J., D. Gutierrez, J. Gutierrez, and V. J. Monserrat. 2007. "An elevational shift in butterfly species richness and composition accompanying recent climate change." *Global Change Biology* 13: 1873–1887.

Winder, M., J. E. Reuter, and S. G. Schladow. 2009. "Lake warming favours small-sized planktonic diatom species." *Proceedings of the Royal Society B: Biological Sciences* 276: 427–435.

Wu, T. J., F. L. Su, H. Y. Han, Y. Du, C. D. Yu, and S. Q. Wan. 2014. "Responses of soil microarthropods to warming and increased precipitation in a semiarid temperate steppe." *Applied Soil Ecology* 84: 200–207.

Zimmerli, S., D. Bernet, P. Burkhardt-Holm, H. Schmidt-Posthaus, P. Vonlanthen, T. Wahli, and H. Segner. 2007. "Assessment of fish health status in four Swiss rivers showing a decline of brown trout catches." *Aquatic Sciences* 69: 11–25.

Zinger, L., B. Shahnavaz, F. Baptist, R. A. Geremia, and P. Choler. 2009. "Microbial diversity in alpine tundra soils correlates with snow cover dynamics." *ISME Journal* 3: 850–859.

Climate Change and Frost Effects in Rocky Mountain Plant Communities
David Inouye

High-altitude ecosystems, including the Rocky Mountains in North America, are experiencing some of the strongest effects of climate change (Imtiaz, Sinsky, and Miller 2013; Pederson, Betancourt, and McCabe 2013; Rangwala and Miller 2012), including both warming of temperatures and changes in precipitation. At 2,900 m (9,500 ft) in the Colorado Rocky Mountains, at the Rocky Mountain Biological Laboratory, the strongest trend for warming has been in April minimum temperatures (3°C change from 1973), which can influence spring snowmelt. Winter snowfall (measured since 1975 and reconstructed from stream runoff in the period 1935–1974) has been highly variable, but there is a declining trend since 1935. The date of snowmelt is similarly variable, for example differing by 40 days between 1980 and 1981, but also trending toward earlier dates, and the year-to-year variation (measured by the moving 3-year range) is increasing. The trends toward both warmer springs and earlier snowmelt dates can have important consequences for the plant community and animal species that interact with plants.

Although snowmelt dates are advancing, the date of the last spring frost has not been changing, still occurring around June 10. If snowmelt is late, plants have not had much time to develop before the frost, so there is little damage. But if snow melts in April, as now sometimes happens, by June many plants have already expanded leaves and initiated flower buds, or even begun to flower. In such years, plants that are sensitive to frost can suffer significant damage. In some cases, only new growth, such as the young leaves on aspen trees or new needles on conifers, or reproductive parts, such as buds or fruits, are sensitive to frost. Some species can develop new leaves to replace the damaged ones, but short-term replacement of reproductive parts does not seem to be an option in this environment. So in those years that do have early snowmelt and late frosts, there may be no flowers, fruits, or seeds produced by frost-sensitive plants.

In addition to the potential effects of loss of flowers, fruits, and seeds for the affected plants, the consequences of the loss of these resources for consumers such as pollinators, seed predators, and seed dispersers can be significant (Inouye 2008). Pollinators will not have access to the pollen and nectar they depend on, and consumers of seeds such as insects and birds will also lose those sources of nutrition. Animals attracted to disperse seeds by the presence of the fruits containing them will also lose access to those resources. Herbivores that have evolved to take advantage of early spring leaves, such as many species of caterpillars, may starve to death before the trees releaf. Although in many cases these consequences are easy to understand, in only a few instances have they been quantified in an ecological context. One example is the long-term study of frost effects on flowers of Erigeron speciosus and on the butterfly species Speyeria mormonia, which is highly dependent on nectar from those flowers in order to reproduce (Boggs and Inouye 2012). In this case 84 percent of the variation in population growth rate of the butterfly was explained by how many flowers avoided frost damage.

Flower buds of the aspen sunflower (Helianthella quinquenervis) are highly sensitive to frost, and in many years close to 100 percent of the buds die off (Figure CS7.1). Flower buds produce extrafloral nectar that attracts ants, which defend them from flies (two species of Tephritidae and one of Agromyzidae) that lay eggs on the heads (Inouye and Taylor 1979). The flowers are visited by bumblebees and flies for nectar and pollen, and the flower heads are popular food for large herbivores such as deer and sheep. Thus, the loss of flower heads to frost affects a variety of consumer species as well as their parasites and predators. The *Erigeron* flower heads that are so important for butterflies for nectar are also used by tephritid flies, whose larvae eat developing seeds, and their populations probably also suffer from the effects of frost on flower abundance.

The pattern of frost damage I have observed in wildflowers in Colorado is also occurring in agriculture there, where the loss of apricots and cherries, which flower early, seems to be happening more frequently. The phenomenon is global, too. For example, a warm spell in early spring 2007 was followed by a hard freeze that caused more than $2 billion in damage to agriculture in the eastern United States (Gu et al. 2008; NOAA/USDA 2008). A similar 2012 event in Michigan (Ault et al. 2013) was responsible for $223 million in losses (Parker 2013). Wine grape harvests were devastated by frost in South Africa in 2013

(the coldest in 126 years; "South African frosts" 2013), and in 2014 in Argentina and Chile (Frank 2013).

These incidents of frost damage to wildflowers and crops may continue to increase in frequency unless the low temperatures in late spring begin to warm significantly. But the fact that some models of climate change predict increased variability suggests that perhaps frost will continue to influence ecological communities and agriculture for quite some time.

REFERENCES

Ault, T. R., G. M. Henebry, K. M. de Beurs, M. D. Schwartz, J. L. Betancourt, and D. Moore. 2013. "The false spring of 2012, earliest in North American record." *Eos*,

Figure CS7.1. Aspen sunflower (*Helianthella quinquenervis*) is a common wildflower near the Rocky Mountain Biological Laboratory, but flower abundance is extremely variable. The plants themselves are long-lived (decades-long) perennials, but the flower buds are sensitive to frost. As snowmelt gets earlier (with the date of the last frost still around June 10), the frequency of frost kills of buds is becoming more common. These data for counts of unfrosted flowers come from two permanent plots at RMBL. (Courtesy of David W. Inouye, data collected with funding from NSF grant DEB-1354104.)

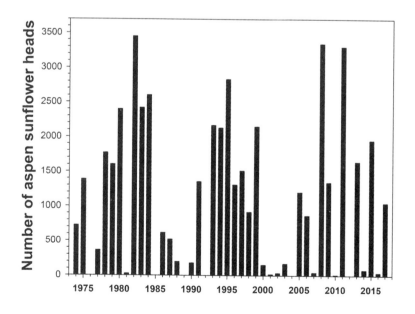

Transactions American Geophysical Union 94 (20): 181–182. https://doi.org/10.1002/2013eo200001.

Boggs, Carol L., and David W. Inouye. 2012. "A single climate driver has direct and indirect effects on insect population dynamics." *Ecology Letters* 15 (5): 502–508. https://doi.org/10.1111/j.1461-0248.2012.01766.x.

Frank, M. 2013. "Chilean wine producers hit by frost." *Wine Spectator*. http://www.winespectator.com/webfeature/show/id/49055.

Gu, Lianhong, Paul J. Hanson, W. Mac Post, Dale P. Kaiser, Bai Yang, Ramakrishna Nemani, Stephen G. Pallardy, and Tilden Meyers. 2008. "The 2007 eastern US spring freeze: Increased cold damage in a warming world?" *BioScience* 58: 253–262.

Inouye, David W. 2008. "Effects of climate change on phenology, frost damage, and floral abundance of montane wildflowers." *Ecology* 89 (2): 353–362.

Inouye, David W., and Orley R. Taylor. 1979. "A temperate region plant-ant-seed predator system: Consequences of extrafloral nectar secretion by *Helianthella quinquenervis*." *Ecology* 60 (1): 1–7.

NOAA/USDA. 2008. "The Easter freeze of April 2007: A climatological perspective and assessment of impacts and services." https://www1.ncdc.noaa.gov/pub/data/techrpts/tr200801/tech-report-200801.pdf.

Parker, R. 2013. "Michigan fruit losses documented in 2012 USDA report." Michigan State University. http://www.mlive.com/news/kalamazoo/index.ssf/2013/02/michigan_fruit_losses_document.html.

Pederson, Gregory T., Julio L. Betancourt, and Gregory J. McCabe. 2013. "Regional patterns and proximal causes of the recent snowpack decline in the Rocky Mountains, U.S." *Geophysical Research Letters* 40 (9): 1811–1816. https://doi.org/10.1002/grl.50424.

Rangwala, Imtiaz, and James R. Miller. 2012. "Climate change in mountains: A review of elevation-dependent warming and its possible causes." *Climatic Change* 114 (3–4): 527–547. https://doi.org/10.1007/s10584-012-0419-3.

Rangwala, Imtiaz, E. Sinsky, and J. R. Miller. 2013. "Amplified warming projections for high altitude regions of the northern hemisphere mid-latitudes from CMIP5 models." *Environmental Research Letters* 8: 024040.

"South African frosts mean job and volume loss for grape sector." 2013. FreshFruitPortal.com. https://www.freshfruitportal.com/news/2013/10/16/south-african-frosts-mean-job-and-volume-loss-for-grape-sector.

CHAPTER EIGHTEEN

Climate Change: Final Arbiter of the Mass Extinction of Freshwater Fishes

LES KAUFMAN

INTRODUCTION

Freshwater ecosystems and their constituent species are exquisitely vulnerable to climate change. The reason is simple. Each of the four horsemen of global climate change—warming, volatility, sea-level rise, and acidification—directly affects freshwater fishes in some profound way. With anthropogenic climate change already in progress we are seeing desiccation, warming, deoxygenation, and violent extremes of flood and drought. Sea-level rise backs up estuaries and pushes salt water into freshwater habitats that harbor unique species, such as in the Everglades and coastal plain ponds. These conditions eliminate some freshwater habitats, make others unlivable, and decouple still others from their marine counterparts (Lin et al. 2017). Climate change brings too much water to some places, or too much within a short time. The elevated violence of cyclonic storms and the concentration of rainfall into extreme precipitation events aggravate natural fluxes in flow and water level, bringing them repeatedly outside the envelope of conditions that aquatic species can deal with.

Hansen et al. (2013) and many others have considered the impacts of climate change on freshwater systems, but the research required for a comprehensive treatment is still under way. The magnitude of expected freshwater fish extinctions that are linked directly to climate change can be estimated computationally by using tools such as the diversity-flow relationship in riverine systems (Xenopoulos et al. 2005) and niche envelope models (Comte and Grenouillet 2013). No matter the approach taken, the data shout for themselves: the global freshwater fish fauna is facing mass extinction, with climate change a major contributing factor. The signal is evident even if we exclude the loss in whole or part of lacustrine fish species flocks such as those in Lakes Victoria (Africa), Lanao (Philippines), and Tana (Ethiopia). For example, using a flow model, Xenopoulos et al. (2005) estimated that up to 75 percent of freshwater fish species are likely to be negatively affected, directly or indirectly, by climate change, with the lion's share of species occurring in developing world nations. Added to the freshwater extinctions already under way due to dams, pollution, and other insults, the prospect from a mass extinction of freshwater species is all but ensured (e.g., Winemiller et al. 2016). The degree to which any particular species will be affected is a function of its evolutionary history, the location and topography of the landscape in which it lives, its life history, and the degree of cumulative human impacts within its range.

There may be no freshwater fish that is threatened only by climate change, but likely no freshwater fish will escape at least some impact from climate change. Climate change effects are superimposed on other anthropogenic stressors, a situation similar to that for coral reefs: climate change constitutes an existential threat, but it acts by amplifying the threats of pollution, overfishing, and habitat destruction. In freshwater systems an added—and indeed, principal—human impact is hydroengineering to stabilize and maximize water availability for human use and to optimize its distribution in time and space to suit human needs. We draw enormous quantities of water from lakes, rivers, and aquifers for domestic, agricultural, and industrial applications, so much that water tables fall, rivers dry before reaching the sea, and deep ancient stores of fossil water are depleted. As freshwater becomes scarce, people arrest control of what remains through technology—damming or diverting it, fighting wars over it, storing or releasing it, driving it into the ground to squeeze the last bits of fossil fuel

out, and triggering floods, droughts, and even earthquakes on a schedule dictated by human, not whole, systems. Hydropower and irrigation projects have been completed for ages as the human population and global economy mushroomed. Yet the key to understanding climate change impacts on freshwater fishes is to first grasp the ricochet dynamic of climate and society.

Climate change most obviously threatens biodiversity hotspots, a few of which have been classified on every continent except Antarctica. According to a recent analysis (Newbold et al. 2016), biodiversity hotspots harbor 55 percent of the world's freshwater fishes, 29 percent as endemics. A different analysis would be required to assess the conservation of evolutionary uniqueness, as many spots with high freshwater fish endemism are dominated by recent radiations (e.g., Lakes Victoria and Malawi, desert springs). Hotspots aside, climate change also threatens species-poor but highly valued fish assemblages of high-latitude lakes and rivers.

Watersheds are broadly similar in function and so are the ways in which they respond to climate change. Differences arise from interactions among precipitation, latitude, topography, and soils, interactions that determine vegetation patterns and so fish habitats as structured by plants growing both above and below the waterline. The impacts of climate change on freshwater fishes in any one watershed cannot be predicted accurately—it is difficult to downscale climate models, and local processes, particularly species interactions, are fraught with stochasticity—but the kinds of climate-related impact are generalizable (Table 18.1).

THREE FISHY BRIEFS: AMERICA, EAST AFRICA, AND EAST ASIA

Freshwater fish diversity is distributed unevenly, but each of the world's great landmasses hosts at least one area of high species diversity. Most of these areas are home to

Table 18.1. Examples of climate change effects on freshwater habitats and fishes

1. *Loss of alpine snowpack and glaciers.* Reduced precipitation at high altitude and increased melting and sublimation cause net loss of snowpack and glaciers. This reduces the seasonal pulse flow on which migratory fishes key their upstream spawning runs, reduces downstream drift important in delivering fry to first-feeding grounds, and can cause intermittent to zero dry-season flow through all or portions of a watershed.

2. *Changes in tundra ponds.* Tundra ponds on permafrost are changing in size and connectivity (mostly shrinking), altering Arctic fish communities.

3. *Reduction in base flow.* A product of reduced precipitation or fog capture anywhere in a watershed but particularly in the headwaters. Effects similar to No. 1, especially in causing ever-flowing streams to become seasonal, intermittent, or completely dry.

4. *Reduction in maximum flood levels.* All freshwater systems occupy some position along an axis from zero to high flood-pulse amplitude. In classic flood-pulse systems like the Mekong and Amazon Rivers, submerged basin at peak flood serves as a crucial young-of-year nursery area, with a food web partially powered by bacterial activity associated with decomposing terrestrial vegetation. If the seasonal floods are reduced (as by regional drying and dams) and flow stabilized overall (as by dams) the contribution of flood-pulse dynamics to fish recruitment and growth are eliminated.

5. *High-altitude habitat "squeeze."* Warming shifts habitat zones to higher elevations, eliminating the highest as they run out of mountainside to be on. This will eliminate low-latitude mountaintop refugia for high-latitude species and threatens distinctive formations such as tropical cloud forest.

6. *Mixing of historically isolated populations and species.*

   a. Lowering of mountain passes. These are famously "higher in the tropics" (i.e., the range in climate from valley to peak is greatest where valley climate is the warmest). Climate warming may increase connectivity of adjacent freshwater systems, reducing isolation.

   b. Wets increasing connectivity in swamp systems, dries reconnecting island and mainland populations. Extreme floods and droughts will drive rapid shifts in population structure and species composition (e.g., spread of exotic species) in subtropical and tropical watershed and lake systems.

7. *Loss of habitat area.*

   a. Species compression: species interactions in shallow lakes and wetlands (and some river systems) will be intensified in drought and relaxed in flood (Itzkowitz paper pupfish/gambusia, Lake Chapala), causing species shuffling and potentiating extinctions.

   b. Habitat reduction and loss. Fishes of ephemeral water bodies (e.g., "annual" cyprinodontiforms) such as the rivuline killifishes will be threatened as the size, abundance, and duration of seasonal ponds are reduced due to increased evaporation.

(continued)

Table 18.1.   (continued)

8. *Ecosystem phase shift.*

   a. Warming of lake surface waters will stabilize stratification and amplify deoxygenation of the hypolimnion, reducing habitable volume for fishes and elevating wind-driven fish kills due to episodic turnover and suffocation.

9. *Disrupted seasonal choreographies.* Fishes are adapted to take advantage of temporal synergies, such as the correlations linking spring flood, nutrient supply, and spring phytoplankton blooms near the pycnocline in estuarine headwaters. Climate change can lead to inappropriate flow rates and thus decouple productivity drivers. For example, spring flow that is too high or low can cause anadromous fish larvae to arrive on first-feeding grounds too late or too early to coincide with the peak in food availability. Or, too-strong spring flows can purge a watershed of nutrients, leaving little for when the sunlight arrives in force to power photosynthesis.

10. *Increased storm intensity.* Warm surface ocean waters strengthen cyclonic storms in some areas, leading to more powerful flash floods. These can wash out riverine habitats and increase sediment loads to rivers and lakes, affecting both water quality and primary production. More powerful storms also generate higher storm surge, pushing salt water inland into normally freshwater coastal habitats.

11. *Saline incursion into estuaries.* Rising sea level and amplified storm surge will drown estuaries, squeezing freshwater species into reduced upper estuary habitat. Modified estuarine flow patterns will disrupt larval transport, delivery to nursery areas, year class strength, and productivity of diadromous fishes.

12. *Acidification.* Acidification may alter behavior and survivorship of diadromous fishes, influencing their abundance and role in freshwater systems. Continued burning of fossil fuel fuels acid rain and lethal acidification of freshwater systems.

---

numerous threatened fishes (and some species are extinct already); about one-third of freshwater species are listed by the International Union for Conservation of Nature (IUCN) as vulnerable to critically endangered. Threat status aside, even species ranked as "least concern" may be at risk from climate change.

North America

Although it is mostly high-latitude and low-species density, North America boasts three concentrations of freshwater fish diversity. One is in Mexico, a country that spans the temperate-tropical divide, whereas the other two are a curious study in contrasts: the southern Appalachian temperate broadleaf forests and the arid Southwest. In the Appalachians, narrow endemics swim beside broadly distributed species (largely of the Mississippi Basin fauna), yielding both high alpha (in any one spot) and beta (across all habitats) diversity plus an unparalleled regional (γ) diversity for a temperate ecosystem: 493 fish species, or about 62 percent of the United States' total. Bivalve mollusks are also phenomenally diverse in this region, with about 269 species, constituting 91 percent of the US tally. Many of these bivalves are gravely endangered, and nearly all depend on fishes as hosts and dispersal agents for their glochidia larvae (one species uses a large salamander, the mudpuppy *Necturus maculosus*, as its host). Species densities are lower on the vast southeastern coastal plain, although there is appreciably high endemism (e.g., among pygmy sunfishes of the genus *Elassoma* and killifishes of the Cyprinodontidae).

Climate change plays a role in what has been called an "extinction vortex" in the southeastern United States (Freeman et al. 2012). We know that the initial trigger for the observed declines in freshwater species is the often-combined impacts of deforestation, mining, sedimentation, hydroengineering (chiefly dams), and pollution. Climate change overlaid on these other factors is expected to include a warming of 3°C–4°C. There may be an increase in precipitation, mostly on the coastal plain, concentrated into more intense and clustered summer storms. Increased evapotranspiration will decrease runoff and stream flow, except in Florida and along the Gulf Coast, where precipitation may exceed evapotranspiration and perhaps increase stream flow, some of which will present as surges and floods. Both spike against a lower baseline flow and extended droughts are likely. Warmer waters, reduced water quality, lower oxygen tensions during the summer, and intense flushing add impetus to incipient mass extinctions in these areas, which host the most species-rich of all the world's temperate freshwater fish communities.

South America

Freshwater fishes in South America, and Brazil in particular, are imperiled by dams, pollution, deforestation, exotic introductions, and rapid industrial and urban growth. In the west are the vulnerable (or extinct) fishes endemic to altiplano lakes and streams in Chile, Peru, and Bolivia, the Galaxiids (southern smelts) of Argentina and Chile, and the poorly known fishes of the western Andean slopes. To the east is the Amazon, with the highest gamma diversity (regional species pool) of any freshwater system; about 8 percent of all the fish species on earth occur there. Even more at risk, the fishes of the Atlantic Forest, reduced to 7 percent of its original extent, are poorly known as a result of incomplete sampling, particularly in northern portions of the forest domain. Few South American

freshwater fishes have been assessed; the IUCN Red List is woefully incomplete for the entire continent.

Loss of fish species in South America can be attributed to climate change of two sorts: regional (caused by mass deforestation) and global. These act together and are supercharged by strengthening climate cycles such as that of El Niño–Southern Oscillation. Global climate change will result in warming and possibly drying of watercourses in the region. Regional climate change driven by deforestation and consequent loss of moisture recycling is particularly important in the Amazon.

The Amazon Basin exhibits an extreme annual cycle of hydrology; the entire basin is a massive flood-pulse system, and fish life histories are geared to it. Any change in the timing and volume of available water, or in the rate of change in its availability, will affect a large fraction of the more than 2,500 fish species that occur there. Such changes come naturally with every El Niño cycle, but recent droughts have been of historic severity. Droughts were so bad in 2005, 2010, and 2014 that the upper reaches of many tributaries dried out completely—obviously a problem for fishes, but also a cause of immense human suffering as lines of transportation and commerce disappeared, literally, into thin air. The entire region encompassing the eastern Amazon and Mata Atlântica (Atlantic Forest) is exquisitely sensitive to climate change. The severity of recent droughts has been tied to the combined effects of deforestation and warming (Xenopoulos et al. 2005).

Evapotranspiration from the Amazon forest is the wellspring of an aerial current of water vapor, or "flying river," which contributes to life-giving precipitation in other parts of South America (Arraut and Nobre 2012). As trees fall, so do the levels of reservoirs that feed São Paulo, Rio de Janeiro, and other heavily populated southeastern Brazilian cities, and agricultural lands in Brazil and other nations to the west and the south. Unique assemblages of fish species isolated in the last glacial period have diversified through allopatric speciation in these watersheds, placing a wealth of endemic species at risk of extinction in the same mountains that provide the water for the great cities of Brazil. It is important to maintain forest cover in the Amazon to maintain regional climate stabilization of the Amazonian forest, which can reduce the likelihood of a downward spiral of drying in response to global climate change.

East Africa

East Africa's Great Lakes boast the highest concentrations of endemic vertebrate species on earth; most of these endemics are fishes, with amphibians likely running a close second, and most of the endemic fishes are cichlids. The three largest lakes alone host a species pool near equal to those of the whole of the Amazon Basin or Mekong Basin; with Tanganyika at 280, Malawi at approximately 1,000, and Victoria at about 600. Levels of endemism range from about 80 percent to 95 percent. Lakes Turkana, Albert, Edward, and George also claim appreciable diversity, as do the swamps, satellite, and soda lakes associated with them. The three greatest lakes sit atop three great watersheds, the Nile (Victoria), the Congo (Tanganyika), and the Zambezi (Malawi); each is home to endemic fishes and diverse fish species assemblages. Miniradiations exist in Lake Tana (the Blue Nile) and in hypersaline soda lakes astride the Rift, such as Lake Natron. Ironically, Africa is a high, dry continent: arid conditions prevail over most it. Microendemic species occur in the tiny places where water is found, as in the deserts of North America, the Middle East, and Australia, complete with an endemic desert sinkhole cichlid (*Tilapia guinasana*) and ephemeral pond rivulines (e.g., *Nothobranchius* spp.). All are at risk from changes in the frequency and intensity of precipitation.

The African Great Lakes would seem to be among those places least vulnerable to warming because their contained water volumes are immense, yet these lakes and their diverse fishes are varyingly vulnerable to climate change (Hecky et al. 2010). This variation in vulnerability owes much to the way endemics evolved and to interactions with nonclimate stressors, such as demands for freshwater by people seeking electricity, food, and other economic returns. In Lake Victoria, climate-related shifts in surface temperature, seasonality, and winds aggravate a history of overfishing, exotic introductions, and eutrophication that collectively have led to the mass extinction of at least half of the lake's endemic species of 500–600 haplochromine cichlids. The primary climate-related impacts to Africa's Great Lakes are lake-level flux and stratification. Intensive rains can cause a rapid rise in lake level, or transgression, that submerges new habitats. Some of these habitats, such as rocky reefs, can harbor founder populations of philopatric fishes that subsequently speciate; mouth-brooding cichlids are famous for this. Even so, rising waters may drown shallow-water species, which may lose important shallow-water habitat as lake levels rise and spill over other kinds of bottom (e.g., Kaufman 1997, 2003; Sturmbaur et al. 2000). Falling lake levels can also pose a conservation challenge, leaving fish populations stranded in peripheral or satellite lakes (Greenwood 1965; Chapman et al. 1996).

In the Lake Kyoga Basin of central Uganda, a system of interconnected papyrus swamp lakes are the most important refugium from the threat of mass extinction in Lake Victoria precipitated by climate change, habitat destruction, and invasive species (Mwanja et al. 2001). A rise in Lake Kyoga caused it to merge with adjacent Lake Nawampassa, exposing endangered cichlids to the possibility of invasion and predation by the introduced Nile perch (*Lates niloticus*). Connectivity was brief and Nawampassa was has not been invaded . . . yet. Comparison between lakes in this system inhabited by, as compared to free of, perch show how the introduction of perch could result in the extinction of endemic fishes (Chapman et al. 1996; Schwartz et al. 2006).

It is natural for lakes to stratify during warm periods and turn over in high winds or cooler weather, but global warming can heat surface waters and lock down stratification. If accompanied by eutrophication, the result will be a shallow, hyperproductive epilimnion and a deadly, hypoxic hypolimnion, a recipe for fish die-offs from the upwelling of deep, oxygen-poor water to the shallows. Lake Victoria, not a rift lake, comparatively shallow (87 m at most), and set in a broad, gradually sloping basin, was dry as recently as 12,500 years ago, and even modest lake-level flux can shift shoreline position. In the deeper and steeper rift lakes, Tanganyika, Malawi, Edward, Albert, and Turkana, the notion of an entire lake drying up completely is unlikely, but these lakes have only modest inputs and outputs, gaining water predominantly from rainfall and losing it mostly through evaporation. Consequently, the fishes they harbor likely will respond to climate change, particularly species that occupy the littoral zone of lakes in shallow basins or lakes with shallow shelf areas.

CONCLUSION

These three fishy stories exemplify four conclusions that can be drawn regarding the role of anthropogenic climate change in determining the conservation status of freshwater fishes: (1) Human demand for water for personal consumption, agriculture, industry, and waste disposal by itself poses an existential threat to a large percentage of surviving freshwater fish species. (2) Anthropogenic climate change is altering the distribution, quality, and availability of surface freshwater habitats and by itself threatens a large percentage of freshwater fish species. (3) Although the percentage of threatened or endangered freshwater fishes attributable to water use versus climate change is difficult or im-

possible to estimate, the two interact in a nonadditive manner. This compounds systemic ills that arise from lack of attention to our dependence on natural systems. Indeed, even if anthropogenic climate change were arrested immediately, freshwater fishes would still be ailing. Hence (4) the systemic nature of the problem challenges species-centric conservation efforts, implying that the most realistic and practical strategy for averting the global mass extinction of freshwater fish species would be ecosystem based. That is, all proposed developments must be reviewed in the context of an ecosystem-based management plan for human activities, rational limits to growth, and most important, a climate adaptation strategy, particularly for the freshwater supply.

Freshwater fish conservation requires action at three scales—species, habitat, and ecosystem—but the ecosystem-level needs have been consistently ignored. Most current attention is directed toward the species level, ranging from protective legislation to breeding and reintroduction programs. Examples of protective legislation include the Convention on International Trade in Endangered Species, special requirements for possession or distribution, and national and local endangered species acts. Legislation is useful in avoiding unintended harm and in preventing trade from becoming a hindrance to recovery. Captive breeding programs now exist for the desert fishes of North America, the Lake Victoria cichlids, Cambodia's dragonfish, Appalachian stream fishes, and quite a few others. We have learned that such programs can work to reverse extinction from the wild, that they are immensely difficult and expensive to complete from first captivity to final release, that they only make sense where intact habitat remains in order to justify the considerable expense of propagation and reintroduction, and that we need a great many such programs in play, way too many to make this a practical primary approach to safeguarding freshwater fish diversity.

Most concentrations of threatened freshwater fish diversity such as those discussed in this chapter occupy areas that are already heavily populated and/or undergoing rapid development. This greatly elevates the importance that development be carefully planned, monitored, and guided in an adaptive fashion. There are very few examples in the world, or in history, of a human society behaving in this manner. Conservation organizations, watershed management councils, and place-based systems of governance must focus their efforts on learning how to do this properly. Otherwise, all the well-intended legislative and captive breeding programs are doomed to fail. In the immediate future the question is one of how much of the wealth of freshwater fishes we can carry through the hysteresis in global climate change. Scientists have written a clear prescription for sustainable, ecosystem-based approaches to human development. These are ways of living that seek to maintain species, restore ecosystem services, and put the lid on the runaway growth of human demands and on the economies that stoke these demands beyond reason. The medicine is sugarcoated with strong logic, powerful incentives, and the love of our children. We have but to take it to heart.

REFERENCES

Arraut, M. A., C. Nobres, H. M. J. Barbosa, G. Obregon and J. Marengo. 2012. "Aerial rivers and lakes: Looking at large-scale moisture transport and its relation to Amazonia and to subtropical rainfall in South America." *Journal of Climate* 25: 543–556.

Chapman, L. J., C. A. Chapman, R. Ogutu-Ohwayo, M. Chandler, L. Kaufman, and A. E. Keiter. 1996. "Refugia for endangered fishes from an introduced predator in Lake Nabugabo, Uganda." *Conservation Biology* 10: 554–561.

Comte, L., and G. Grenouillet. 2013. "Do stream fishes track climate change? Assessing distribution shifts in recent decades." *Ecography* 36 (11): 1236–1246.

Downing, A. S., E. Van Nes, J. Balirwa, J. Beuving, P. Bwathondi, L. J. Chapman, I. J. M. Cornelissen, et al. 2014. "Coupled human and natural system dynamics as key to the sustainability of Lake Victoria's ecosys-

tem services." *Ecology and Society* 19 (4): 31. http://dx.doi.org/10.5751/ES-06965-190431.

Freeman, M. C., G. R. Buel, L. E. Hay, W. B. Hughes, R. B. Jacobson, J. W. Jones, S. A. Jones, et al. 2012. "Linking river management to species conservation using dynamic landscape-scale models." *River Research and Applications* 29: 906–918.

Greenwood, P. H. 1965. "The cichlid fishes of Lake Nabugabo, Uganda." *Bulletin of the British Museum* 12 (9).

Hansen, J., P. Kharecha, M. Sato, V. Masson-Delmotte, F. Ackerman et al. 2013. "Dangerous climate change: Required reduction of carbon emissions to protect young people, future generations, and nature." *PLOS One* 8 (10): 1–26.

Hecky, R., R. Mugidde, P. S. Ramlal, M. R. Talbot, and G. W. Kling. 2010. "Multiple stressors cause rapid ecosystem change in Lake Victoria." *Freshwater Biology* 55 (S1): 19–42.

Hulsey, C. D., J. Marks, D. A. Hendrickson, C. A. Williamson, A. E. Cohen, and M. J. Stephens. 2006. "Feeding specialization in *Herichthys minckleyi*: A trophically polymorphic fish." *Journal of Fish Biology* 68: 1–12.

Kaufman, L. S. 1992. "Catastrophic change in species-rich freshwater ecosystems: The lessons of Lake Victoria." *Bioscience* 42: 846–858.

Kaufman, L. S. 1997. "Asynchronous taxon cycles in haplochromine fishes of the greater Lake Victoria region." *South African Journal of Science* 93: 601–606.

Kaufman, L. S. 2003. "Evolutionary footprints in ecological time: Water management and aquatic conservation in African lakes." In *Conservation, Ecology, and Management of African Freshwaters*, ed. T. L. Crisman, L. J. Chapman, C. A. Chapman, and L. S. Kaufman, 460–490. University Press of Florida.

Lin, Hsien-Yung, Alex Bush, Simon Linke, Hugh P. Possingham, and Christopher J. Brown. 2017. "Climate change decouples marine and freshwater habitats of a threatened migratory fish." *Diversity and Distributions* 23 (7): 751–760.

Lundberg, J. G., J. P. Sullivan, R. Rodiles-Hernandez, and D. A. Hendrickson. 2007. "Discovery of African roots for the Mesoamerican Chiapas catfish, *Lacantunia enigmatica*, requires an ancient intercontinental passage." *Proceedings of the Academy of Natural Sciences of Philadelphia* 156: 39–53. https://doi.org/10.1635/0097-3157(2007)156 [39: DOARFT]2.0.CO;2.

Mulholland, P. J., G. R. Best, C. C. Coutant, G. M. Hornberger, J. L. Meyer, P. J. Robinson, J. R. Steinberg, R. E. Turner, F. Vera-Herrera, R. G. Wetzel. 1997. "Effects of climate change on freshwater ecosystems of the southeastern United States and the Gulf coast of Mexico." *Hydrological Processes* 11: 949–970.

Mwanja, W. W., A. S. Armoudlian, S. B. Wandera, L. Kaufman, L. Wu, G. C. Booton, and P. A. Fuerst. 2001. "The bounty of minor lakes: The role of small satellite water bodies in evolution and conservation of fishes." *Hydrobiologia* 458: 55–62.

Newbold, Tim, Lawrence N. Hudson, Andrew P. Arnell, Sara Contu, Adriana De Palma, Simon Ferrier, and Samantha L. Hill. 2016. "Has land use pushed terrestrial biodiversity beyond the planetary boundary? A global assessment." *Science* 353 (6296): 288–291.

Schwartz, J. D. M., M. J. Pallin, R. H. Michener, D. Mbabazi, and L. S. Kaufman. 2006. "Effects of Nile perch, *Lates niloticus*, on functional and specific fish diversity in Uganda's Lake Kyoga system." *African Journal of Ecology* 44 (2): 145–156.

Shapiro, J. 2001. *Mao's War Against Nature: Politics and the Environment in Revolutionary China.* Cambridge University Press.

Van Zwieten, P., J. Kolding, M. J. Plank, R. E. Hecky, T. B. Bridgeman, S. MacIntyre, O. Seehausen, and G. M. Silsbe. 2015. "The Nile perch invasion in Lake Victoria: Cause or consequence of the haplochromine decline?" *Canadian Journal of Fisheries and Aquatic Sciences.* https://doi.org/10.1139/cjfas-2015-0130.

Winemiller, Kirk O., P. B. McIntyre, L. Castello, E. Fluet-Chouinard, T. Giarrizzo, S. Nam, I. G. Baird, et al. 2016. "Balancing hydropower and biodiversity in the Amazon, Congo, and Mekong." *Science* 351 (6269): 128–129.

Xenopoulos, M., D. M. Lodge, J. Alcamo, M. Marker, K. Schultze, and D. P. Van Vuurens. 2005. "Scenarios of freshwater fish extinction from climate change and water withdrawal." *Global Change Biology* 11: 1–8.

CHAPTER NINETEEN

The *Asymmetrical* Impacts of Climate Change on Food Webs

LAUREN JARVIS, KEVIN MCCANN, AND
MARY O'CONNOR

INTRODUCTION

All living organisms are thermally driven entities vulnerable to changes in their external environment via suites of temperature-dependent traits (Gillooly et al. 2001; Gilbert et al. 2014; Fussmann et al. 2014; Kraemer et al. 2016). That climate change affects a suite of organismal traits makes the dynamic responses of individuals to ecosystems contingent on how traits, collectively, are affected. Interacting species often experience a habitat differently by virtue of their mobility. The aim of this chapter is to explore the role of asymmetrical thermal responses within food webs to highlight potential ways to understand climate impacts on species in diverse ecosystems. In what follows, temperature is the primary focus, but these same principles could be extended to other facets of climate change (e.g., extreme weather events, rising CO_2 levels). There are two asymmetries: first is the asymmetrical trait response to climate change, in which the response of a consumer-resource (C-R) interaction is driven by how temperature differentially changes important suites of resource and consumer traits, and second is the asymmetrical compartment response, in which the effects of climate warming are different across connected habitat divisions (compartments or subwebs). Although webs are amazingly complex, insight provided by considering asymmetric thermal responses may enable theory to predict the dynamic consequences of climate warming on food webs.

CONSUMER-RESOURCE DYNAMICS AND THE CONSEQUENCES OF ASYMMETRICAL-TRAIT RESPONSES

Temperature affects nearly every biological process. Scientists relate effects of temperature on the outcomes of cellular metabolic processes to whole animal functions using thermal performance curves (TPCs) (Huey and Stevenson 1979). TPCs depict how biological traits respond across a range of relevant temperatures, describing the breadth of thermal tolerance and providing estimates of "optimal" temperatures that yield maximal trait performance (Huey and Kingsolver 1989). Asymmetries arise at the trait level when multiple traits differentially respond to temperature (Dell et al. 2013). For example, a study of experimental algae (*Chlorella vulgaris*) populations found that the intrinsic per capita growth rate (r) showed a hump-shaped response to temperature, whereas carrying capacity (K) remained unchanged across biologically relevant temperature ranges (Jarvis et al. 2016). In this example, growth rate is the dominant trait to act on the population at any given temperature, although multiple traits, with various TPCs, may interact to yield a variety of population dynamics under changing thermal stress. Other empirical studies have identified asymmetries in consumer and resource growth, activity rates, metabolism, ingestion, and feeding rates—asymmetries that shift interactions over a thermal gradient (Rall et al. 2010; O'Connor 2009; Rose and Caron 2007; Dell et al. 2011; Dell et al. 2013; Monaco et al. 2017). Recognizing this dynamic, Gilbert et al. (2014) developed a simple theory to identify how asymmetrical trait responses influence the dynamics and composition of consumer-resource interactions.

In consumer-resource interactions, warming affects the production of the resource (growth rate r and carrying capacity K), the consumption and assimilation of the resource (attack rate *a* and conver-

sion efficiency *e*) by the consumer, and the consumer mortality rate (m) (Figure 19.1). Hereafter, these parameters are referred to as functions of temperature to emphasize that it is their response to temperature that collectively determines biomass and dynamic responses to climate warming. Individual trait performances manifest in the overall expression of each aspect to determine how energy and material flow through the interaction and thus determine dynamic outcomes of climate warming. Climate-driven asymmetries make a trait-based approach seem daunting; however, the theoretical framework proposed by Gilbert et al. (2014) uses bioenergetic reasoning to aggregate these key traits to assess the temperature dependence of trophic interactions, and thus the structural (biomass ratios) and dynamic (stability) implications of change. An understanding of which dynamics drive changes in abundance is essential to incorporate temperature-dependent rates into estimates of how consumers and resource abundance or biomass will be altered by future climate.

Imagine that the biomass of the consumer in Figure 19.1 is a balloon. The size of this balloon (i.e., biomass of the consumer) depends on three things: how much air is blown into the balloon; the elasticity of the balloon (i.e., how easily it expands); and the size of the leak in the balloon. Note that production (via growth rate r, or carrying

Figure 19.1. Three key components of production (R growth), energy transfer (from R to C) and loss (mortality or metabolic costs) contribute to the size of the consumer biomass or the size of the consumer "balloon."

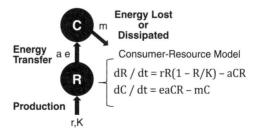

Energy Lost or Dissipated

Energy Transfer a e

Production

r,K

Consumer-Resource Model

$$dR / dt = rR(1 - R/K) - aCR$$
$$dC / dt = eaCR - mC$$

capacity K), energy transfer (via attack rate a or conversion efficiency e) and energy loss (i.e., mortality m) respectively govern each of these attributes. As this chapter proceeds to navigate the theory behind trait (r, K, a, e, and m) responses to climate, keep in mind that if relatively more air is entering the "consumer biomass" balloon than is leaving it, the balloon will expand, and vice versa.

Understanding the effects of production, energy transfer, and energy loss on consumer biomass can help explain how the implications of climate warming ultimately depend on whether the traits are affected symmetrically (Figure 19.2A) or asymmetrically (Figure 19.2B). There is no reason to

expect a priori that one parameter will always dominate; similarly, there is no reason to expect all parameters to be affected iden-

Figure 19.2. Hypothetical cases of trait responses to temperature (left) and subsequent changes to C-R interactions (right). In (A), the thermal performances of ae (solid line) and m (dotted line) are equal, thus the influx of energy to the consumer (ae) is similar to the energy lost (m) across increasing temperatures and temperature has no impact on the C-R interaction. (B) is showing a case where both ae and m have the same optimum temperature, but warming has a stronger effect on ae than m such that higher temperatures increase the C-R interaction. In the final case (C), ae and m have differential temperature responses in that they peak at different temperatures, causing stronger interactions at intermediate temperatures and weaker interactions at relatively high temperatures.

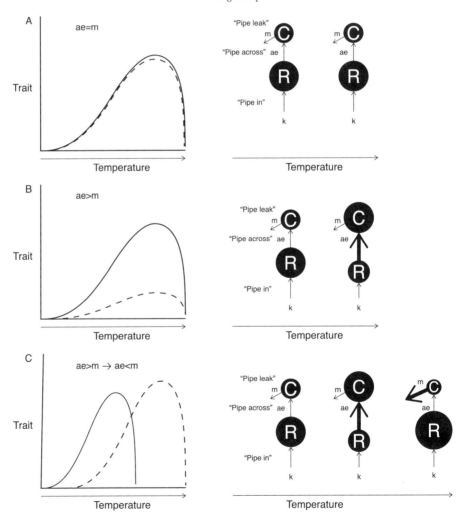

tically. Figure 19.2 outlines three different scenarios.

To put asymmetrical trait responses into context, a recent study by Bost et al. (2015) found that temperature had an impact on the population sizes of king penguins (*Aptenodytes patagonicus*) by altering their foraging behavior (and thus their attack rate). Temperature shifts associated with tropical anomalies pushed prime foraging habitats farther south and increased the depth of the ocean's thermocline (the depth at which temperature decreases rapidly). These shifts forced the penguins to travel farther and dive deeper for food, consequently increasing the total time spent foraging. Though not part of the study, subsequent reductions in attack rate and conversion efficiency (recall energy transfer) means less energy is available to be allocated to reproduction. Assuming that mortality rates remain constant, warmer temperatures asymmetrically altered consumer traits (reduced attack rates but not mortality) and, in this case, decreased population sizes.

This framework suggests that understanding the relationship between temperature dependence of consumer traits, $a(Temp)$ and $e(Temp)$, and consumer mortality rates, $m(Temp)$, provides insight into how dynamic systems respond to warming, even without information on values (a, e, m, r, and K) or thermal dependence of each parameter ($a(Temp)$, $e(Temp)$, $m(Temp)$, $r(Temp)$, and $K(Temp)$). Broadly synthesized empirical evidence suggests that across species, mortality and attack rates both increase similarly with temperature (McCoy and Gillooly 2008; Englund et al. 2011), although Dell et al. (2011) suggest a moderate asymmetry in which mortality is less sensitive to warming. For the outcome of dynamic systems, the relative temperature dependences of these rates will reflect not only broad metabolic constraints but also species-specific temperature performance curves for a given system (Englund et al. 2011; Dell et al. 2013). Within C-R systems, how the ratio of energy gain to energy loss (i.e., $a[Temp]$

$e[Temp]/m[Temp]$) varies with temperature, and whether asymmetries are common in one direction or another, is poorly understood. Increases in consumer standing stocks may often be an early but not lasting signal of warming (Rose and Caron 2007; O'Connor et al. 2011).

The example in Figure 19.1 illustrates how understanding relative thermal performance of traits can explain how the abundance of a species in a C-R interaction responds to temperature. Abundance or biomass typically and tractably are measured in experiments or in the field, and changes through time are often used to infer future trajectories of persistence and extinction, although a dynamic approach reminds us that temporal trajectories in abundance alone are insufficient to predict the future. For example, compare the patterns of relative abundance of consumers and resources in Figures 19.2. The decrease in consumer biomass at high temperatures (Figure 19.2) reflects the reversal in the temperature dependence of the underlying ratio of ae/m, even though the C-R biomass ratio (C/R) at lower temperatures was indistinguishable from the pattern in Figure 19.2. The bioenergetic framework outlined here, with information on the thermal asymmetry of ae/m, could be used to predict temporal trends in C/R with warming through time.

So far the effects of temperature on interactions have been considered without the consideration of temperature-dependent resource population growth (r and K), each of which may vary with temperature; and there is little reason to think they are thermally symmetric with consumer rates (a and m). For plant-grazer (or algal-grazer) C-R interactions, primary production tends to be less sensitive to warming than secondary production, which suggests an asymmetry between consumer and resource rates (Allen et al. 2005; O'Connor et al. 2011). Another source of consumer-resource thermal asymmetry is an animal's velocity (Dell et al. 2011, 2013). Differences in the mobility of consumers and their prey

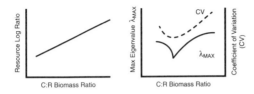

Figure 19.3. In a simple C-R model (e.g., type I functional response), the C:R biomass ratio correlates to increased top-down control. Here, shown by measuring
the change in R density (called log ratio; see Shurin et al.
2002) with and without the consumer (A). A top-heavy
biomass ratio drives a large cascade. Similarly, high C:R
biomass ratios also tend to drive heightened instability
(B). For more details on these results in numerous models, see McCann (2012) and Gilbert et al. (2014).

can generate thermal asymmetries in the
rates underlying consumption, and when
these differ from population growth rates
the framework here suggests shifts in interaction outcomes.

Several important ecological properties are altered with biomass accumulation
changes. First, increasing consumer biomass in simple C-R theory tends to equate
with the strength of top-down control
(Figure 19.3A). That is, the larger the C:R
biomass ratio, the greater the release of
resource biomass (represented by the resource log ratio; Shurin et al. 2002) when
the consumer is removed. In other words,
with increased C:R ratios, the strength of
the trophic cascade and top-down control
grows (Gilbert et al. 2014; DeLong et al.
2015). Such a change in consumer biomass
affects the stability of the interaction (Figure 19.3B): there is a strong tendency for
top-heavy interactions (i.e., high C:R biomass ratios) to be less stable (McCann 2012;
Gilbert et al. 2014).

It is evident that specific impacts of
warming are context dependent; however,
two important points are addressed here:
First, the relative response of entire suites
of traits govern general dynamic outcomes
of climate change, and second, climate
change can alter the structure or architectural framework of systems and subsystems
via interacting traits. To simplify the vexing problem of simultaneously responding

traits, Gilbert et al. (2014) aggregated key
ecological traits into consumer rates (ae/m)
and production (r, K) to form ratios that determine biomass flux across temperature
gradients, thus enabling us to track the
flow of energy between interacting species
and the impact of change on relative abundances and stability.

CLIMATE CHANGE AND ASYMMETRIC COMPARTMENT RESPONSES

The theoretical exercise suggested a relatively straightforward way to understand
how a given C-R interaction can be affected
by warming. To scale up to whole food
webs is difficult. Certain properties inherent in real food webs may allow an informative macroscopic glimpse of how climate
warming ought to affect the energy flow of
whole webs. However, different habitats often have unique characteristics (i.e., organismal sizes, habitat conditions) that likely
mediate the impacts of climate change differently. If true, then climate change may
fundamentally alter energy flow through
whole ecosystems.

Empirical results suggest that food webs
typically have a gross architecture in which
lower trophic level habitats are compartmentalized (species relatively isolated) and
ultimately are coupled through foraging to
higher-trophic-level predators, which tend
to be more mobile (Figure 19.4). Mobile
organisms tend to be generalist foragers
capable of consuming a variety of distinct
prey types because, all else equal, mobility exposes consumers to a greater number
of habitats and prey types (i.e., encounter
probability is positively related to mobility; Pyke et al. 1977). This correlation between size and generalism suggests that
each progressively higher trophic position
will increasingly use different habitats
(Pimm and Lawton 1977). Increased generalism at higher trophic levels increases
the likelihood for omnivory, defined here
as feeding at more than one trophic level

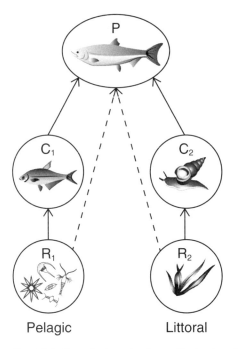

Pelagic   Littoral

Figure 19.4. General schematic of gross food-web architecture. Here, the top predator (P) is a generalist across habitat compartments and trophic levels. That is, the predator couples distinct energy channels (subscripts 1 and 2), selectively feeding on consumers (C) and resources (R) in either habitat.

(increasing omnivory is used to mean increasing consumption of lower-trophic-level prey).

Integration by upper trophic levels of lower-level compartmentalization has been demonstrated in aquatic and terrestrial systems and at different habitat scales from microhabitat to ecosystem (McMeans et al. 2013; Rooney et al. 2008) (Figure 19.4). This energetic integration by trophic level is therefore a theoretically motivated and empirically appropriate characterization of major carbon-flow pathways through food webs (McCann 2012). Energy integration across trophic levels is a fundamental component to the assessment of adaptive capacity of food webs in a variable world. Climate change may fundamentally alter this gross stabilizing structure and in doing so threaten critical ecosystem functions.

ASYMMETRICAL FOOD-WEB COMPARTMENT RESPONSES TO CLIMATE CHANGE

There are several reasons to believe that these different habitat compartments may be differentially affected by climate change (e.g., one habitat altered more than another) and so drive major changes to the energy pathways that maintain food webs. "Food-web compartments" may be differentially affected by climate change because they are compartmentalized by habitats with different sensitivities to environmental variability. Such differential impacts on habitats are not confined to lakes. In general different properties of different habitat compartments (e.g., grassland vs. forest, aboveground vs. belowground) suggest, if anything, that climate change ought to affect habitats asymmetrically. There is little reason to believe that they will be affected identically.

Habitat compartments can be expected, through evolution in different habitats, to include organisms with unique or habitat-dependent attributes. For example, at identical trophic levels the body sizes of littoral organisms tend to be much larger than the body sizes of pelagic organisms (Rooney et al. 2008). The littoral is a much more two-dimensional habitat than the open-water pelagic—an environment that favors small primary producers that float (enabling them to use the Sun's energy). Such different attributes tend to cascade through the ecosystem (Rooney et al. 2008) and so may alter the way a whole compartment of organisms respond to climate change, thus driving asymmetric responses at the whole energy pathway scale.

Finally, generalists that couple across compartments or channels can be expected to alter their behavior under asymmetric habitat impacts. As an example, if a top predator fish is a cold-water fish that couples pelagic and littoral pathways, then the differential heating of one pathway (i.e., the littoral) may dramatically affect the "coupler's" behavior. Altered behavior could fundamentally alter the flow of energy even if things like compartmental productivity or energy transfers between trophic levels do not change.

Taken together, climate change is expected to have asymmetric impacts on different food web compartments that could drastically change energy flow. Below are some simple empirical examples—one aquatic and one terrestrial—to show that some emerging data agree with this food-web-scale perspective on climate change.

EMERGING EMPIRICAL EXAMPLES OF ASYMMETRIC WHOLE FOOD-WEB IMPACTS

In lake ecosystems, as air temperature increases, shallow, near-shore (littoral) habitats warm more quickly than deep, offshore (pelagic) habitats. Tunney et al. (2014) found that cold-adapted predatory lake trout (*Salvelinus namaycush*) are increasingly restricted from accessing the littoral carbon pool when temperatures exceed that of optimal foraging. Across 50 boreal lake ecosystems, warming had a "decoupling" effect, meaning that lake trout diet became increasingly derived from pelagic habitats in warmer lakes, independent of prey abundance in either habitat (Figure 19.5A; see Guzzo et al. 2017 for similar patterns found temporally). This differential warming and subsequent shifts in species behavior are not confined to aquatic ecosystems. Barton and Schmitz (2009) documented a similar behavioral shift for a predatory spider species in a grassland food web with a vertical temperature differential in the vegetation canopy (temperatures increase from the ground surface to the top of the canopy). They examined two spider species that differed in foraging strategy (ambush sit-and-wait vs. active hunting) and principal habitat (the ambush species, *Pisaurina mira*, resided lower in the canopy than the active hunter, *Phidippus rimator*). Experimental warming in-

creased the absolute temperature at each elevation of the vegetation canopy and set new spatial constraints, thus altering the habitat domain of the thermally sensitive active hunting spider (vulnerable to the increased temperatures at higher elevations) and causing overlap of the two species. This resulted in intraguild predation and extinction of the active species (Figure 19.5B).

Climate warming drives asymmetries in compartmentalized food webs by differentially warming alternate habitats and affecting trait-mediated responses of species at both the physiological and the behavioral levels. Warming can alter the architectural framework of food webs via changes in biomass production (e.g., increased growth rates, thermal refugia for prey species) and top-down control.

DISCUSSION

Context-dependent, TPC-driven response of C-R interactions to climate change can yield complex outcomes (see Figure 19.2C). This suggests that warming associated with climate change acts on asymmetric thermal responses of individual species traits to alter whole "compartment" responses that affect the gross architecture of entire food webs due to habitat heterogeneity and the physiology and behavior of resident species (Figure 19.4).

Empirical patterns in gross food-web energy flows suggest that webs are often compartmentalized at lower trophic levels and coupled by the behavior of higher-order generalists. This structure is often "adaptive" and stabilizing to changing conditions, as one compartment serves as a buffer to the pressures that act on another (e.g., overconsumption, seasonal fluxes). Evolution has equipped species for habitat-specific conditions, and this compartmentalized or channeled food web structure sets up the possibility for food webs to be affected asymmetrically by change and so potentially altered in terms of their gross

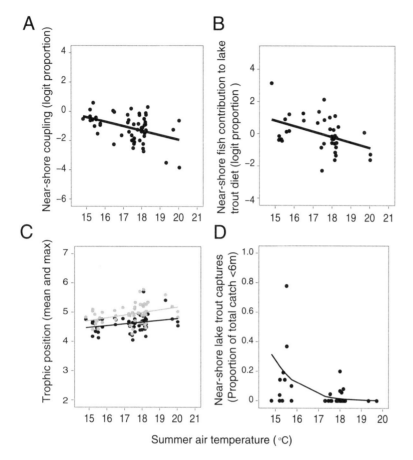

Figure 19.5. Empirical data from Tunney et al. (2014), showing the impacts of climate warming on food web structure. (A) shows that there is evidence for a decrease in near- and offshore coupling by lake trout, (B) shows a decrease in the contribution of nearshore fish to lake trout diet, (C) shows an increase in trophic position, and (D) shows a decrease in lake trout presence in nearshore habitats with increasing temperatures.

gross architecture of many aquatic and terrestrial webs suggests that their inherently spatial structuring can generate different responses that completely reroute energy and carbon flow in whole ecosystems.

energy flow patterns (due to changes in, for example, primary production or consumption rates). Changes can alter the behavior of larger, mobile species, potentially restricting accessibility to alternate habitats and eliminating the buffering or stabilizing properties of compartmentalized webs.

Food webs are complex, and the fate of our ecosystems in response to climate change depends not only on individual species or individual interactions but also on the collective response of the whole system to change. This chapter shows that a first glimpse of the

REFERENCES

Allen, A. P., J. F. Gillooly, and H. Brown. 2005. "Linking the global carbon cycle to individual metabolism." *Functional Ecology* 19: 202–213. https://doi.org/10.1111/j.1365-2435.2005.00952.

Barton, B. T., and O. J. Schmitz. 2009. "Experimental warming transforms multiple predator effects in a grassland food web." *Ecology Letters* 12: 1317–1325. https://doi.org/10.1111/j.1461-0248.2009.01386.x.

Bost, C. A., C. Cotté, P. Terray, C. Barbraud, C. Bon, K. Delord, O. Gimenez, Y. Handrich, Y. Naito, C. Guinet, and H. Weimerskirch. 2015. "Large-scale climatic anomalies affect marine predator foraging behaviour and demography." *Nature Communications* 6. https://doi.org/10.1038/ncomms9220.

Dell, A. I., S. Pawar, and V. M. Savage. 2013. "Temperature dependence of trophic interactions are driven by asymmetry of species responses and foraging strategy." Journal of Animal Ecology 83: 70–84. https://doi.org/10.1111/1365-2656.12081.

Dell, A. I., S. Pawar, and V. M. Savage. 2011. "Systematic variation in the temperature dependence of physiological and ecological traits." Proceedings of the National Academy of Sciences 108: 10591–10596. https://doi.org/10.1073/pnas.1015178108.

Englund, G., G. Ohlund, C. L. Hein, and S. Diehl. 2011. "Temperature dependence of the functional response." Ecology Letters 14: 914–921. https://doi.org/10.1111/j.1461-0248.2011.01661.x.

Fussmann, K. E., F. Schwarzmuller, U. Brose, A. Jousset, and B. C. Rall. 2014. "Ecological stability in response to warming." Nature Climate Change 4: 206–210. https://doi.org/10.1038/nclimate2134.

Gilbert, B., T. D. Tunney, K. S. McCann, J. P. DeLong, D. A. Vasseur, V. M. Savage, J. B. Shurin, A. Dell, B. T. Barton, C. D. G. Harley, P. Kratina, J. L. Blanchard, C. Clements, M. Winder, H. S. Greig, and M. I. O'Connor. 2014. "A bioenergetics framework for the temperature dependence of trophic interaction strength." Ecology Letters 17: 1–12. https://doi.org/10.1111/ele.12307.

Gillooly, J. F., J. H. Brown, G. B. West, V. M. Savage, and E. L. Charnov. 2001. "Effects of size and temperature on metabolic rate." Science 293: 2248–2251. https://doi.org/10.1126/science.1061967.

Guzzo, M., P. J. Blanchfield, and M. D. Rennie. "Behavioural responses to annual temperature variation alter the dominant energy pathway, growth, and condition of a cold-water predator." PNAS. https://10.1073/pnas.1702584114.

Huey, R. B., and J. G. Kingsolver. 1989. "Evolution of thermal sensitivity of ectotherm performance." Tree 4: 131–135. https://doi.org/10.1016/0169-5347(89)90211-5.

Huey, R. B., and R. D. Stevenson. 1979. "Integrating thermal physiology and ecology of ectotherms: A discussion of approaches." American Zoologist 19: 357–366.

Jarvis, L., K. S. McCann, T. Tunney, G. Gellner, and J. Fryxell. 2016. "Early warning signals detect critical impacts of experimental warming." Ecology and Evolution 6: 6097–6106. https://doi.org/10.1002/ece3.2339.

Kraemer, B. M., S. Chandra, A. I. Dell, M. Dix, E. Kuusisto, D. M. Livingstone, S. G. Schladow, E. Silow, L. M. Sitoki, R. Tamatamah, and P. B. McIntyre. 2016. "Global patterns in lake ecosystem responses to warming based on the temperature dependence of metabolism." Global Change Biology. https://doi.org/10.1111/gcb.13459.

McCann, K. S. 2012. Food Webs. Princeton University Press.

McCoy, M. W., and J. F. Gillooly. 2008. "Predicting natural mortality rates of plants and animals." Ecology Letters 11: 710–716. https://doi.org/10.1111/j.1461-0248.2008.01190.x.

McMeans, B. C., N. Rooney, M. T. Arts, and A. T. Fisk. 2013. "Food web structure of a coastal Arctic marine ecosystem and implications for stability." Marine Ecology Progress Series 482: 17–28. https://doi.org/10.3354/meps10278.

Monaco, C. J., C. D. McQuaid, and D. Marshall. 2017. "Decoupling of behavioural and physiological thermal performance curves in ectothermic animals: A critical adaptive trait." Oecologia 185: 583–593. https://doi.org/10.1007/s00442-017-3974-5.

O'Connor, M. I. 2009. "Warming strengthens an herbivore-plant interaction." Ecology 90: 388–398. https://doi.org/10.1890/08-0034.1.

O'Connor, M. I., B. Gilbert, and C. J. Brown. 2011. "Theoretical predictions for how temperature affects the dynamics of interacting herbivores and plants." American Naturalist 178: 626–638. https://doi.org/10.1086/662171.

Pimm, S. L., and J. H. Lawton. 1977. "Number of trophic levels in ecological communities." Nature 268: 329–331. https://doi.org/10.1038/268329a0.

Pyke, G. H., H. R. Pulliam, and E. L. Charnov. 1977. "Optimal foraging: A selective review of theory and tests." Quarterly Review of Biology 52: 137–154. http://www.jstor.org/stable/2824020.

Rall, B. C., O. Vucic-Pestic, R. B. Ehnes, M. Emmerson, and U. Brose. 2010. "Temperature, predator-prey interaction strength and population stability." Global Change Biology 16: 2145–2157. https://doi.org/10.1111/j.1365–2486.2009.02124.x.

Rooney, N., K. S. McCann, and J. C. Moore. 2008. "A landscape theory for food web architecture." Ecology Letters 11: 867–881. https://doi.org/10.1111/j.1461-0248.2008.01193.x.

Rooney, N., K. S. McCann, G. Gellner, and J. C. Moore. 2006. "Structural asymmetry and the stability of diverse food webs." Nature 442: 265–269. https://doi.org/10.1038/nature04887.

Rose, J. M., and D. A. Caron. 2007. "Does low temperature constrain the growth rates of heterotrophic protists? Evidence and implications for algal blooms in cold waters." Limnology and Oceanography 52: 886–895. https://doi.org/10.4319/lo.2007.52.2.0886.

Shurin, J. B., E. T. Borer, E. W. Seabloom, K. Anderson, C. A. Blanchette, B. Broitman, S. D. Cooper, and B. S. Halpem. 2002. "A cross-ecosystem comparison of trophic cascades." Ecology Letters 5: 785–791. https://doi.org/10.1046/j.1461-0248.2002.00381.x.

Tunney, T., K. S. McCann, N. Lester, and B. Shuter. 2014. "Effects of differential habitat warming on complex communities." Proceedings of the National Academy of Sciences. 111: 8077–8082. https://doi.org/10.1073/pnas.1319618111.

Vadeboncouer, Y. M., M. J. Vander Zanden, and D. M. Lodge. 2002. "Putting the lake back together: Reintegrating benthic pathways into lake food web models." BioScience 52: 44–54. https://doi.org/10.1641/0006-3568(2002)052[0044:PTLBTR]2.0.CO;2.

Dynamic Spatial Management in an Australian Tuna Fishery

Jason R. Hartog and Alistair J. Hobday

Global oceans have warmed over the past 100 years by an average of approximately 0.6°C, however, ocean temperatures are not increasing homogeneously, as some areas are warming even more rapidly than the average (Hobday and Pecl 2014). These rapidly warming areas are concentrated against coastal margins where western boundary currents move water poleward. This long-term warming moves isotherms poleward and shifts ocean habitats—as a result, many changes in species' distribution have been reported from these regions. In areas with a strong north-south current axis, this "climate scale" habitat shift also occurs seasonally, with the advance and retreat of currents. For example, the seasonal variation in the East Australian Current off southeastern Australia is superimposed on the long-term southward extension, where warming is about four times the global average. Such dynamic and fast-warming areas represent an obvious location for developing and testing climate-proof management approaches (see Hobday et al. 2016).

In the ocean, spatial management continues to be an important tool for managing marine resource extraction, as well as for meeting conservation objectives. However, in regions with both short- and long-term change, static spatial management may not represent the best solution when there are competing goals for ocean use (protection or exploitation), as oceanic habitats are mobile, and static protection often requires large areas to cover all the habitat locations over some time period. Instead, dynamic spatial management may be a suitable alternative, provided that species movements are predictable and suitable incentives exist. Dynamic spatial management thus also represents a climate-proof strategy that will be robust to shifting habitats and species under climate change (see Hobday et al. 2016).

The Australian Eastern Tuna and Bill-fish Fishery operates along much of the East Coast inside and beyond the exclusive economic zone with effort concentrated in the dynamic portion of the East Australian Current. This multispecies longline fishery is managed by integrating single species assessments, catch-limit trigger points, harvest strategies, and gear restrictions in a whole-of-area management approach. Fishers often target different species, such as yellowfin (*Thunnus albacares*), bigeye (*T. obesus*), and southern bluefin (*T. maccoyii*) tuna; striped marlin (*Kajikia audax*, formerly *Tetrapturus audax*); and swordfish (*Xiphias gladius*), depending on seasonal availability and prevailing ocean conditions. Fishers are also subjected to management decisions that alter their fishing behavior.

In this fishery region, where a range of other activities also occurs, dynamic ocean management has been used to reduce unwanted bycatch of southern bluefin tuna. Southern bluefin tuna are managed by the Commission for the Conservation of Southern Bluefin Tuna. As part of this management process there is an agreed global quota of which Australia has an allocation (full details can be found at http://www.ccsbt.org). Managing the catch of southern bluefin tuna within these limits (while still allowing capture of other species) is an important fisheries management goal for the Australian Fisheries Management

Authority. The distribution of likely southern bluefin tuna (SBT) habitat, which can change rapidly with the movement of the East Australian Current, has been used to dynamically regulate fisher access to East Coast fishing areas, by dividing the ocean into a series of zones based on the expected distribution of southern bluefin tuna. A habitat preference model was used to provide near-real-time advice to management about likely tuna habitat, delivered in the form of regular reports (every 1–2 weeks) to the authority during the fishing season (Hobday et al. 2010). Managers used these habitat preference reports to frequently update spatial restrictions to fishing grounds. These restrictions limit unwanted interactions by fishers that do not hold SBT quota (SBT cannot be landed without quota and in that situation must be discarded) and allow access to those that do have SBT quota to operate efficiently (Hobday et al. 2010), and they are enforced by vessel monitoring systems and fisheries observers.

The data requirements for this dynamic habitat prediction system are information on the temperature preference profile from SBT in the study area provided from electronic tags, near-real-time sea surface temperature from satellites, and near-real-time subsurface temperatures from an operational ocean model. Efficient computing systems support real-time delivery, for example, with temperature preference profiles obtained from a tagging database of up-to-date data from electronic tags deployed on southern bluefin tuna in the study area. The habitat prediction system integrates these components to produce a two-dimensional map of the habitat preference in near real-time (see Plate 7). Following quality control, reports on the current southern bluefin tuna habitat prediction in the fishery area are provided to fisheries managers (see Plate 7), who make a decision about changing management zones within hours of receiving the reports. The ocean data products

that underpin this dynamic management approach are not mission-critical-supported products, and occasional failures have prevented on-time delivery of southern bluefin tuna habitat reports. This risk is mitigated by close engagement with the authority to ensure that, in the event of disruption, historical patterns in the distribution of the habitat zones can be coupled with the seasonal forecasts and the management system is not compromised.

In operational use by the Australian Fisheries Management Authority from 2003 to 2015, this habitat model evolved from one based on surface temperature to an integrated surface and subsurface model, before finally including a seasonal forecasting element to aid managers and fishers in planning for future changes in the location of the habitat zones (Hobday et al. 2016). The continual improvement and adaptation of the system saw new oceanographic products tested and included in the operational model. Stakeholder engagement has been critical to the development, validation, and refinement of the system (Hobday et al. 2010). For example, incorporating a seasonal forecasting component was an important step in informing and encouraging managers and fishers to think about decisions on longer time scales (Hobday et al. 2016). The dynamic approach reduced the need for closures of large areas while still meeting the management goal, but required that more flexible fishing strategies be developed, including planning vessel movements, home-port selection, and quota purchase. From 2015, the authority has opted to use video monitoring of the fleet to ensure accurate reporting of catch of all species, SBT included, and reliance on the habitat model has declined.

This example from Australia's East Coast is one of a growing set of dynamic ocean management applications for fisheries, conservation, and shipping. Under dynamic management, rather than closing off large areas, managers can integrate information from habitat models, recent interactions, and historical knowledge to dynamically close smaller areas for shorter periods of time. At longer time scales, similar strategies can be implemented to minimize overlaps of protected and desired fish species. Climate-aware conservation planners may also consider the use of mobile protected areas, defined by dynamic habitat models, to afford protection to species changing their distribution in response to long-term ocean change.

REFERENCES

Hobday, A. J., J. R. Hartog, T. Timmis, and J. Fielding. 2010. "Dynamic spatial zoning to manage southern bluefin tuna capture in a multi-species longline fishery." *Fisheries Oceanography* 19 (3): 243–253.

Hobday, A. J., and G. T. Pecl. 2014. "Identification of global marine hotspots: Sentinels for change and vanguards for adaptation action." *Reviews in Fish Biology and Fisheries* 24: 415–425. https://doi.org/410.1007/s11160-11013-19326-11166.

Hobday, A. J., C. M. Spillman, J. P. Eveson, and J. R. Hartog. 2016. "Seasonal forecasting for decision support in marine fisheries and aquaculture." *Fisheries Oceanography* 25 (S1): 45–56.

CHAPTER TWENTY

Invasive Species and Climate Change

ELIZABETH H. T. HIROYASU
AND JESSICA J. HELLMANN

INTRODUCTION

Invasive species have been identified as a major threat to ecosystem integrity (e.g., Bailie et al. 2004; Millennium Ecosystem Assessment 2005). Coupled with the impacts of climate change, invasive species are a major driver of biodiversity change (Sala et al. 2000; Early et al. 2016). Evidence suggests that climate change undermines the resilience of ecosystems and exacerbates disturbance, thus creating favorable conditions of invasive species. However, it is difficult to predict how species will respond to changes in disturbance regimes, weather events, and novel species assemblages. We expect climate change to alter current distributions, survivorship, and life-history traits of species. This information is crucial for effective invasive species management and control.

INVASIVE SPECIES AND CLIMATE CHANGE

Invasive species often possess traits that allow them to persist and succeed in a changing environment, including broad environmental tolerance, short juvenile periods, and the ability to disperse long distances (Hellmann et al. 2008). These biological traits suggest that many invasive species may be more likely to be successful under climate change. We see in a variety of cases that invasive species are often better equipped to be successful under changing conditions than their native competitors. With changes in precipitation levels, for example, introduced cheatgrass (*Bromus tectorum*) range is expected to expand by up to 45 percent in the western United States, increasing the risk of wildfires across the continent and reducing native biodiversity (Bradley 2009). In India, the geographical distribution of the invasion range of the giant African snail (*Achatina fulica*) is not expected to change, but already-invaded areas are expected to be more prone to future invasions (Sarma et al. 2015). In the oceans, international shipping has been an important vector for invasive species, with 84 percent of marine ecoregions reporting marine invasions (Molnar et al. 2008). Climate change impacts on global shipping routes may increase the propagule pressure in the oceans, exacerbating current invasion problems.

Although these impacts have been clearly documented for select invasive species, the impacts of climate change for others are less clear. Invasive ant populations are expected to respond in a variety of ways to climate change, with climate suitability models predicting that some species will expand their ranges and others will contract (Bertelsmeier et al. 2014; Figure 20.1). Ecological niche models predict that the potential of weedy European plant species to invade new areas will increase across the Southern Hemisphere but decrease in the Northern Hemisphere across four emissions scenarios (Peterson et al. 2008), suggesting that climate change will not necessarily work synergistically with invasive species everywhere. These variable responses to climate change mean that management strategies will have to be flexible, adaptive, and largely region-specific in order to respond to changing invasion pressures.

In this chapter, we present possible consequences of climate change for invasive species impacts, using the stages of invasion as a guide to evaluate invasion risk. According to Hellmann et al. (2008), climate change is expected to impact the success of invasive species in a variety of ways, including (1) altering the mechanisms of transport and introduction, (2) altering climatic constraints on invasive species, (3) altering

the distribution of existing populations of invasive species, (4) altering the impacts of existing populations of invasive species, and (5) altering the effectiveness of management of invasive species. We explore each of five dimensions proposed by Hellmann et al. (2008) in turn, with an eye to literature emerging since 2008. Specifically, we explore new evidence, counterexamples, and more nuanced expectations that have arisen for each dimension since Hellmann et al. (2008) was published. This chapter is intended to provide readers with an introduction to the impacts of invasive species and climate change on biodiversity; it is not a comprehensive review. We conclude the chapter with some possible solutions for mitigating the climate change–related impacts of invasive species.

ALTERED INTRODUCTION AND TRANSPORT

Climate change may alter the way that invasive species are introduced to novel environments through changes in transport patterns or novel geographic linkages (Hellmann et al. 2008). As climate change alters local weather patterns, changes to commerce and tourism routes may change both the introduction rate and the local survival probabilities of invasive species.

In their global assessment of marine invaders, Molnar et al. (2008) found that international shipping was the major means of introduction of marine invaders. Several recent studies have revealed new trade routes that may emerge under climate change. For example, Smith and Stephenson (2013) found that historic lows in Arctic summer sea ice will allow for increased and new routes across the North Pole for vessel shipping. This new route will increase ship traffic, increasing the introduction rates of nonnative species to novel habitats through ballast water and on hulls of ships. Further, Ware et al. (2016) found that current ballast water management practices in

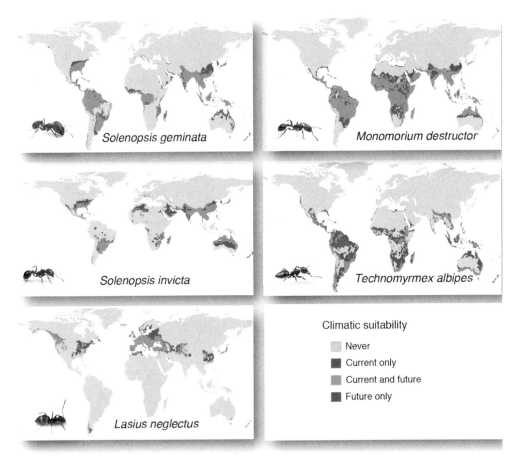

Figure 20.1. Spatial shift of suitable areas for five different ant species. Darkest gray: suitable currently, but not in the future. Lighter gray: suitable currently and in the future (to 2080). Medium gray: suitable only in the future. (Reproduced from Bertelsmeier et al. 2014.)

the Arctic do not prevent the introduction of nonindigenous zooplankton and marine invertebrates. In Europe, nonnative spider introduction has been shown to track with shipping routes, with the highest introductions coming from the areas with the highest amounts of trade (Kobelt and Nentwig 2008).

Increased propagule pressure and warming temperatures may also make these novel environments more hospitable for nonnatives. Ware et al. (2014) found that the Arctic ecosystem is at greater risk for invasion due to increased sea surface temperature and environmental similarity to more ports. Increasing temperatures and temperature oscillations have allowed for the range expansion of previously extant species in the Arctic (Berge et al. 2005), suggesting that conditions are becoming more favorable for warmer adapted species.

Intentional introductions for recreation or conservation purposes may also facilitate the transfer of invasive species in novel habitats. With the increased ability to ship plants and animals in the global pet trade, there is an increased probability of introductions through intentional releases. Carrete and Tella (2008) found a link between increasing demand for exotic birds in the pet trade and increased invasion risk of wild-caught bird species. Similarly, Strecker et al. (2011) found that approximately 2,500 exotic fish are released into Puget Sound, a temperate aquatic system, annually.

Invasive species proliferation may also be an unintended consequence of other policies aimed at mitigating the effects of climate change. Recent studies suggest that plants meeting the criteria for ideal biofuels often also have traits that can be classified as "weedy" (Mainka and Howard 2010); with the increased demand for biofuels, we may be inadvertently introducing new invaders into our agricultural systems. California's low carbon fuel standard (LCFS), part of the state's Assembly Bill 32, provides benefits to the transportation industry through the use of biofuel blends but does not provide guidelines for mitigating potential invasive species issues from biofuel production (Pyke et al. 2008; Chapman et al. 2016).

Finally, climate change may allow for extended recreational seasons or open new areas to recreational activities. These changes in recreation patterns may lead to an increase in introduction opportunities for invasive species. Zebra mussels (Dreissena polymorpha) have been introduced to novel freshwater environments via recreational boaters, resulting in costly damage and mitigation programs. A recent meta-analysis found that abundance and richness of nonnative species are higher in sites with high tourist activity (Anderson et al. 2015), suggesting that sites where people from diverse areas congregate can also be important introduction sites for invasive species.

ALTERED CLIMATIC CONSTRAINTS

Climate change can facilitate invasion in the colonization, establishment, and spread phases of the invasion process. In addition to changes in transport pathways discussed earlier, altered climatic constraints may provide more hospitable environments for introduced propagules or allow currently rare species to establish and colonize an area (Sorte et al. 2013). Milder winter temperatures can increase survival for many

temperate species into higher latitudes (Walther et al. 2009; Chen et al. 2011), thus leading to range expansions. Additionally, increased extreme weather events may facilitate disturbance—a process that creates openings for invaders.

In countries with a high Human Development Index (HDI), invasion threat was found to coincide with areas where climate-driven biome shifts are projected (Early et al. 2016). Invasive cheatgrass (Bromus tectorum) growth and success has been found to increase with temperature (Blumenthal et al. 2016), with implications for the success of native semiarid grasslands as temperatures rise. Bertelsmeier et al. (2014) found that 5 of 15 invasive ant populations are expected to expand their ranges up to 35.8 percent. Sanders et al. (2003) have shown that the invasive Argentine ant (Linepithema humile) has the potential to disassemble native ant communities in northern California by reducing species aggregations and shifting the community from structured to random, consequently reducing biodiversity and altering total community structure. If Argentine ants or cheatgrass expand with climate change, their disruption of community dynamics can drive larger losses in biodiversity.

Extreme weather events—predicted to worsen under climate change scenarios—may reduce the ability of native species to resist establishment of and compete with invaders. Through direct physiological stress as well as indirect mechanisms, climate change has been shown to undermine the resilience of aquatic systems to invaders. In their meta-analysis on different systems, Sorte et al. (2013) found that increases in temperature and CO_2 inhibited native species in aquatic systems. In some cases, climate change is expected to work in tandem with the invasion process, with positive feedbacks for invasive species, which are driven by extreme weather events. This appears to be the case for salt cedar (Tamarix spp.). The species is better able to withstand drought conditions than native flora and is outcompeting natives in areas expressing increased drought (Mainka and Howard 2010). As droughts become more severe, salt cedar is expected to continue to expand its range. The establishment of large, monoculture stands of salt cedar significantly reduces water table levels, further reducing biodiversity (Chapin et al. 2000).

Changes to dispersal abilities and rates of spread may also occur. Here again, new examples suggest that this mechanism is occurring under climate change. Increased temperatures increase flight activity and therefore dispersal ability of the pine processionary moth (*Thaumetopoea pityocampa*) (Battisti et al. 2006). Recent studies show that ornamental, garden-variety plants are shifting in latitude approximately 1,000 km beyond the range of their native counterparts as the climate warms (Van der Veken et al. 2008). In China, Huang et al. (2011) found a positive relationship between the establishment of invasive alien insects and change in temperature over a 100-year period. As the dispersal abilities of species shift with climate change, so, too, will their impacts on associated biodiversity.

ALTERED DISTRIBUTION IN ESTABLISHED POPULATIONS

Climate change has been linked to range shifts (see Chapter 3 in this volume), and invasive species are no exception to this pattern. In particular, warming conditions are expected to result in expansion of invasive species poleward and could also increase reproductive output with longer growing seasons. Described in Chapter 7, pine bark beetles have expanded their range in the western United States, but the beetles are expanding ahead of tree species, with impacts on trees that have not previously been affected by the beetle (Thuiller, Richardson, and Midgley 2007). Because invasive species are already well suited to rapid spread and establishment, they may have a competitive advantage over their native counterparts to take advantage of expanded ranges.

As ocean temperatures rise, organisms with broader thermal limits may be able to expand their range. For example, in the Antarctic and sub-Antarctic, Byrne et al. (2016) found that the carnivorous Arctic sea star (*Asterias amurensis*) has the potential to invade as ocean temperatures warm, with potential cascading effects for native biodiversity in lower trophic levels. It has been suggested that as native species shift their ranges, they too may become invasive. Unlike traditional invasive species, which are transplanted away from their native ranges, these species would be invading habitat at the edges of their current ranges. In Europe and North America, the deer tick *Ixodes scapularis* has expanded its range northward as temperatures warm, with important implications for the expansion of Lyme disease (Brownstein, Holford, and Fish 2005; Jore et al. 2014). Invasive purple loosestrife has been found to undergo rapid evolution to expand its current invaded range even further across the United States (Colautti and Barrett 2013).

Current conservation measures to move species to mitigate the impacts of climate change, known as "managed relocation" (or "assisted migration") have the potential to establish native species as pest species (Hellmann et al. 2008; Mueller and Hellmann 2008; Richardson et al. 2009). Alternatively, moving near natives can intentionally fill open niches and thus set a strategy for mitigating the spread of harmful or undesirable pests that spread and establish on their own (Lunt et al. 2013). The debate about managed relocation continues (e.g., Hewitt et al. 2011; Schwartz et al. 2012; Vitt et al. 2016), but the development of new modeling techniques such as bioclimatic envelope models, species distribution models, and trait-based models can be used to help forecast the dynamics and expansion of introduced populations (Barbet-Massin et al. 2018; Vitt et al. 2016; Jeschke and Strayer 2008).

ALTERED IMPACTS OF EXISTING INVASIVE SPECIES

It is difficult to predict how invasive species impacts will be affected by climate change, because both climate change and invasive species impacts occur on many different ecological scales and properties. Total impact of an invasive species is measured across three axes: spatial extent, average abundance within its current range, and per unit biomass impact on the local environment (Parker et al. 1999). We have discussed how we might expect the first axis, spatial extent, to change. Along the second axis, it seems likely that invasive species will be able to increase their abundance as the climate warms given the biological qualities that allow an organism to be a successful invader. In Thoreau's woods, Willis et al. (2010) identified earlier flowering times in nonnative and invasive species, over their native counterparts. This suggests that climate change can facilitate community invasion, as nonnatives are able to respond more quickly. As species expand their spatial extent, they will also increase their local abundance. In addition, climate change may impact the population dynamics of species such that they increase either survival or reproductive output (Ogden and Radojevic 2014). The generalist tendencies of invasives are likely to enable them to outcompete native species as they become decoupled from the environments that they are best adapted to (Hellmann et al. 2008). Along the third axis, climate change is expected to alter the relative impact of an invasive species in a variety of ways. For example, invasion of nonnative grasses in grassland ecosystems has altered fire regimes, creating a positive feedback loop of increasing the abundance of nonnative grasses and fire frequency (Brooks et al. 2010; Mack and D'Antonio 1998). Increasing temperatures as a result of climate change can also increase the probability of extreme fire events, further promoting the success of invasive species.

As environmental stressors become more pronounced, invasive species may exacerbate these stresses or their impacts may become more severe relative to the climate change impact.

ALTERED MANAGEMENT EFFECTIVENESS

Climate change will undoubtedly alter the effectiveness of management strategies used to prevent establishment and spread of invasive species. Climate change may require more vigilant and increased monitoring for invasive species, because it is unknown exactly where or how invasive species may relocate due to climate change. Changes to temperature fluctuations may also lead to differential effectiveness and nontarget impacts from chemical control (Maino et al. 2017). Therefore, strategies used for mitigating and reducing invasive species may be ineffective as the climate warms. Consideration of location of invasion is important; recent meta-analyses have found that areas at risk of biological invasion are often biodiversity hotspots (Bertelsmeier et al. 2014; Li et al. 2016). Adaptive management will be increasingly important as biocontrols change in effectiveness. Current biocontrols that are in place are likely to be reduced or could become invasive themselves (Lu et al. 2015; Selvaraj, Ganeshamoorthi, and Pandiaraj 2013). Long-term invasive species management plans will need to include climate change as an important factor when determining the sustainability and effectiveness of biocontrols and reliance on natural disturbance regimes to control invasive species. Accounting for sea-level rise will be important for predicting how invasive species might impact ecosystems; for example, changes in tidal height can determine whether invasive cordgrass (Spartina) will alter the state of tidal mudflats (Grosholz et al. 2009). This may alter how management strategies or restoration locations are prioritized.

SOLUTIONS

In addition to shedding new light on the invasion mechanisms proposed by Hellmann et al. (2008), several new ideas related to species invasion and climate change solutions have appeared in the literature in recent years. Courchamp et al. (2017) recently outlined the problems associated with invasive species management and the discipline of invasion biology as a whole. The authors identified four major categories of problems and solutions: understanding, alerting, supporting, and implementation. By identifying these categories of problems, scientists can work in each of these categories to identify solutions. Some examples of more concrete solutions include more coordinated research and restoration, better assessment techniques and policy-driven solutions to go along with them, and proposed impact classification frameworks. The following sections highlight some of the proposed methods for reducing the impacts of invasive species on biodiversity as the climate warms.

Biodiversity Conservation and Restoration

Community ecology theory predicts decreased success of invasion establishment if all niches of an ecosystem are filled; as a result, restoration efforts often focus on filling niches with native species that invasive species might otherwise exploit (Lunt et al. 2013). Restoration projects that focus on managing native species in restored or altered landscapes may mitigate the presence and impacts of invasive species (Funk et al. 2008). Increasing diversity and maturity may help reduce the invasion risk of a community (Levine and D'Antonio 1999; Naeem et al. 2000; Knops et al. 1999), thereby allowing a community to better withstand a release from climatic constraints or changes in extreme weather events. Conversely, overall reductions in biodiversity may lead to decreased resistance to invaders (Bellard et al. 2012). Management that focuses

on biodiversity conservation can help buffer communities from the impacts of both climate change and invasive species. Restoration that incorporates seeds from more southerly locations or lower elevations might be an increasing successful strategy in a warming climate as well (Aitken et al. 2008; Bower et al. 2014; Havens et al. 2014).

Coordinated Research and Policy Efforts

To reduce the impacts of invasive species, more cohesive international, national, and regional policies and research programs are needed. To effectively prevent introductions of invasive species in the first place, efforts will have to be coordinated across introduction pathways, sectors of policy and law (e.g., international trade agreements, national health and safety standards), and geographic scale (Burgiel 2015). Packer et al. (2017) proposed a framework for global network science to coordinate large-scale biogeographic research related to invasive species. Networks such as these are important for providing insights to our bigger-picture understanding of both the invasion process and how to best manage invasions.

Given that many countries have a limited ability to respond to invasions, global efforts will be necessary to reduce the impacts of invasive species (Early et al. 2016). Box 20.1 outlines the gaps and inconsistencies in the international legal framework for managing invasive species. One of the first international lines of defense is through strengthening the World Trade Organization's (WTO) General Agreement on the Application of Sanitary and Phytosanitary Measures (SPS), which sets a framework for countries to protect the health of plants, animals, humans, and the environment without large impacts on trade. Regulation through the WTO SPS would provide the broadest control for prevention of introductions of invasive species at the international level (Burgiel 2015). Increasing information exchange and collaborations between the WTO and countries will help increase

---

Box 20.1   **Gaps and Inconsistencies in the Legal Framework
for Managing Invasive Species**

- Animals (not plant pests), including pets, live bait, live food fish (World Organization for Animal Health [OIE], International Plant Protection Convention [IPPC], World Trade Organization Sanitary and Phytosanitary Measures [WTO/SPS] Agreement, UN Food and Agriculture Organization [FAO])

- Marine biofouling (International Maritime Organization—IMO)

- Civil air transport (International Civil Aviation Organization—ICAO)

- Aquaculture/mariculture (FAO)

- Conveyances (IPPC sea and air containers)

- Interbasin water transfers and canals

- Emergency relief

- Development assistance

- Military activities

(From Shimura et al. 2010.)

---

the effectiveness of these policies. Keller and Springborn (2014) found that imperfect screening is still a better method than an open-door policy to invasive species. Latombe et al. (2016) outlined a modular approach to monitoring invasive species globally through information transfer at the country level. Overall, coordinated policies at the international, national, and regional level are needed to protect biodiversity from invasive species.

Classifying Invasive Species

Blackburn et al. (2014; see also Hawkins et al. 2015) recently developed a framework for classifying invasive species on the basis of their impacts. The Environmental Impact Classification for Alien Taxa (EICAT) has been adopted by the IUCN to classify invasive species according to their predicted impact on the environment. The EICAT scheme has already been applied to a variety of taxa and will be used to assess and compare the impacts of known invasive species by the IUCN (Evans, Kumschick, and Blackburn 2016; Kumschick

et al. 2017). One of the most fascinating things about climate change is how it calls into question the very definition of *native* and *invasive*. As species reorganize under a changing climate—when they can—new species will appear in new areas, and land managers will have to decide whether they are welcome arrivals.

CONCLUSION

Following on a paper published nearly a decade ago, we find that emerging evidence supports earlier claims about emerging pathways and mechanisms of invasion under climate change. Like all other species, invasive species and their impacts are expected to be strongly influenced by climate change. Overall, invasive species are likely to expand their ranges and increase their impacts as the climate warms and species are reorganized. Increasing losses in biodiversity and the increasing probability of extreme weather events suggest that climate change opens niche space and creates disturbed habitats for generalist, disturbance-driven species to thrive. This suggests that climate change may act as a filter to promote a weedier world.

As trade becomes increasingly globalized, better management and policies should be put into place to help mitigate the impacts of invasive species and reduce introduction pressure. Changes to climatic constraints, distributions, and impacts with warming show that much more needs to be done to mitigate and reduce the impacts of invasive species. More research is still needed to reveal the circumstances when climate change enhances versus reduces the risk of invasion, but data from the past decade give considerable credence to the expectation that climate change could increase the abundance and spread of pest species without thoughtful ecosystem management informed by an understanding of the invasion pathway.

REFERENCES

Aitken, Sally N., Sam Yeaman, Jason A. Holliday, Tongli Wang, Sierra Curtis-McLane. 2008. "Adaptation, migration or extirpation: climate change outcomes for tree populations." *Evolutionary Applications*: 95–111.

Anderson, Lucy G., Steve Rocliffe, Neal R. Haddaway, and Alison M. Dunn. 2015. "The role of tourism and recreation in the spread of non-native species: A systematic review and meta-analysis." *PLOS One*. https://doi.org/10.1371/journal.pone.0140833.

Bailie, Jonathan, Craig Hilton-Taylor, and S. N. Stuart, eds. 2004. *2004 IUCN Red List of Threatened Species: A Global Species Assessment*. International Union for Conservation of Nature (IUCN). https://portals.iucn.org/library/node/9830.

Barbet-Massin, Morgane, Quentin Rom, Claire Villemant, and Franck Courchamp. 2018. "Can species distribution models really predict the expansion of invasive species?" *PLOS One* 13 (3): e0193085.

Battisti, Andrea, Michael Stastny, Emiliano Buffo, and Stig Larsson. 2006. "A rapid altitudinal range expansion in the pine processionary moth produced by the 2003 climatic anomaly." *Global Change Biology* 12: 662–671. https://doi.org/10.1111/j.1365-2486.2006.01124.x.

Bellard, Céline, Cleo Bertelsmeier, Paul Leadley, Wilfried Thuiller, and Franck Courchamp. 2012. "Impacts of climate change on the future of biodiversity." *Ecology Letters* 15 (4): 365–377. https://doi.org/10.1111/j.1461-0248.2011.01736.x.

Berge, Jørgen, Geir Johnsen, Frank Nilsen, Bjørn Gulliksen, and Dag Slagstad. 2005. "Ocean temperature oscillations enable reappearance of blue mussels *Mytilus edulis* in Svalbard after a 1000 year absence." *Marine Ecology Progress Series* 303: 167–175. https://doi.org/10.3354/meps303167.

Bertelsmeier, Cleo, Gloria M. Luque, Benjamin D. Hoffmann, and Franck Courchamp. 2014. "Worldwide ant invasions under climate change." *Biodiversity and Conservation* 24 (1): 117–128. https://doi.org/10.1007/s10531-014-0794-3.

Blackburn, Tim M., Franz Essl, Thomas Evans, Philip E. Hulme, Jonathan M. Jeschke, Ingolf Kühn, Sabrina Kumschick, et al. 2014. "A unified classification of alien species based on the magnitude of their environmental impacts." *PLOS Biology* 12 (5). https://doi.org/10.1371/journal.pbio.1001850.

Blumenthal, Dana M., Julie A. Kray, William Ortmans, Lewis H. Ziska, and Elise Pendall. 2016. "Cheatgrass is favored by warming but not CO_2 enrichment in a semi-arid grassland." *Global Change Biology* 22 (9): 3026–3038. https://doi.org/10.1111/gcb.13278.

Bradley, Bethany A. 2009. "Regional analysis of the impacts of climate change on cheatgrass invasion shows potential risk and opportunity." *Global Change Biology* 15 (1): 196–208. https://doi.org/10.1111/j.1365-2486.2008.01709.x.

Brooks, Matthew L., Carla M. D. Antonio, David M. Richardson, James B. Grace, E. Keeley, Joseph M. Ditomaso, Richard J. Hobbs, et al. 2010. "Effects of invasive alien plants on fire regimes." *BioScience* 54 (7): 677–688.

Brower, Andrew D., J. Bradley St. Clair, and Vicky Erickson. 2014. "Generalized provisional seed zones for native plants." *Ecological Applications* 24: 913–919. https://doi.org/10.1890/13-0285.1.

Brownstein, John, Theodore Holford, and Durland Fish. 2005. "Effect of climate change on Lyme disease risk in North America." *Ecohealth* 2 (1): 38–46. https://doi.org/10.1007/s10393-004-0139-x.Effect.

Burgiel, Stanley W. 2015. "From global to local: Integrating policy frameworks for the prevention and management of invasive species." In *Invasive Species in a Globalized World: Ecological, Social, and Legal Perspectives on Policy*, ed. Reuben P. Keller, Marc W. Cadotte, and Glenn Sandiford, 283–302. University of Chicago Press.

Byrne, Maria, Mailie Gall, Kennedy Wolfe, and Antonio Agüera. 2016. "From pole to pole: The potential for the Arctic seastar *Asterias amurensis* to invade a warming Southern Ocean." *Global Change Biology* (March). https://doi.org/10.1111/gcb.13304.

Carrete, Martina, and Jose L. Tella. 2008. "Wild-bird trade and exotic invasions: A new link of conservation concern?" *Frontiers in Ecology and the Environment* 6 (4): 207–211. https://doi.org/10.1890/070075.

Chapin, F. S., E. S. Zavaleta, V. T. Eviner, R. L. Naylor, P. M. Vitousek, H. L. Reynolds, D. U. Hooper, et al. 2000. "Consequences of changing biodiversity." *Nature* 405 (6783): 234–242. https://doi.org/10.1038/35012241.

Chapman, Daniel S., László Makra, Roberto Albertini, Maira Bonini, Anna Páldy, Victoria Rodinkova, Branko Šikoparija, Elżbieta Weryszko-Chmielewska, and James M. Bullock. 2016. "Modelling the introduction and spread of non-native species: International trade and climate change drive ragweed invasion." *Global Change Biology* 22 (9): 3067–3079. https://doi.org/10.1111/gcb.13220.

Chen, I-ching, Jane K. Hill, Ralf Ohlemüller, David B. Roy, and Chris D. Thomas. 2011. "Rapid range shifts of species of climate warming." *Science* 333: 1024–1027.

Colautti, Robert I., and Spencer C. Barrett. 2013. "Rapid adaptation to climate facilitates range expansion of an invasive plant." *Science* 342. https://doi.org/10.1126/science.1242121.

Courchamp, Franck, Alice Fournier, Céline Bellard, Cleo Bertelsmeier, Elsa Bonnaud, Jonathan M. Jeschke, and James C. Russell. 2017. "Invasion biology: Specific problems and possible solutions." *Trends in Ecology and Evolution* 32 (1): 13–22. https://doi.org/10.1016/j.tree.2016.11.001.

Early, Regan, Bethany A. Bradley, Jeffrey S. Dukes, Joshua J. Lawler, Julian D. Olden, Dana M. Blumenthal, Patrick Gonzalez, et al. 2016. "Global threats from invasive alien species in the twenty-first century and national response capacities." *Nature Communications* 7: 12485. https://doi.org/10.1038/ncomms12485.

Evans, Thomas, Sabrina Kumschick, and Tim M. Blackburn. 2016. "Application of the Environmental Impact Classification for Alien Taxa (EICAT) to a global assessment of alien bird impacts." *Diversity and Distributions* 22 (9). https://doi.org/10.1111/ddi.12464.

Funk, Jennifer L., Elsa E. Cleland, Katherine N. Suding, and Erika S. Zavaleta. 2008. "Restoration through reassembly: Plant traits and invasion resistance." *Trends in Ecology and Evolution* 23 (12): 695–703. https://doi.org/10.1016/j.tree.2008.07.013.

Grosholz, E. D., L. A. Levin, A. C. Tyler, and C. Neira. 2009. "Changes in community structure and ecosystem function following *Spartina alterniflora* invasion of Pacific estuaries." In *Human Impacts on Salt Marshes: A Global Perspective*, ed. B. R. Silliman, M. D. Bertness, and E. D. Grosholz, 23–40. University of California Press.

Havens, Kayri, Pati Vitt, Shannon Still, Andrea T. Kramer, Jeremie B. Fant, and Katherine Schatz. 2015. "Seed sourcing for restoration in an era of climate change." *Natural Areas Journal* 35 (1): 122–133.

Hawkins, Charlotte L., Sven Bacher, Franz Essl, Philip E. Hulme, Jonathan M. Jeschke, Ingolf Kühn, Sabrina Kumschick, et al. 2015. "Framework and guidelines for implementing the proposed IUCN Environmental Impact Classification for Alien Taxa (EICAT)." *Diversity and Distributions* 21 (11): 1360–1363. https://doi.org/10.1111/ddi.12379.

Hellmann, Jessica J., James E. Byers, Britta G. Bierwagen, and Jeffrey S. Dukes. 2008. "Five potential consequences of climate change for invasive species." *Conservation Biology* 22 (3): 534–543. https://doi.org/10.1111/j.1523-1739.2008.00951.x.

Hewitt, N., N. Klenk, A. L. Smith, D. R. Bazely, N. Yan, S. Wood, J. I. Maclellan, C. Lipsig-mumme, and I. Henriques. 2011. "Taking stock of the assisted migration debate." *Biological Conservation* 144 (11): 2560–2572. https://doi.org/10.1016/j.biocon.2011.04.031.

Huang, Dingcheng, Robert A. Haack, and Runzhi Zhang. 2011. "Does global warming increase establishment rates of invasive alien species? A centurial time series analysis." *PLOS One* 6 (9). https://doi.org/10.1371/journal.pone.0024733.

Jeschke, Jonathan M., and David L. Strayer. 2008. "Usefulness of bioclimatic models for studying climate change and invasive species." *Annals of the New York Academy of Sciences* 1134: 1–24. https://doi.org/10.1196/annals.1439.002.

Jore, Solveig, Sophie O. Vanwambeke, Hildegunn Viljugrein, Ketil Isaksen, Anja B. Kristoffersen, Zerai Woldehiwet, Bernt Johansen, et al. 2014. "Climate and environmental change drives *Ixodes ricinus* geographical expansion at the northern range margin." *Parasites and Vectors* 7 (11): 1–14.

Keller, Reuben P., and Michael R. Springborn. 2014. "Closing the screen door to new invasions." *Conservation Letters* 7 (3): 285–292. https://doi.org/10.1111/conl.12071.

Knops, M., D. Tilman, N. M. Haddad, S. Naeem, C. E. Mitchell, J. Haarstad, M. E. Ritchie, et al. 1999. "Effects of plant species richness on invasion dynamics, disease outbreaks, insect abundances and diversity." *Ecology Letters* 2: 286–293. https://doi.org/10.1046/j.1461-0248.1999.00083.x.

Kobelt, Manuel, and Wolfgang Nentwig. 2008. "Alien spider introductions to Europe Supported by global trade." *Diversity and Distributions* 14 (2): 273–280. https://doi.org/10.1111/j.

Kumschick, Sabrina, G. John Measey, Giovanni Vimercati, F., Andre de Villiers, Mohlamatsane M. Mokhatla, Sarah J. Davies, Corey J. Thorp, Alexander D. Rebelo, Tim M. Blackburn, and Fred Kraus. 2017. "How repeatable is the Environmental Impact Classification of Alien Taxa (EICAT)? Comparing independent global impact assessments of amphibians." *Ecology and Evolution* 7 (8): 2661–2670. https://doi.org/10.1002/ece3.2877.

Latombe, Guillaume, Petr Pyšek, Jonathan M. Jeschke, Tim M. Blackburn, Sven Bacher, César Capinha, Mark J. Costello, et al. 2016. "A vision for global monitoring of biological invasions." *Biological Conservation* (June). https://doi.org/10.1016/j.biocon.2016.06.013.

Levine, Jonathan M., and Carla M. D'Antonio. 1999. "Elton revisited: A review of evidence linking diversity and invasibility." *Oikos* 87 (1): 15–26.

Li, Xianping, Xuan Liu, Fred Kraus, Reid Tingley, and Yiming Li. 2016. "Risk of biological invasions is concentrated in biodiversity hotspots." *Frontiers in Ecology and the Environment* 14 (8): 411–417. https://doi.org/10.1002/fee.1321.

Lu, Xinmin, Evan Siemann, Minyan He, Hui Wei, Xu Shao, and Jianqing Ding. 2015. "Climate warming increases biological control agent impact on a non-target species." *Ecology Letters* 18: 48–56. https://doi.org/10.1111/ele.12391.

Lunt, Ian D., Margaret Byrne, Jessica J. Hellmann, Nicola J. Mitchell, Stephen T. Garnett, Matt W. Hayward, Tara G. Martin, Eve McDonald-Maddden, Stephen E. Williams, and Kertin K. Zander. 2013. "Using assisted colonization to conserve biodiversity and restore ecosystem function under climate change." *Biological Conservation* 157: 172–177.

Mack, Michelle C., and Carla M. D'Antonio. 1998. "Impacts of biological invasions on disturbance regimes." *Trends in Ecology and Evolution* 13 (5): 195–198. https://doi.org/10.1016/S0169-5347(97)01286-X.

Mainka, Susan A., and Geoffrey W. Howard. 2010. "Climate change and invasive species: Double jeopardy." *Integrative Zoology* 5 (2): 102–111. https://doi.org/10.1111/j.1749-4877.2010.00193.x.

Maino, James L. Paul A. Umina, Ary A. Hoffmann, and Arndt Hampe. 2017. "Climate contributes to the evolution of pesticide resistance." *Global Ecology and Biogeography* 27: 223–232.

Millenium Ecosystem Assessment. 2005. "Ecosystems and human well-being: Synthesis." World Resources Institute.

Molnar, Jennifer L., Rebecca L. Gamboa, Carmen Revenga, and Mark D. Spalding. 2008. "Assessing the global threat of invasive species to marine biodiversity." *Frontiers in Ecology and the Environment* 6 (9): 485–492. https://doi.org/10.1890/070064.

Mueller, J. M., and J. J. Hellmann. 2008. "An assessment of invasion risk from assisted migration." *Conservation Biology* 22: 562–567.

Naeem, Shahid, Johannes M. H. Knops, David Tilman, Katherine M. Howe, Theodore Kennedy, and Samuel Gale. 2000. "Plant diversity increases resistance to invasion in the absence of covarying extrinsic factors." *Oikos* 91 (1): 97–108. https://doi.org/10.1034/j.1600-0706.2000.910108.x.

Ogden, Nicholas H., and Milka Radojevic. 2014. "Estimated effects of projected climate change on the basic reproductive number of the Lyme disease vector *Ixodes scapularis*." *Environmental Health Perspectives* 122 (6): 631–638.

Packer, Jasmin G., Laura A. Meyerson, David M. Richardson, Giuseppe Brundu, Warwick J. Allen, Ganesh P. Bhattarai, Hans Brix, et al. 2017. "Global networks for invasion science: Benefits, challenges and guidelines." *Biological Invasions* 19 (4): 1081–1096. https://doi.org/10.1007/s10530-016-1302-3.

Parker, I. M., D. Simberloff, W. M. Lonsdale, K. Goodell, M. Wonham, P. M. Kareiva, M. H. Williamson, B. von Holle, P. B. Moyle, J. E. Byers, and L. Goldwasser. 1999. "Toward a framework for understanding the ecological effects of invaders." *Biological Invasions* 1: 3–19. https://doi.org/10.1023/A:1010034312781.

Peterson, A. Townsend, Aimee Stewart, Kamal I. Mohamed, and Miguel B. Araújo. 2008. "Shifting global invasive potential of European plants with climate change." *PLOS One* 3 (6): 1–7. https://doi.org/10.1371/journal.pone.0002441.

Pyke, Christopher R., Roxanne Thomas, Read D. Porter, Jessica J. Hellmann, Jeffrey S. Dukes, David M. Lodge, and Gabriela Chavarria. 2008. "Current practices and future opportunities for policy on climate change and invasive species." *Conservation Biology* 22 (3): 585–592. https://doi.org/10.1111/j.1523-1739.2008.00956.x.

Richardson, D. M., J. J. Hellmann, J. S. McLachlan, D. F. Sax, M. W. Schwartz, P. Gonzalez, E. J. Brennan, A. Camacho, T. L. Root, O. E. Sala, S. H. Schneider, D. M. Ashe, J. R. Clark, R. Early, J. R. Etterson, E. D. Fielder, J. L. Gill, B. A. Minteer, S. Polasky, H. D. Safford, A. R. Thompson, and M. Vellend. 2009. "Multidimensional evaluation of managed relocation." *Proceedings of the National Academy of Sciences* 106: 9721–9724.

Sala, Osvaldo E., F. Stuart Chapin III, Juan J. Armesto, Eric Berlow, Janine Bloomfield, Rodolfo Dirzo, Elisabeth Huber-Sanwald, et al. 2000. "Global biodiversity scenarios for the year 2100." *Science* 287 (5459): 1770–1775. https://doi.org/10.1126/science.287.5459.1770.

Sanders, Nathan J., Nicholas J. Gotelli, Nicole E. Heller, and Deborah M. Gordon. 2003. "Community disassembly by an invasive species." *Proceedings of the National Academy of Sciences* 100 (5): 2474–2477. https://doi.org/10.1073/pnas.0437913100.

Sarma, Roshmi Rekha, Madhushree Munsi, and Aravind Neelavara Ananthram. 2015. "Effect of climate change on invasion risk of giant African snail (*Achatina fulica ferussac*, 1821: Achatinidae) in India." *PLOS One* 10 (11): 1–16. https://doi.org/10.1371/journal.pone.0143724.

Selvaraj, S., P. Ganeshamoorthi, and T. Pandiaraj. 2013. "Potential impacts of recent climate change on biological control agents in agro-ecosystem: A review." *Journal of Biodiversity and Conservation* 5 (12): 845–852. https://doi.org/10.5897/IJBC2013.0551.

Smith, Laurence C., and Scott R. Stephenson. 2013. "New trans-Arctic shipping routes navigable by midcentury." *Proceedings of the National Academy of Sciences* 110 (13): E1191–E1195. https://doi.org/10.1073/pnas.1214212110.

Sorte, Cascade J. B., Ines Ibáñez, Dana M. Blumenthal, Nicole A. Molinari, Luke P. Miller, Edwin D. Grosholz, Jeffrey M. Diez, et al. 2013. "Poised to prosper? A cross-system comparison of climate change effects on native and non-native species performance." *Ecology Letters* 16 (2): 261–270. https://doi.org/10.1111/ele.12017.

Strecker, Angela L., Philip M. Campbell, Julian D. Olden, Angela L. Strecker, Philip M. Campbell, Julian D. Olden, and Philip M. Campbell. 2011. "The aquarium trade as an invasion pathway in the Pacific Northwest." *Fisheries* 36 (2): 74–85. https://doi.org/10.1577/03632415.2011.10389070.

Thuiller, Wilfried, D. M. Richardson, and G. F. Midgley. 2007. "Will climate change promote alien plant invasions?" In *Biological Invasions*, ed. W. Nentwig, 193, 197–211. Springer Verlag. http://www.springerlink.com/index/G66R623223012856.pdf.

Veken, Sebastiaan Van der, Martin Hermy, Mark Vellend, Anne Knapen, Martin Hermy, Mark Veliend, Anne Knapenl, and Kris Verheyen. 2008. "Garden plants get a head start on climate change." *Frontiers in Ecology and the Environment* 6 (4): 212–216. https://doi.org/10.1890/070063.

Vitt, Pati, Pairsa N. Belmaric, Rily Book, and Melissa Curran. 2016. "Assisted migration as a climate change adaptation strategy: Lessons from restoration and plant reintroductions." *Israel Journal of Plant Sciences* 63: 250–261.

Walther, Gian-reto, Alain Roques, Philip E. Hulme, Martin T. Sykes, Petr Pys, Christelle Robinet, and Vitaliy Semenchenko. 2009. "Alien species in a warmer world: Risks and opportunities." *Trends in Ecology and Evolution* 24 (12). https://doi.org/10.1016/j.tree.2009.06.008.

Ware, Chris, Jørgen Berge, Anders Jelmert, Steffen M. Olsen, Loïc Pellissier, Mary Wisz, Darren Kriticos, Georgy Semenov, Sławomir Kwaśniewski, and Inger G. Alsos. 2016. "Biological introduction risks from shipping in a warming Arctic." *Journal of Applied Ecology* 53: 340–349. https://doi.org/10.1111/1365-2664.12566.

Ware, Chris, Jørgen Berge, Jan H. Sundet, Jamie B. Kirkpatrick, Ashley D. M. Coutts, Anders Jelmert, Steffen M. Olsen, Oliver Floerl, Mary S. Wisz, and Inger G. Alsos. 2014. "Climate change, non-indigenous species and shipping: Assessing the risk of species introduction to a high-Arctic archipelago." *Diversity and Distributions* 20: 10–19. https://doi.org/10.1111/ddi.12117.

Willis, Charles G., Brad R. Ruhfel, Richard B. Primack, Abraham J. Miller-Rushing, Jonathan B. Loso, and Charles C. Davis. 2010. "Favorable climate change response explains non-native species' success in Thoreau's woods." *PLOS One* 5 (1): e8878.

CHAPTER TWENTY-ONE

Climate Change and Disease

LINDSAY P. CAMPBELL, A. TOWNSEND
PETERSON, ABDALLAH M. SAMY, AND
CARLOS YAÑEZ-ARENAS

Rumination about biological implications began soon after the realization that greenhouse gases were increasing in the atmosphere; a prominent element was that disease transmission would increase (e.g., Epstein et al. 1998). A 1995 Intergovernmental Panel on Climate Change (IPCC) report (Watson et al. 1996) speculated that the potential geographic distribution of malaria could increase "from approximately 45 percent of the world population to 60 percent by the latter half of the next century," although subsequent IPCC reports became more measured in their disease projections and emphasized human socioeconomic dimensions more clearly (McCarthy et al. 2001).

Considering disease transmission as the product of interactions among suites of species and the environmental landscapes where they are distributed (Peterson 2014), basic implications of warming climate for disease transmission are clear: ranges will generally shift to higher latitudes and elevations, and transmission phenomena will likely occur earlier in spring and later in fall. There are different geographic and environmental dimensions of climate change effects on species. Many of these expectations are already being fulfilled (see other contributions in this volume). Still, evidence needed to pinpoint climate change causally is difficult to assemble, such that climate change effects may often go unappreciated.

Here, we review models that have been used to explore climate change effects on disease. These distinct approaches offer some possibility of anticipation of potential changes in disease transmission across landscapes. We explore a case study—malaria in highland East Africa—to illustrate complexities inherent in the question.

TRANSMISSION MODELS

Disease transmission models were developed to understand and predict disease dynamics. Transmission models are mathematical models that mimic circulation of pathogens at the population level (Keeling and Rohani 2007). SIR models, which partition populations into classes of susceptible (S), infected (I), and recovered (R), have seen considerable development (Hethcote 2000). These models are used to predict and explain changes in disease transmission.

Ronald Ross and George MacDonald were pioneers in this field. Ross (1904) developed a first mathematical model to describe adult mosquito movements and the spatial scale of larval control that would best reduce mosquito populations, thus contributing to elimination of disease transmission. Ross's first model did not address the question of malaria transmission directly, but it certainly earned him a place in history. Later, he developed malaria models incorporating differential equations. In 1950–1956, MacDonald used field-collected data to test Ross's models on infant malaria rates and sporozoite prevalences; finally, in 1957, he published a book *Epidemiology and Control of Malaria* (MacDonald 1957), summarizing factors limiting malaria transmission. MacDonald proposed expression of the basic reproductive rate (R_0) for malaria as the mean number of secondary infections produced when an infected individual is introduced into a host population, assuming that every other individual in the population is susceptible; Keeling and Rohani (2007) estimated R_0 for several diseases and found values ranging from ~5 (polio) to ~100 (malaria).

Such models incorporated environmental drivers in malaria transmission. Lindsay and Birley (1996) used a simple mathematical framework to study temperature effects on vectorial capacity of *Anopheles maculipennis* in transmitting *Plasmodium vivax*; they found that small changes in temperature can change malaria transmission dramati-

cally. Martens et al. (1999) analyzed effects of temperature and precipitation on biting rates and adult emergence in *Anopheles* under scenarios derived from HadCM2 and HadCM3 climate models and concluded that a 1°C increase could increase malaria transmission in low-temperature regions. Ogden et al. (2014) evaluated climate change effects on R_0 for Lyme disease and projected increases of 150 percent–500 percent in North America.

Other transmission models have focused on climate change effects on vectorial capacity, the integration of feeding rate, survival, and incubation period. Kearney et al. (2009) evaluated how climate change may expand distributional potential of the mosquito *Aedes aegypti* in Australia (Figure 21.1); Liu-Helmersson et al. (2014) examined effects of climate change on *Aedes aegypti* vectorial capacity in dengue transmission. In both, mean temperature and diurnal temperature range were key factors. Another example compared development of *Plasmodium falciparum* at forested and deforested sites (Afrane et al. 2008): increased temperature and solar radiation at deforested sites elevates infection rate and vectorial capacity of the mosquito *Anopheles gambiae*, and *Plasmodium* sporozoites developed more rapidly than at forested sites, suggesting combined effects of deforestation and climate change on malaria dynamics.

Transmission models are thus powerful in permitting "experimentation" with specific effects of particular parameters, though not without limitations. Keeling and Rohani (2007) noted frequent information gaps for specific model parameter values: full model implementation requires estimates of all parameters, but population- or species-specific estimates are frequently unavailable or imprecise. Even when detailed estimates are available, interactions among parameters may be poorly understood; assumptions of independence can bias results substantially. Models do not necessarily reconstruct reality; rather, they should be used as heuristics that instruct about the behavior of complex systems.

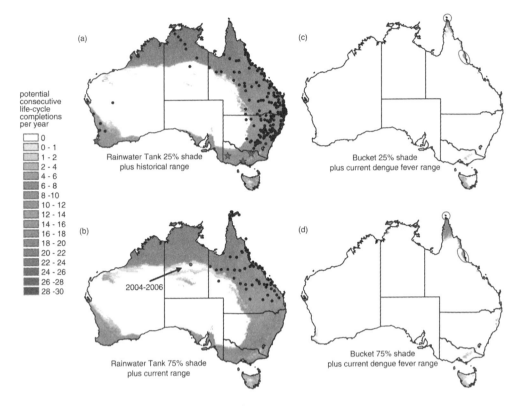

Figure 21.1. Predictions of number of consecutive life-cycle completions per year that could emerge from two simulated breeding containers (rainwater tanks, buckets) under different levels of shade. The model assumes that urbanized areas with containers that could sustain *Aedes aegypti* are present. (A) The historical distribution is imposed, (B) the present distribution is imposed, and (C) and (D) the locations of recent (since 1990) dengue outbreaks are imposed. (From Kearney et al. 2009.)

CORRELATIVE MODELS

Another approach to understanding climate change effects on disease transmission has been correlative ecological niche modeling (Peterson et al. 2011); these approaches are also referred to as species distribution models. The approach relates occurrences of a species or disease to raster data sets summarizing environmental variables to identify environments under which transmission occurs (or under which a species can maintain populations). These putatively suitable environments (i.e., ecological niches) can be used to identify regions where the species can maintain populations. Transferring

niche estimates onto future climate scenarios allows for estimating future potential distributions of disease-related phenomena.

Early examples applying ecological niche models (ENMs) to questions of climate change and disease include Rogers and Randolph (2000), who assessed climate effects on human malaria cases and future expansions in some areas being offset by contractions elsewhere (Figure 21.2). Peterson and Shaw (2003) focused on climate change effects on distributions of *Lutzomyia* sandfly species (vectors of cutaneous leishmaniasis) in Brazil and anticipated southward expansion of distributions. Many subsequent studies have used ENMs to assess potential effects of climate change on disease geography, and many have found significant distributional implications.

Recent studies have refined techniques in model calibration to deal with limited dispersal and uneven sampling, which affect model results (Peterson et al. 2011). Costa et al. (2014) included explicit consideration

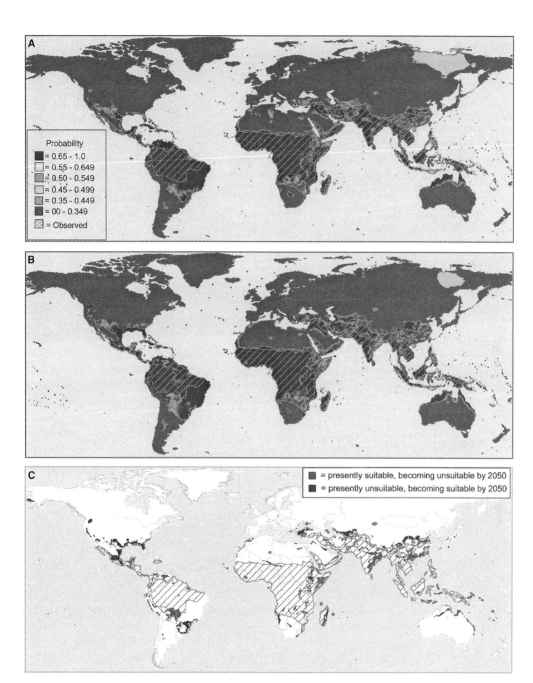

Figure 21.2. Early example of application of correlative ecological niche modeling to questions of disease transmission risk and climate change. (A) Current global map of malaria caused by *Plasmodium falciparum* (hatching) and predicted distribution (light gray areas indicate no prediction). (B) Discriminating criteria from the current situation were applied to equivalent climate surfaces from the high scenario from HadCM2: hatching and the probability scale are the same as in (A). (C) The difference between predicted distributions in (A) and (B), showing areas where malaria is predicted to disappear (i.e., probability of occurrence decreases from >0.5 to <0.5) (in medium gray) or invade (i.e., probability of occurrence increases from <0.5 to >0.5) (in dark gray) by the 2050s in relation to the present situation. The hatching is the current global malaria map shown in hatching in (A). (From Rogers and Randolph 2000.)

of accessible areas in model calibration to characterize future potential distributions of species in the *Triatoma brasiliensis* species complex (Chagas disease vectors). Campbell et al. (2015) summarized overall sampling of *Aedes* mosquitoes to characterize environmental variation across the calibration area. These methodological steps affect model outcomes; ignoring them compromises model quality seriously (Peterson 2014). Another recent improvement is use of multiple general circulation models to understand uncertainty in future predictions, as exemplified by Roy-Dufresne et al. (2013) in an assessment of the future potential distribution of the mouse *Peromyscus leucopus*, a Lyme disease reservoir.

ENMs are a practical option for assessment of climate change effects on disease geography for several reasons. First, data required are commonly available via online digital databases. Second, the conceptual framework of ecological niche modeling can incorporate diverse processes of abiotic, biotic, accessibility, and sampling variables (Peterson et al. 2011). Third, numerous options in model development and selection permit powerful steps in model calibration to meet diverse inferential challenges. Finally, new generations of ENMs cross diverse spatial and temporal scales to incorporate multiscalar phenomena (Kearney et al. 2009).

However, despite these strengths, several limitations must be taken into account. ENMs depend strongly on amounts, quality, and relevance of input data. The data used in ENMs are often "found" data not collected directly for the purpose of the study, which adds biases. Data documenting absences of species, specifically absences directly relevant to estimating niche dimensions, are frequently lacking. ENMs have generally ignored spatial autocorrelation, both in the sampling process and in the structure of environmental landscapes, notwithstanding the fundamental effects of such processes; spatially explicit approaches incorporating spatial autocorrelation exist but are not yet in broad use. ENMs associate occurrences with environmental characteristics; these "models" therefore must be interpreted with caution because occurrence-environment associations manifest fundamental ecological niches only *after* effects of limited dispersal (the "existing fundamental niche") and biotic interactions (Peterson et al. 2011). The latter point is important as regards disease studies because biotic effects may be pronounced in disease systems (realized niche; Peterson et al. 2011). Accessibility and movement effects often place disease-relevant taxa outside of their long-term geographic ranges. Finally, ENMs have been criticized for their correlational nature, linking geographic occurrences and environmental parameters without pondering underlying mechanisms explicitly.

EXAMPLES

Given the analytical frameworks described here, we explore examples linking disease transmission to weather and climate. These examples are invariably partial and incomplete, as warming trends are only a few decades old, causation is almost always multifactorial, and data are invariably incomplete. Still, these examples are suggestive and exploration is informative.

Several studies have linked variation in disease transmission to extreme weather, in particular to El Niño events. Perhaps best documented is Sin Nombre virus (a hantavirus) in the southwestern United States (Yates et al. 2002): elevated rainfall and mild winters may lead to high rodent populations, and rodents spill over into anthropogenic habitats when resources become scarce again in subsequent years. Similar linkages have been documented for plague (Stapp, Antolin, and Ball 2004) and Rift Valley fever (Anyamba et al. 2009). These studies, however, are linked to weather patterns rather than longer-term climate change.

Siraj et al. (2014) examined distribution of malaria cases in highland Ethiopia and

Colombia in relation to long-term temperature change. They derived median elevation values from yearly cumulative distributions of malaria incidence across elevational gradients and assessed elevational shifts through time. The spatial distribution of malaria cases extended to higher elevations with higher temperatures.

Another study that examined spatial and temporal trends of disease transmission examined tularemia and plague transmission in North America over the late twentieth century and tested hypotheses of climate causation of those shifts (Nakazawa et al. 2007). Specifically, this study tested whether spatial pattern of distributional shifts in these diseases was consistent with known climate changes. Both diseases showed shifts consistent with climate change expectations, although other causal factors probably were acting.

Vibrio cholaerae, which occurs in coastal and estuarine waters, causes human cholera (Lipp, Huq, and Colwell 2002). Cholera is seasonal, with detection and cases concentrated in months when water temperatures are warmest (Lipp, Huq, and Colwell 2002). Warming temperatures may affect transmission to humans directly by accelerating bacteria production, expanding the host plankton species' range, and sea-level rise could introduce estuarine bacteria farther inland (Lipp, Huq, and Colwell 2002). Correlations between cholera cases and warming temperatures have already been found in endemic countries in Asia and Africa (Vezzulli, Colwell, and Pruzzo 2013). Cholera epidemics in South America and Asia have been associated with El Niño events that produce unseasonal and variable weather patterns (Cash et al. 2014).

Wildlife diseases have also begun to undergo climate-related shifts. *Plasmodium* species represent threats not only to humans but also to wildlife, particularly birds native to Hawaii and New Zealand. Avian malaria has rather minor effects on most wild birds, but island bird populations appear to be particularly vulnerable (LaPointe et al. 2012). Although opinion has long held that some species may have high-elevation strongholds above the elevational range of the vector mosquitoes, individual dispersal and movement patterns may nonetheless expose much of even those refugial populations (Guillaumet et al. 2017). Garamszegi (2011) conducted a meta-analysis of avian malaria prevalence among bird species in relation to temperature anomalies over the past 70 years: controlling for species-specific effects, *Plasmodium* prevalence in birds was strongly related to positive temperature anomalies, with a 1°C increase corresponding to two- to threefold increases in prevalence; effects varied by continent, with Africa and Europe exhibiting the most pronounced increases.

Climate change has been implicated in driving shifts or potentially driving future shifts in several other wildlife diseases (Harvell et al. 2002; Lips et al. 2008; Descamps et al. 2016). By a similar token, climate change has been implicated in driving emergence of several crop diseases, such as coffee leaf rust (Bebber et al. 2016), rice leaf blast and sheath blight (Kim and Cho 2016), and wheat leaf rust (Junk et al. 2016), among others. Quite simply, global climate change affects any species that has a significant portion of its life history under conditions tied to climate conditions, which will involve many disease systems.

Highland Malaria in East Africa

Perhaps the best-studied example of climate change effects on disease transmission is the apparent resurgence of malaria in the highlands of East Africa in recent decades. Complex interactions among vectors, hosts, environments, landscapes, and anthropogenic factors (e.g., public health interventions, socioeconomic factors) have made direct linkages between climate change and disease risk challenging to document, leading to debate among researchers regarding a causal role of climate change in recent malaria epidemics. We present this case in detail to illustrate these complexities and to emphasize the interconnected, interdependent nature of the panoply of factors influencing disease transmission.

Highland malaria regions are areas near the maximum elevational limit of malaria transmission, where transmission is unstable, resulting in epidemic or episodic outbreaks. Endemic malaria is not generally present above 1,500 m in East Africa, although cases have been recorded above 2,000 m; a minimum temperature threshold of <18°C exists at which *Plasmodium falciparum* transmission is not stable, which was the mean temperature at ~2,000 m in East Africa in 2006 (Githeko et al. 2006). Periodically, human and environmental conditions become suitable for malaria transmission in these regions, often following anomalous weather events.

Malaria transmission in the western highlands of Kenya can be traced to completion of a railway system in 1901 that moved infected mosquitoes and laborers into the region; epidemics began around 1918, lasting until widespread mosquito control and antimalarial programs were implemented in the 1950s, particularly as regards the dominant vector species, *Anopheles gambiae*. In the late 1980s, however, epidemic malaria reappeared in this region, with outbreaks occurring nearly every year. Understanding how disturbances of different duration could affect transmission equilibrium began to define scientific research in this area: the following is a summary of the ongoing dialogue and debate.

A major factor in the debate is differences in how climate change was characterized in analyses. Although several studies analyzed changes in mean values, others focused on variability and emphasized that climate change is not restricted to simple, general warming trends (Patz et al. 2002). For example, Malakooti, Biomndo, and Shanks (1998) analyzed temporal relationships between malaria epidemics and temperature and precipitation in 1990–1997; results suggested that disease incidence fluctuated but

that significant changes in mean precipitation or temperature had not occurred in the region and thus could not be a causal factor in increasing case rates. Shanks et al. (2002) also argued that climate change was not a significant factor in increased malaria incidence in the 1990s because significant changes in mean ambient temperature and mean precipitation were not found across 1965–1995.

Alternatively, Githeko (2001) investigated effects of temperature and precipitation anomalies on incubation periods of *Plasmodium falciparum* in *Anopheles gambiae s.s.*, and *A. funestus*, creating an epidemic prediction model based on their results. Positive temperature anomalies from mean monthly climate trends were, they found, a precursor to malaria epidemics when mean monthly precipitation was above 150 mm; the authors suggested that monthly temperature data are too general to detect important malaria risk signals.

As studies became more sophisticated, the debate grew. A series of analyses of malaria incidence at four high-elevation sites in Kenya, Uganda, Rwanda, and Burundi highlight differences in inferences that can be obtained from a single climate data set. Hay et al. (2002) investigated meteorological trends, with results indicating no significant changes in mean temperature or vapor pressure during 1911–1995 at any of the study locations, despite warmer and wetter seasonal trends. A separate analysis of the same data set (Patz et al. 2002) found that warming trends did, in fact, correspond to increased malaria incidence at specific sites, noting methods used to downscale the climate data by the original authors as a likely driver of the incongruity in results. Additionally, malaria incidence corresponded more closely to climate anomalies, so the authors emphasized that mean warming trends were not sufficient as indicators of climate change, echoing the results of Githeko (2001).

A third analysis added 5 years of daily climate data to the data set (Pascual et al.

2006). The authors used a nonparametric singular spectrum analysis and a parametric seasonal autoregressive moving average (SARMA) model to examine data for significant warming trends. A significant warming trend of 0.5°C was identified, contradicting Hay et al. (2002). The warming trend began in 1980, coinciding with the rise in malaria incidence at these locations; although other factors (e.g., drug resistance, land-cover change, their interactions) could not be excluded as important factors, changing climate could not be excluded as a contributing factor.

Zhou et al. (2004) analyzed climate variable interactions and their effects on malaria incidence at highland locations in Ethiopia, Kenya, and Uganda. Whereas only two of seven sites exhibited statistically significant increases in mean temperature (1989–1998 vs. 1978–1988), variance in maximum monthly temperature increased significantly at five of seven sites, and all seven sites showed significant positive effects of interactions on malaria transmission.

Results from more recent analyses continue to support significant climate change impacts on malaria incidence in the region. Alonso, Bouma, and Pascual (2011) created a coupled human-malaria model that incorporated temperature and precipitation time series data, again using malaria data from the Kericho tea estate, and identified a nonlinear positive relationship between mean warmer temperatures and more cases from the 1970s to the 1990s. Pascual et al. (2006) implemented a stage-structured biological model for *Anopheles gambiae*, incorporating climate data from the 1980s to the present and found amplification of mosquito population dynamics of a magnitude much greater than changes in the environmental variables, such that subtle climatic changes can have major biological implications.

Clearly, the relationship between malaria and climate change in the highland East Africa is complex, with multiple human, biological, and environmental factors affecting transmission. However, several studies have found statistically significant signals relating climate factors and their interactions to recent increases in malaria incidence in the region. These findings do not negate effects of population movements, drug resistance, land-use change, population immunity, or other factors, but changing climate appears to make a significant contribution.

CONCLUSIONS

Links between climate change and shifts in the distribution, timing, and intensity of disease transmission are complex. The early IPCC reports painted a dark picture in which climate change would somehow "activate" diseases and create a wave of new problems. Twenty or so years later, the picture is more nuanced: whereas some diseases are indeed beginning to shift distributionally, those shifts are invariably subtle and multifactorial. What is more, increases and expansions in transmission may be balanced in other regions, where transmission may be reduced. In sum, then, climate change will indeed change disease transmission patterns, regionally and for particular disease systems, but it may not augment or reduce the overall magnitude of disease transmission worldwide.

Two points are worth highlighting. First, as diseases are the product of interactions among biological species (e.g., pathogens, vectors, hosts), their transmission areas and intensities depend on the ecological niches of those species and their interactions, which will make for considerable complexity in responses to climate change. Second, is more of a comment on human perspectives: early commentaries on effects of climate change on disease were written from a very "developed world" or "northern" perspective, such that transformation of temperate zone climates to subtropical climates at the southern borders of those regions and consequent arrival of "tropical" diseases might seem a dominant expectation of warming climates. The reality, of course, is that people the world over are experiencing these changes, so a more varied suite of responses will be experienced: in some areas, disease transmission will increase, whereas in others it may decrease.

As can be appreciated from the discussions above, however, the picture of climate change effects on disease transmission patterns remains far from clear, and causal chains have not been documented definitively in any disease system. In terms of tools with which to identify and anticipate these changes, mechanistic and correlative models offer complementary perspectives. The few direct comparisons between the two approaches indicate that they converge on similar projections, but clearly much more experimentation, exploration, and comparison is needed to understand the two approaches in greater detail.

The data on which conclusions about climate change and disease are based are crucial, and yet have not been curated or conserved in any detail (Peterson 2008). That is, baseline data to which present and future transmission patterns can be compared to document changes are sorely and woefully lacking. As a consequence, a crucial priority is organization, integration, and documentation of existing primary data regarding disease geography (i.e., data that place occurrences of pathogens, vectors, and hosts in particular places at particular times), and making those data openly available to the broader scientific and public health communities. Such retrospective data curation can be paired with efficient monitoring systems put in place explicitly to detect future changes: permanent monitoring networks can characterize transmission patterns over the long term, being sampled at regular intervals (e.g., years, decades, centuries), such that ideal data exist with which to understand these complex processes.

ACKNOWLEDGMENTS

We thank various agencies for support during the development of this contribution: CONACyT for support of CYR, the C-CHANGE IGERT program for support of LPC, the Egyptian Fulbright Mission Program (EFMP) for support of AMS, and a grant from the Inter-American Institute to Tulane University, which supported (partially) ATP's work.

REFERENCES

Afrane, Y. A., T. J. Little, B. W. Lawson, A. K. Githeko, and G. Y. Yan. 2008. "Deforestation and vectorial capacity of *Anopheles gambiae* mosquitoes in malaria transmission, Kenya." *Emerging Infectious Diseases* 14: 1533–1538. https://doi.org/10.3201/eid1410.070781.

Alonso, D., M. J. Bouma, and M. Pascual. 2011. "Epidemic malaria and warmer temperatures in recent decades in an East African highland." *Proceedings of the Royal Society B* 278: 1661–1669.

Anyamba, A., J.-P. Chretien, J. Small, C. J. Tucker, P. B. Formenty, J. H. Richardson, S. C. Britch, D. C. Schnabel, R. L. Erickson, and K. J. Linthicum. 2009. "Prediction of a Rift Valley fever outbreak." *Proceedings of the National Academy of Sciences* 106: 955–959.

Bebber, D. P., Á. Delgado Castillo, and S. J. Gurr. 2016. "Modelling coffee leaf rust risk in Colombia with climate reanalysis data." *Philosophical Transactions of the Royal Society B* 371: 20150458.

Campbell, L. P., C. Luther, D. Moo-Llanes, J. M. Ramsey, R. Danis-Lozano, and A. T. Peterson. 2015. "Climate change influences on global distributions of dengue and chikungunya virus vectors." *Philosophical Transactions of the Royal Society B* 370: 20140135.

Cash, B. A., X. Rodo, M. Emch, M. Yunus, A. S. Faruque, and M. Pascual. 2014. "Cholera and shigellosis: Different epidemiology but similar responses to climate variability." *PLOS One* 9: e107223.

Costa, J., L. L. Dornak, C. E. Almeida, and A. T. Peterson. 2014. "Distributional potential of the *Triatoma brasiliensis* species complex at present and under scenarios of future climate conditions." *Parasites and Vectors* 7: 238.

Descamps, S., J. Aars, E. Fuglei, K. M. Kovacs, C. Lydersen, O. Pavlova, Å. Ø. Pedersen, V. Ravolainen, and H. Strøm. 2016. "Climate change impacts on wildlife in a High Arctic archipelago—Svalbard, Norway." *Global Change Biology* 23: 490–502.

Epstein, P. R., H. F. Diaz, S. Elias, G. Grabherr, N. E. Graham, W. J. M. Martens, E. Mosley-Thompson, and J. Susskind. 1998. "Biological and physical signs of climate change: Focus on mosquito-borne diseases." *Bulletin of the American Meteorological Society* 79 (3): 409–417. https://doi.org/10.1175/1520-0477(1998)079<0409:Bapsoc>2.0.Co;2.

Garamszegi, L. Z. 2011. "Climate change increases malaria risk in birds." *Global Change Biology* 17: 1751–1759.

Githeko, A. K. 2001. "Predicting malaria epidemics in the Kenyan highlands using climate data: A tool for decision makers." *Global Change and Human Health* 2: 54–63.

Githeko, A. K., J. M. Ayisi, P. K. Odada, F. K. Atieli, B. A. Ndenga, J. I. Githure, and G. Yan. 2006. "Topography and malaria transmission heterogeneity in western Kenya highlands: Prospects for focal vector control." *Malaria Journal* 5: 107.

Guillaumet, A., W. A. Kuntz, M. D. Samuel, and E. H. Paxton. 2017. "Altitudinal migration and the future of an iconic Hawaiian honeycreeper in response to climate change and management." *Ecological Monographs.* https://doi.org/10.1002/ecm.1253.

Harvell, C. D., C. E. Mitchell, J. R. Ward, S. Altizer, A. P. Dobson, R. S. Ostfeld, and M. D. Samuel. 2002. "Climate warming and disease risks for terrestrial and marine biota." *Science* 296: 2158–2162.

Hay, S. I., J. Cox, D. J. Rogers, S. E. Randolph, D. I. Stern, G. D. Shanks, M. F. Myers, and R. W. Snow. 2002. "Climate change and the resurgence of malaria in the East African highlands." *Nature* 415: 905–909.

Hethcote, H. W. 2000. "The mathematics of infectious diseases." *SIAM Review* 42: 599–653.

Junk, J., L. Kouadio, P. Delfosse, and M. El Jarroudi. 2016. "Effects of regional climate change on brown rust disease in winter wheat." *Climatic Change* 135: 439–451.

Kearney, M., W. P. Porter, C. Williams, S. Ritchie, and A. A. Hoffmann. 2009. "Integrating biophysical models and evolutionary theory to predict climatic impacts on species' ranges: The dengue mosquito *Aedes aegypti* in Australia." *Functional Ecology* 23: 528–538. https://doi.org/10.1111/j.1365-2435.2008.01538.x.

Keeling, M. J., and P. Rohani. 2007. *Modeling Infectious Diseases in Humans and Animals.* Princeton University Press.

Kim, K.-H., and J. Cho. 2016. "Predicting potential epidemics of rice diseases in Korea using multi-model ensembles for assessment of climate change impacts with uncertainty information." *Climatic Change* 134: 327–339.

LaPointe, D. A., C. T. Atkinson, and M. D. Samuel. "Ecology and conservation biology of avian malaria." *Annals of the New York Academy of Sciences* 1249: 211–226.

Lindsay, S. W., and M. H. Birley. 1996. "Climate change and malaria transmission." *Annals of Tropical Medicine and Parasitology* 90: 573–588.

Lipp, E. K., A. Huq, and R. R. Colwell. 2002. "Effects of global climate on infectious disease: The cholera model." *Clinical Microbiology Reviews* 15: 757–770.

Lips, K. R., J. Diffendorfer, J. R Mendelson III, M. W. Sears. 2008. "Riding the wave: Reconciling the roles of disease and climate change in amphibian declines." *PLOS Biology* 6: e72.

Liu-Helmersson, J., H. Stenlund, A. Wilder-Smith, and J. Rocklov. 2014. "Vectorial capacity of *Aedes aegypti*: Effects of temperature and implications for global dengue epidemic potential." *PLOS One* 9: e89783.

MacDonald, G. 1957. *The Epidemiology and Control of Malaria.* Oxford University Press.

Malakooti, M. A., K. Biomndo, and G. D. Shanks. 1998. "Reemergence of epidemic malaria in the highlands of western Kenya." *Emerging Infectious Diseases* 4: 671–676.

Martens, P., R. S. Kovats, S. Nijhof, P. de Vries, M. T. J. Livermore, D. J. Bradley, J. Cox, and A. J. McMichael. 1999. "Climate change and future populations at risk of malaria." *Global Environmental Change* 9: S89–S107.

McCarthy, J. J., O. F. Canziani, N. A. Leary, D. J. Dokken, and K. S. White, eds. 2001. *Climate Change 2001: Impacts, Adaptation, and Vulnerability; Contribution of Working Group II to the Third Assessment Report of the Intergovernmental Panel on Climate Change.* Cambridge University Press.

Nakazawa, Y., R. Williams, A. T. Peterson, P. Mead, E. Staples, and K. L. Gage. 2007. "Climate change effects on plague and tularemia in the United States." *Vector Borne and Zoonotic Diseases* 7: 529–540.

Ogden, N. H., M. Radojevic, X. Wu, V. R. Duvvuri, P. A. Leighton, and J. P. Wu. 2014. "Estimated effects of projected climate change on the basic reproductive number of the Lyme disease vector *Ixodes scapularis.*" *Environmental Health Perspectives* 122: 631–638.

Pascual, M., J. A. Ahumada, L. F. Chaves, X. Rodo, and M. Bouma. 2006. "Malaria resurgence in the East African highlands: Temperature trends revisited." *Proceedings of the National Academy of Sciences* 103: 5829–5834.

Patz, J. A., M. Hulme, C. Rosenzweig, T. D. Mitchell, R. A. Goldberg, A. K. Githeko, S. Lele, A. J. McMichael, and D. Le Sueur. 2002. "Climate change: Regional warming and malaria resurgence." *Nature* 420: 627–628.

Peterson, A. T., and J. J. Shaw. 2003. "*Lutzomyia* vectors for cutaneous leishmaniasis in southern Brazil: Ecological niche models, predicted geographic distributions, and climate change effects." *International Journal of Parasitology* 33: 919–931.

Peterson, A. T. 2008. "Improving methods for reporting spatial epidemiologic data." *Emerging Infectious Diseases* 14: 1335–1336.

Peterson, A. T. 2014. *Mapping Disease Transmission Risk in Geographic and Ecological Contexts.* Johns Hopkins University Press.

Peterson, A. T., J. Soberón, R. G. Pearson, R. P. Anderson, E. Martínez-Meyer, M. Nakamura, and M. B. Araújo. 2011. *Ecological Niches and Geographic Distributions.* Princeton University Press.

Rogers, D. J., and S. E. Randolph. 2000. "The global spread of malaria in a future, warmer world." *Science* 289: 1763–1766.

Ross, R. 1904. "The anti-malaria experiment at Mian-Mir." *British Medical Journal* 2: 632–635.

Roy-Dufresne, E., T. Logan, J. A. Simon, G. L. Chmura, and V. Millien. 2013. "Poleward expansion of the white-footed mouse (*Peromyscus leucopus*) under climate change: Implications for the spread of Lyme disease." *PLOS One* 8: e80724.

Siraj, A. S., M. Santos-Vega, M. J. Bouma, D. Yadeta, D. R. Carrascal, and M. Pascual. 2014. "Altitudinal changes in malaria incidence in highlands of Ethiopia and Colombia." *Science* 343: 1154–1158. https://doi.org/10.1126/science.1244325.

Stapp, P., M. F. Antolin, and M. Ball. 2004. "Patterns of extinction in prairie dog metapopulations: Plague outbreaks follow El Nino events." *Frontiers in Ecology and the Environment* 2: 235–240.

Vezzulli, L., R. R. Colwell, and C. Pruzzo. 2013. "Ocean warming and spread of pathogenic vibrios in the aquatic environment." *Microbial Ecology* 65: 817–825.

Watson, R. T., M. C. Zinyowera, R. H. Moss, and Intergovernmental Panel on Climate Change, Working Group II. 1996. *Climate Change, 1995: Impacts, Adaptations, and Mitigation of Climate Change: Scientific-Technical Analyses; Contribution of Working Group II to the Second Assessment Report of the Intergovernmental Panel on Climate Change.* Cambridge University Press.

Yates, T. L., J. N. Mills, C. A. Parmenter, T. G. Ksiazek, R. R. Parmenter, J. R. Vande Castle, C. H. Calisher, S. T. Nichol, K. D. Abbott, J. C. Young, M. L. Morrison, B. J. Beaty, J. L. Dunnum, R. J. Baker, J. Salazar-Bravo, and C. J. Peters. 2002. "The ecology and evolutionary history of an emergent disease: Hantavirus pulmonary syndrome." *BioScience* 52: 989–998.

Zhou, G., N. Minakawa, A. K. Githeko, and G. Yan. 2004. "Association between climate variability and malaria epidemics in the East African highlands." *Proceedings of the National Academy of Sciences* 191: 2375–2380.

PART V

How Can Conservation
and Policy Respond?

CHAPTER TWENTY-TWO

Protected–Area Management and Climate Change

PABLO A. MARQUET, JANETH LESSMANN,
AND M. REBECCA SHAW

INTRODUCTION

Protected-area networks aim to adequately represent and protect biodiversity (Margules and Pressey 2000; Possingham et al. 2006). This objective is seldom fulfilled in practice, as most protected-area networks are inefficient in representing the full diversity of species (e.g., Rodrigues et al. 2004; Tognelli, de Arellano, and Marquet 2008). The challenge is made greater in the context of a changing climate, which decreases the effectiveness of protected areas for biodiversity conservation (Peters and Darling 1985; Hannah et al. 2008). There is a very high risk that representation in protected areas will decrease under changing environmental conditions, as the expected response of many species will be to either shift their geographical distribution to track suitable climates or perish if dispersal abilities or other physical barriers impede their migration (e.g., Thomas and Gillingham 2015; Thomas et al. 2004). Both responses will affect the species' representation within protected areas (Peters and Darling 1985; Hannah et al. 2005). With the high likelihood of species extinction, it is imperative to analyze how existing protected areas networks can be improved, enhanced, and managed in the face of climate change (Plate 8).

We believe that robust protected area design and management in an era of climate change depends on considering four components:

1. The magnitude, rate, timing, and intensity of climate change and its impact on conservation targets

2. Gaps in protected-area design and management

3. A simple theory that can effectively accommodate the biophysical dynamics important to species persistence

4. Innovative protected area design and management approaches that translate the theory into effective practice for conservation

We discuss these components below and provide an assessment of emerging approaches to address the adaptation of protected areas to climate change.

THE MAGNITUDE, RATE, TIMING, AND INTENSITY OF CLIMATE CHANGE AND PROTECTED AREAS

In 2018, we passed an average of 1°C degrees of warming globally, and we expect to reach an average of 2°C warming by 2050. With just 1°C warming, changes in species are documented across the globe (e.g., Parmesan 2006), including distributional shifts, changes in timing of biological phenomena (e.g., flowering, breeding, migration), and decoupling of coevolved species interactions. Species responses have resulted in observed range shifts both poleward and upward along elevational gradients (Parmesan 2006).

Predicting species- and ecosystem-level responses to future warming is difficult, as warming has not been, and will not be, uniform across the globe. As we head to a 2°C increase in global average temperature by midcentury, we expect an increase of more than 4°C in some geographies and less than 0.5°C in others (IPCC 2014). In general, we expect greater warming at the poles than at the equator. This variation in the magnitude of change will manifest in significant variations in the rate of change. On average, the velocity of the climate change (i.e., the instantaneous local velocity along Earth's surface needed to maintain a constant temperature) is 0.42 km each

year (Loarie et al. 2009); that rate can be much higher, depending on the latitude, altitude, and topography of a given location. The velocity of change will be lowest in mountainous biomes, as a result of topographic variability, and higher in lowlands. Lowland species will need to travel greater distances and at faster rates to track their suitable climates than will their mountain-dwelling counterparts. This suggests that mountainous regions may be critically important to maintaining biodiversity in some regions by facilitating species movement to suitable climate.

Recent studies in paleoecology suggest that species occupying discrete microclimates played an important role in species responses to rapid climate change during the Last Glacial Maximum (Stewart et al. 2010). The effective role of microrefugia, holdouts, and stepping stones in species-range shift dynamics depends on the magnitude and velocity of climate change (Hannah 2014), as well as direct and indirect impacts of extreme climatic events. Even so, it is possible that some species will be able thrive in protected areas if they are able to find microenvironments that will allow them to adapt to changing climate conditions, even if only temporarily (Thomas and Gillingham 2015), thus buying time for more significant management intervention. The possibility for microrefugia or temporary stepping stones within and between existing protected areas will be very important in the successful adaptation of some species.

Just as we are beginning to fully consider the potential conservation value of microclimates for facilitating biodiversity persistence, we are beginning to consider the potential consequences of the less predictable extreme events for the protection of biodiversity. As is evident by the increasing frequency and magnitude of severe weather events worldwide, climate change is characterized not simply by shifts in mean global temperature and precipitation alone but also by changes in the pattern of extreme climatic and climate-induced events. To challenge the survival of species even further, extreme climatic events will deliver punctuated impacts in time and space that are likely to magnify the influence of average climatic trends and other stressors. That is, the character and severity of impacts from climate extremes depend not only on the extremes themselves but also on the background exposure and vulnerability of the species.

GAPS IN PROTECTED-AREA DESIGN AND MANAGEMENT

Effective policy, design, and management of protected areas in the context of climate change require relevant science to inform decision making. In the past decade there has been a surge in the number of scientific articles focusing on the effects of climate change on protected areas, which usually provide recommendations for management and design of protected areas to support adaptation (Heller and Zavaleta 2009). Most of these studies do not address the implications of the full range of direct and indirect climate-induced impacts; therefore, they have not fully explored the range of protected-area design and management approaches to facilitate species persistence in the future. The scientific literature has important gaps, which limit the applicability of the design and management approaches into different situations and regions across the world. These gaps include the science-management gap, the geographic gap, and the ecosystem gap.

The Science-Management Gap

Although many studies focus on basic scientific issues emphasizing theoretical and/or general principles of climate change biology, only a few studies provide practical guidance for protected areas design and management and adaptation planning (Heller and Zavaleta 2009). There is a pau-

city of information about the desirability, feasibility, and effectiveness of adaptation and mitigation options (Lemieux et al. 2011; Hannah 2008). In addition, much of this research aimed to solve ecological and biological problems (e.g., loss of species' representation) caused by climate change in protected areas, whereas impacts on local communities or their participation in the management solution are rarely discussed. It is the internal institutional publications, or the "gray" literature, that are beginning to deliver an integrated vision (i.e., including social and political aspects) of the role of protected areas under climate change (e.g., Hansen et al. 2003; Dudley et al. 2010).

The Geographic Gap

There are gaps in the geographic focus of studies on the impact, vulnerability, and adaptation of protected areas to climate change, with a significant bias toward North America and Europe (Heller and Zavaleta 2009). The urgency to address the impact of climate change on biodiversity is important worldwide, but the gap in information to support action is much higher in developing countries, where species extinction risk rates are high and financial and technical capacities can be very low (Hannah et al. 2002). Although some of the knowledge that is generated in the scientific literature is transferable to multiple regions, it is necessary to address the geographic gap and focus future research on less studied and more vulnerable regions to attend their specific contexts and needs. For example, the consequences of climate change for the representation of species in protected areas in lowland tropical rainforest are poorly documented (Thomas and Gillingham 2015). Furthermore, several strategies developed for protected areas under climate change are "data hungry," demanding high-quality information about geographical distribution, vulnerability, and dispersal capacity of the target species. Clearly, these strategies are very difficult to implement in protected areas located in regions of low capacity where detailed information about species is scarce and species inventories are incomplete.

The Ecosystem Gap

Finally, there are gaps in the type of the ecosystems studied (e.g., terrestrial, marine, freshwater). Most of the research on climate change effects on biodiversity and their implications for conservation planning has concentrated on the terrestrial and marine systems, whereas studies that consider freshwater diversity have been limited, despite the high vulnerability of freshwater diversity to climate change (Bush et al. 2014). The combined species distribution-climate models assessments for freshwater needs to be improved rapidly to adequately address freshwater conservation planning that incorporates both biodiversity and human needs in the face of climate change. To do this, it will be important to understand the effects of climate change on the unique properties of freshwater systems, such as regime flow and intercatchment connectivity (Bush et al. 2014). A broader analysis of all ecosystems is needed because the effects of climate change and requirements for biodiversity conservation differ across systems.

THEORY OF CLIMATE CHANGE AND PROTECTED-AREA DESIGN AND MANAGEMENT

One of the pressing challenges to protecting biodiversity in the future is the extent to which protected-area networks will be able to retain their biodiversity, provide connectivity for migrating species, and serve as refuges for new ones. In general, this amounts to providing an answer to the following questions: If a given protected area possesses S_0 species today, how many species will it likely have by the end of the century? How different will these species be from the original assemblage? Will this new assemblage maintain key ecosystem services? The complexity of the answers to these simple questions illustrates the vexing nature of the issue at hand. First, we need to explore which theories of species persistence can inform our answer. There are two with which to begin, island biogeography and metapopulation theory, complemented by a third emerging one, graph theory or network theory. These theories tell us that the expected number of species in the protected area will depend on the rate at which species colonize (through in-migration or speciation) and leave (through emigration or extinction) the protected areas. While the former rate depends on the size of the pool and the degree of isolation of the protected area or how reachable it is, the latter depends on area, shape, topography, and condition of the surrounding agricultural, urban, and natural land use matrix. In particular, it will depend on whether the matrix will allow for the species to leave the protected area, as many species are very sensitive to movement across matrix habitats, making it likely that they will become increasingly isolated within protected areas. This is, by no means, the whole story as rates will also depend on the number of species already present in the protected area, their interaction, and the ability of protected areas to support viable populations for different species.

In Figure 22.1 we show the relationship between body size and home-range size for marine and terrestrial species and, superimposed over it, the median size of marine and terrestrial protected areas. It is striking that most protected areas are smaller than the area required by an average individual of most marine and approximately 50 percent of the terrestrial vertebrate species considered. Further, since the median size of minimum viable populations (MVPs) is in the range of 1,000 to 4,000 individuals (Brook, Traill, and Bradshaw 2006), Figure 22.1 implies that very few protected areas will be able to support species over the long term unless connectivity is enhanced and

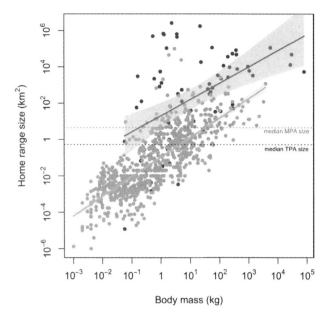

Figure 22.1. Relationship between the scale of habitat use per individual in marine and terrestrial species. The dotted lines demarcate the current median size of all marine (MPA) and terrestrial (TPA) protected areas. (Redrawn from McCauley et al. 2015.) The median in terrestrial PAs was calculated by considering all PAs described in the World Protected Area Data Base (IUCN and UNEP-WCMC, 2015) that do not include marine habitats.

the matrix in between protected areas becomes more biodiversity friendly. This figure highlights the deep disconnect between the scale at which species interact with their environments and how we manage land use to enhance species persistence (McCauley et al. 2015). Thus, everything else being equal, even if a well-connected network of protected areas will do better than a small isolated one, the average size of protected areas is so small that many species risk extinction if the hostility of the surrounding matrix increases as a result of climate change.

Available evidence shows that lack of connectivity in fragmented landscapes has negative impacts upon biodiversity conservation and functioning and that species do expand their ranges out of or into protected areas while responding to climate change (Araújo et al. 2004; Thomas et al. 2012). This suggests that connectivity is essential in ensuring resilience. However, it

should be borne in mind that it is very difficult to assess the number of species that have currently failed to expand because of lack of connectivity between protected areas and surrounding habitats, and so it is difficult to assess its importance in comparative terms, and improving connectivity may not be the best option in all places and times. Indeed, the usual strategy of restoring or generating habitat corridors within landscapes may have unintended negative effects associated with the spread of pathogens, disturbances (e.g., fire), or increasing synchrony in local population fluctuations (Simberloff and Cox 1987), thus leading to metapopulation extinction due to regional stochasticity. Investments in connectivity should be analyzed in the context of potential positive and negative impacts as well as the expected benefits that may be accrued by investing in alternative actions to build resilience, such as increasing the number, area, habitat quality, and heterogeneity of protected areas.

When it comes to the theoretical understanding of climate change impacts on species, it is essential to bear in mind that the problem of persistence is a problem of scale commensurability. That is, between the rate at which a landscape or habitat changes and the changes that a species can tolerate or adapt to, as the saying goes, it takes two to tango. Habitats change because of changes in their physical and chemical properties but also because of changes in the biological processes and species that are found in those habitats; under climate change both rates will increase. Unfortunately, little theory has been developed to understand the issue of persistence under climate change, or in dynamic landscapes, and how it depends on the climate change itself but in interaction with the scale at which species perceive and interact with their environments (but see Marquet and Velasco Hernandez 1997). Similarly, the impact of climate change on networks of interacting species is similarly challenging and difficult to tackle, first because interactions will likely change before species go extinct, and second because multiple anthropogenic drivers affect them at the same time, which undermines the ability to make predictions. In recent years, simple static approaches (e.g., Araújo et al. 2011) and dynamic models have been developed to make predictions on the impact of climate change on networks of interacting species. However, the complexity of the issue still defies the identification of first principles (e.g., Marquet et al. 2014) that can be translated into simple guidelines for management. In many ways, this is an empirical problem as most networks are not monitored in time or at different locations in space, which hinders our capacity to understand how they change and how resilient they are. The challenge is also theoretical, as we do not fully understand how richness, interaction type, and interaction strengths affect stability and resilience; nor do we know how to include these principles in conservation planning and the design of protected-area networks. Simple theory and empirical evidence show that, in general, species diversity and population diversity can bring, through portfolio effects, stability to ecosystems and the services they provide (e.g., Hilborn et al. 2003)

and help buffer the impact of climate fluctuation (Schindler et al. 2010).

INNOVATIVE APPROACHES TO ADAPT PROTECTED AREAS TO CLIMATE CHANGE

Several actions can be taken to improve the role of protected area networks in protecting biodiversity under climate change. These range from maintaining current conservation successes to innovative responses specific to climate change.

Ensure the Permanence and Quality of Existing Protected Areas

Protected areas are an important tool to conserve biodiversity and adapt to climate change. The current protected-area network provides a platform from which to build. It is critical that the existing protected area network be maintained and strengthened as current trends and future scenarios of land use may threaten their persistence and viability. Probably most important, shifting climate conditions will affect where food and energy crops can be grown to meet the demands of a growing human population, thus creating the potential for land-use conflicts. Hannah et al. (2013) studied the changing geographic suitability of regions for wine grape production under a changing climate and highlighted the potential for a shift into existing protected areas and important matrix habitat for species of concern across the globe. Predicted shifts in agricultural suitability such as these can create competition between conservation and food production for land and will have a significant impact on protected areas and habitat connectivity for wide-ranging species.

Expand or Modify Protected Areas and Networks

As species ranges shift in response to changes in climatic factors, many are projected to move outside the boundaries of existing protected lands and waters. There will need to be the modification of boundaries of existing protected areas, or establishment of new protected areas to cover lands and waters likely to be important to these species in the future. Even with new protected areas, it is predicted that by 2050, 20 percent of species will have disappeared from reserves (Hannah et al. 2007); thus, it is important to keep in mind that the function of protected areas will not continue under business-as-usual emissions trajectories (Thomas and Gillingham 2015), and more innovative and aggressive management of species and protected areas will be necessary.

Use Protected Areas to Enhance Connectivity between Present and Future Populations

Regardless of how large an individual protected area is, if it is an island of habitat surrounded by inhospitable habitat, it will be difficult or impossible for species to migrate as climate shifts. Enhancing connectivity with special attention to microclimate stepping stones among protected areas to enable species movements should be employed to ensure resilience of protected areas over time (e.g., Belote et al. 2016). For climate change it is especially important to connect present and future populations of species, which will not often be the same as connecting protected areas. Although connections between protected areas for large mammals may provide connectivity for climate change, connecting present and future populations of plants and nonvertebrates may involve different areas and be less area demanding.

Protect Enduring Geophysical Features

Protecting geophysical settings will be important for protecting species diversity in the future (Anderson and Ferree 2010; Brost and Beier 2012). The relationship between species diversity and geophysical setting (elevation, slope, substrate, and aspect) sug-

gests that protecting the "ecological stage" for processes as well as preserving particular "actors" (species, ecosystems) may be important in protecting species diversity over time.

Assist Migration

Assisted migration is a strategy designed to help species overcome dispersal barriers associated with tracking climate shifts by physically relocating them outside their historical range to climatic zones considered suitable. The outcome of assisted migration is not qualitatively different from enhancing habitat connectivity (Lawler and Olden 2011), and although this may be the only strategy that will be effective for achieving the persistence of some species, many are opposed to it given the poor performance of past efforts (Ricciardi and Simberloff 2009).

Enhance Resilience

Resilience has emerged as perhaps the dominant paradigm for addressing climate change impacts on protected areas. Resilience can refer to the ability of a system to return to its original state following a perturbation, to maintain functionality during the course of a transition, or to self-organize in maintaining basic system traits and functioning (Folke et al. 2004). In the practice of biodiversity conservation, the concept emphasizes the notion that robust protected-area networks will be better able to ensure persistence of species across the landscape in the context of climate change. How to do this requires a multifaceted approach including abating not only climate change but also other global change drivers and fostering legislation and education to better preserve the seminatural matrix or landscapes where human uses coexist with protected areas.

Create Stepping Stones

Existing conservation tools for creating protected areas like purchase fee or ease-

ments, though valuable, are static and insufficient to maintain biodiversity over
time. To encourage investment at the size
and rate required, we must develop conservation tools that are adaptive and dynamic over time, in response to the changing resource. New conservation tools such
as habitat exchange or bird return are designed to achieve spatially and temporally
significant landscape-scale conservation
through participation from private landowners. The Habitat Exchange allows de-
velopers to offset their impacts on habitat
and species by purchasing credits generated through conservation actions of private landowners. The Birds Returns program typically pays farmers to flood their
fields at certain levels and certain times of
the year to provide habitat for birds. Both
of these are dynamic, incentive-based approaches that provide economic opportunity for private landowners while delivering scientifically robust benefits to species
and habitat.

Enhance Synergies between Protected-Area Design and Management to Mitigate Climate Change

Despite the described negative effects of
climate change on natural reserve systems, protected areas may be an effective
response to climate change mitigation
and adaptation (Thomas and Gillingham
2015; Hannah et al. 2007; Hole et al. 2009).
Protected areas contribute directly to climate change mitigation by avoiding deforestation and thus reducing greenhouse
gas emissions into the atmosphere, by
sequestering carbon and by offering opportunities for restoration (Soares-Filho et
al. 2010). Ecosystems represented within
global terrestrial protected areas store over
312 GtC, or 15 percent of the terrestrial
carbon stock (Campbell et al. 2008). The
protection and management offered on the
ground by these reserves will be essential
to reduce carbon fluxes.

Enhance Synergies between Protected-Area Design and Management for the Adaptation and Resilience of Human Communities to Climate Change

Some of the impacts of climate change are
unavoidable, but protected areas may offer
options for biodiversity and society adaptation to these impacts (Heller and Zavaleta
2009; Welch 2005). By helping maintain
natural ecosystems and enhancing their resilience to climate change, protected areas
can reduce the risk of disasters such as hurricanes, droughts, floods, and landslides,
which will likely increase as a consequence
of increasing temperature and deforestation. In the same way, the protection of
natural ecosystems can contribute to securing the supply of ecosystem services to local
communities and providing additional in-
surance against the predicted instability of
agriculture, fisheries, and water resources.

CONSERVATION IN A CHANGING CLIMATE

Climate change inspires us to rethink the
role of protected areas in biodiversity conservation and challenges us to systematically
think through their design and modification
in the context of realistic climate change
(Shaw et al. 2012). Many are doing just
that. Several researchers have identified the
need for additional and connected reserves
to maintain current levels of representation and to compensate for future potential
losses (Araújo et al. 2004, 2011; Hannah et
al. 2007; Shaw et al. 2012). Some are beginning to shift to thinking about process-
oriented conservation goals, such as maintaining ecological processes that sustain
the ecosystem services, migration patterns,
species flow, and/or gene flow (Anderson
and Ferree 2010; Stein and Shaw 2013),
which may promote species diversification
as well as the emergence of novel ecosystems. Novel ecosystems are combinations
and relative abundances of species that have

not previously occurred (Seastedt, Hobbs, and Suding 2008). Although challenging for managers under current conservation goals in which we attempt to restore such systems back to natural conditions, novel ecosystems can be managed to promote biodiversity and the delivery of key ecosystem services (Jackson and Hobbs 2009). Given the magnitude, rate, timing, and intensity of projected changes, some existing goals may no longer be relevant or attainable, and the role of protected areas may need to change. What will be needed is a commitment to undertaking the challenging and psychologically demanding task of reconsidering our conservation goals and the role of protected areas in achieving them.

REFERENCES

Anderson, Mark G., and Charles E. Ferree. 2010. "Conserving the stage: Climate change and the geophysical underpinnings of species diversity." PLOS One 5 (7): e11554. https://doi.org/10.1371/journal.pone.0011554.

Araújo, Miguel B., Mar Cabeza, Wilfried Thuiller, Lee Hannah, and Paul H. Williams. 2004. "Would climate change drive species out of reserves? An assessment of existing reserve-selection methods." Global Change Biology 10 (9): 1618–1626. https://doi.org/10.1111/j.1365-2486.2004.00828.x.

Araújo, Miguel B., Alejandro Rozenfeld, Carsten Rahbek, and Pablo A. Marquet. 2011. "Using species co-occurrence networks to assess the impacts of climate change." Ecography 34 (6): 897–908. https://doi.org/10.1111/j.1600-0587.2011.06919.x.

Belote, R. Travis, Matthew S. Dietz, Brad H. McRae, David M. Theobald, Meredith L. McClure, G. Hugh Irwin, Peter S. McKinley, Josh A. Gage, and Gregory H. Aplet. 2016. "Identifying corridors among large protected areas in the United States." PLOS One 11 (4): e0154223. https://doi.org/10.1371/journal.pone.0154223.

Brook, Barry W., Lochran W. Traill, and Corey J. A. Bradshaw. 2006. "Minimum viable population sizes and global extinction risk are unrelated." Ecology Letters 9 (4): 375–382. https://doi.org/10.1111/j.1461-0248.2006.00883.x.

Brost, Brian M., and Paul Beier. 2012. "Use of land facets to design linkages for climate change." Ecological Applications 22 (1): 87–103.

Bush, Alex, Virgilio Hermoso, Simon Linke, David Nipperess, Eren Turak, and Lesley Hughes. 2014. "Freshwater conservation planning under climate change:

Demonstrating proactive approaches for Australian Odonata." Journal of Applied Ecology 51 (5): 1273–1281. https://doi.org/10.1111/1365-2664.12295.

Campbell, Alison, Valerie Kapos, Igor Lysenko, Jorn Scharlemann, Barney Dickson, Holly Gibbs, Matthew Hansen, and Lera Miles. 2008. Carbon Emissions from Forest Loss in Protected Areas. UNEP World Conservation Monitoring Centre.

Dirzo, Rodolfo, Hillary S. Young, Mauro Galetti, Gerardo Ceballos, Nick J. B. Isaac, and Ben Collen. 2014. "Defaunation in the Anthropocene." Science 345 (6195): 401–406. https://doi.org/10.1126/science.1251817.

Dudley, Nigel, Sue Stolton, Alexander Belokurov, Linda Krueger, Nik Lopoukhine, Kathy MacKinnon, Trevor Sandwith, and Nik Sekhran. 2010. Natural Solutions: Protected Areas Helping People Cope with Climate Change. World Commission of Protected Areas of the International Union for Conservation of Nature (IUCN-WCPA), Nature Conservancy (TNC), UN Development Programme (UNDP), Wildlife Conservation Society (WCS), World Bank, and World Wide Fund for Nature (WWF).

Folke, Carl, Steve Carpenter, Brian Walker, Marten Scheffer, Thomas Elmqvist, Lance Gunderson, and C. S. Holling. 2004. "Regime shifts, resilience, and biodiversity in ecosystem management." Annual Review of Ecology Evolution and Systematics 35: 557–581. https://doi.org/10.1146/annurev.ecolsys.35.021103.105711.

Franks, Steven J., and Ary A. Hoffmann. 2012. "Genetics of climate change adaptation." Annual Review of Genetics 46: 185–208. https://doi.org/10.1146/annurev-genet-110711-155511.

Hannah, L., G. F. Midgley, T. E. Lovejoy, W. J. Bond, M. Bush, J. C. Lovett, D. Scott, and F. I. Woodwards. 2002. "Conservation of biodiversity in a changing climate." Conservation Biology 16: 264–268. https://doi.org/10.1046/j.1523-1739.2002.00465.x.

Hannah, Lee, Guy Midgley, Sandy Andelman, Miguel Araújo, Greg Hughes, Enrique Martinez-Meyer, Richard Pearson, and Paul Williams. 2007. "Protected area needs in a changing climate." Frontiers in Ecology and the Environment 5 (3): 131–138. https://doi.org/10.1890/1540-9295(2007)5[131: PANIAC]2.0.CO;2.

Hannah, Lee, Patrick R. Roehrdanz, Makihiko Ikegami, Anderson V. Shepard, M. Rebecca Shaw, Gary Tabor, Lu Zhi, Pablo A. Marquet, and Robert J. Hijmans. 2013. "Climate change, wine, and conservation." Proceedings of the National Academy of Sciences 110 (17): 6907–6912. https://doi.org/10.1073/pnas.1210127110.

Hannah, Lee, Lorraine Flint, Alexandra D. Syphard, Max A. Moritz, Lauren B. Buckley, and Ian M. McCullough. 2014. "Fine-grain modeling of species' response to climate change: Holdouts, stepping-stones, and microrefugia." Trends in Ecology & Evolution 29 (7): 390–397.

Hansen, L. J., J. L. Biringer, and J. R. Hoffman. 2003. In Buying Time: A User's Manual for Building Resistance and Resilience to Climate Change in Natural Systems. World Wildlife Fund.

Heller, Nicole E., and Erika S. Zavaleta. 2009. "Biodiversity management in the face of climate change:

A review of 22 years of recommendations." *Biological Conservation* 142 (1): 14–32. https://doi.org/10.1016/j .biocon.2008.10.006.

Hilborn, Ray, Thomas P. Quinn, Daniel E. Schindler, and Donald E. Rogers. 2003. "Biocomplexity and fisheries sustainability." *Proceedings of the National Academy of Sciences* 100 (11): 6564–6568. https://doi.org/10.1073/pnas .1037274100.

Hole, David G., Stephen G. Willis, Deborah J. Pain, Lincoln D. Fishpool, Stuart H. M. Butchart, Yvonne C. Collingham, Carsten Rahbek, and Brian Huntley. 2009. "Projected impacts of climate change on a continent-wide protected area network." *Ecology Letters* 12 (5): 420–431. https://doi.org/10.1111/j.1461-0248 .2009.01297.x.

IPCC. 2014. *Climate Change 2014: Synthesis Report. Contribution of Working Groups I, II and III to the Fifth Assessment Report of the Intergovernmental Panel on Climate Change*, ed. Core Writing Team, R. K. Pachauri, and L. A. Meyer. IPCC.

Jackson, Stephen T., and Richard J. Hobbs. 2009. "Ecological restoration in the light of ecological history." *Science* 325 (5940): 567–569. https://doi.org/10.1126/ science.1172977.

Lawler, Joshua J., and Julian D. Olden. 2011. "Reframing the debate over assisted colonization." *Frontiers in Ecology and the Environment* 9 (10): 569–574. https://doi.org/ 10.1890/100106.

Lawson, Callum R., Jonathan J. Bennie, Chris D. Thomas, Jenny A. Hodgson, and Robert J. Wilson. 2014. "Active management of protected areas enhances metapopulation expansion under climate change." *Conservation Letters* 7 (2): 111–118. https://doi.org/10.1111/conl .12036.

Lemieux, Christopher J., Thomas J. Beechey, Daniel J. Scott, and Paul A. Gray. 2011. "The state of climate change adaptation in Canada's protected areas sector." *Canadian Geographer/Geographe Canadien* 55 (3): 301–317. https://doi.org/10.1111/j.1541-0064.2010.00336.x.

Lenoir, J., J. C. Gégout, P. A. Marquet, P. de Ruffray, and H. Brisse. 2008. "A significant upward shift in plant species optimum elevation during the 20th century." *Science* 320 (5884): 1768–1771. https://doi.org/10 .1126/science.1156831.

Loarie, Scott R., Philip B. Duffy, Healy Hamilton, Gregory P. Asner, Christopher B. Field, and David D. Ackerly. 2009. "The velocity of climate change." *Nature* 462 (7276): 1052–1055. https://doi.org/10.1038/ nature08649.

Margules, Chris R., and Robert L. Pressey. 2000. "Systematic conservation planning." *Nature* 405: 243–253. https://doi.org/10.1038/35012251.

Marquet, Pablo A., Andrew P. Allen, James H. Brown, Jennifer A. Dunne, Brian J. Enquist, James F. Gillooly, Patricia A. Gowaty, Jessica L. Green, John Harte, Steve P. Hubbell, James O'Dwyer, Jordan G. Okie, Annette Ostling, Mark Ritchie, David Storch, and Geoffrey B. West. 2014. "On theory in ecology." *BioScience* 64 (8): 701–710. https://doi.org/10.1093/biosci/biu098.

Marquet, Pablo A., and Jorge X. Velasco Hernandez. 1997. "A source-sink patch occupancy metapopulation model." *Revista chilena de historia natural* 70 (3): 371–380.

McCauley, Douglas J., Malin L. Pinsky, Stephen R. Palumbi, James A. Estes, Francis H. Joyce, and Robert R. Warner. 2015. "Marine defaunation: Animal loss in the global ocean." *Science* 347 (6219). https://doi.org/ 10.1126/science.1255641.

Parmesan, Camille. 2006. "Ecological and evolutionary responses to recent climate change." *Annual Review of Ecology Evolution and Systematics* 37: 637–669. https://doi .org/10.1146/annurev.ecolsys.37.091305.110100.

Peters, Robert L., and Joan D. S. Darling. 1985. "The greenhouse effect and nature reserves." *BioScience* 35 (11): 707–717. https://doi.org/10.2307/1310052.

Pressey, Robert L., Mar Cabeza, Matthew E. Watts, Richard M. Cowling, and Kerrie A. Wilson. 2007. "Conservation planning in a changing world." *Trends in Ecology and Evolution* 22 (11): 583–592. https://doi.org/10.1016/ j.tree.2007.10.001.

Ricciardi, Anthony, and Daniel Simberloff. 2009. "Assisted colonization is not a viable conservation strategy." *Trends in Ecology and Evolution* 24 (5): 248–253. https://doi.org/10.1016/j.tree.2008.12.006.

Rodrigues, Ana S. L., Sandy J. Andelman, Mohamed I. Bakarr, Luigi Boitani, Thomas M. Brooks, Richard M. Cowling, Lincoln D. C. Fishpool, Gustavo A. B. da Fonseca, Kevin J. Gaston, Michael Hoffmann, Janice S. Long, Pablo A. Marquet, John D. Pilgrim, Robert L. Pressey, Jan Schipper, Wes Sechrest, Simon N. Stuart, Les G. Underhill, Robert W. Waller, Matthew E. J. Watts, and Xie Yan. 2004. "Effectiveness of the global protected area network in representing species diversity." *Nature* 428: 640–643. https://doi.org/10.1038/ nature02422.

Schindler, Daniel E., Ray Hilborn, Brandon Chasco, Christopher P. Boatright, Thomas P. Quinn, Lauren A. Rogers, and Michael S. Webster. 2010. "Population diversity and the portfolio effect in an exploited species." *Nature* 465 (7298): 609–612. https://doi.org/10 .1038/nature09060.

Seastedt, Timothy R., Richard J. Hobbs, and Katharine N. Suding. 2008. "Management of novel ecosystems: Are novel approaches required?" *Frontiers in Ecology and the Environment* 6 (10): 547–553. https://doi.org/10.1890/ 070046.

Shaw, M. Rebecca, Kirk Klausmeyer, D. Richard Cameron, Jason Mackenzie, and Patrick Roehrdanz. 2012. "Economic costs of achieving current conservation goals in the future as climate changes." *Conservation Biology* 26 (3): 385–396. https://doi.org/10.1111/j.1523 -1739.2012.01824.x.

Shaw, M. Rebecca, Linwood Pendleton, D. Richard Cameron, Belinda Morris, Dominique Bachelet, Kirk Klausmeyer, Jason MacKenzie, David R. Conklin, Gregory N. Bratman, James Lenihan, Erik Haunreiter, Christopher Daly, and Patrick R. Roehrdanz. 2011. "The impact of climate change on California's ecosystem services." *Climatic Change* 109 (1): 465–484. https://doi.org/10.1007/s10584-011-0313-4.

Simberloff, Daniel, and James Cox. 1987. "Consequences and costs of conservation corridors." *Conservation Biology* 1 (1): 63–71. https://doi.org/10.1111/j.1523-1739.1987.tb00010.x.

Soares-Filho, Britaldo, Paulo Moutinho, Daniel Nepstad, Anthony Anderson, Hermann Rodrigues, Ricardo Garcia, Laura Dietzsch, Frank Merry, Maria Bowman, Leticia Hissa, Rafaella Silvestrini, and Claudio Maretti. 2010. "Role of Brazilian Amazon protected areas in climate change mitigation." *Proceedings of the National Academy of Sciences* 107 (24): 10821–10826. https://doi.org/10.1073/pnas.0913048107.

Staudinger, Michelle D., Nancy B. Grimm, Amanda Staudt, Shawn L. Carter, F. Stuart Chapin III, Peter Kareiva, Mary Ruckelshaus, and Bruce A. Stein. 2012. *Impacts of Climate Change on Biodiversity, Ecosystems, and Ecosystem Services: Technical Input to the 2013 National Climate Assessment.* Cooperative Report of the 2013 National Climate Assessment.

Stein, Bruce A., and M. Rebecca Shaw. 2013. "Biodiversity conservation in a climate-altered future." In *Successful Adaptation to Climate Change: Linking Science and Practice in a Rapidly Changing World*, ed. S. C. Moser and M. T. Boykoff, 50–66. Routledge.

Stewart, John R., Adrian M. Lister, Ian Barnes, and Love Dalén. 2010. "Refugia revisited: Individualistic responses of species in space and time." *Proceedings of the Royal Society B: Biological Sciences* 277 (1682): 661–671. https://doi.org/10.1098/rspb.2009.1272.

Thomas, Chris D., and Phillipa K. Gillingham. 2015. "The performance of protected areas for biodiversity under climate change." *Biological Journal of the Linnean Society* 115: 718–730. https://doi.org/10.1111/bij.12510.

Thomas, Chris D., Alison Cameron, Rhys E. Green, Michel Bakkenes, Linda J. Beaumont, Yvonne C. Collingham, Barend F. N. Erasmus, Marinez Ferreira de Siqueira, Alan Grainger, Lee Hannah, Lesley Hughes, Brian Huntley, Albert S. van Jaarsveld, Guy F. Midgley, Lera Miles, Miguel A. Ortega-Huerta, A. Townsend Peterson, Oliver L. Phillips, and Stephen E. Williams. 2004. "Extinction risk from climate change." *Nature* 427 (6970): 145–148. https://doi.org/10.1038/nature02121.

Tognelli, Marcelo F., Pablo I. Ramirez de Arellano, and Pablo A. Marquet. 2008. "How well do the existing and proposed reserve networks represent vertebrate species in Chile?" *Diversity and Distributions* 14 (1): 148–158. https://doi.org/10.1111/j.1472-4642.2007.00437.x.

Turner, Bryan M. 2009. "Epigenetic responses to environmental change and their evolutionary implications." *Philosophical Transactions of the Royal Society B: Biological Sciences* 364 (1534): 3403–3418. https://doi.org/10.1098/rstb.2009.0125.

Extinction Risk from Climate Change

Guy Midgley and Lee Hannah

Extinction risk from climate change first made a major impact on international climate change policy dialogue in 2004, following publication of the first global estimates of possible species losses due to climate change (Thomas et al. 2004). Following the publication of these initial estimates, extinction risk was discussed in climate change debates in the House of Lords and was the subject of US Senate hearings. While these and other, more multilateral debates (as in the Convention for Biological Diversity and UN Framework Convention on Climate Change) finally began to grapple with this issue, the concept of extinction risk itself had much deeper roots in climate change biology.

One of the most important insights from paleoecology is that species' ranges shift in response to climate change. Such shifts have demonstrably been the overwhelming response of both plant and animal species to climate change for several millions of years. The process can even lead to speciation as ranges are fragmented (e.g., Hewitt 2000). However, these range shifts can also result in extinction when species ranges run into barriers or are in competition with new species, or they can involve catastrophic loss of range. In a corollary to this, regions that experienced relatively less recent paleoclimate change tend to have higher biodiversity (Dynesius and Janssen 2013).

The lesson from paleoecology is that context matters. Major extinction events in the past 500 million years are associated, directly or indirectly, with climate change. But some major climatic events are not associated with megaextinction, although they may have driven major shifts in dominant taxa. The Paleocene-Eocene Thermal Maxima (a rapid warming period), for instance, is associated with the rise of primates and many modern mammals but with few global extinctions. More recently, there was a wave of extinctions as the world cooled into the ice ages but few extinctions associated with the glacial and interglacial transitions themselves. An explanation for this is that the regime shift caused extinctions, but then the remaining species were preselected to cope with cooler conditions and climate transitions. But all these Paleo events offer at best hints of what the future may hold. There is no exact past analog for the warming that is now unfolding.

Beginning to quantify the possible scale of the extinction problem associated with human-caused climate change was possible only with the evolution of species distribution models (ecological niche models) prior to the 2004 landmark paper. Large-scale modeling of thousands of species across continental or subcontinental domains emerged in the late 1990s, enabling authors to assess extinction risk and entire floras and faunas. A group of these researchers pooled several large-scale modeling efforts to produce the 2004 estimates.

Although the first estimates of extinction risk from climate change attracted much public policy attention, they also garnered substantial scientific criticism. However, the general magnitude of extinction risk in these initial estimates has now been corroborated through several independent studies (Malcolm et al. 2006; Sinervo 2010; Urban 2015) and in theory (Deutsch et al. 2008; see Plate 4). There is now little expert disagreement that climate change poses a major extinction risk, perhaps to hundreds of thousands of species or more around the world, which appears to be strongly dependent on the rate of climate change (e.g., O'Neill et al. 2017).

The Intergovernmental Panel on Climate Change (IPCC) published a consensus estimate of species extinction (Fischlin et al. 2007) that used several of the studies included in the 2004 Thomas study, together with several studies published subsequently. It estimated that "approximately 20 to 30 percent of plant and animal species assessed so far (in an unbiased sample) are likely to be at increasingly high risk of extinction as global mean temperatures exceed a warming of 2°C to 3°C above pre-industrial levels (medium confidence)" (Climate Change 2007, 213). This study was an advance on Thomas et al. (2004) in the sense that the estimate was made after all local studies had been referenced to a global temperature increase, allowing for impacts to be directly comparable on this temperature scale. This was a significant effort with the tools of the day. IPCC authors deliberately steered away from the phrase "committed to extinction," and the species area curve technique, using rather the raw projections of species losses due to projected total range loss. The phrase "at increasingly high risk of extinction" was used to indicate a significant threat to biodiversity. It was possible to focus on a policy relevant global warming range of 1.5°C–2.5°C above a 1980–1999 average baseline (2°C–3°C above preindustrial average baseline) temperatures because of a concentration of studies in that range, allowing the authors to average a number of studies. A simple linear regression on the results revealed a possible linear response of extinction risk with warming, a trend that was described in a subsequent CBD report as a "roughly 10 percent increase with every degree C warming" (Convention on Biological Diversity 2009).

Urban (2015) published a revised estimate of extinction risk that used more studies and employed a superior weighting system based on numbers of species per study. This study reduced the estimate of species extinction risk over the temperature range of 2°C–3°C above preindustrial levels but identified a logarithmic increase with increasing temperature, indicating accelerating risk of extinction. More detailed work on several key taxonomic groupings and using even finer-grained species and climate data (Warren et al. 2018) supports a finding of an accelerating trend of projected species geographic range loss. They report that the number of species projected to lose over half their range doubles between 1.5°C and 2°C above the preindustrial baseline, an increase of more than 8 times with 4.5°C of warming.

Whatever the proportional loss of species projected, it must be multiplied by the number of species in the world to estimate a global extinction risk estimate. This is what Thomas et al. did in the press release for the 2004 study, to translate it for the media and the public. With 6 million–10 million species in the world, conservatively, 30 percent risk translates to between 1.8 million and 3 million species. The majority of these are insects, so what happens to insects is critical to these estimates. Few modeling studies have addressed large numbers of insects, but we know that most insect diversity is narrowly distributed in tropical forests, where recent upticks in drought and fire associate with or are compounded by climate change, are of great concern.

This risk of extinction is further intensified because of extensive human modification of landscapes. The combination of species ranges moving in response to climate change and human-dominated landscapes full of hard edges is what the late Stephen Schneider referred to as "a no-brainer for an extinction spasm." In its most recent report, IPCC (Settele et al. 2015) concluded that "a large fraction of species faces increased extinction risk due to climate change during and beyond the 21st century, especially as climate change interacts with other pressures . . . (high confidence)" (Climate Change 2014, 275). This study explicitly quantified how most plant species would be unable to naturally shift their geographical ranges fast enough to keep pace with projected rates of climate change in most landscapes. Most small mammals and freshwater mollusks would also not be able to keep pace at rates projected under a RCP4.5 emissions scenario in flat landscapes.

The question now is not so much whether there is major extinction risk from climate change, but rather how to go about addressing it. Cash-strapped conservation agencies in the tropics, where much of global biodiversity resides, are often too busy fighting urgent deforestation and habitat loss to battle the long-term effects of climate change. People living around protected areas may suffer when local land-use practices are overturned by climate change, bringing them into conflict with protected areas. People benefiting from tourism in protected areas may see incomes and job opportunities erode as habitat quality and species compositions are affected by climate change. For all these reasons, and for all these beneficiaries, it is important that the Global Climate Fund prioritize adaptation of biodiversity and prevention of extinctions due to climate change.

Fully understanding extinction risk from climate change nonetheless presents a rich and important research challenge. An accelerating increase in risk (e.g., Urban 2015) should theoretically result because faunas and floras have evolved under the relative cool conditions of the Pleistocene. Global biodiversity has for several million years not been exposed to levels of warmth approaching that projected for the next few decades. It might be expected that concentrations of biodiversity of relatively recent origin would show the highest susceptibility.

Fully integrated effects resulting from shifts in disturbance regimes, and the direct effects of rising CO_2 especially on plants, are not accounted for in niche-based models, but they may also represent a significant shock to biodiversity (e.g., Midgley and Bond 2015). The potential of such effects for extinction risk has not yet been calculated.

REFERENCES

Deutsch, Curtis A., Joshua J. Tewksbury, Raymond B. Huey, Kimberly S. Sheldon, Cameron K. Ghalambor, David C. Haak, and Paul R. Martin. 2008. "Impacts of climate warming on terrestrial ectotherms across latitude." *Proceedings of the National Academy of Sciences* 105 (18): 6668–6672.

Fischlin, A., G. F. Midgley, J. T. Price, R. Leemans, B. Gopal, C. Turley, M. D. A. Rounsevell, O. P. Dube, J. Tarazona, A. A. Velichko. 2007. "Ecosystems, their properties, goods, and services." *Climate Change 2007: Impacts, Adaptation and Vulnerability. Contribution of Working Group II to the Fourth Assessment Report of the Intergovernmental Panel on Climate Change*, ed. M. L. Parry, O. F. Canziani, J. P. Palutikof, P. J. van der Linden, and C. E. Hanson, 211–272. Cambridge University Press.

Hewitt G. 2000. "The genetic legacy of the Quaternary ice ages." *Nature* 22: 907.

Malcolm, Jay R., Canran Liu, Ronald P. Neilson, Lara Hansen, and Lee Hannah. 2006. "Global warming and extinctions of endemic species from biodiversity hotspots." *Conservation Biology* 20 (2): 538–548.

Secretariat of the Convention on Biological Diversity. 2009. *Connecting Biodiversity and Climate Change Mitigation and Adaptation: Report of the Second Ad Hoc Technical Expert Group on Biodiversity and Climate Change*. Technical Series No. 41. Montreal: Secretariat of the Convention on Biological Diversity.

Settele, J., R. Scholes, R. Betts, S. Bunn, P. Leadley, D. Nepstad, J. T. Overpeck, and M. A. Taboada. 2014. "Terrestrial and inland water systems." In *Climate Change 2014: Impacts, Adaptation, and Vulnerability. Part A: Global and Sectoral Aspects. Contribution of Working Group II to the Fifth Assessment Report of the Intergovernmental Panel on Climate Change*, ed. C. B. Field, V. R. Barros, D. J. Dokken, K. J. Mach, M. D. Mastrandrea, T. E. Bilir, M. Chatterjee, et al., 271–359. Cambridge University Press.

Sinervo, Barry, Fausto Mendez-De-La-Cruz, Donald B. Miles, Benoit Heulin, Elizabeth Bastiaans, Maricela Villagrán-Santa Cruz, Rafael Lara-Resendiz, et al. 2010. "Erosion of lizard diversity by climate change and altered thermal niches." *Science* 328 (5980): 894–899.

Thomas, Chris D., Alison Cameron, Rhys E. Green, Michel Bakkenes, Linda J. Beaumont, Yvonne C. Collingham, Malcolm, Guy Midgley, Lee Hannah, et al. 2004. "Extinction risk from climate change." *Nature* 427 (6970): 148.

Urban, Mark C. 2015. "Accelerating extinction risk from climate change." *Science* 348 (6234): 571–573.

Warren, R., J. Price, J. VanDerWal, S. Cornelius, and H. Sohl. 2018. "The implications of the United Nations Paris Agreement on climate change for globally significant biodiversity areas." *Climatic Change* 147 (3–4): 395–409.

Warren, R., J. Price, J. VanDerWal, E. Graham, and N. Forstenhauesler. (in press). *Science*.

CHAPTER TWENTY-THREE

Ecosystem–Based Adaptation

CAITLIN LITTLEFIELD, ERIK NELSON,
BENJAMIN J. DITTBRENNER,
JOHN WITHEY, KATIE K. ARKEMA,
AND JOSHUA J. LAWLER

Human well-being is inextricably linked to ecosystem processes. The success of societies is predicated on past, current, and future states of the natural environment, and humans have struggled to adapt their systems to changes in ecosystem processes for millennia. Periods of unsuccessful adaptation have led to societal distress (Parker 2013). Functioning ecosystems and the services they provide may, in many cases, supply humans with the best opportunities to adapt under climate change.

Ecosystem services are the benefits that people obtain from ecosystems. Although any classification scheme belies the interconnectedness of these services, they are most frequently identified as supporting services that underpin all others (e.g., primary production, nutrient cycling), provisioning services (e.g., providing food, fibers, natural medicines), regulating services (e.g., climate regulation, water purification), and cultural services (e.g., recreational opportunities, spiritual importance; Millennium Ecosystem Assessment 2003). Some services can be replaced by technology at low cost, whereas other large-scale services have no feasible substitutes. Biodiversity is a contributing factor to sustainable delivery of these services, and redundancy in ecosystem functionality that accompanies healthy, biodiverse ecosystems further ensures both natural and human system stability.

Ecosystem-based adaptation (EbA) strategies "harness the capacity of nature to buffer human communities against the adverse impacts of climate change through the sustainable delivery of ecosystem services" (Jones et al. 2012). EbA leverages and aims to protect ecosystem services to build adaptive capacity, resistance, and resilience into human systems (Box 23.1). EbA can

Box 23.1   **Glossary**

Agroecosystem: A holistic agricultural system that is typically small scale and that leverages natural ecosystem processes (e.g., nutrient cycling, energy flows, biotic interactions between diverse species) in management practices in a way that minimizes synthetic inputs.

Carbon sequestration: The removal and storage of carbon dioxide from the atmosphere and into carbon sinks (e.g., oceans, vegetation, geologic formations deep underground) through physical or biological processes.

Green infrastructure: Any of a variety of stormwater-management techniques, installations, or systems that use vegetation, soils, and natural processes as compared to engineered water collection systems of storm drains and pipes.

Hard adaptation approaches: Strategies for adapting to climate change that tend to use specific technologies, infrastructure, and actions that may require more capital goods and be more permanent than other adaptation approaches (e.g., sea walls to ward against sea-level rise).

Resilience: The capacity of a human or natural system to regain the essential components and processes that characterize the system after a perturbation or various stressors. Thresholds or tipping points are crossed when a system does not return to its characteristic state following a perturbation.

Resistance: The capacity of a human or natural system to maintain its essential components and processes despite a perturbation or various stressors.

Soft adaptation approaches: Strategies for helping communities adapt to climate change that primarily relate to social systems, knowledge transfer, and human behavior (e.g., livelihood diversification, establishment of early warning systems).

Soil organic carbon (SOC): The pool of carbon occurring in organic form in the soil, usually contained in soil organic matter (e.g., dead plant and animal tissue, decomposition by-products, soil microbial biomass). It is the primary source of energy for soil microorganisms and serves as a good proxy for soil biodiversity.

Sea-level rise (SLR): The global and local rise in sea level due to a change in ocean volume. A volume change can result from an increase in the amount (i.e., mass) of water in the oceans (e.g., due to melting ice caps) and from the thermal expansion of ocean water as its temperature rises. Changes in salinity may also have an impact on sea level.

Urban greenspace: Lands in urban areas that are primarily covered by vegetation. Greenspaces can be publically or privately owned, and include a variety of maintenance and management regimes from golf courses and cemeteries to protected natural forests.

Urban heat island (UHI): The common phenomenon of warmer air and surface temperatures in urban areas compared to nearby rural areas. The pavement and dark building materials of urban areas have lower albedo—they do not reflect as much solar energy. The temperature differential, typically 1°C–3°C on an annual basis for a city of 1 million, can be as high as 12°C at night.

complement "soft" adaptation approaches such as livelihood diversification or replace "hard" adaptation approaches, which use specific technologies and capital goods and are often engineered, infrastructure-based interventions (Jones et al. 2012; Morris et al. 2018).

Like ecosystem services, EbA approaches transcend rigid categorization; many overlap and reinforce one another. Here, for illustrative purposes, we focus on five broadly defined environments: agricultural landscapes, urban areas, coastal zones, freshwater, and forests. We identify several ecosystem services generated in each environment, how climate change has and will affect these services, and potential EbA strategies to maintain them. We also identify related EbA co-benefits and alternative interventions. We conclude with a discussion of both the opportunities and potential pitfalls of EbA.

MAINTAINING AGRICULTURAL PRODUCTIVITY

Climate change is expected to lengthen growing seasons, alter rainfall patterns, increase the frequency of extreme weather events, and shift both pest and pest-predator ranges (Wheeler and von Braun 2013). Given current agricultural production technology, these changes are expected to make global agricultural production more variable over space and time (Asseng et al. 2015) and apply downward pressure on average yields around the world (Rosenzweig et al. 2014). Technological improvements in agricultural production will ameliorate some of climate change's negative impacts on yields (Abberton et al. 2016). In addition, production farmers (as opposed to subsistence farmers) may change seed varieties, adjust irrigation and fertilizer amounts, and modify planting and harvesting dates to better adapt production to the changing climate.[1] Finally, farmers may also leverage ecosystem services to maintain yields in the face

of climate change, for example, by increasing soil biodiversity, maintaining pollinator habitat, and ensuring sustainable water provisioning.

Farmers can improve yields by increasing biodiversity in their soils. For example, wheat yield in the Scania region of Sweden increased by 3.2 Mg ha^{-1} when soil organic carbon (SOC) content—a proxy indicator of soil biodiversity (de Vries et al. 2013)—was increased from 7.9 g kg^{-1} of soil to 19 g kg^{-1} of soil (Brady et al. 2015). An examination of the relationship between 2009 crop yield, growing season weather, and SOC across all of Europe for multiple crops corroborates the positive impact SOC can have on yields at the margin, especially when growing season weather is not ideal (see shaded leaf nodes in Figure 23.1).

Activities that increase SOC, such as adding manure to the soil and including cover crops in rotations (Alvarez 2005), generate both direct and opportunity costs (e.g., a rotation of cover crops means no marketable yield that season or year). If, however, the increase in yield from enhanced SOC outweighs costs, farmers will have an incentive to increase SOC. Furthermore, the societal co-benefits generated by greater SOC levels, including reduced need for chemical fertilizers (and therefore reduced eutrophication) and reduced atmospheric CO_2 concentrations, may make it optimal for governments to subsidize SOC investments.

Increasing pollination capacity is another EbA strategy to buttress food production from adverse climate change impacts (Hannah et al. 2017). Globally, 75 percent of all human-consumed crops require insect pollination (Klein et al. 2007), yet widespread declines in pollinator abundance, mostly due to habitat conversion, are compromising the quality and quantity of food production. Ensuring pollination services requires coordinated action across agricultural landscapes, as ecosystem services mediated by mobile organisms like pollinators are affected by management at scales larger than individual farms (Cong et al.

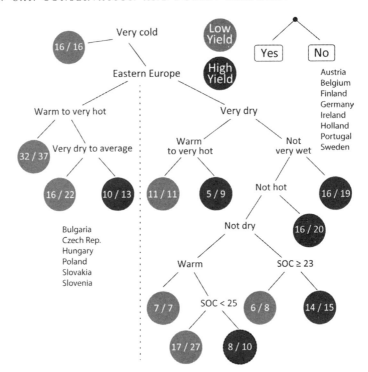

Figure 23.1. Classification tree for 2009 crop yields in Europe as explained by growing season weather and SOC. The tree represents the partitioning of 2009 crop-specific growing season weather and SOC across Europe that best explains a sample of observed 2009 yields. Observed yields are either placed in the lower-yield bin (a bottom 50th percentile yield observation for the given crop) or the higher-yield bin (a top 50th percentile yield observation for the given crop). At each node a yes to the weather or soil characteristic means a move to the left best fits the data and a no means a move to the right best fits the data (Loh 2011; Varian 2014). A yes on the Eastern Europe nodes indicates that the observation is from Eastern Europe. Eastern European farms tend to use less chemical inputs and are less capitalized than their Western European counterparts. At the end of each branch, called a leaf, the first number gives the count of observations on that branch that are in the given yield bin and the second is the number of observations on that branch.

For example, the higher-yield "8/10" leaf (with dashed black border at bottom) indicates that eight Western Europe yield observations with higher yields did not experience very extreme growing season weather and had an SOC of 25 g kg^{-1} or more. In contrast, only two lower Western European yields had similar weather and soil conditions. In other words, observations on this branch are predicted to have a top half yield. The other SOC node, SOC ≥ 23, indicates that observations with that branch's growing season weather profile are much more likely to have a top half yield if SOC < 23 g kg^{-1}. In other words, in this leaf's particular growing season weather profile, too much SOC is associated with lower yields. All in all, SOC impacts yield at the margin while growing season weather and agricultural investment (crudely represented by Eastern vs. Western European observations) are the main drivers of observed yields. See http://www.bowdoin.edu/faculty/e/enelson/ for more details on this analysis.

2014). Economists have demonstrated that small payments encourage farmers to provide pollinator habitat when accompanied by larger fines for any subsequent habitat destruction (Cong et al. 2014).

As global demand for food grows and rainfall patterns change, farmers will increasingly look to irrigation as an adaptation measure. For example, northern China is projected to become drier while southern China gets wetter (Piao et al. 2010). Such changes will force northern farmers to adopt more drought-tolerant crops or, less likely, devise ways to transport water from south to north (Piao et al. 2010). In other places, climate change may cause increased precipitation in the winter and decreased precipitation during the growing season.

Storing non-growing-season precipitation in networks of constructed retention ponds and restored wetlands could ensure water availability during increasingly dry growing seasons (Baker et al. 2012) while also reducing flood risk and providing wildlife habitat.

Farmers of agroecosystems, subsistence farmers, and small-scale farmers generally tend to rely much more heavily on ecosystem services to manage uncertainty and environmental variability than do large-scale production farmers (Tengö and Belfrage 2004). Grazing animals on crop fields and including grasses in crop rotations are two tactics still widely used by subsistence farmers to maintain SOC and biodiversity. Intercropping and maintenance of pollinator and natural pest habitat also help maintain acceptable yields. Although these agricultural systems are much more sustainable than the high-yield systems described above, their low productivity means that they will not contribute significantly to global food supplies. Instead, if the global agriculture system is to become more resilient to climate change, the more input-intensive farmers will have to adopt agroecosystem techniques that are compatible with high-yield farming.

MAKING URBAN LIVING MORE SUSTAINABLE

According to the United Nations (2014), 66 percent of the world's population will live in urban areas by 2050, which means that the majority of the world's people will directly experience climate change and attempt to adapt to it in the urban environment. Warmer temperatures will exacerbate the urban heat island (UHI) effect and the frequencies of extreme weather events are expected to increase in urban areas. Extreme heat and precipitation events in cities increase human mortality rates and hamper the ability of infrastructure to perform adequately (IPCC 2014). In many coastal cities, sea-level rise (SLR) and storm surge are direct consequences of climate change that require management. (We address SLR management specifically in the coastal-zone section.) Green infrastructure and urban greenspace are two related EbA approaches that can be used to address many of these issues related to heat and water management.

Green infrastructure alternatives to the hard or "gray" infrastructure typically used for stormwater management (e.g., drains, pipes) include green roofs, bioswales, rain gardens, and constructed retention ponds and wetlands. The vegetation and soils associated with green infrastructure intercept precipitation and reduce the rate and volume of runoff (Gill et al. 2007; Pappalardo et al. 2017). For example, a project in Seattle, Washington, that included bioswales, retention ponds, and a series of stepped pools retained 99 percent of wet-season runoff (Horner et al. 2004). Similarly, a review of green roofs in German cities showed that intensive green roofs (with substrate >150 mm) could retain 75 percent of annual runoff (Mentens et al. 2006). In other words, green infrastructure can mitigate urban flooding by storing some of the excess water created by a storm. Further, any urban drought after an extreme precipitation event can be alleviated by the slow release of this excess water.

Ground and rooftop plantings in lieu of conventional dark materials and pavement reduces the UHI in general and may reduce mortality impact of a heat wave specifically (Li et al. 2014). Alternatively, "white" and "cool" roofs, or even lighter-colored pavements that have higher albedo, can reduce the UHI and the effects of heat waves, although these approaches do not moderate stormwater runoff. Co-benefits of green infrastructure include water-quality improvements due to filtration of natural pollutants and a reduction of water temperatures, aesthetic values of plantings, the potential for wildlife and pollinator habitat provisioning, and opportunities for small-scale urban agriculture.

Expanding and conserving existing urban greenspaces in and around cities—from pocket parks to vegetated corridors to protected forests—provide myriad ecosystem services and inherently entail EbA. Forested lands and street trees increase evaporative cooling and shade pavement, which help reduce the magnitude of the UHI, the impacts of heat waves, and energy consumption (e.g., for air conditioning; Parmova et al. 2012). For example, adding just 10 percent green cover to urbanized parts of Greater Manchester, in the United Kingdom, is projected to keep maximum summer surface temperatures at 29°C in the 2080s, compared to 32°C with current cover and 35°C with a 10 percent loss of green cover (Gill et al. 2007).

Restoring native species or planting drought-resistant species or hybrids can help ensure urban tree health while also promoting biodiversity in plantings, which helps prevent outbreaks of species-specific diseases and pests (Alvey 2006; Paap et al. 2017), particularly if urban trees are already experiencing climate-related stress. Co-benefits of increased greenspaces, especially urban forests, include carbon sequestration, improved air quality, psychological benefits, and opportunities for recreation, in addition to other benefits enumerated above and in the forest discussion below.

COASTAL STORM PROTECTION AND FISHERIES PRODUCTION

Coastal and marine ecosystems provide a diversity of benefits. Fish and shellfish are important sources of sustenance and protein, and many coastal habitats protect infrastructure from storm surge and offer opportunities for recreation and aesthetic enjoyment. However, ocean acidification and warming are leading to shifts in the distribution of economically and ecologically important ecosystems and species, while more frequent and intense storms increase the threat to coastal infrastructure

(IPCC 2014). Conserving existing ecosystems, restoring degraded ones, and pursuing integrated management to reduce the cumulative risks from local stressors all have the potential to enhance and maintain the functioning of coastal and marine ecosystems and the benefits they provide (Ruckelshaus et al. 2014; Morris et al. 2018). Below we discuss these EbA strategies in light of two important services: coastal storm protection and fisheries production.

Rising seas and potential increases in the intensity and frequency of storms pose risks to the 200 million people living in coastal regions worldwide (IPCC 2014). By attenuating waves and storm surge, coastal and marine ecosystems such as wetlands, coral reefs, and coastal forests can help reduce the impacts of such hazards (Shepard et al. 2011; Arkema et al. 2013). Conserving existing habitats and restoring degraded ones in regions with low-lying sandy and muddy coastlines (e.g., the US east and gulf coasts) may effectively halve the number of people at high risk under climate change (Arkema et al. 2013) (Figure 23.2). Funded by the US federal government in the wake of the 2012 storm Hurricane Sandy, several innovative projects involve building reefs to serve as "natural breakwaters" to attenuate waves and reduce erosion while providing habitat for fish, shellfish, and lobsters.[2] Leveraging ecosystems for coastal defense is often less costly to implement and maintain than hard infrastructure approaches including seawalls and levees (Morris et al. 2018).

Increasing ocean temperatures and changing water chemistry are disrupting the delivery of marine food and the livelihoods that facilitate this service (Pinsky and Mantua 2014; Weatherdon et al. 2016). For example, in the northwestern Atlantic, 24 out of 36 commercially exploited fish showed significant range (latitudinal and depth) shifts between 1968 and 2007 due to warming water (Nye et al. 2009). Changes in water chemistry affect the calcification rates of marine organisms (IPCC 2014), many of which are an important food

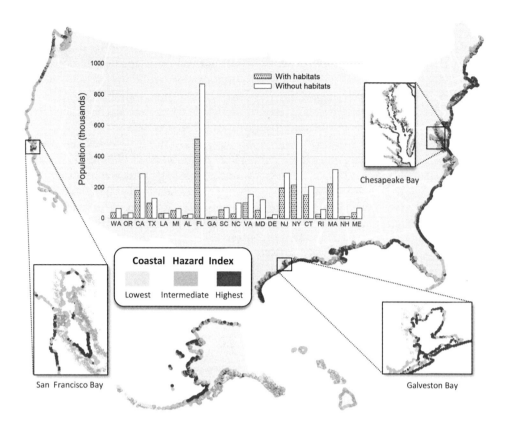

Figure 23.2. Exposure of the US coastline and coastal populations to sea-level rise in 2100 (A2 scenario) and storms. Darker colors indicate regions with more exposure to coastal hazards (index >3.36). The bar graph shows the population living in areas most exposed to hazards (darkest 1 km² coastal segments in the map) with protection provided by habitats (dotted bars) and the increase in population exposed to hazards if habitats were lost owing to climate change or human impacts (white bars). Letters on the x-axis represent US state abbreviations. This figure first appeared in the journal *Nature Climate Change* (Arkema et al. 2013).

source (e.g., oysters) or provide important nursery and adult habitats for fishes (e.g., corals). Coral conservation is particularly challenging in tropical systems where pollution, sedimentation, and unsustainable fishing stressors have not been addressed as successfully as in other parts of the world (Ruckelshaus et al. 2014). Efforts such as the multilateral Coral Triangle Initiative draw on ecosystem-based fisheries management

and marine protected areas (MPAs) to safeguard biodiverse sites and sustain fish stocks to ensure food security in coastal areas.[3] Although alternative approaches to marine food production such as aquaculture may offset some climate impacts on natural systems, these alternatives come with risks of disease, habitat destruction, and pollution (Ruckelshaus et al. 2014).

Conservation, restoration, and integrated management of coastal and marine ecosystems have numerous co-benefits. Seagrasses and mangroves sequester carbon, such that degradation and conversion of these systems globally release 0.15–1.02 Pg CO_2 annually and result in damages of US$6–$42 billion annually (Pendleton et al. 2012). Healthy coastal and marine ecosystems also provide tourism opportunities and support livelihoods in this sector. Given these additional co-benefits, EbA approaches are often superior to other options, but they do have their challenges. In the case of coastal protection services, perceptions of risk can be higher with green infrastructure than with built systems, and thus challenging to implement. Likewise, fisheries management can require coordination among many entities, making it particularly challenging.

MAINTAINING FRESHWATER RESOURCES

Climate change is projected to substantially and nonlinearly affect surface water and groundwater resources. These impacts are likely to progress more quickly in heavily populated areas (Gerten et al. 2013). Geographic and temporal shifts in precipitation regimes will interact with changing temperatures to create more prolonged drought or, conversely, flooding associated with extreme weather events (IPCC 2014), impacts that will be felt most acutely in heavily populated areas (Gerten et al. 2013).

Traditional flood management strategies, such as straightening river channels and building dikes and levees, are vulnerable to failure and largely pass potential flooding problems downstream. EbA strategies for reducing the impact of floods enable excess water to spread into side channels and beyond riverbanks, where it slows and infiltrates soils, thereby reconnecting rivers with floodplains, and holding excess water in natural areas above population centers (Palmer et al. 2009). Stream and riparian restoration that increases large woody debris, structural habitat, and channel complexity encourages incised channels to aggrade (Palmer et al. 2009) and improves overall ecosystem health (Thompson et al. 2018). Beyond the stream channel, afforestation and reforestation increase evapotranspiration and can prolong snow cover, thereby reducing downstream flooding by up to 54 percent in drier areas and 15 percent in more humid areas, as demonstrated in four South American case studies (Tra-

bucco et al. 2008). These approaches also increase soil moisture and groundwater recharge, thereby providing direct benefits for agriculture, facilitating downstream groundwater withdrawals, and reducing fire risk (Ellison et al. 2017), among other forest-related co-benefits identified in the forest discussion below.

It is likely that climate change will further reduce water availability in areas currently suffering from water scarcity (Gosling and Arnell 2016). Indeed, 1.3 billion people already live in water-scarce regions, and global warming of 2°C, 3.5°C, and 5°C are projected to expose an additional 8 percent, 11 percent, and 13 percent of the world population to greater water scarcity, respectively (Gerten et al. 2013). Beyond the problem of limited water for direct human consumption and agricultural purposes, water shortages may result in habitat loss for pollinators and other species of economic importance, increased threat of forest fires, and saltwater intrusion into overdrawn aquifers. Substantial efficiencies can be achieved in water use by adopting practices such as upgrading leaky water-delivery systems, recycling wastewater, and implementing more efficient agricultural practices (IPCC 2014).

Some of these solutions, however, may be prohibitively expensive and will offer little benefit if natural ecosystems and processes that augment or ensure water provisioning are not protected. As noted in the forest discussion, forest ecosystems are critical for the sustainable water delivery to over a third of the world's largest cities (Dudley and Stolten 2003). The protection of high-elevation wetlands and peatlands—such as the Andean *bofedales*—may be critical for the persistence of pastoral native communities in otherwise inhospitable environments; these fragile ecosystems are extremely sensitive to climatic changes (Squeo et al. 2006). Elsewhere, natural and constructed wetlands have been effectively used to retain surface water, recharge groundwater, and filter out pollutants; these systems can be more cost-effective and permanent than treatment facilities (Jones et al. 2012). Natural ecosystem engineers like beavers (*Castor canadensis*) have been used to increase water retention and hydrologic stability through their creation of wetland complexes where wetlands would otherwise not exist (Dittbrenner et al. 2018). Although these EbA strategies are promising, realization of large-scale adaptation goals will require watershed assessments, long-term planning, synchronization of multiple cross-watershed entities, and close coordination with local populations to ensure stakeholder buy-in.

MAINTAINING THE BENEFITS THAT FORESTS PROVIDE

Humans derive many benefits from forests, from local to global scales. For example, one-third of the world's largest cities obtain a significant proportion of their drinking water directly from protected forests (Dudley and Stolten 2003). Conserving the biodiversity and processes of forest ecosystems and restoring the integrity and resilience of degraded ones are primary strategies for leveraging forests for climate change adaptation.

Restoration of coastal forests can minimize inland flooding and coastal storm-surge events, which are projected to become more frequent and greater in magnitude (IPCC 2014). From riparian forests to coastal mangroves, vegetation structure physically slows water flow, attenuates wave and tidal energy, and stores water through plant uptake, thus minimizing the threat of flooding. Indeed, fewer lives were lost in coastal communities with healthy mangrove forests than those without during the 2004 Indian Ocean tsunami (Das and Vincent 2009). Mangroves additionally capture nutrient-rich sediments in their root structures and maintain important habitat for birds, fish, and other marine species. Similarly, through shading and inputs of in-

stream large woody debris, riparian forests maintain thermal refugia for temperature-sensitive species, such as spawning salmon (Palmer et al. 2009). Although conservation and restoration of these natural coastal and inland buffering systems can require multiscale and cross-sector coordination for successful maintenance, the co-benefits and cost savings compared to dams, levees, and other shoreline defense approaches can be considerable (Jones et al. 2012).

As described above, changes in precipitation regimes may necessitate greater water storage capacity and filtration—services that forest vegetation and soils afford. Already many municipalities have realized significant cost savings from investing in the forested watershed conservation (e.g., the Catskill-Delaware watershed north of New York City) instead of water purification infrastructure, which requires initial capital and continued maintenance. Elsewhere, tropical montane cloud forests are prime candidates for continued conservation in a changing climate, as they play an important role in water supplies: water vapor condenses on foliage and flows into streams, significantly augmenting water availability from rainfall in drier, low-elevation areas (Ellison et al. 2017). By contrast, deforestation throughout Amazonia increases runoff and water discharge at local scales and—as demonstrated with simulations—affects the water balance, hydrology, and surface temperatures across the entire Amazon Basin and likely globally (Foley et al. 2007).

Along with deforestation and degradation, a major threat to ecosystem services from forests is forest fires. From Australia to North America, synergistic effects of climate change–related drivers are fueling more catastrophic forest fires. For example, earlier snowmelt and drought conditions in the western United States are interacting with increasingly widespread mountain pine beetle (*Dendroctonus ponderosae*) outbreaks as the species' range expands (Loehman et al. 2016). Combined with a management legacy of fire suppression and selective harvesting, these climate-related stressors may contribute to larger fires and longer fire seasons (Westerling et al. 2006). Managers in fire-prone areas can ameliorate this stress and reestablish system resilience by restoring the patterns and processes typical of healthy, fire-prone forests—for example, through thinning and prescribed burning to reduce fuel loads and minimize the threat of catastrophic fires (Calkin et al. 2014). This ecosystem-based approach in turn decreases the tremendous costs associated with firefighting, both in dollars and in lives.

Beyond the ways in which forest conservation—particularly in the tropics—can help communities and society adapt to climate change, co-benefits associated with these adaptation strategies are innumerable. For example, non-timber-forest products including food, fiber, and fuel are especially important for subsistence livelihoods. With regard to human health, higher levels of tree cover in upstream watersheds are associated with lower probabilities of childhood diarrheal disease in downstream rural communities across 35 developing nations (Herrera et al. 2017). It has also been found that tropical forest cover can moderate the spread of infectious disease through regulating pathogen populations and their hosts. For example, in the Peruvian Amazon, the biting rates of mosquitoes that are the primary malaria vector in South America were found to be 278 times higher in deforested areas than in areas that remained forested (Vittor et al. 2006). Furthermore, biodiversity within these systems has yielded myriad medicinal natural products (Foley et al. 2007). Finally, the climate change mitigation potential of forests is immense. The global forest carbon sink rate is estimated to be 2.4 Pg C/yr (Pan et al. 2011), although emissions from tropical deforestation and degradation effectively halve this rate. In terms of climate change adaptation strategies, there are no fathomable human-built alternatives to tropical forest conservation that deliver comparable co-benefits.

CONCLUSION

From agricultural lands to cities, from rivers and oceans to forests, the EbA approaches discussed above create a deep, diverse suite of co-benefits that technological adaptation measures do not. Whether the values of these co-benefits are high enough to make EbA strategies preferable to technological adaptations is an ongoing question. EbA approaches are usually more holistic and proactive in design than conventional interventions, which may be more reactive and focused on singular goals (e.g., levees for flood control; Jones et al. 2012). However, quantifying the future benefits of EbA can be challenging given the difficulty of accurately modeling ecosystems and a lack of consensus on how to place values on non-marketed ecosystem services for comparison purposes. These methodological difficulties can make engineered approaches more compelling because the costs and outcomes of such interventions are more easily quantified (Morris et al. 2018). Furthermore, the time frame over which the primary benefits of EbA measures materialize may not always coincide with more immediate adaptation needs. Thus, successful selection of adaptation measures requires identifying the contexts in which a given measure provides competitive adaptation options, even if primary services are not delivered for many years (Jones et al. 2012). Such accounting must also consider that EbA strategies—especially conservation and restoration ones—may be self-renewing and are inherently plastic whereas hard infrastructure and engineering solutions may end up mismatched to future conditions.

In the best of circumstances, EbA approaches coincide with and reinforce human health and poverty alleviation goals. For example, global health experts have hypothesized that climate change, deforestation, poverty, and civil unrest interacted to set the stage for the 2014 West African Ebola outbreak (Bausch and Schwarz 2014). A prolonged dry season, linked to extreme deforestation and climate trends, may have driven the rural poor deeper into remaining forests in search of food and wood. As they expanded their geographic range and the variety of species they hunted, their risk of exposure to Ebola and other zoonotic pathogens increased. As exemplified in this case, strategies for poverty alleviation and forest restoration for EbA may reinforce one another, although the danger remains that such EbA efforts may unintentionally undermine development efforts (e.g., conservation schemes that disenfranchise or exclude local peoples; Tallis et al. 2008).

Despite these potential shortcomings and trade-offs, positive synergies between EbA approaches, other climate change mitigation efforts, and conservation are highly likely. This is particularly true because EbA inherently seek to improve or augment the resilience of systems, thereby reducing the risks of crossing tipping points and shifting to unmanageable or unrecoverable states (Jones et al. 2012). Achieving such system resilience and harnessing nature to buffer communities from climate change impacts requires cross-scale coordination. Natural processes and ecosystem services do not conform to political boundaries, human institutions, or specific landscapes. Relying on and safeguarding the ecosystem service of pollination, for example, requires individual and collective actions—perhaps mediated by top-down government incentives (Cong et al. 2014)—from agricultural landscapes to cities. Lastly, in undertaking any EbA effort, we must evaluate the potential for benefits to persist through time and across space. After all, the very systems and processes we seek to leverage through EbA are also subject to unforeseen climate impacts.

NOTES

1. In some cases the optimal response to climatic changes will be to change the types of crops planted. However, national agriculture policies may make such welfare-enhancing crop swaps difficult. For example, many national biofuel policies that subsidize maize or sugar production may incentivize farmers to continue planting crops not best suited for the emerging climate. Unless subsidy policies are sensitive to climate change, regulatory inflexibility is likely to make global agriculture less resilient to climate change.

2. See http://www.rebuildbydesign.org/our-work/sandy -projects. In the aftermath of Hurricane Sandy in 2012, President Obama launched Rebuild by Design, a design competition, to stimulate the development of innovative solutions for addressing the affected region's most complex needs. Funded projects are showcased at this webpage.

3. See http://www.coraltriangleinitiative.org. The Coral Triangle Initiative is a multilateral partnership of six countries working together to sustain marine and coastal resources by addressing crucial issues including food security, climate change, and marine biodiversity.

REFERENCES

Abberton, M., J. Batley, A. Bentley, J. Bryant, H. Cai, J. Cockram, A. Costa de Oliveira, L. J. Cseke, H. Dempewolf, C. De Pace, et al. 2016. "Global agricultural intensification during climate change: A role for genomics." *Plant Biotechnology Journal* 14: 1095–1098.

Alvarez, R. 2005. "A review of nitrogen fertilizer and conservation tillage effects on soil organic carbon storage." *Soil Use and Management* 21: 38–52.

Alvey, A. A. 2006. "Promoting and preserving biodiversity in the urban forest." *Urban Forestry and Urban Greening* 5: 195–201.

Arkema, K. K., G. Guannel, G. Verutes, S. A. Wood, A. Guerry, M. Ruckelshaus, P. Kareiva, M. Lacayo, and J. M. Silver. 2013. "Coastal habitats shield people and property from sea-level rise and storms." *Nature Climate Change* 3 (10): 913–918.

Asseng, S., F. Ewert, P. Martre, R. P. Rötter, D. B. Lobell, et al. 2015. "Rising temperatures reduce global wheat production." *Nature Climate Change* 5 (2): 143–147.

Baker, J. M., T. J. Griffis, and T. E. Ochsner. 2012. "Coupling landscape water storage and supplemental irrigation to increase productivity and improve environmental stewardship in the US Midwest." *Water Resources Research* 48: W05301.

Bausch D. G., and L. Schwarz. 2014. "Outbreak of Ebola virus disease in Guinea: Where ecology meets economy." *PLOS Neglected Tropical Disease* 8 (7): e3056.

Brady M. V., K. Hedlund, R. G. Cong, L. Hemerik, S. Hotes, S. Machado, L. Mattsson, E. Schulz, and I. K. Thomsen. 2015. "Valuing supporting soil ecosystem services in agriculture: A natural capital approach." *Agronomy Journal* 107 (5): 1809–1821.

Calkin, D. E., J. D. Cohen, M. A. Finney, and M. P. Thompson. 2014. "How risk management can prevent future wildfire disasters in the wildland-urban interface." *Proceedings of the National Academy of Sciences* 111 (2): 746–751.

Cong, R. G., H. G. Smith, O. Olsson, M. Brady. 2014. "Managing ecosystem services for agriculture: Will landscape-scale management pay?" *Ecological Economics* 99: 53–62.

Das, S., and J. R. Vincent. 2009. "Mangroves protected villages and reduced death toll during Indian super cyclone." *Proceedings of the National Academy of Sciences* 106: 7357–7360.

de Vries, F. T., E. Thébault, M. Liiri, K. Birkhofer, M. A. Tsiafouli, L. Bjørnlund, H. B. Jørgensen, et al. 2013. "Soil food web properties explain ecosystem services across European land use systems." *Proceedings of the National Academy of Sciences* 110: 14296–14301.

Dittbrenner, B. J., M. M. Pollock, J. W. Schilling, J. D. Olden, J. J. Lawler, and C. E. Torgersen. 2018. "Modeling intrinsic potential for beaver (*Castor canadensis*) habitat to inform restoration and climate change adaptation." *PLOS One* 13 (2): e0192538.

Dudley, N., and S. Stolton. 2003. *Running Pure: The Importance of Forest Protected Areas to Drinking Water*. World Bank and World Wildlife Fund Alliance for Forest Conservation and Sustainable Use.

Foley, J. A., G. P. Asner, M. H. Costa, M. T. Coe, R. De-Fries, H. K. Gibbs, E. A. Howard, S. Olson, J. Patz, and N. Ramankutty. 2007. "Amazonia revealed: Forest degradation and loss of ecosystem goods and services in the Amazon Basin." *Frontiers in Ecology and the Environment* 5: 25–32.

Gerten, D., W. Lucht, S. Ostberg, J. Heinke, M. Kowarsch, H. Kreft, Z. W. Kundzewicz, J. Rastgooy, R. Warren, and H. J. Schellnhuber. 2013. "Asynchronous exposure to global warming: Freshwater resources and terrestrial ecosystems." *Environmental Research Letters* 8: 34032.

Gill, S. E., J. F. Handley, A. R. Ennos, and S. Pauleit. 2007. "Adapting cities for climate change: The role of the green infrastructure." *Built Environment* 33: 115–133.

Gosling, S. N., and N. W. Arnell. 2016. "A global assessment of the impact of climate change on water scarcity." *Climatic Change* 134 (3): 371–385.

Hannah, L., M. Steele, E. Fung, P. Imbach, L. Flint, and A. Flint. 2017. "Climate change influences on pollinator, forest, and farm interactions across a climate gradient." *Climatic Change* 141 (1): 63–75.

Herrera, D., A. Ellis, B. Fisher, C. D. Golden, K. Johnson, M. Mulligan, A. Pfaff, T. Treuer, and T. H. Ricketts. 2017. "Upstream watershed condition predicts rural children's health across 35 developing countries." *Nature Communications* 8: 811.

Horner, R. R., H. Lim, and S. J. Burges. 2004. *Hydrological Monitoring of the Seattle Ultra-Urban Stormwater Management Projects: Summary of the 2000–2003 Water Years*. Water Resources Series Technical Report No. 181. Department of Civil and Environmental Engineering, University of Washington.

Intergovernmental Panel on Climate Change (IPCC). 2014. *Climate Change 2014: Impacts, Adaptation, and Vulnerability. Contribution of Working Group II to the Fifth Assessment Report of the Intergovernmental Panel on Climate Change*, ed. C. Field, V. Barros, D. Dokken, K. Mach, M. Mastrandrea, T. Bilir, M. Chatterjee, K. Ebi, Y. Estrada, R. Genova, B. Girma, E. Kissel, A. Levy, S. MacCracken, P. Mastrandrea, and L. White. Cambridge University Press.

Jones, H. P., D. G. Hole, and E. S. Zavaleta. 2012. "Harnessing nature to help people adapt to climate change." *Nature Climate Change* 2: 504–509.

Klein, A. M., B. E. Vaissière, J. H. Cane, I. Steffan-Dewenter, S. A. Cunningham, C. Kremen, and T. Tscharntke. 2007. "Importance of pollinators in changing landscapes for world crops." *Proceedings of the Royal Society B: Biological Sciences* 274: 303–313.

Li, D., E. Bou-Zeid, and M. Oppenheimer. 2014. "The effectiveness of cool and green roofs as urban heat island mitigation strategies." *Environmental Research Letters* 9 (5): 055002.

Loehman, R. A., R. E. Keane, L. M. Holsinger, and Z. Wu. 2017. "Interactions of landscape disturbances and climate change dictate ecological pattern and process: Spatial modeling of wildfire, insect, and disease dynamics under future climates." *Landscape Ecology* 32 (7): 1447–1459.

Loh, W. 2011. "Classification and regression trees." *Wiley Interdisciplinary Reviews: Data Mining and Knowledge Discovery* 1: 14–23.

Mentens, J., D. Raes, and M. Hermy. 2006. "Green roofs as a tool for solving the rainwater runoff problem in the urbanized 21st century?" *Landscape and Urban Planning* 77: 217–226.

Millennium Ecosystem Assessment. 2003. *Ecosystems and Human Well-Being: A Framework for Assessment*. Island Press.

Morris, R. L., T. M. Konlechner, M. Ghisalberti, and S. E. Swearer. 2018. "From grey to green: Efficacy of eco-engineering solutions for nature-based coastal defense." *Global Change Biology* 24 (5): 1827–1842.

Nye, J. A., J. S. Link, J. A. Hare, and W. J. Overholtz. 2009. "Changing spatial distribution of fish stocks in relation to climate and population size on the Northeast United States continental shelf." *Marine Ecology Progress Series* 393: 111–129.

Paap, T., T. I. Burgess, and M. J. Wingfield. 2017. "Urban trees: Bridge-heads for forest pest invasions and sentinels for early detection." *Biological Invasions* 19 (12): 3515–3526.

Palmer, M. A., D. P. Lettenmaier, N. L. Poff, S. L. Postel, B. Richter, and R. Warner. 2009. "Climate change and river ecosystems: Protection and adaptation options." *Environmental Management* 44: 1053–1068.

Pan, Y., R. A. Birdsey, J. Fang, R. Houghton, P. E. Kauppi, W. A. Kurz, O. L. Phillips, A. Shvidenko, S. L. Lewis, J. G. Canadell, P. Ciais, R. B. Jackson, S. W. Pacala, A. D. McGuire, S. Piao, A. Rautiainen, S. Sitch, and D. Hayes. 2011. "A large and persistent carbon sink in the world's forests." *Science* 333: 988–993.

Pappalardo, V., D. La Rosa, A. Campisano, and P. La Greca. 2017. "The potential of green infrastructure application in urban runoff control for land use planning: A preliminary evaluation from a southern Italy case study." *Ecosystem Services* 26: 345–354.

Parker, G. 2013. *Global Crisis: War, Climate Change and Catastrophe in the Seventeenth Century.* Yale University Press.

Parmova, E., B. Locatelli, H. Djoudi, and O. A. Somorin. 2012. "Forests and trees for social adaptation to climate variability and change." *Wiley Interdisciplinary Reviews: Climate Change* 3: 581–596.

Pendleton, L., D. C. Donato, B. C. Murray, S. Crooks, W. A. Jenkins, S. Sifleet, C. Craft, et al. 2012. "Estimating global 'blue carbon' emissions from conversion and degradation of vegetated coastal ecosystems." *PLOS One* 7 (9): e43542.

Piao, S., P. Ciais, Y. Huang, Z. Shen, S. Peng, J. Li, L. Zhou, et al. 2010. "The impacts of climate change on water resources and agriculture in China." *Nature* 467: 43–51.

Pinsky, M. L., and N. J. Mantua. 2014. "Emerging adaptation approaches for climate-ready fisheries management." *Oceanography* 27 (4): 146–159.

Rosenzweig, C., J. Elliott, D. Deryng, A. C. Ruane, C. Müller, A. Arneth, K. J. Boote, C. Folberth, M. Glotter, N. Khabarov, et al. 2014. "Assessing agricultural risks of climate change in the 21st century in a global gridded crop model intercomparison." *Proceedings of the National Academy of Sciences* 111 (9): 3268–3273.

Ruckelshaus, M., P. Kareiva, and L. Crowder. 2014. "The future of marine conservation." In *Marine Community Ecology and Conservation*, ed. M. Bertness and B. Silliman. Sinauer Associates.

Shepard, C. C., C. M. Crain, and M. W. Beck. 2011. "The protective role of coastal marshes: A systematic review and meta-analysis." *PLOS One* 6 (11): e27374.

Squeo, F. A., B. G. Warner, R. Aravena, and D. Espinoza. 2006. "Bofedales: High altitude peatlands of the central Andes." *Revista chilena de historia natural* 79: 245–255.

Tallis, H., P. Kareiva, M. Marvier, and A. Chang. 2008. "An ecosystem services framework to support both practical conservation and economic development." *Proceedings of the National Academy of Sciences* 105 (28): 9457–9464.

Tengö, M., and K. Belfrage. 2004. "Local management practices for dealing with change and uncertainty: A cross-scale comparison of cases in Sweden and Tanzania." *Ecology and Society* 9: 4.

Thompson, M. S., S. J. Brooks, C. D. Sayer, G. Woodward, J. C. Axmacher, D. M. Perkins, and C. Gray. 2018. "Large woody debris 'rewilding' rapidly restores biodiversity in riverine food webs." *Journal of Applied Ecology* 55 (2): 895–904.

Trabucco, A., R. J. Zomer, D. A. Bossio, O. van Straaten, and L. V. Verchot. 2008. "Climate change mitigation through afforestation/reforestation: A global analysis of hydrologic impacts with four case studies." *Agriculture Ecosystems and Environment* 126: 81–97.

UN Department of Economic and Social Affairs, Population Division. 2014. *World Urbanization Prospects: The 2014 Revision* (ST/ESA/SER.A/352). United Nations.

Varian, H. R. 2014. "Big data: New tricks for econometrics." *Journal of Economic Perspectives* 28: 3–27.

Vittor, A. Y., R. H. Gilman, J. Tielsch, G. Glass, T. Shields, W. S. Lozano, V. Pinedo-Cancino, and J. A. Patz. 2006. "The effect of deforestation on the human-biting rate of *Anopheles darlingi*, the primary vector of falciparum malaria in the Peruvian Amazon." *American Journal of Tropical Medicine and Hygiene* 1 (74): 3–11.

Weatherdon, L. V., A. K. Magnan, A. D. Rogers, U. R. Sumaila, and W. W. Cheung. 2016. "Observed and projected impacts of climate change on marine fisheries, aquaculture, coastal tourism, and human health: An update." *Frontiers in Marine Science* 3: 48.

Westerling, A. L., H. G. Hidalgo, D. R. Cayan, and T. W. Swetnam. 2006. "Warming and earlier spring increase western US forest wildfire activity." *Science* 313 (5789): 940–943.

Wheeler, T., and J. von Braun. 2013. "Climate change impacts on global food security." *Science* 341: 508–513.

CHAPTER TWENTY-FOUR

Climate Change Mitigation Using Terrestrial Ecosystems: Options and Biodiversity Impacts

MONIKA BERTZKY, REBECCA C. BROCK, LERA MILES, AND VALERIE KAPOS

INTRODUCTION

To reduce the effects of climate change on humans and biodiversity, there is a need not only to adapt to climate change impacts that cannot be avoided but also to mitigate climate change, that is, to reduce the sources and enhance the sinks of greenhouse gases (GHGs). Mitigation can be undertaken in a wide variety of ways, targeting all the different sources of GHG emissions. Land-use change is one of the largest sources of GHGs. Land management for climate change mitigation, here referred to as ecosystem-based mitigation, can substantially reduce carbon emissions, and thus provides an important tool for countries' contributions to achieving the objectives of the Paris Climate Agreement (Griscom et al. 2017). Some mitigation options, such as protecting forests, retain and enhance carbon stocks and make a positive contribution to biodiversity conservation. However, some mitigation practices can affect biodiversity in ways that may compound the more direct impacts of climate change on biodiversity and ecosystem services (outlined in previous chapters). This chapter explores the role of ecosystems in climate change mitigation, the impacts of mitigation policies on ecosystems and biodiversity, and the importance of safeguards in the endeavor to balance different policy objectives.

ROLE OF ECOSYSTEMS IN MITIGATION

The degradation or destruction of ecosystems can cause carbon stored in their biomass and soils to be released as GHGs, such as carbon dioxide (CO_2). The quantity of GHGs emitted depends on the original and the "new" use of the land as well as on longer-term land management. Some ecosystems have particularly high potential to contribute to climate change mitigation. These include forests, wetlands, coastal ecosystems, and drylands. Management of agricultural ecosystems offers great potential for mitigation.

Mitigation Measures Targeting Forest: REDD+

The largest emissions from land-use change result from conversion of mature tropical forest to agricultural land to produce both food and bioenergy crops, and from the burning and drainage of tropical peat-swamp forests and soils (van der Werf et al. 2009). In 2015, about 30.6 percent of the world's terrestrial surface (just under 4 billion ha) was covered by forests, storing about 296 gigatons of biomass carbon (Food and Agriculture Organization of the United Nations [FAO] 2016). The global rate of net annual forest loss more than halved from 7.2 million ha in the 1990s, and the current net annual decrease in forest area is still very large: about 3.3 million ha, averaged over the period 2010–2015 (FAO 2016). In addition, roughly 20 percent of the world's remaining forests are degraded (Potapov, Laestadius, and Minnemeyer 2011), often through selective timber harvesting and damage caused to surrounding vegetation (Pearson, Brown, and Casarim 2014). Over the decade 2007–2016, CO_2 emissions from land use change (including deforestation, afforestation, logging and associated forest degradation, shifting cultivation, and regrowth of forests following wood harvest or abandonment of agriculture) accounted for about 12 percent of the world's total carbon emissions (Le Quéré et al. 2018; see also Figure 24.1).

The UN Framework Convention on Climate Change (UNFCCC) has agreed on key components to provide financial incentives to reduce forest-related GHG emissions in developing countries, and some funders are making resources available to support this. Such incentives would cover five activities, collectively referred to as REDD+: (1) reducing emissions from deforestation and (2) from forest degradation, plus (3) conservation of forest carbon stocks, (4) sustainable management of forest, and (5) enhancement of forest carbon stocks.

REDD+ has the potential to contribute substantially to climate change mitigation and thus to achievement of the goals in the Paris Agreement and to help meet the objectives of the New York Declaration on Forests (halving natural forest loss by 2020, and striving to end it by 2030), endorsed by world leaders at the UN Secretary-General's Climate Summit in 2014. Although the five REDD+ activities are not further defined, they are widely interpreted as applying to (1) forests under direct deforestation pressure, (2) forests under direct degradation pressure, (3) forests not under direct pressure but acting as important carbon sinks, (4) forests used for their resources, and (5) areas that provide opportunities for forest restoration (including degraded forest), reforestation, or afforestation. REDD+ thus addresses the different sources of GHG emissions, and opportunities for sequestration, in the forest sector and from deforestation. Ecosystems and biodiversity play a role in each of the REDD+ activities. Intact natural forests are more resilient to climate change than degraded, often less biodiverse forests (Miles et al. 2010). Conserving these areas thus helps maintain not only important forest carbon reservoirs but also the ecosystem's natural resilience and biodiversity.

WIDER ECOSYSTEM-BASED MITIGATION

Management of Non-Forest Ecosystems

Ecosystems such as peatlands, littoral zones, and dryland ecosystems have great potential to contribute to climate change mitigation.

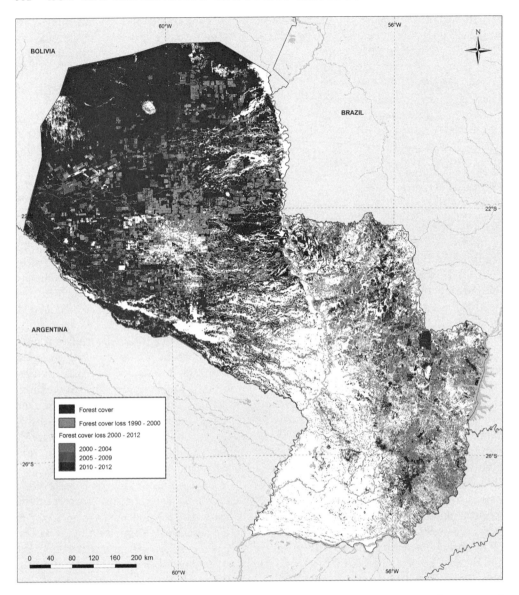

Figure 24.1. Forest cover loss in Paraguay between 2000 and 2012 (*map*) and emissions/removals of CO$_2$ equivalents associated with net forest change over the same period of time (*graph*). *Sources*: Walcott et al. 2015 (using data from PNC-ONU REDD+ Paraguay 2011, The Global Land Cover Facility 2006) by permission of UNEP-WCMC; Hansen et al. 2013 (*map*), and FAOSTAT 2014 (*graph*).

Whereas their aboveground biomass may be lower than that of forests, the carbon stored in their belowground biomass (roots, soils, and sediment) can be substantial. Overall, the carbon stored in the world's soils is much larger than that in its aboveground biomass (Scharlemann et al. 2014). Peat soils and coastal sediments are particularly rich in carbon given their density, depth, and anaerobic conditions, so CO$_2$ emissions from soil and sediment are of particular concern in these ecosystems. The world's wetlands are also the largest natural source of methane (Turetsky et al. 2014), a GHG that has a much stronger global warming potential than CO$_2$, at least for the first six or seven decades after its emission.

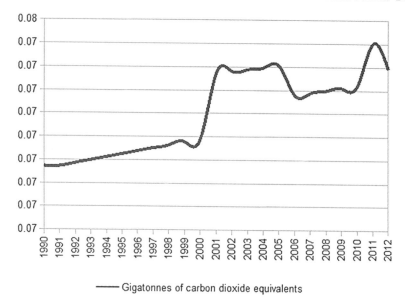

0.08
0.07
0.07
0.07
0.07
0.07
0.07
0.07
0.07
0.07
0.07

1990 1991 1992 1993 1994 1995 1996 1997 1998 1999 2000 2001 2002 2003 2004 2005 2006 2007 2008 2009 2010 2011 2012

——— Gigatonnes of carbon dioxide equivalents

Figure 24.1. (continued)

Peatlands occur on all continents from the tropics to the Arctic, and from low-lying to high alpine locations. They are being converted at large scale for agriculture, forestry, and grazing, and in some parts of the world, peat is harvested for fuel. Some of the most carbon-dense peatland areas are tropical peat-swamp forests, where improved management may be supported under REDD+. Nevertheless, peatland outside of forest deserves attention. Whereas emissions from forest clearance are largely instantaneous, those from peatland drainage and from subsurface fire continue for as long as the land remains drained and the peat can continue to oxidize (Joosten 2015b).

The role of coastal ecosystems in climate change mitigation, including mangroves, salt marshes, and seagrass beds, has received growing attention of late. These habitats cover an area one to two orders of magnitude smaller than that covered by terrestrial forests, but their potential contribution per unit area to long-term carbon sequestration is much greater (Mcleod et al. 2011). Conversely, drylands have relatively low biomass carbon density, but because of their large surface area, dryland carbon storage

and sequestration have global significance. Drylands include most of the temperate, subtropical, and tropical grasslands, savannas, shrublands, and deserts (UN Convention to Combat Desertification, UNCCD, definition). Their soil organic carbon stocks account for 27 percent of global soil carbon reserves (Millennium Ecosystem Assessment 2005). They have suffered considerably from degradation and conversion, with more than 90 percent of the world's drylands converted to pasture and croplands (Safriel et al. 2005). Around 0.3 Gt C is estimated to be lost from dryland soils every year as a result of unsustainable agricultural and pastoral practices (Joosten 2015a). In these ecosystems, avoidance of further destruction and degradation, restoration of degraded areas, and sustainable management of those areas in productive use can help to avoid or reduce GHG emissions and enhance their removals.

Agricultural Practices

In 2009, about 37 percent of the terrestrial area of the world was used for agriculture, and this percentage continues to increase (FAO 2013). How agriculture is managed and the resulting impacts on soils can influence emissions. In addition to CO_2

emissions from agricultural soils, crop and livestock production account for half of the anthropogenic emissions of methane and two-thirds of those of nitrous oxide. The technical mitigation potential from global agriculture (except fossil-fuel offsets from biomass), including climate-smart agricultural practices, is estimated at 5.5–6 gigatons of CO_2 equivalents per year by 2030 (Smith 2012).

IMPACT OF MITIGATION ON ECOSYSTEMS AND BIODIVERSITY

Ecosystem-based and other types of mitigation can have large impacts, both positive and negative, on biodiversity. A complete understanding of all impacts of mitigation options on biodiversity requires a full life-cycle analysis that considers all stages entailed in implementing the respective measures, including their immediate and long-term, direct and indirect, and on- and off-site effects. Such analyses are not yet available for most mitigation options, and it is beyond the scope of this chapter to detail all possible impacts on ecosystems and biodiversity of the range of different mitigation options. Instead, examples are provided to highlight key biodiversity impacts of ecosystem-based mitigation measures, and the role of emerging safeguards and standards is discussed.

Forest and Nonforest Ecosystems

By reducing deforestation and degradation, strengthening efforts to conserve forest carbon, establishing and improving sustainable management of forests, and restoring and reforesting areas where appropriate, there is great potential for REDD+ to benefit forest biodiversity (Miles and Kapos 2008; Harvey, Dickson, and Kormos 2010). However, there is also potential for REDD+ to cause harm (Gardner et al. 2012; Harvey, Dickson, and Kormos 2010; Miles and Kapos 2008), including where (1) it causes the displacement of conversion or degradation either to other ecosystems in that country or to countries that do not participate in REDD+ ("leakage"), (2) afforestation is undertaken in nonforest areas that are highly biodiverse, (3) REDD+ leads to the conversion of naturally regenerating forests or nonforest systems to plantations, and (4) REDD+ activities are not designed with due consideration of biodiversity and conservation targets. For example, in Vietnam deforestation is much reduced and tropical forest cover increasing. Yet about 39 percent of the regrowth that occurred between 1987 and 2006 was achieved by displacement of forest extraction to other countries, and about half of the wood imports in the same period were illegal (Meyfroidt and Lambin 2009). Biodiversity impacts depend on the exact location of the displaced deforestation, which are difficult to identify but are assumed to be substantial.

Degraded lands, which in some countries include degraded forests, may host biodiversity of importance for grazing animals, wild foods, medicines, artisanal products, and wildlife but could be perceived as suitable areas for forest plantations, resulting in establishment of ecosystems that are less biodiverse and unable to provide similar functions. In general, forest restoration using a range of native species to enhance connectivity between protected areas will have larger biodiversity benefits than reforestation using monoculture plantations of fast-growing exotic species (Lamb, Erskine, and Parrotta 2005).

The "Cancún Agreements" of the UNFCCC formally recognize the concerns over potential environmental harm from REDD+ by including guidance and safeguards for policy approaches and positive incentives (Appendix I of Decision 1/CP.16 in UNFCCC 2011). The Cancún safeguards cover both social and environmental issues. The inclusion of social issues is essential—without addressing potential social harm, governance, acceptance of local and indigenous peoples rights and needs, REDD+ is unlikely to achieve either environmental or social benefits. The main reference to biodiversity is in safeguard (e), which stipulates that REDD+ "actions are consistent with the conservation of natural forests and biological diversity, ensuring that the [REDD+ actions] are not used for the conversion of natural forests, but are instead used to incentivize the protection and conservation of natural forests and their ecosystem services, and to enhance other social and environmental benefits." To support the implementation of these safeguards, additional guidance has been produced by a number of organizations, such as the UN-REDD Programme (the United Nations' collaborative initiative on REDD+).

REDD+ is only one policy mechanism that targets ecosystem-based climate change mitigation in forest areas and focuses on developing countries. Land-use change emissions are included within the GHG emissions inventories that Annex 1 (developed) country parties submit annually to the UNFCCC. Categories of climate change mitigation activities for all ecosystems are similar to those identified as part of REDD+, in that they include conversion avoided, restoration, and sustainable management, and thus have high potential to provide biodiversity benefits. For example, rewetting of peatlands reduces emissions but also is used by several Nordic and Baltic countries and the United Kingdom as a strategy to restore biodiversity (Joosten 2015b; Peh et al. 2014).

It is likely that there will be more specific climate change mitigation policies that target wetlands, drylands, and coastal ecosystems in the future. For example, developing countries have agreed to establish nationally appropriate mitigation actions (NAMAs), which may include avoided conversion as well as conservation and restoration activities in these habitats. As different mitigation approaches expand, there may be a need for specific biodiversity safeguards, similar to those developed for REDD+. Additionally, monitoring to assess the impacts of mitigation policies is critical both to improve their effectiveness (e.g., through amendments to close loopholes) and to avoid harm.

Agriculture and Biofuels

The climate change mitigation potential of adjusting agricultural practices has only in recent years gained attention in international climate talks. Mitigation measures discussed include reducing soil erosion, reducing leaching of nitrogen and phosphorous from fertilizer, conserving soil moisture, increasing the diversity of crop rotations, and establishing agroforestry systems where appropriate. These measures, which can also help to conserve biodiversity, overlap with those of conservation agriculture, which aims to reduce negative environmental impacts. The development of specific policies that address climate change mitigation on agricultural land is still in its infancy, and an understanding of how these measures will affect biodiversity is still developing.

One area that has received more attention is the cultivation of biofuel feedstocks. Biofuels, produced from living organisms such as plants and algae, have been promoted as a key option to mitigate climate change. Many national emission reduction policies include mandates to increase the share of biofuels in their energy mix. Estimates suggest that between 44 million and 118 million ha of additional cropland will be required for biofuel production by 2030 (Lambin and Meyfroidt 2011). Most governments, however, fail to account correctly for the emissions from clearing land for biofuel feedstock cultivation or for burning biofuels. Some 60 percent of renewable energy in the European Union is from bioenergy that is subsidized the same as zero-emission sources such as wind and solar because it is erroneously considered "carbon neutral" (Brack 2017). In fact, forest bioenergy (wood) for producing electricity is a growing source of forest loss. In the United States and Canada the attempt to meet demand for wood pellets that are replacing coal for electric power generation has caused significant deforestation.

In some tropical areas of the world, such as Brazil and Indonesia, biofuel expansion is a major emerging threat to biodiversity (Stromberg et al. 2010). Biodiversity impacts are feedstock specific. For example, oil palm (mainly *Elaeis guineensis* but also *Elaeis oleifera* and *Attalea maripa*) cultivation often entails conversion of virgin forest, whereas cassava (*Manihot esculenta*) and Jatropha (*Jatropha curcas*) can be cultivated on grassland or land already used for agriculture. A comparison of studies from Indonesia suggests that total vertebrate species richness in oil palm plantations is less than half that of tropical forests, with great differences in community composition (Figure 24.2). Some of the main components of tropical forests, such as forest trees, lianas, epiphytic orchids, and indigenous palms, are absent, so the flora in plantations is impoverished (Danielsen et al. 2009).

There is increasing concern about the impact of oil palm cultivation on great apes in Southeast Asia and Africa. Models indicate that between 1999 and 2015 half the population of the Bornean orangutan (*Pongo pygmaeus*) was affected by logging, deforestation, or industrialized plantations (Voigt et al. 2018). More than half the oil palm concessions for which data were available in Africa overlap with great ape distributions, and about 42 percent of the geographic distribution of great apes overlaps with areas suitable for oil palm production (Wich et al. 2014). Studies have found that orangutans disperse into mature oil palm plantations only where natural forest patches are nearby (Ancrenaz et al. 2014). Large-scale conversion for monoculture oil palm plantations both in Southeast Asia and Africa, which is likely to be increased by demand for biofuel, could severely affect the distribution and the conservation status of great apes.

Although difficult to trace, it can be assumed that indirect land-use change from biofuel production occurs at substantial extent. For example, in Brazil, area for both biofuel production and cattle ranches has expanded over recent years. In the Amazon region more than 90 percent of soybean plantations planted after 2006 replaced cattle ranches (Lapola et al. 2010), which in turn were displaced into areas of forest and *cerrado* (tropical savanna ecoregion of Brazil). Land-use change scenarios for Europe for the period 2000–2030 have shown that

Figure 24.2. Impact on fauna of replacing forest with oil palms. Mean number of invertebrate species (patterned bar) and mammal, bird, and reptile species (solid bar) recorded in oil palm (including new colonists of oil palm) as a proportion of those recorded in forest. Meta-analysis sample sizes are provided in parentheses; error bars are 95% confidence intervals. (Redrawn from Danielsen et al. 2009.)

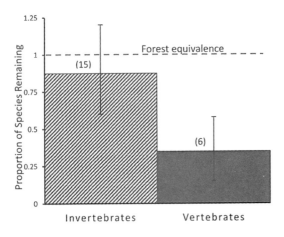

indirect effects of the European Union's biofuel directive on land use and biodiversity in Europe are much larger than the direct effects (Hellmann and Verburg 2010).

Ways to reduce the impact of biofuel production on ecosystems and biodiversity are being explored. For example, growing low-input, high-diversity mixtures of native grassland perennials on degraded agricultural land could provide more usable energy, greater GHG reductions, and less agrichemical pollution than conventional biofuel crop cultivation without displacing food production or harming biodiversity (Tilman, Hill, and Lehman 2006; Verdade, Piña, and Rosalino 2015). Agricultural intensification has been suggested as another alternative to continued growth in land use by the biofuels sector.

The known risks to biodiversity of biofuel production have triggered the development of safeguards and standards for this sector. The EU Renewable Energy Directive (RED), for example, includes comprehensive sustainability criteria that aim to prevent the conversion of areas of high biodiversity and high carbon stock for the production of biofuel feedstocks. Biofuels must comply with these criteria if they are to be counted toward the 10 percent renewable energy target for the transport sector or to receive state financial support. Given the limited contribution of the EU to global biofuel consumption, only a small share of global biofuel production is directly subject to these targets and standards. Nonetheless, concerns about biofuels impacts have prompted development of relevant voluntary safeguards and standards initiatives that may be more widely applied, including the Roundtable on Sustainable Biomaterials, the Roundtable on Sustainable Palm Oil, and the Better Sugarcane Initiative.

Other Renewable Energy Options

The use of non-ecosystem-based renewable energy options, such as direct solar energy, geothermal, hydropower, ocean energy, and wind, is continuously increasing, spurred by government policies and subsidies, the decreasing cost of renewable energy technologies, and other factors (IPCC 2011). Such options presumably can have large impacts on ecosystems and biodiversity, although knowledge of these impacts is limited for some options.

In general, impacts fall into five main groups: (1) direct environmental footprint of the hard structures (including habitat alteration, fragmentation, isolation of protected areas, and loss); (2) noise of construction, decommissioning, and disturbance by human presence; (3) collision with operating infrastructure; (4) pollution and disturbance from operating infrastructure; and (5) provision of new substrate for establishment of new communities, including invasive species (amended from Bertzky et al. 2010; Hoetker, Thomsen, and Jeromin 2006).

Wind farms can harm biodiversity through enhanced collision risks for birds and bats as well as disturbance leading to displacement, for example, of waders nesting on open ground. However, impacts are species specific, and some species seem to benefit from wind farms, possibly because the agricultural use of land in the immediate vicinity often is suspended. Limitations to fishing in the vicinity of offshore wind farms can have similar positive impacts (Bergström et al. 2014). The species affected depend on the location, such as near to shore, on mountain ridges, or close to forest (Hoetker, Thomsen, and Jeromin 2006). Appropriate siting and the use of blinking lights have reduced collisions of birds with wind towers.

Hydropower is a major renewable energy source for a number of countries worldwide. Environmental impacts of dam construction include the destruction of habitat and altered water-flow patterns, sedimentation, discharge change, and heavy metal pollution. It is anticipated that the hundreds of new dams planned for construction in the major tropical rivers will disrupt fish migration and thus endanger a large number of endemic species, as well as harm local livelihoods (Winemiller et al. 2016). The impacts of renewables such as dams are not always instantaneous. A life-cycle impact assessment of the dam construction on the upper Mekong River suggests that the initial impact accounts for only about 30 percent of the total environmental impact (Chen, Chen, and Fath 2015). Existing dams in the Amazon region have already fragmented the networks of six of eight major Andean Amazon river basins, while another 160 dams are proposed. Such development not only threatens biodiversity; over time it is also likely to drastically alter river channel and floodplain geomorphology and associated ecosystem services due to changes in sediment deposition (Anderson et al. 2018).

Biodiversity effects of wave and tidal energy installations largely depend on the site selection for placing the energy converters. In certain areas, such as fish-spawning areas or highly biodiverse sites, their placement may harm marine habitats and species (Witt et al. 2012). Similarly, possible direct impacts on biodiversity of solar power can be expected, especially where the aboveground vegetation is cleared for construction and landscapes are fragmented. Indirect impacts may also occur in a wide variety of ways including from changes in microclimate and local hydrology, electromagnetic field effects, erosion, water, and soil contamination (Hernandez et al. 2014).

Governments and funding agencies often require strategic environmental assessments at the planning stage of renewable energy development. Results of such assessments can be used to identify ways to reduce negative environmental impacts, where feasible, even if the stringency of assessment varies between countries and the negative impacts identified by assessments do not alter the implementation of plans. Increasingly, conservation organizations (e.g., BirdLife International) advocate for strong biodiversity and social safeguards to be included in lending decisions and development priorities.

Other Mitigation Options

Mitigation can also be achieved through increased energy efficiency and geoengineering. Increased energy efficiency is considered to be of relatively low ecological risk, although the production and disposal of some energy-saving material, such as mercury (used in compact fluorescent bulbs), can have negative environmental impacts (Cusack et al. 2014). These lamps are being replaced with even more efficient mercury-free light-emitting diodes. Climate geoengineering proposals include solar radiation management and engineered carbon capture and storage (the removal of atmospheric CO_2 and its storage in long-lived reservoirs). Many questions persist on the impacts of such endeavors on ecosystems and biodiversity. However, because they can involve purposeful alteration of ecological processes, such as the transfer of the Sun's energy, at unprecedented scales, consideration of possible environmental impacts is vital (Russell et al. 2012). An interdisciplinary assessment of different climate engineering strategies concluded that both solar radiation management and carbon capture and storage carry ecological risks of potentially serious magnitude and duration (Cusack et al. 2014). For example, accidental leakage of concentrated liquid CO_2 from deep-ocean reservoirs could cause extreme ocean acidification, with serious effects on marine species and ecosystem services (Cusack et al. 2014). A further concern is that implementation and termination of some forms of geoengineering may lead to rapid shifts in climatic conditions. Sudden termination of solar geoengineering, for example, would increase the threats to biodiversity from climate change considerably (Trisos et al. 2018).

CONCLUSIONS

Different ecosystems can play an important role in climate change mitigation. Mitigation policies targeting these ecosystems are partly in place and increasingly under development, but their potential as a contribution to delivering on the Paris Climate Agreement is far from fully realized. Such policies not only can use ecosystems to help mitigate the climate but also help protect biodiversity and ecosystem services. The benefits and costs of these policies depend on the translation of the policies to action on the ground as well as on the interpretation and application of safeguards. Where mitigation policies are implemented without due consideration of ecosystems and biodiversity, the risk of harm is high and the potential to achieve multiple benefits, for the climate as well as for biodiversity, may not be realized. Biodiversity safeguards are a crucial tool to help avoid negative impacts on biodiversity and increase synergies between mitigation and conservation policies. Nevertheless, safeguards are not yet in place for all of the discussed mitigation options. Their application is often voluntary, and they leave a lot of room for interpretation or do not provide comprehensive geographical coverage. However, if the goal is forest ecosystem restoration, a full menu of ecosystem services may be achieved (e.g., Ren et al. 2017). The regenerative system of forests and related ecosystems that has kept carbon dioxide in balance for the past 300 million years can be restored and expanded to absorb and store carbon in plants and soils, while contributing to conserving systems that support the greatest amount of terrestrial biodiversity. Further work on implementation of safeguards and how to ensure environmental risks are reduced and benefits maximized will therefore be crucial in the coming years.

Apart from the existing synergies between climate change mitigation and conservation policies, more such synergies can be identified, such as with climate change adaptation policies (see previous chapter). While in the planning phase, consideration of a range of synergies between policies may appear to complicate implementation, the extra effort may nonetheless help avoid implementation of conflicting measures and thus increase overall policy efficiency and effectiveness.

The role of ecosystems and biodiversity in mitigation as well as the success of climate change mitigation policies and their environmental impacts depend on one more fundamental influence: climate change itself. Each of the ecosystems highlighted here as having potential to contribute to mitigation can be affected by climate change (see previous chapters) and may lose some capacity for carbon sequestration and storage. For mitigation policies that target these ecosystems, it will be important to monitor changes and adapt mitigation measures according to site-specific conditions to be able to achieve their objectives.

ACKNOWLEDGMENTS

The authors are particularly grateful to Corinna Ravilious and Xavier de Lamo for reproducing the map as part of Figure 24.1, to Finn Danielsen for his help with Figure 24.2, to Neil Burgess and William Moomaw for their input, and to all anonymous reviewers for their comments.

REFERENCES

Ancrenaz, Marc, Felicity Oram, Laurentius Ambu, Isabelle Lackman, Eddie Ahmad, Hamisah Elahan, Harjinder Kler, Nicola K. Abram, and Erik Meijaard. 2015. "Of Pongo, palms and perceptions: A multidisciplinary assessment of Bornean orang-utans *Pongo pygmaeus* in an oil palm context." *Oryx* 49: 465–472.

Anderson, Elizabeth P., Clinton N. Jenkins, Sebastian Heilpern, Javier A Maldonado-Ocampo, Fernando M. Carvajal-Vallejos, Andrea C. Encalada, and Juan Francisco Rivadeneira. 2018. "Fragmentation of Andes-to-Amazon connectivity by hydropower dams." *Science Advances* 4: 1–8. https://doi.org/10.1126/sciadv .aao1642.

Bergström, Lena, Lena Kautsky, Torleif Malm, Rutger Rosenberg, Magnus Wahlberg, Nastassja Åstrand Capetillo, and Dan Wilhelmsson. 2014. "Effects of offshore wind farms on marine wildlife: A generalized impact assessment." *Environmental Research Letters* 9

(3): 34012. https://doi.org/10.1088/1748-9326/9/3/034012.

Bertzky, Monika, Barney Dickson, Russell Galt, Eleanor Glen, Mike Harley, Nikki Hodgson, Igor Lysenko, et al. 2010. *Impacts of Climate Change and Selected Renewable Energy Infrastructures on EU Biodiversity and the Natura 2000 Network: Summary Report.* European Commission and International Union for Conservation of Nature.

Brack, Duncan. 2017. *The Impacts of the Demand for Woody Biomass for Power and Heat on Climate and Forests.* Royal Institute of International Affairs, Chatham House.

Chen, Shaoqing, Bin Chen, and Brian D. Fath. 2015. "Assessing the cumulative environmental impact of hydropower construction on river systems based on energy network model." *Renewable and Sustainable Energy Reviews* 42: 78–92. https://doi.org/10.1016/j.rser.2014.10.017.

Cusack, Daniela F., Jonn Axsen, Rachael Shwom, Lauren Hartzell-Nichols, Sam White, and Katherine R. M. Mackey. 2014. "An interdisciplinary assessment of climate engineering strategies." *Frontiers in Ecology and the Environment* 12 (5): 280–287.

Danielsen, Finn, Hendrien Beukema, Neil D. Burgess, Faizal Parish, Carsten A. Brühl, Paul F. Donald, Daniel Murdiyarso, et al. 2009. "Biofuel plantations on forested lands: Double jeopardy for biodiversity and climate." *Conservation Biology* 23 (2): 348–358. https://doi.org/10.1111/j.1523-1739.2008.01096.x.

FAO. 2013. *FAO Statistical Yearbook 2013.* Food and Agriculture Organization of the United Nations.

FAO. 2016. *Global Forest Resources Assessment 2015.* FAO Forestry. Food and Agriculture Organization of the United Nations. https://doi.org/10.1002/2014GB005021.

FAOSTAT. 2014. "Emissions/removals of Carbon Dioxide Associated with Forest and Net Forest Conversion. Paraguay." Food and Agriculture Organization of the United Nations Statistics Division. http://faostat3.fao.org/browse/G2/GF/E.

Gardner, Toby A., Neil D. Burgess, Naikoa Aguilar-Amuchastegui, Jos Barlow, Erika Berenguer, Tom Clements, Finn Danielsen, et al. 2012. "A framework for integrating biodiversity concerns into national REDD+ programmes." *Biological Conservation* 154: 61–71. https://doi.org/10.1016/j.biocon.2011.11.018.

Global Land Cover Facility. 2006. Forest Cover Change in Paraguay, Version 1.0. University of Maryland Institute for Advanced Computer Studies. College Park, MD, 1990–2000.

Griscom, Bronson W., Justin Adams, Peter W. Ellis, Richard A. Houghton, Guy Lomax, Daniela A. Miteva, William H. Schlesinger, et al. 2017. "Natural climate solutions." *Proceedings of the National Academy of Sciences* 114 (44): 11645–11650. https://doi.org/10.1073/pnas.1710465114.

Hansen, M. C., P. V. Potapov, R. Moore, M. Hancher, S. A. Turubanova, A. Tyukavina, D. Thau, et al. 2013. "High-resolution global maps of 21st-century forest cover change." *Science* 342 (6160): 850–853. http://science.sciencemag.org/content/342/6160/850.abstract.

Harvey, Celia A., Barney Dickson, and Cyril Kormos. 2010. "Opportunities for achieving biodiversity conservation through REDD." *Conservation Letters* 3 (1): 53–61. https://doi.org/10.1111/j.1755-263X.2009.00086.x.

Hellmann, Fritz, and Peter H. Verburg. 2010. "Impact assessment of the European biofuel directive on land use and biodiversity." *Journal of Environmental Management* 91 (6): 1389–1396. https://doi.org/10.1016/j.jenvman.2010.02.022.

Hernandez, Rebecca R., Shane B. Easter, Michelle L. Murphy-Mariscal, Fernando T. Maestre, M. Tavassoli, Edith B. Allen, Cameron W. Barrows, et al. 2014. "Environmental impacts of utility-scale solar energy." *Renewable and Sustainable Energy Reviews* 29: 766–779. https://doi.org/10.1016/j.rser.2013.08.041.

Hoetker, Hermann, Kai-Michael Thomsen, and Heike Jeromin. 2006. *Impacts on Biodiversity of Exploitation of Renewable Energy Sources: The Example of Birds and Bats—Facts, Gaps in Knowledge, Demands for Further Research, and Ornithological Guidelines for the Development of Renewable Energy Exploitation.* Michael-Otto-Institut of Naturschutzbund Deutschland.

IPCC. 2011. *Special Report on Renewable Energy Sources and Climate Change Mitigation.* Edited by Ottmar Edenhofer, Rafael Pichs-Madruga, Youba Sokona, Kristin Seyboth, Patrick Matschoss, Susanne Kadner, Timm Zwickel, et al. Cambridge University Press.

Joosten, Hans. 2015a. "Current soil carbon loss and land degradation globally—Where are the hotspots and why there?" In *Soil Carbon: Science, Management and Policy for Multiple Benefits,* ed. Steven A. Banwart, Elke Noellemeyer, and Elinor Milne, 224–234. CAB International.

Joosten, Hans. 2015b. *Peatlands, Climate Change Mitigation and Biodiversity Conservation: An Issue Brief on the Importance of Peatlands for Carbon and Biodiversity Conservation and the Role of Drained Peatlands as Greenhouse Gas Emission Hotspots.* Nordic Council of Ministers. http://norden.diva-portal.org/smash/get/diva2:806688/FULLTEXT01.pdf.

Lamb, David, Peter D. Erskine, and John A. Parrotta. 2005. "Restoration of degraded tropical forest landscapes." *Science* 310 (5754): 1628–1632. https://doi.org/10.1126/science.1111773.

Lambin, Eric F., and Patrick Meyfroidt. 2011. "Global land use change, economic globalization, and the looming land scarcity." *Proceedings of the National Academy of Sciences* 108 (9): 3465–3472. https://doi.org/10.1073/pnas.1100480108.

Lapola, David M., Ruediger Schaldach, Joseph Alcamo, Alberte Bondeau, Jennifer Koch, Christina Koelking, and Joerg A. Priess. 2010. "Indirect land-use changes can overcome carbon savings from biofuels in Brazil." *Proceedings of the National Academy of Sciences* 107 (8): 3388–3393. https://doi.org/10.1073/pnas.0907318107.

Le Quéré, Corinne, Robbie M. Andrew, Pierre Friedlingstein, Stephen Sitch, Julia Pongratz, Dan Zhun Zhu Korsbakken, Glen P. Peters, Josep G. Canadell, Robert B. Jackson, Thomas A. Boden, et al. 2018. "The global carbon budget 2017." *Earth System Science Data Discussions* 405–448. https://doi.org/10.5194/essdd-2017-123.

Mcleod, Elizabeth, Gail L. Chmura, Steven Bouillon, Rodney Salm, Mats Björk, Carlos M. Duarte, Catherine E. Lovelock, William H. Schlesinger, and Brian R. Silliman. 2011. "A blueprint for blue carbon: Toward an improved understanding of the role of vegetated coastal habitats in sequestering CO_2." Frontiers in Ecology and the Environment 9 (10): 552–560. https://doi .org/10.1890/110004.

Meyfroidt, Patrick, and Eric F. Lambin. 2009. "Forest transition in Vietnam and displacement of deforestation abroad." Proceedings of the National Academy of Sciences 106 (38): 16139–16144. https://doi.org/10.1073/pnas .0904942106.

Miles, Lera, Emily Dunning, Nathalie Doswald, and Matea Osti. 2010. "A safer bet for REDD+: Review of the evidence on the relationship between biodiversity and the resilience of forest carbon stocks." Working Paper v2, Multiple Benefits Series 10. UNEP World Conservation Monitoring Centre.

Miles, Lera, and Valerie Kapos. 2008. "Reducing greenhouse gas emissions from deforestation and forest degradation: Global land-use implications." Science 320 (5882): 1454–1455. https://doi.org/10.1126/science .1155358.

Millennium Ecosystem Assessment. 2005. Ecosystems and Human Well-Being: Desertification Synthesis. World Resources Institute.

Pearson, Timothy R. H., Sandra Brown, and Felipe M. Casarim. 2014. "Carbon emissions from tropical forest degradation caused by logging." Environmental Research Letters 9 (3): 34017. https://doi .org/10.1088/1748-9326/9/3/034017.

Peh, Kelvin S. H., Andrew Balmford, Rob H. Field, Anthony Lamb, Jennifer C. Birch, Richard B. Bradbury, Claire Brown, et al. 2014. "Benefits and costs of ecological restoration: Rapid assessment of changing ecosystem service values at a U.K. Wetland." Ecology and Evolution 4 (20): 3875–3886. https://doi.org/10.1002/ ece3.1248.

PNC ONU-REDD+ Paraguay. 2011. Mapa de bosque/no bosque. Inventario Forestal Nacional. Asunción, Paraguay: PNC ONU-REDD+.

Potapov, Peter, Lars Laestadius, and Susan Minnemeyer. 2011. Global Map of Forest Landscape Restoration Opportunities. World Resources Institute. http://www.wri.org/forest -restoration-atlas.

Ren, Yanjiao, Yihe Lü, Bojie Fu, and Kun Zhang. 2017. "Biodiversity and ecosystem functional enhancement by forest restoration: A meta-analysis in China." Land Degradation & Development 28 (7): 2062–2073. https://doi .org/10.1002/ldr.2728.

Russell, Lynn M., Philip J. Rasch, Georgina M. Mace, Robert B. Jackson, John Shepherd, Peter Liss, Margaret Leinen, et al. 2012. "Ecosystem impacts of geoengineering: A review for developing a science plan." AMBIO 41 (4): 350–369. https://doi.org/10.1007/s13280 -012-0258-5.

Safriel, Uriel, Zafar Adeel, David Niemeijer, Juan Puigdefabregas, Robin White, Rattan Lal, Mark Winslow, et al. 2005. "Dryland systems." In Ecosystems and Human Well Being: Current State and Trends, ed. Rashid Hassan, Robert Scholes, and Neville J. Ash, 623–662. Island Press.

Scharlemann, Jörn P. W., Edmund V. J. Tanner, Roland Hiederer, and Valerie Kapos. 2014. "Global soil carbon: Understanding and managing the largest terrestrial carbon pool." Carbon Management 5 (1): 81–91. https://doi.org/10.4155/cmt.13.77.

Smith, Pete. 2012. "Agricultural greenhouse gas mitigation potential globally, in Europe and in the UK: What have we learnt in the last 20 years?" Global Change Biology 18 (1): 35–43. https://doi .org/10.1111/j.1365-2486.2011.02517.x.

Stromberg, Per M., Alexandros Gasparatos, Janice S. H. Lee, John Garcia-Ulloa, Lian Pin Koh, and Kazuhiko Takeuchi. 2010. Impacts of Liquid Biofuels on Ecosystem Services and Biodiversity. United Nations University Institute of Advanced Studies.

Tilman, David, Jason Hill, and Clarence Lehman. 2006. "Carbon-negative biofuels from low-input high-diversity grassland biomass." Science 314 (5805): 1598–1600. https://doi.org/10.1126/science.1133306.

Trisos, Christopher H., Giuseppe Amatulli, Jessica Gurevitch, Alan Robock, Lili Xia, and Brian Zambri. 2018. "Potentially dangerous consequences for biodiversity of solar geoengineering implementation and termination." Nature Ecology & Evolution 2 (3): 475–482. https:// doi.org/10.1038/s41559-017-0431-0.

Turetsky, Merritt R., Agnieszka Kotowska, Jill Bubier, Nancy B. Dise, Patrick Crill, Ed R. C. Hornibrook, Kari Minkkinen, et al. 2014. "A synthesis of methane emissions from 71 northern, temperate, and subtropical wetlands." Global Change Biology 20 (7): 2183–2197. https://doi.org/10.1111/gcb.12580.

van der Werf, Guido R., Douglas C. Morton, Ruth S. DeFries, Jos G. J. Olivier, Prasad S. Kasibhatla, Robert B. Jackson, G. Jim Collatz, and James T. Randerson. 2009. "CO_2 emissions from forest loss." Nature Geoscience 2: 737–738.

Verdade, Luciano M., Carlos I. Piña, and Luís Miguel Rosalino. 2015. "Biofuels and biodiversity: Challenges and opportunities." Environmental Development 15: 64–78. https://doi.org/10.1016/j.envdev.2015.05.003.

Voigt, Maria, Serge A. Wich, Marc Ancrenaz, Erik Meijaard, Nicola Abram, Graham L. Banes, Gail Campbell-Smith, et al. 2018. "Global demand for natural resources eliminated more than 100,000 Bornean orangutans." Current Biology 28 (5): 761–769. https://doi .org/10.1016/J.CUB.2018.01.053.

Walcott, Judith, Julia Thorley, Valerie Kapos, Lera Miles, Stephen Woroniecki, and Ralph Blaney. 2015. Mapping Multiple Benefits of REDD+ in Paraguay: Using Spatial Information to Support Land-Use Planning. UNEP-WCMC. http://www .un-redd.org/tabid/5954/Default.aspx.

Wich, Serge A., John Garcia-Ulloa, Hjalmar S. Kühl, Tatanya Humle, Janice S. H. Lee, and Lian Pin Koh. 2014. "Will oil palm's homecoming spell doom for Africa's great apes?" *Current Biology* 24 (14): 1659–1663. https://doi.org/10.1016/j.cub.2014.05.077.

Winemiller, K. O., P. B. McIntyre, L. Castello, E. Fluet-Chouinard, T. Giarrizzo, S. Nam, I. G. Baird, et al. 2016. "Balancing hydropower and biodiversity in the Amazon, Congo, and Mekong." *Science* 351 (6269): 128–129. https://doi.org/10.1126/science.aac7082.

Witt, Matthew J., Emma V. Sheehan, Stuart Bearhop, Annette C. Broderick, Daniel C. Conley, Stephen P. Cotterell, E. Crow, et al. 2012. "Assessing wave energy effects on biodiversity: The wave hub experience." *Philosophical Transactions Series A: Mathematical, Physical, and Engineering Sciences* 370 (1959): 502–529. https://doi.org/10.1098/rsta.2011.0265.

Connectivity by Design: A Multiobjective Ecological Network for Biodiversity That Is Robust to Land Use and Regional Climate Change

Andrew Gonzalez, Cecile Albert, Bronwyn Rayfield, and Maria Dumitru

The accelerated pace of both land use and climate change is closing the window of opportunity to design and manage landscapes for biodiversity. The paradigm of connectivity conservation emphasizes that biodiversity and ecosystem services can be maintained by optimizing the spatial composition and configuration of ecosystem and habitat types in the landscape. Hundreds of habitat-network projects that implement this paradigm are being established around the world, but the application of connectivity concepts for biodiversity in changing landscapes remains a challenge.

In 2009 we received a request from the Ministry of Sustainable Development, Environment, and Parks of the Quebec government to identify a robust habitat network for the long-term conservation of forest biodiversity for the region surrounding the city of Montreal in southern Quebec, Canada (27,500 km² centered on 45°40′N, 73°15′W). Extending from the Appalachian Mountains in the southeast to the Laurentian Mountains in the northwest, the study area encompasses ~40 percent forest, which is heavily fragmented due to the predominance of productive agroecosystems in the St. Lawrence lowlands. Ongoing urban sprawl to the north and south of the city alters forest patch area, isolation, shape, quality, and connectivity. Projections of future forest fragmentation, derived from the current rate, and scenarios of potential climate change, derived from regional climate models, make conservation of ecological connectivity an imperative.

FRAMEWORK AND METHODS

Three aspects are crucial to the effective design of habitat networks. First, they must meet the distinct habitat requirements of a wide array of species throughout their life cycles. Second, habitat networks must accommodate different movement types, including short-range connectivity relevant to the persistence of metapopulations within the networks and long-range connectivity relevant to climate-driven range shifts across the networks. Third, habitat networks must be robust to habitat loss and changing connectivity. Future land-use and climate change scenarios are uncertain, so network design should allow for the broadest range of future conditions.

Here we adopt a multiobjective spatial optimization approach to the design of a habitat network that meets the movement needs of multiple species at several spatial scales and that is robust to land use and climate change (Albert et al. 2017). Our methodology (see Figure CS10.1) produces a spatial prioritization of forest habitats that maintains metapopulation connectivity and regional traversability to aid migration across the St. Lawrence central lowlands. Here we provide an example for five focal species. The inclusion of scenarios for future land use and climate change out to 2050 ensures that the conservation priorities reflect projected habitat and connectivity loss for the region.

RESULTS

Different elements of the forest network promote different types of movement, from

323

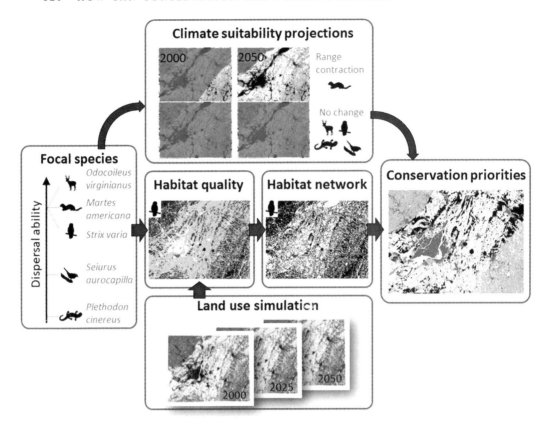

Figure CS10.1. The six sequential and parallel steps (structured as a loop) of the general framework to identify spatial conservation priorities: 1) In this example, we selected five focal species based on their dispersal ability—as a proxy for vulnerability to fragmentation—among birds (*Seiurus aurocapilla, Strix varia*), amphibians (*Plethodon cinereus*), and mammals (*Odocoileus virginianus, Martes americana*). 2) Assess species-specific habitat quality and resistance layers (quality shown by tone of gray). 3) Habitat network: Each species network is composed of nodes and links with higher to lower (grayscale) probabilities of being used. Habitat nodes are connected from edge-to-edge via least-cost links into a stepping-stone structure. 4) Climate suitability was derived for each species from an ensemble forecasting approach. Models were calibrated for the period 1971–2000 and projected for the period 2041–2070 (type 1, decrease; 2, no change; 3, increase; gray layer over map = climatically suitable areas); model changes in regional climate suitability. 5) Spatial prioritization: We used the software Zonation v3.1 to produce a spatial ranking of forest conservation priorities (grayscale with darkest tones = highest priority) based on habitat quality and complementary measures of short- and long-range connectivity. 6) Land-use change simulations: We simulated land-use change over the period 2000–2050 with a model combining a top-down demand for new urban and agricultural areas, bottom-up constraints (e.g., slope, soil quality), and spatial processes (e.g., diffusion). Land-use change simulations involved probabilistic transitions from forest to agricultural lands or urban land cover. (For further methodological details, see Albert et al. 2017.)

short to long range, such as well-connected large patches, matrix areas that channel flow from one forest patch to another, and strategically positioned patches that form a stepping-stone pathway between the Laurentian and Appalachian Mountains. The spatial ranking of conservation priorities was derived from connectivity and habitat-quality maps in the years 2000, 2025, and 2050 under land-use and climate change scenarios.

Highest conservation priority is given to the series of stepping-stone patches to the north that promotes traversability between the mountain ranges (Figure CS10.1). Many small forest fragments in the agriculturally intensive zone are also of high conservation value, providing connectivity within the St. Lawrence lowlands, particularly in the southwestern and northeastern portions.

CONCLUSIONS

We quantified and ranked the contribution of all forest fragments to the region's forest network. Our approach to network design stresses the value of different dimensions of habitat connectivity needed to mitigate the impacts of land use and climate change on the region's biodiversity. Networks optimized for multiple movement objectives can allow a diverse set of species to move among habitat patches in the short term and adjust their distributions in the longer term as climate and land use change. We recommend the use of realistic scenario-based projections of climate and land-use change so as to future-proof the structure of the habitat network being prioritized. Given the investment required to acquire, restore, and protect habitat within a network, it is important for the network to be robust to as many likely scenarios of change as possible.

The strong stakeholder relationships we developed with government ministries, municipalities, nongovernmental organizations, and landowners mean that our results are being used in the design of Montreal's greenbelt project; measures for the implementation of the habitat network include new protected areas, large-scale tree planting, and conservation easements for landowners. Ongoing work involves strengthening the economic argument for investing in the greenbelt for the range of ecosystem services the forest delivers to the people that inhabit the region.

REFERENCES

Albert, C., B. Rayfield, M. Dumitru, and A. Gonzalez. 2017. "Applying network theory to prioritize multi-species habitat networks that are robust to climate and land-use change." *Conservation Biology* 31: 1383–1396.

CHAPTER TWENTY-FIVE

Regreening the Emerald Planet: The Role of Ecosystem Restoration in Reducing Climate Change

THOMAS E. LOVEJOY

It is clear from the preceding chapters that the biosphere and biological diversity are more sensitive to climate change than had previously been imagined, and that it is important to try to limit climate change and impacts to a manageable level. With the 1.0°C increase in global temperature, there already are some ecosystems beyond the point of adapting naturally (Box 25.1).

That conclusion leads quite naturally to the question of whether there are ways to reduce the amount of impending climate change. There are two ways to do it: increase the albedo of the planet so that more incoming radiation is reflected back to space, or reduce the atmospheric CO_2 concentrations so that less radiant heat is trapped to warm the planet.

Reducing albedo can include modest changes to the built environment such as painting roofs white as well as schemes, often called geoengineering. An example would be the introduction of sulfates into the atmosphere to reflect incoming radiation, mimicking the known effects from major volcanic eruptions such as that of Mt. Pinatubo in 1991. Those kinds of schemes have the serious disadvantage that they address temperature (the symptom) but not elevated greenhouse gas levels (the cause). Also, the moment such an intervention ceases, planetary temperature jumps back to what it would have been without the intervention. In addition, they do nothing to address ocean acidification.

Because most of such geoengineering schemes are planetary in scale, whatever downsides they have will similarly be planetary in scale. In addition, the National Academy of Sciences deemed "geoengineering" an inappropriate term because it is only possible to engineer systems that are

Box 25.1   **Two Degrees Is Too Much**

For a long time much of the international negotiation and discussion has been around not exceeding a target of 2°C of warming. This was largely based on the perception that it was achievable from an energy perspective rather than having intrinsic merit. At the Paris COP 21, 1.5°C, long championed by James Hansen (Hansen et al. 2013), emerged as an important topic largely at the behest of indigenous peoples and small island developing states.

That 2°C is too much should have emerged far earlier for the simple reason that the last time the world was two degrees warmer, the oceans were four to six meters higher (Kopp et al. 2009). That would imply the elimination of low-lying nations and huge coastal areas involving the considerable human populations clustered on them. In a sense the planet has already done "the experiment," and the only questions are about rates of rise, not about the endpoint.

Impacts on biodiversity also point to 1.5° being far preferable to 2°C (Midgley 2018; Warren et al. 2018). As this book reveals, the impacts of climate change on biodiversity are essentially ubiquitous. The actual language of the convention reads:

> The ultimate objective of the Convention is to stabilize greenhouse gas concentrations "at a level that would prevent dangerous anthropogenic (human induced) interference with the climate system." It states that "such a level should be achieved within a time-frame sufficient to allow ecosystems to adapt naturally to climate change, to en-

sure that food production is not threatened, and to enable economic development to proceed in a sustainable manner."

The impacts already observed of coral bleaching events and pervasive mortality of coniferous forests (because the balance has been tipped in favor of pine bark beetles) cannot be viewed as "adapting naturally." In both cases what is occurring was not predictable with climate models and vegetation models because both come down to specific biological relationships: between the coral animals and algae species in the case of the reefs, and the trees and native beetles in the second.

It can be argued that some species of corals are more resilient in the face of higher temperatures and that even the sensitive ones may switch to a more resistant symbiont. Nonetheless bleaching events will occur with greater and greater frequency before such adaptations can take place. There is the additional factor of increasing acidity that will surely play a stressful role.

These are effects observed at 0.9°C, and it is logical to assume there will be many more as the planet warms to 1.5°C. The conclusion is the biology of the planet is very sensitive to climate change. The more that climate change can be limited, the easier it will be to manage the consequences, and beyond 1.5°C the biology of the planet basically will become unmanageable.

understood (National Academy of Sciences 2015).

To the extent that it is possible, reducing atmospheric CO_2 can reduce the problem directly. CO_2 injection into oil wells is a time-tested technique of the oil industry that extends the productive lifetime of the wells. Carbon capture and storage are still

328 HOW CAN CONSERVATION AND POLICY RESPOND?

in early experimental phases. Localized iron fertilization of the ocean replicates a natural phenomenon that stimulates phytoplankton blooms; what is still unclear is whether the CO_2 remains sequestered for long periods of time.

Largely overlooked until recent decades is the substantial amount of CO_2 in the atmosphere from degradation and destruction of modern ecosystems, mostly in the past two centuries. Although there has correctly been a strong effort to reduce emissions from fossil fuel use, limiting emissions especially from deforestation has also been a justified part of the global environmental (carbon cycle management) agenda. Indeed, the gross figure for deforestation is about 30 percent of all emissions (bigger than the transportation sector) (Houghton et al. 2015), a hidden number when the net emissions (about 10 percent) are reported, as is usually the case.

The amount of carbon lost to the atmosphere from terrestrial ecosystems has long been estimated to be 200–300 Pg (1 petagram = 1 trillion kilograms = 1 gigaton) (Houghton 2012). A more conservative estimate is about 150 Pg (Shevliakova et al. 2009). Part of the range of uncertainty reflects inadequate knowledge of soil carbon (belowground carbon). Soil carbon is probably distributed patchily in any given soil type, and knowledge about that distribution is sketchy. Even with this imprecision it has been clear for some time that the number is large (Lal et al. 2012). Recently Sanderman et al. (2017) have estimated the amount of soil carbon lost to the atmosphere from human landuse to be on the order of 133 Pg C itself.

The most recent estimate of the loss of CO_2 from ecosystem destruction and degradation is that only 450 Pg remain in extant ecosystems out of the 913 Pg prior to human impact (Erb et al. 2018). In other words the amount of carbon in the atmosphere from destroyed and degraded ecosystems is essentially equal to what remains.

Reaching a given atmospheric CO_2 concentration does not result in instant conse-

quent warming, because it takes time for the CO_2 to trap the corresponding amount of radiant heat. Reducing the atmospheric CO_2 concentration before that heat is trapped can therefore reduce the amount of global warming.

Reducing the atmospheric CO_2 burden by 1 ppm is equivalent to removing 7.7 Pg of carbon. To lower global CO_2 concentrations to close to 350 ppm from the current level of 400 ppm would mean removal of 388.5 Pg of carbon (Shukla, personal communication). That is at the same order of magnitude as the estimated loss (450+ Pg) from ecosystem destruction and degradation

The magnitude of the potential to reduce CO_2 concentrations and associated global warming focusing on tropical forests alone is considerable. Elimination of current deforestation and degradation of forests and reforesting 500 million ha could together reduce annual emissions by 4.4 Pg/yr (Houghton et al. 2015). A further complication is that removing one gigaton does not lower CO_2 concentration by exactly 1 ppm because of a very slow release back from the oceans, which of course produces a small reduction of acidity.

This kind of carbon dynamic is not new in the history of life on Earth. Twice there have been extremely high atmospheric concentrations of CO_2. In both instances they were brought down to preindustrial levels by natural processes (Figure 25.1) (Royer 2006; Beerling 2008). In the Cretaceous the arrival of plants on land and a major opportunity for additional photosynthesis resulted in a major CO_2 sequestration. There was also a significant CO_2 drawdown contribution from soil creation. Although soil formation is essentially an abiotic process, soil biota also contribute significantly. Mycorrhizal processes played a magnifying role.

The second major drawdown occurred in the Paleogene and the Neogene because of major engagement of modern flowering plants. In both cases the process took tens of millions of years—a luxury we do not

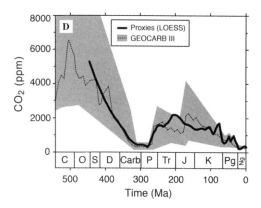

Figure 25.1. During prior geological periods there were two times when very high CO_2 concentrations were brought down to preindustrial levels by natural processes (mostly terrestrial photosynthesis and soil formation, both physical and biological). The first reduction coincided with the arrival of higher plants on land and the second with the arrival of modern flowering plants. (From Royer 2006.)

have. We can, however, proactively restore ecosystems (Lal et al. 2012; Lovejoy 2014; Griscom et al. 2017) as part of a global effort to recarbonize the biosphere by promoting ecological restoration.

A particularly important source of anthropogenic carbon emissions from natural ecosystems is from peatlands. When peatlands are drained, the peat oxidizes and burns readily. In Southeast Asia there are extensive peat forests, which if deforested render the underlying peat deposits very vulnerable to burning, especially in the dry El Niño years. During those years, fires have been so extensive in Indonesia and Borneo as to cause health-threatening air pollution even in adjacent mainland (South) Asia.

There should be a concerted effort globally to maintain all peatland as wet environments. Carbon accretes slowly in peatlands, so they do not have a major role to play in carbon sequestration through restoration, but it is important that their integrity as carbon stocks be protected.

The Woods Hole Research Center, a world authority on the role of forests in the carbon cycle, is currently preparing a global map of potential ecosystem restora-

tion. There are obviously places where the potential is limited (e.g., urban centers), but there is an enormous area of degraded land where ecosystem restoration would be of value in itself.

Reforestation has an important role to play because forests hold so much carbon per unit area, but there are other benefits as well, including biodiversity conservation and watershed functions. Reforestation in the Amazon would secure the important hydrological cycle that maintains that forest and provides moisture south of the Amazon even as far as northern Argentina (Martinez and Dominguez 2014; Lovejoy and Nobre 2018). Worldwide, reforestation is important in restoration of watershed function. These forests need not all be natural, but they could include production forests of various kinds as well. The carbon and ecosystem service values are not substantially affected by harvest of minor forest products (e.g., rattan), which can provide additional economic return.

Degraded grasslands and grazing lands represent considerable potential for carbon sequestration, not just in the plant systems but also in the soils. Beyond the carbon value, restoration would result in better grazing or other uses of grasslands, simultaneously improving productivity.

Degraded lands encompass abandoned agricultural lands and grazing lands that were originally grasslands or former forested lands. Globally degraded lands are between 1 billion and 6 billion ha in extent because of considerable uncertainty in many instances (Gibbs and Salmon 2014). Whatever the number, the potential to sequester carbon and simultaneously restore productivity is substantial.

Agroecosystems also have tremendous potential to sequester carbon instead of leaking carbon, as is so often the case in current agricultural practice. An important benefit of sequestering carbon is improvement in soil fertility. Interestingly, even though industrial agriculture practices would seem counter to that, there are places in the

American Midwest where no-till and low-till practices have taken hold spontaneously. An important aspect is to restore riparian vegetation that improves water quality and prevents soil erosion. That in turn adds natural connectivity in the landscape—so important for normal dispersal of plants and animals, and even more so under climate change, which will drive many organisms to move and track their climatic envelope.

Only relatively recently has it been recognized that coastal wetlands, which total 49 million ha worldwide, have an important role to play in the carbon cycle (Pendleton et al. 2012). Mangroves, coastal marshes, and seagrasses have experienced considerable destruction and degradation but are being recognized as important buffers against storm surge. The Mississippi Delta, where such ecosystems have been degraded and destroyed, is undergoing priority restoration for those reasons alone, but there are significant carbon sequestration benefits as well.

Ecosystem restoration has yet to emerge as the global priority it deserves. Still, the 2015 New Climate Economy report produced for the United Nations concluded there can be healthy economic growth while rapidly reducing emissions, if there is a strong land use component and all three "Rio conventions" (i.e., UN Framework Convention of Climate Change, Convention on Biological Diversity, and UN Convention to Combat Desertification) include restoration as part of their core agendas. Many of the independently determined national commitments made at the 2015 Paris Climate Change Conference of the Parties focused not only on forest protection but also on forest restoration. The Nature Conservancy has developed a major portfolio around land management for climate change mitigation.

Extensive restoration is not simple to achieve on a planet already experiencing climate change and a large, still-growing human population. The large amount of carbon sequestered in the Amazon forest could be lost if climate change causes Amazon dieback (transformation to a savanna vegetation) in the eastern and southeastern Amazon (Nobre et al. 2016). Southern edges of Northern Hemisphere boreal forest could also die back, although they might be offset by new boreal forests on the northern edge of the current distribution. With more warming, there could be massive release of methane as permafrost melts. Basically, the ability to use ecosystem restoration to sequester carbon depends on not straying too far from current climatic regimes, or at least understanding the dynamics of possible climatic change and integrating it into planning.

The success of ecosystem restoration for climate management will require a more integrated and sustainable approach to development, especially in the face of additional billions of people and the need to feed and provide for them adequately. That can occur only with a more coordinated approach to development (see Chapter 28): one that avoids isolated decision making and puts a premium on sustainable infrastructure and protection or conservation of natural ecosystems. It is a vision congruent with Edward O. Wilson's (2016) Half-Earth, which argues for half of the planet to be set aside primarily for nature.

At a finer and more local scale, management of ecosystems can have important carbon value implications. Changes in trophic structure can directly affect carbon values, as in the example of old fields and different dominant species of predators (Case Study 11). In blue-carbon ecosystems, predators are important for maintaining high carbon stocks (Atwood et al. 2015).

The term "Anthropocene" was coined to recognize a new geological age in which human impacts became planetary and geological in scale, like the giant meteor that ushered out the dinosaurs. It could also come to mean a new age in which we learn to manage ourselves for better outcomes for humans and other forms of life.

When ecosystem restoration and management attain their rightful place in the climate change agenda, it will transform public perception of the planet on which we live. It will require recognition that the Earth functions not just as a physical system but also as a combined biological and physical system—and that we in fact live on, and are part of, a living planet. Such a view would profoundly alter our sense of our place in nature and what is appropriate behavior.

Similarly, it can empower individuals who previously could see no way in which they individually can make a difference about climate change because it is such a gigantic problem. In a simplistic way every individual can plant a tree or help restore a coastal wetland, much as people of all walks of life were widely involved in victory gardens during World War II.

REFERENCES

Atwood, T. B., R. M. Connolly, E. G., Ritchie, C. E. Lovelock, M. R. Heithaus, G. C. Hays, J. W. Fourqurean, and P. I. Macreadie. 2015. "Predators help protect carbon stocks in blue carbon ecosystems." *Nature Climate Change* 5: 1038–1045.

Beerling, D. J. 2008. *The Emerald Planet*. Oxford University Press.

Erb, K. H., T. Kastner, C. Plutzar, A. S. Bais, N. Carvalhais, T. Fetzel, S. Gingrich, H. Haberl, C. Lauk, M. Neidertscheider, J. Pongratz, M. Turner, and S. Luyssaert. 2018. "Unexpectedly large impact of forest management and grazing on global vegetation biomass." *Nature* 553: 73–76.

Gibbs, H. K., and J. M. Salmon. 2014. "Mapping the world's degraded lands." *Applied Geography* 57: 12–21.

Griscom, B. W., J. Adams, P. W. Ellis, R. A. Houghton, G. Lomax, D. A. Miteva, W. H. Schlesinger, et al. 2017. "Natural climate solutions." *Proceedings of the National Academy of Sciences* 114 (44): 11645–11650.

Hansen, James, et al. 2013. "Assessing 'dangerous climate change': Required reduction of carbon emissions to protect young people, future generations and nature." *PLOS One* 8: 81648.

Houghton, R. A. 2012. "Historic changes in terrestrial carbon storage." In *Recarbonization of the Biosphere: Ecosystems and the Global Carbon Cycle*, ed. R. Lal, K. Lorenz, R. F. Hüttl, B. U. Schneider, and J. von Braun, 59–82. Springer. https://doi.org/10.1007/978-94-007-4159-1_4.

Houghton, R. A., B. Byers, and A. A. Nassikas. 2015. "A role for tropical forests in stabilizing atmospheric CO_2." *Nature Climate Change* 5: 1022–1023.

Kopp, R. E., F. J. Simons, J. X. Mitrovica, A. C. Maloof, and M. Oppenheimer. 2009. "Probabilistic assessment of sea level during the last interglacial stage." *Nature* 462: 863–867.

Lal, R., K. Lorenz, R. F. Huttle, B. U. Schneider, and J. von Braun, eds. 2012. *Recarbonization of the Biosphere: Ecosystems and the Global Carbon Cycle*. Springer. https://doi.org/10.1007/978-94-007-4159-1_4.

Lovejoy. T. E. 2014. "A natural proposal for addressing climate change." *Ethics and International Affairs* 28: 359–363.

Lovejoy, T. E., and C. Nobre. 2018. "Amazon tipping point." *Science Advances* 4: eaat2340.

Martinez, J., and F. Dominguez. 2014. "Sources of atmospheric moisture for the La Plata River basin." *Journal of Climate* 27: 6737–6753.

Midgley, G. 2018. "Narrowing pathways to a sustainable future." *Science* 360 (6390): 714–715. https://doi.org/10.1126/science.aat6671.

National Academy of Science. 2015. *Climate Intervention: Carbon Dioxide Removal and Reliable Sequestration and Climate Intervention: Reflecting Sunlight to Cool Earth*. National Academies Press. https://doi.org/10.17226/18988.

Nobre, C. A., G. Sampaio, L. Borma, J. C. Castilla-Rubio, J. S. Silva, and M. Cardoso. 2016. "The fate of the Amazon forests: Land-use and climate change risks and the need of a novel sustainable development paradigm." *Proceedings of the National Academy of Sciences* 113 (39): 10759–10768.

Pendleton, L., et al. 2012. "Estimating global 'blue carbon' emissions from conversion and degradation of vegetated coastal ecosystems." *PLOS One* 7: 1–7.

Royer, D. A. 2006. "CO_2-forced climate thresholds during the Phanerozoic." *Geochimica* 70: 5665–5675.

Sanderman, J., T. Hengl, and G. J. Fiske. 2017. "Soil carbon debt of 12,000 years of human land use." *Proceedings of the National Academy of Sciences* 114: 9575–9580.

Shevliakova, E., et al. 2009. "Carbon cycling under 300 years of land use change: Importance of the secondary vegetation sink." *Global Biogeochemical Cycles* 23 (GB2022): 1–16.

Warren, R., J. Price, E. Graham, N. Forstenhaeusler, and J. VanDerWal. 2018. "The projected effect on insects, vertebrates, and plants of limiting global warming to 1.5°C rather than 2°C." *Science* 360 (6390): 791–795. https://doi.org/10.1126/science.aar3646.

Enlisting Ecological Interactions among Animals to Balance the Carbon Budget

Oswald J. Schmitz

Efforts to mitigate climate change have tended to focus on managing human activities like fossil-fuel burning and land clearing, and on innovating technologically, to reduce CO_2 emissions to the atmosphere (Pacala and Socolow 2004). But this particular focus on controlling atmospheric inputs represents only one side of the equation for balancing the carbon budget. A potentially important yet largely untapped opportunity is to actively enlist natural ecological processes to recapture carbon within terrestrial and aquatic ecosystems, thereby offsetting atmospheric buildup. Indeed, ecosystems globally are removing as much as half of the CO_2 emitted to the atmosphere each year from anthropogenic activities (Schmitz et al. 2014). But it may be possible to do more through strategic management of the functional composition of living organisms—especially animals—and their interactions within ecosystems, thereby enhancing the nature and magnitude of carbon exchange and storage (Schmitz et al. 2014). Recent estimates suggest that the magnitudes of animal effects can be profound, holding much promise in protecting and managing biodiversity as a way to help manage the global carbon budget.

Animals can determine the fate of carbon in ecosystems through trophic interactions that directly and indirectly influence the abundance and chemical composition of plant and animal biomass and its fate as soil organic matter. Top predators especially can trigger knock-on effects that modulate the amount of carbon exchanged between ecosystem reservoirs (e.g., soils and plants) and the atmosphere (Schmitz et al. 2014). Predators do this in at least two ways. They may have consumptive effects that reduce herbivore density (and hence herbivory), in turn enabling more plant-based carbon to enter soil organic matter storage pools as uneaten senescent plant biomass, relative to conditions where predators are absent (Schmitz et al. 2014). Their presence within ecosystems may also cause nonconsumptive fear effects that cause herbivores to shift their foraging preferences from nitrogen-rich plants that support growth and high reproduction to plants with higher content of soluble carbon to support heightened metabolic costs caused by chronic stress from fear of predation (Schmitz et al. 2017). Such a diet shift can alter the spe-

cies composition of the plant community as well as the amount of carbon in senescent plant matter entering the soil storage pool (Schmitz et al. 2017). Whether predator effects on herbivores are largely consumptive or nonconsumptive generally depends on their hunting mode: active-hunting predators generally have consumptive effects, whereas sit-and-wait predators have nonconsumptive effects.

Demonstration of these two predation effects at play come from empirical analyses examining the link among spider predators, grasshopper prey, and grasses and herbs among old-field ecosystems in Connecticut, United States. The important players in these ecosystems are three functional groups of plants represented by the herb *Solidago rugosa* (hereafter *Solidago*), a variety of other old-field herb species, and grasses; a dominant generalist grasshopper herbivore *Melanoplus femurrubrum*; and the sit-and-wait spider predator *Pisaurina mira* and active-hunting predator *Phidippus clarus*. In these systems, the plant functional groups compete asymmetrically (see Figure CS11.1), with *Solidago* competitively dominating the other two functional groups of plants (Schmitz et al. 2017).

Solidago competitive dominance is altered, however, by hunting-mode-dependent effects of predators on grasshoppers, which in turn shape the composition of the plant community predominantly through changing effects on *Solidago* and grasses (Figure CS11.1). In the absence of predators, grasshoppers prefer nitrogen-rich grasses to maximize survival and reproduction. Sit-and-wait spiders are persistently present and thus trigger grasshoppers to modify foraging in favor of *Solidago* because it is a key source of dietary soluble carbohydrate to fuel heightened grasshopper respiration caused by chronic stress from perceived predation risk (Schmitz et al. 2017). The potential for predator-prey encounter is less frequent with widely roaming active-hunting spiders. Consequently risk of predation is episodic for any individual grasshopper, making it energetically inefficient for grasshoppers facing active-hunting spiders to be chronically stressed. Hence, grasshoppers facing active-hunting predators favor grasses over *Solidago* (Figure CS11.1).

As a consequence, when active-hunting spiders dominate, *Solidago*, along with its relatively higher carbon content, flourishes and dominates the plant community. When sit-and-wait predators dominate, *Solidago* is suppressed, favoring a more equitable representation of the three functional plant groups in the plant community with less total carbon. Intermediate combinations of the two spiders should result in intermediate levels of plant functional group abundances. Sampling among 15 old fields across Connecticut revealed that the amount of carbon retained in soils among these old fields is related to the relative abundance of active-hunting and sit-and-wait predators within the fields, mediated via changes in plant community composi-

tion. This relationship improves when accounting for the strength of the ecological interactions of predators on the plant community (Schmitz et al. 2017). Most striking is that shifting from a dominance of sit-and-wait predators to a dominance of active-hunting predators can result in a doubling of soil carbon retention. This range in per unit area of carbon retained in soils rivals or exceeds the range achieved through conventional management of soil carbon and land use (Post and Kwan 2000; Guo and Gifford 2002). More important, several other candidate factors, such as live plant biomass, insect diversity, soil arthropod decomposers, degree of land-use development around the fields, field age, and soil texture, did not significantly explain soil carbon retention among the fields (Schmitz et al. 2017).

These ecological principles are also scalable. The loss of important predators—from wolves in boreal forests to sharks in seagrass meadows—can lead to growing populations of terrestrial and marine herbivores, whose widespread grazing can and does reduce the ability of ecosystems to absorb and protect hundreds of years of built-up soil and sediment carbon (Atwood et al. 2015; Wilmers and Schmitz 2016). These encouraging new insights support expanding consideration of biodiversity management to balance carbon budgets. Humans already manage populations of many animals for other purposes. So, while representing a shift in management goals, this would not involve a radical shift in management approach. But individual animal species tend to occur regionally, not globally. This requires shying away from finding the single, global-scale, home-run solution to seek more regionally nuanced alternatives. Animal management has the advantage that it can take into account the values and preferences of local societies and would allow for regional players to reconcile their particular concerns and values with broader climate solutions. The many local and regional

Trophic interactions

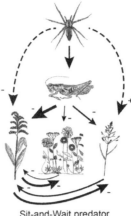 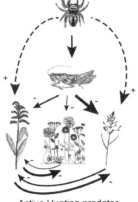

Sit-and-Wait predator   Active Hunting predator

Plant community composition

Sit-and-Wait predators dominate   Active Hunting predators dominate

Soil carbon retention

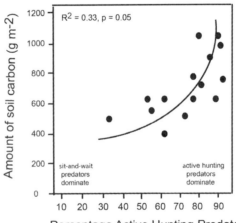

animal management strategies could then add up globally to create a portfolio of solutions that, together with other mitigation efforts (Pacala and Socolow 2004), can meaningfully help slow climate change.

REFERENCES

Atwood, T. B., R. M. Connolly, E. G. Ritchie, C. E. Lovelock, M. R. Heithaus, G. C. Hays, J. W. Fourqurean, and P. I. Macreadie. 2015. "Predators help protect carbon stocks in blue carbon ecosystems." *Nature Geoscience* 5: 1038–1045.

Guo, L. B., and R. M. Gifford. 2002. "Soil carbon stocks and land use change: A meta-analysis." *Global Change Biology* 8: 345–360.

Pacala, S., and R. Socolow. 2004. "Stabilization wedges: Solving the climate problem for the next 50 years with current technologies." *Science* 305: 968–972.

Post, W. M., and K. C. Kwan. 2000. "Soil carbon sequestration and land use change: Processes and potential." *Global Change Biology* 6: 317–327.

Schmitz, O. J., R. W. Buchkowski, J. R. Smith, M. Telthorst, and A. E. Rosenblatt. 2017. "Predator community composition is linked to soil carbon retention across a human land use gradient." *Ecology* 98: 1256–1265.

Schmitz, O. J., P. A. Raymond, J. A. Estes, W. A. Kurz, G. W. Holtgrieve, M. E. Ritchie, D. E. Schindler, et al. 2014. "Animating the carbon cycle." *Ecosystems* 7: 344–359.

Wilmers, C. C., and O. J. Schmitz. 2016. "Effects of wolf-induced trophic cascades on ecosystem carbon cycling." *Ecosphere* 7: e01501.

Figure CS11.1. Case example of how spider predators with different hunting modes influence food-web interactions in old-field ecosystems, with attendant effects on plant community composition and soil carbon retention. *Trophic interactions*: Depiction of direct (solid arrows) and indirect (dashed arrows) interactions among hunting spiders, grasshopper herbivores, and three functional groups of plants (left to right: carbon-rich *Solidago*, less carbon-rich herbs and grasses). The strength of grasshopper direct effects on a plant functional group, via diet selection (depicted as arrow thickness), depends on which hunting mode of spider predator dominates and thus mediates competition among the plant functional groups. Dominance of active-hunting predators like *Phidippus clarus* leads to grasshopper preference for grasses, thereby abetting *Solidago* competitive dominance (positive indirect effect of predators on *Solidago*). Dominance of sit-and-wait predators like *Pisaurina mira* leads to grasshopper preference for Soli- dago, thereby suppressing *Solidago* competitive dominance (negative indirect effect of effect of predators on *Solidago*). *Plant community composition*: When sit-and-wait spiders dominate, the plant community is equitably comprised of the three plant functional groups. When active-hunting spiders dominate, *Solidago* dominates the plant community. Mixed communities of active-hunting and sit-and-wait spiders result in intermediate levels of the three plant functional groups. Hence sit-and-wait spider dominance leads to lower biomass density of carbon in the plant community than when active-hunting spiders dominate. *Soil carbon retention*: Sampling of soil carbon among 15 old-field ecosystems reveals that the knock-on effects of spiders on carbon density of the plant community carry further to determine the amount of carbon retained in soils, which is related to the relative abundance of active-hunting and sit-and-wait predators within an ecosystem. (Courtesy of Oswald Schmitz, Yale University.)

CHAPTER TWENTY-SIX

Increasing Public Awareness and Facilitating Behavior Change: Two Guiding Heuristics

EDWARD MAIBACH

Everything should be made as simple as possible, but no simpler.
—Quote commonly attributed to Albert Einstein

If there is a single aspiration that unifies the professionals who work on the challenges associated with climate change and biodiversity, it is likely their desire to see policy makers, business managers, and members of the public make decisions that are better informed by the realities of what we know about how to stabilize the climate, conserve biodiversity, and prevent needless harm to people and ecosystems. If there is a single emotion that unifies them, it is likely angst—as a result of feeling that, collectively, we are falling far short of our aspirations.

This calls out an obvious question: What can we do to more effectively promote wise decision making and actions by decision makers? Many excellent books (Hornik 2002; Moser and Dilling 2007; Whitmarsh et al. 2011; McKenzie-Mohr 2011; Crow and Boykoff 2014; Marshall 2014) and articles (Maibach et al. 2007; Holmes and Clark 2008; Ryder et al. 2010) offer important insights and partial answers to the question, but none offers a simple, clear answer that working scientists—and science institutions—will find to be practical.

In this chapter I offer my best shot at a practical answer. It is by no means the only answer, or the definitive answer, but it is—by design—the simplest answer I can offer while still staying true to the best available evidence on the science of science communication. Moreover, my answer is intended to be equally helpful both to individual scientists—in any relevant discipline, at any stage in her or his career—and to the full

Note: This essay builds on a previously published essay by the author (Maibach and Covello 2016).

range of science and science-based institutions that strive to share current scientific insights about the physical world with decision makers and the public (e.g., National Academy of Science, National Science Foundation, professional societies, science museums, media producers).

My answer also aims to be useful regardless of which category of decision makers is most relevant in a given situation—community leaders, national leaders, business leaders, people in a specific profession (e.g., building contractors), or even individuals and families who are trying to manage their own lives in the best possible manner. All of these people have important climate- and biodiversity-related decisions to make, whether they currently know it or not. Individual scientists and the scientific community can be of value in helping these people make wise decisions and take wise actions.

The question posed above includes two related yet distinct challenges. The first challenge is helping decision makers make wise decisions; the second challenge is helping them take wise actions. To help people make wise decisions, we must effectively bring the issue to their attention, suggest the need to make decisions, clarify the nature of the problems and opportunities, and make available the best science-based information—in appropriate formats—for decision makers to consider. In short, we must effectively share what we know.

Helping people take wise actions involves a different set of activities. If it were easy for people to convert their good decisions (i.e., their good intentions) into effective actions, the proverb "the road to hell is paved with good intentions" would never have arisen. Fortunately, steps can be taken to help people convert their good intentions into effective actions.

My answer to the question, therefore, has two parts. To effectively share what we know, we need simple clear messages, repeated often, by a variety of trusted sources. To help people convert their good intentions into effective actions, we need to do everything we can to make the behaviors we are promoting easy, fun, and popular. I refer to each of these as "heuristics," in the sense that they organize a relatively large amount of prescriptive information into a relatively easy-to-use method or process.

In the remainder of this chapter, I unpack these two heuristics, with the aim of making them practical for all readers. I assume that most readers of this chapter will be scientists and allied professionals—and I therefore tailor my comments to them—but the recommendations are equally relevant to anyone seeking to improve climate change and biodiversity outcomes in ways that are grounded in scientific evidence.

SHARING WHAT WE KNOW

Scientists are highly trained to share what they know, but primarily to colleagues in their own scientific discipline. Typically, this process begins with our research, where we develop and test ideas. If an idea proves to have merit, we take steps to share it with our colleagues—at professional meetings, in journal articles, and in books. Perhaps if we are really excited by the idea, we might make extra efforts to share it more broadly—possibly with scientists in other disciplines (e.g., by giving talks at their meetings) or with the general public (e.g., by working with our institution's press office to issue a press release), although these efforts tend to be fleeting. These approaches to sharing what we know work reasonably well with colleagues in our own discipline, less well with scientists in other disciplines, and not well enough with policy makers, business leaders, and members of the public. Metaphorically, these approaches are akin to tossing what we know "over the transom" or out the window of our lab, and into the outside world, expecting relevant people to pick up our knowledge, consider it, and know what to do with it.

There is a better way: simple clear messages, repeated often, by a variety of trusted sources. This is not a "magic bullet," and it will not solve our biodiversity and climate challenges overnight (or anytime soon), but the approach is evidence based (based in the science of science communication), reasonably easy (once you understand it, it is no harder than what you are already doing), and ethical (it involves providing people with information they are likely to find helpful). The heuristic itself has three elements—simple messages, message repetition, and trusted messengers—each of which is based on empirical evidence and offers practical guidance.

The Importance of Message Simplicity

Most people, in most situations, do not deal well with complex information; complex information is cognitively taxing, and most people (even very bright people), in most situations, are unwilling to make the effort. Instead, people typically use mental shortcuts to avoid cognitively difficult tasks, and when they do, they often end up reaching conclusions that differ from those intended by the information provider (Kahneman 2011). The risk communication expert Baruch Fischhoff (1989) trenchantly summarized the situation—and his prescription about how best to manage it—in the following manner: "People simplify." Our job (as risk and science communicators) is to help people simplify appropriately.

So, what can we—as communicators—do to help people simplify appropriately? We can develop "messages" about the information we wish to share that are specifically intended to help people simplify complex information appropriately. (For readers who disfavor the term "messages"—equating it to persuasive intent—the term "brief statements" is an acceptable synonym.) Audience research is a powerful tool for developing such messages. Through audience research, we can systematically collect data to assess audience members' preexisting knowledge, attitudes, and values, and test their responses to draft messages. In this

manner, we stand a much better chance of designing messages that illuminate, rather than messages that alienate. Admittedly, conducting audience research is not always feasible (perhaps because of lack of time, funds, or expertise), but that is not an excuse for failure to seriously consider how members of the target audience are likely to understand the information we wish to share with them. Making the effort to consider our audience members' views—and values—is helpful in that it forces us as communicators to look beyond ourselves and to think carefully about both what is most worth saying and how best to say it.

An approach that all science communicators can use to improve their messages involves anticipating the questions that people are likely to ask, and drafting messages that attempt to proactively answer those questions. For example, when the issue pertains to a risk, people are likely to ask some variation of the following questions: What is the problem? How will it affect me (and my people)? How serious is it? Who or what is causing the problem? What are the options for dealing with it? What, if anything, can I do about it?

To test your success in drafting simple, clear messages in response to likely audience questions, share the messages with a few members of your target audience, one person at a time. After each person has had a chance to consider your messages, ask each one to explain—in his or her own words—what the messages mean and how he or she would explain them to a friend. Also ask, "What questions, if any, do you have for me about this information?" When members of your audience can adequately explain your messages in their own words, and when your messages help them ask good questions, you have succeeded in writing simple, clear messages.

The Third U.S. National Climate Assessment (NCA3) provides a useful example of simple clear messages (Melillo et al. 2014). Although the full report is more than 1,000 pages, for the NCA3 each set of chapter authors was asked to identify key messages for their chapter. In addition, the federal advisory committee that oversaw the development of NCA3 developed key messages for the report as a whole. Authors and advisory committee members were well aware of audience research showing that most Americans understand that climate change is happening but see it as a distant threat—distant in space (i.e., the problems will primarily manifest elsewhere, not in the United States), distant in time (i.e., the problems will start in the future and are not happening yet), and distant from humans (i.e., the problems will be primarily felt by plants, penguins, and polar bears—not people; Leiserowitz 2005; Leiserowitz et al. 2017). As a result, NCA3 authors developed key messages intended to correct the misperception of climate change as a distant threat. The opening words on the NCA3 report website are the following: "The National Climate Assessment provides an in-depth look at climate change impacts on the U.S. It details the multitude of ways climate change is already affecting and will increasingly affect the lives of Americans. Explore how climate change affects you and your family" (Melillo et al. 2014). This theme is also clearly in evidence in the report's introduction. These opening words, key messages, and even chapter names (e.g., extreme weather, human health, water, agriculture, oceans) were all intended to help all readers—even those giving the information only a quick glance—to understand the most important, overarching, finding of the assessment: climate change is happening here, now, in every region of America.

The Importance of Message Repetition

One of the most robust findings to have ever emerged from communication research is that "repetition is the mother of all learning" (Lang 2013)—an expression that comes to us originally from an ancient Latin proverb (*repetitio est mater studiorum*). Repetition increases message persuasive-

ness cognitively (by increasing salience and availability of the information) and affectively (by increasing positive feelings about the message) (Batra and Ray 1986; Pechman and Stewart 1988; Chong and Druckman 2013). Although truthfulness is of paramount importance in science communication, truthful messages amount to little without adequate message repetition, a point nicely illustrated by this quote from a political consultant: "You take the truth, and I'll take repetition; I'll beat you every time" (Castellanos, personal communication, 2010).

The importance of message repetition is something that every politician learns in her first political campaign, and every business executive learns in his first marketing course, but it is a lesson rarely taught to scientists. Admittedly, repetition is boring to most communicators, especially scientists. As scientists, novelty, innovation, and controversy are what excite us—and what we like to talk about—but what excites us is not a relevant consideration in determining how best to share what we know with decision makers. Moreover, like all people, scientists suffer the "curse of knowledge" (Heath and Heath 2007); we forget that most people do not know what we know, and as a result we end up making assumptions in our communication that inadvertently excludes the very people we are seeking to share our knowledge with. The discipline of message repetition—repeating the messages that we have designed for the express purpose of helping audience members simplify appropriately—forces us to stay true to our plan for sharing the information that is most helpful to members of our target audience (rather than sharing the information that most interests us).

Fortunately, message repetition is not the sole burden of any one individual or organization; message repetition works best when many different messengers repeat the same set of messages, consistently, over time. Individuals and organizations working on climate change and biodiversity issues should develop the discipline to work together to design, use, and repeat—at every communication opportunity—a shared set of messages specifically intended to help audience members reach appropriate conclusions about the complex problems they are urging them to engage with.

Reach—that is, reaching members of your intended target—is an important consideration, too. Messages that are repeated often but fail to reach their intended audience will have no benefit for that audience. Consumer brands typically strive to achieve both message reach and frequency with a combination of paid advertising, earned media (i.e., outreach to news media and bloggers), social media, endorsements, paid placements, and other means. Climate change and biodiversity professionals and organizations rarely have the opportunity to achieve reach through paid placements (e.g., advertising), but through the kind of collaboration suggested in the paragraph above, they can strive to maximize both message reach and frequency (i.e., repetition), especially to the extent that they succeed in bringing other trusted messengers into the communication mix. For further elaboration of this important idea, read on.

The Importance of Trusted Messengers

Quite simply, where there is no trust, there can be no learning. As a group, scientists are highly trusted. For example, scientists are trusted "a lot" by two-thirds of American adults—tied with medical doctors and second only to members of the military and teachers (Pew Research Center 2013). However, when target audience members do not know the specific scientist who is attempting to communicate with them—personally or by reputation—their trust in that messenger is likely to be superficial, provisional, and vulnerable, and communication mistakes (e.g., unclear messages, seemingly evasive answers, lack of empathy) can rapidly undermine trust (Maibach and Covello 2016).

Climate change and biodiversity communicators can earn the trust of their target, and leverage the impact of their communication, by recruiting additional trusted voices—people who are known by the target audience, personally or by reputation—to embrace, repeat, and thereby validate their simple clear messages. These additional trusted voices need not necessarily be from the science community. Indeed, we can and should cultivate communication partnerships with individuals and organizations outside the realm of climate change and biodiversity who are highly trusted by members of our audience—for example, leaders in the faith community or in the business community—because doing so is a way of demonstrating one's trustworthiness, and of maximizing one's message reach and frequency.

The most effective endorsements come from people that our target audience trusts the most, regardless of their level of expertise. On the issue of climate change, for example, people typically trust most the people they know the best—their family members, friends, and coworkers (Figure 26.1; Leiserowitz et al. 2009). Scientists are highly trusted, too, but it is the rare individual who places greater trust in a scientist (whom he or she has never met) than in his or her own family and friends. This is precisely why the best test of the simplicity and clarity of a science-based message is whether members of the target audience are willing and able to convey the message to their family, friends, and coworkers. Ultimately, that should be the aim of our communication—to motivate and enable members of our target audience to share our messages with one another.

Feeling overwhelmed at the prospect of designing and communicating simple, clear messages, repeated often, by a variety of trusted sources? Don't be. In his article "Communicating about Matters of Greatest Urgency: Climate Change," Baruch Fischhoff (2007) made a strong case for improving the effectiveness of science communication by

Americans Trust Climate Scientists, Friends & Family Most As Sources Of Information About Global Warming
- % of Americans who strongly or somewhat trust... -

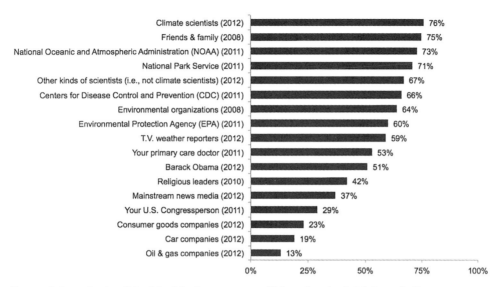

Source	%
Climate scientists (2012)	76%
Friends & family (2008)	75%
National Oceanic and Atmospheric Administration (NOAA) (2011)	73%
National Park Service (2011)	71%
Other kinds of scientists (i.e., not climate scientists) (2012)	67%
Centers for Disease Control and Prevention (CDC) (2011)	66%
Environmental organizations (2008)	64%
Environmental Protection Agency (EPA) (2011)	60%
T.V. weather reporters (2012)	59%
Your primary care doctor (2011)	53%
Barack Obama (2012)	51%
Religious leaders (2010)	42%
Mainstream news media (2012)	37%
Your U.S. Congressperson (2011)	29%
Consumer goods companies (2012)	23%
Car companies (2012)	19%
Oil & gas companies (2012)	13%

How much do you trust or distrust the following as a source of information about global warming?

Base: Americans 18+.

4C
George Mason University
Center for Climate Change Communication

Figure 26.1. Americans' trust in various possible sources of information about global warming. Data from Yale/George Mason University Climate Change in the American Mind surveys conducted between 2008 and 2012 (Leiserowitz et al. 2012).

approaching it as a team sport rather than as solo sport in which every scientist is expected to be master of the art and science of communication. Specifically, Fischhoff encourages the development of science communication teams that include three distinct types of expertise, which can be provided by a minimum of three people: a content scientist (i.e., a person with expertise on the risk or the issue), a social scientist (i.e., a person with expertise on how people interpret information), and a communications practitioner (i.e., a person with expertise in creating communications opportunities). These three types of professionals each bring unique knowledge and skills to the process of developing simple, clear messages and in working to ensure that those messages are conveyed often, by a variety of trusted sources.

By way of example, I (as a social scientist) helped organize a team of climate scientists, social scientists, and communications practitioners at various universities (George Mason and Yale), nonprofit organizations (Climate Central, American Meteorological Society), and government agencies (National Oceanic and Atmospheric Administration, National Aeronautics and Space Adminstration) to develop and distribute to TV weathercasters broadcast-quality materials to help them report on the local impacts of climate change in their area. Called *Climate Matters*, the collaboration started with a successful pilot test at a single TV station in Columbia, South Carolina (Zhao et al. 2014), and has expanded to a national network with more than 500 participating weathercasters, and growing (Placky et al. 2016; Maibach et al. 2016).

In addition to sustained collaborations, as described in the *Climate Matters* example, this team-based approach to science communication is also practical for ad hoc communications opportunities. For example, climate scientists—who will soon be publishing findings with important implications for decision makers—can ask a social science colleague and a member of the media relations team at their institution to help them craft messages and develop and implement a communication plan through which to communicate the messages.

INFLUENCING BEHAVIOR

Effective communication is important, although it is often not sufficient to change people's behavior (Hornik 2002; McKenzie-Mohr 2011). Even after people decide to take action, many will not, or they will not persevere long enough to succeed. Consider, for example, your most recent New Year's resolution.

Social marketing—the use of marketing methods to promote behaviors that benefit society—is a method developed specifically to help address this problem (Maibach et al. 2002). Many excellent texts lay bare the principles of social marketing, including two that specifically explore its application to environmental challenges (McKenzie-Mohr 2011; McKenzie-Mohr et al. 2012). I particularly encourage readers of this chapter, however, to watch a TED Talk by Bill Smith (2011)—one of social marketing's pioneers. In his talk, Smith lays out a simple heuristic to guide the implementation of social marketing programs: make the behavior you are promoting easy, fun, and popular. Although it sounds cheeky, the heuristic is based on a large body of empirical research, and it offers important, practical guidance.

The Importance of Making the Behavior Easy

Social scientists have long known that there is often a large gap between people's attitudes toward a behavior (e.g., vegetables are very good for you) and their behavior (e.g., "I'll have the cheeseburger and fries, please"). One of the most effective means of reducing this attitude-behavior gap is to make the recommended behavior easier to perform (e.g., "Would you like carrot sticks, an apple, or fries with your burger?").

People are likely to perform easy behaviors that they believe to be in their best interest, but they often defer—and never get around to—behaviors they find more difficult. To save money on utility bills, for example, a homeowner may switch her lights over time from incandescent bulbs to LEDs (because doing so is relatively easy), but she may not take steps to weatherize her home (because doing so is harder), despite the fact that the cost savings from the latter are considerably larger.

Many important actions are not easy to perform. Steps can be taken, however, to make them easier to perform. In his excellent book *Fostering Sustainable Behavior*, Doug McKenzie-Mohr (2011) recommends taking an engineering-like approach to the task of making behaviors easier. The first step in the process is to conduct audience research for the purpose of identifying the barriers that impede people's performance of a behavior of whose value they are already convinced. These barriers might include a lack of knowledge about how to perform the action (e.g., "I can't remember which kinds of fish are sustainable"), a lack of skills necessary to perform the behavior well (e.g., "I don't know how to cook that kind of fish"), a lack of necessary resources (e.g., "Sustainably caught fish is too expensive"), concern about the negative consequences of performing the behavior incorrectly (e.g., "My kids won't eat it if they don't like it"), and so on.

The next step in McKenzie-Mohr's approach is to develop and pilot test ways of reducing—or ideally eliminating—the barriers found to be particularly common. The Seafood Watch app developed by the Monterey Bay Aquarium is a good example of a program intended to reduce at least one barrier to purchasing sustainably caught fish—not knowing which fish are sustainably caught. If the pilot-test results are promising, efforts can be made to encourage widespread adoption of the approach. The Marine Stewardship Council's Certified Sustainable Seafood program is an example of a program that has achieved considerable success through adoption by large food companies and retailers.

Another important way to make behaviors easier is to have members of the target audience demonstrate to other audience members how they perform the behavior, live or on video (Bandura 2004). Modeling demonstrations of this type are particularly effective when the models make explicit the necessary steps to perform the behavior successfully, the pitfalls to avoid, and the benefits of performing the behavior. Both of these approaches—reducing barriers and modeling the behavior—will increase people's sense of self-efficacy (self-confidence) to perform the recommended action, which increases the odds that people will try, persevere, and eventually succeed in performing the behavior (Bandura 2004).

In their terrific book *Switch*, Chip and Dan Health (2010) lay out a host of practical ways to make behavior change easier and to make behavior change programs more successful. Drawing on the metaphor of a rider (to represent people's thoughts), an elephant (to represent people's emotions), and their path (to represent the social and physical environment in which people are operating), the Heaths recommend setting a clear (i.e., unambiguous) goal, charting milestones so that progress made toward the goal will be positively reinforcing, and "tweaking the environment" (i.e., modifying or removing personal, social, or environmental barriers to performance of the behavior).

The *Climate Matters* program provides an example of how my colleagues and I have sought to make behavior change easier for TV weathercasters. Our audience research with weathercasters identified several key barriers to their ability to report on local climate change impacts stories, including lack of time to prepare stories, lack of access to data on local impacts, and lack of access to appropriate graphics and visuals to support their reporting. In response, to make the recommended behavior easier for weathercasters, each week our team produces and distributes broadcast-quality graphics, customized to each participating weathercasters' media market, which often feature data on the local impacts of climate change in their area; see Figure 26.2 for an example. (All current and past *Climate Matters* materials are available online at the website http://www.climatecentral.org/climate-matters.) To model use of these materials, and thereby increase participating weathercasters' sense of self-efficacy in using the materials on-air, the *Climate Matters* Facebook page includes examples of how other weathercasters have used the materials (see http://www.facebook.com/climate.matters/videos).

The Importance of Making the Behavior Fun

Climate change and biodiversity experts recommend behaviors not because the behaviors are fun, but because they offer important benefits. Regardless, experts should not lose sight of two important facts: people are more likely to perform behaviors that are fun than behaviors that are not; and receiving benefits is fun, while incurring costs is not.

People are often willing to incur costs to secure benefits that they value (including but certainly not limited to fun). The most attractive offers, however, are those that deliver valued benefits to people at the same time as—or even before—they are required to incur the costs (e.g., "No money down . . . take the car today"). Conversely, the least attractive offers are those that require incurring costs up front and receiving the benefits only much later (Rothschild 1999). Many actions recommended to prevent climate change and species loss are seen as

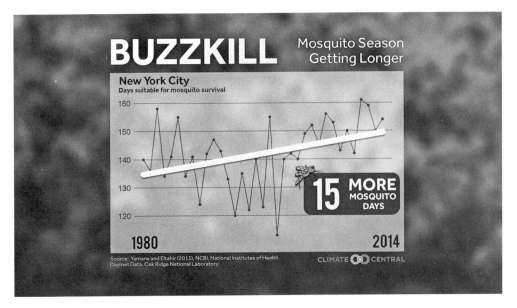

Figure 26.2. An example of broadcast-ready *Climate Matters* graphics that are produced by Climate Central and distributed to TV weathercasters. (Courtesy of Climate Central.)

requiring costs up front while delivering benefits only in the future, possibly the distant future.

To enhance the odds that people will adopt behaviors they have already decided are in their best interest, climate change and biodiversity professionals should consider two important questions: What can I do to make the behavior easier to perform? And what can I do to help decision makers get immediate benefits from the behavior, especially benefits they care most about?

Rare, a biodiversity conservation and behavior change organization—based in the United States but working worldwide—provides a great example. Rare has developed a highly successful model for enhancing fun, making recommended behaviors easier, and delivering valued benefits immediately— called "pride campaigns" (Boss 2008; Butler et al. 2013). Pride can be a powerful motivator (Patrick et al. 2009). Rare's approach centers on cultivating local pride in a community or in a nation—in their land, in their culture, and in their willingness to rally behind a local iconic species that is threatened

by current practices. This iconic species becomes the mascot of their campaign, a campaign that embraces and embodies fun, and offers people immediate benefits in the form of positive reinforcement for participating and the satisfaction of contributing to something of value to entire community.

The Importance of Making the Behavior (at Least Seem) Popular

People are highly sensitive to social norms. The more common (or normative) a behavior is perceived to be, the more likely people are to perform it (Cialdini 2006). There are two distinct types of social norms, both of which exert subtle but powerful influences on people's behavior. Descriptive norms are people's perceptions of how common a behavior (or attitude) is among people like themselves (e.g., friends in their social network, members of their "tribe," and/ or citizens of their community). Injunctive norms, conversely, are people's perceptions of the degree to which other relevant people (friends, "tribal" members, citizens) approve or disapprove approve of the behavior or attitude.

The most useful way to harness the influence of social norms depends, in part, on the degree to which a behavior being promoted is currently normative. When seeking to promote a behavior that is currently uncommon among members of the target audience, one can draw attention to specific notable people who are already performing the behavior, to their reasons for performing the behavior, and to the benefits they are enjoying as a result. Shining a light on these behavioral models makes the behavior appear more descriptively normative than it might otherwise seem, and as described above, it can also highlight the behavior's benefits and promote self-efficacy among decision makers who see the modeling.

Uncommon behaviors can quickly become popular when opinion leaders within a target audience embrace and endorse the behavior, thereby exerting their powerful social influence through injunctive norms. Sustainability professionals can seek out and recruit opinion leaders in their target population as a strategy for accelerating uptake of behaviors are recommending (Valente 2012).

If the recommended behavior is gaining in popularity but is not yet normative, efforts can be made to highlight its growing popularity—in the news, in entertainment media, and in social media—as a means of reinforcing the growing norm. Such efforts are particularly likely to be effective when they highlight notable respected individuals who are embracing the behavior (e.g., Warren Buffett), especially if most people would not expect those individuals to embrace the behavior.

Opower provides an excellent example. Opower is an American corporation that harnesses social science research on the power of social norms to help utility companies reduce consumer demand for electricity (http://opower.com/designprinciples). A study by some of the company's behavioral science advisers (Schultz et al. 2007) found that when shown on their monthly utility bill their own energy use data relative to the neighborhood's average energy use, above-

average households subsequently decreased their energy use, and below-average households subsequently increased their energy use, thereby demonstrating the power of descriptive norms. The boomerang effect among below-average households was prevented, however, by adding a smiley-face image on the utility bill to signal the utility company's approval of energy conservation—thereby demonstrating the power of injunctive norms. This simple insight about the power of social norms has led to a thriving business that is helping utility companies in a half dozen nations reduce their need to generate electricity.

In conclusion, communications efforts that use simple, clear messages, repeated often by a variety of trusted sources, and behavior change efforts that strive to make the behavior you are promoting easy, fun, and popular hold considerable promise in helping translate the insights of environmental science into more sustainable civilizations across the globe.

ACKNOWLEDGMENTS

This material is based on work supported by the National Science Foundation under Grant Numbers DRL-1422431 and DRL-1713450. Any opinions, findings, and conclusions or recommendations expressed in this material are those of the author(s) and do not necessarily reflect the views of the National Science Foundation.

REFERENCES

Bandura, Albert. 2004. "Health promotion by social cognitive means." *Health Education and Behavior* 31: 143–164.

Batra, Rajeev, and Michael Ray. 1986. "Situational effects of advertising repetition: The moderating influence of motivation, ability and opportunity to respond." *Journal of Consumer Research* 12: 432–445.

Boss, Suzie. 2008. "The cultural touch: Environmental non-profit RARE tailors its programs to local cultures and needs." *Stanford Social Innovation Review*. https://ssir .org/articles/entry/the_cultural_touch.

Butler, Paul, Kevin Green, and Dale Galvin. 2013. *The Principles of Pride: The Science Behind the Mascots.* Rare. http://rare.org/publications.

Chong, Dennis, and James Druckman. 2013. "Counterframing effects." *Journal of Politics* 75: 1–16.

Cialdini, Robert. 2006. *Influence: The Psychology of Persuasion.* William Morrow.

Crow, Desiree, and Maxwell Boykoff. 2014. *Culture, Politics and Climate Change: How Information Shapes Our Common Future.* Routledge.

Fischhoff, Baruch. 1989. "Risk: A guide to controversy." In *Improving Risk Communication,* ed. Institute of Medicine, 211–319. National Academies Press.

Fischhoff, Baruch. 2007. "Non-persuasive communication about matters of the greatest urgency: Climate change." *Environmental Science and Technology* 41: 7204–7208.

Heath, Chip, and Dan Heath. 2010. *Switch: How to Change Things When Change Is Hard.* Broadway Books.

Holmes, John, and Rebecca Clark. 2008. "Enhancing the use of science in environmental policy-making and regulation." *Environmental Science and Policy* 11: 702–711.

Hornik, Robert. 2002. *Public Health Communication: Evidence for Behavior Change.* Lawrence Erlbaum Associates.

Lang, Annie. 2013. "Discipline in crisis? The shifting paradigm of mass communication research." *Communication Theory* 23: 10–24.

Leiserowitz, Anthony. 2005. "American risk perceptions: Is climate change dangerous?" *Risk Analysis* 23: 1433–1442.

Leiserowitz, Anthony, Edward Maibach, and Connie Roser-Renouf. 2009. *Climate Change in the American Mind: American's Climate Change Beliefs, Attitudes, Policy Preferences, and Actions.* Yale Project on Climate Change Communication.

Leiserowitz, Anthony, Edward Maibach, Connie Roser-Renouf, Seth Rosenthal, Matt Cutler, and John Kotcher. 2017. "Climate change in the American mind: October, 2017." Yale Project on Climate Change Communication. https://www.climatechangecommunication.org/wp-content/uploads/2017/11/Climate-Change-American-Mind-October-2017-min.pdf.

Maibach, Edward, Bernadette Placky, Joe Witte, Keith Seitter, Ned Gardiner, Teresa Myers, Sean Sublette, and Heidi Cullen. 2016. "TV meteorologists as local climate educators." *Oxford Research Encyclopedia, Climate Science.* https://doi.org/10.1093/acrefore/9780190228620.013.505.

Maibach, Edward, and Vincent Covello. 2016. "Communicating environmental health." In *Environmental Health: From Global to Local,* 3rd ed., ed. Howard Frumkin, 769–791. Jossey-Bass.

Maibach, Edward, Michael Rothschild, and William Novelli. 2002. "Social marketing." In *Health Behavior and Health Education,* 3rd ed., ed. K. Glanz, Barbara Rimer, and Fran Marcus Lewis, 437–461. Jossey-Bass.

Maibach, Edward, Lorien Abroms, and Mark Marosits. 2007. "Communication and marketing as tool to cultivate the public's health: A proposed 'people and places' framework." *BMC Public Health* 7: 88.

Marshall, G. 2014. *Don't Even Think about It: Why Our Brains Are Wired to Ignore Climate Change.* Bloomsbury.

McKenzie-Mohr, Doug. 2011. *Fostering Sustainable Behavior.* 3rd. ed. New Society Publishers.

McKenzie-Mohr, Doug, Nancy Lee, Wesley Schultz, and Phillip Kotler. 2012. *Social Marketing to Protect the Environment.* SAGE Publications.

Moser, Suzanne, and Lisa Dilling. 2007. *Creating a Climate for Change.* Cambridge University Press.

Patrick, Vanessa, HaeEun Chun, and Deborah MacInnis. 2009. "Affective forecasting and self-control: When anticipating pride wins over anticipating shame in a self-regulation context." *Journal of Consumer Psychology* 19: 537–545.

Pechman, Cornelia, and David Stewart. 1988. "Advertising repetition: A critical review of wear-in and wear-out." *Critical Issues and Research in Advertising* 11: 285–329.

Pew Research Center. *Public Esteem for Military Still High.* http://www.pewforum.org/2013/07/11/public-esteem-for-military-still-high/.

Placky, Bernadette, Edward Maibach, Joe Witte, Bud Ward, Keith Seitter, Ned Garniner, David Herring, and Heidi Cullen. 2015. "Climate Matters: A comprehensive educational resource for broadcast meteorologists." *Bulletin of the American Meteorological Society.* http://dx.doi.org/10.1175/BAMS-D-14-00235.1.

Rothschild, Michael. 1999. "Carrots, sticks and promises: A conceptual framework for the management of public health and social issue behaviors." *Journal of Marketing* 63: 24–37.

Ryder, Dennis, Moya Tomlinson, Ben Gawne, and Gene Likens. 2010. "Defining and using 'best available science': A policy conundrum for the management of aquatic ecosystems." *Marine and Freshwater Research* 61: 821–828.

Schultz, Wesley, Jessica Nolan, Robert Cialdini, Noah Goldstein, and Vladas Griskevicius. 2007. "The constructive, destructive and reconstructive power of social norms." *Psychological Science* 18: 429–434.

Smith, William. 2011. *Reinventing Social Marketing.* TedX Penn Quarter. https://www.youtube.com/watch?v=IECY9LJvTf4.

Valente, Tom. 2012. "Network interventions." *Science* 337: 49–53.

Whitmarsh, Lorraine, Sophie O'Neill, and Irene Lorenzoni. 2011. *Engaging the Public with Climate Change.* Earthscan.

CHAPTER TWENTY-SEVEN

Climate Change, Food, and Biodiversity

CARY FOWLER AND
OLA TVEITEREID WESTENGEN

INTRODUCTION

Most human nutrition is derived from domesticated plants. Significant factors controlling the productivity of agricultural
crops are under human control; fundamentally their evolution is in our hands. As
Sir Otto Frankel once remarked, we have
acquired "evolutionary responsibility" for
them. A discussion about the impact of climate change on agriculture cannot, therefore, be about climate's effects on only
plants and agricultural ecosystems. It must
also concern how we discharge our evolutionary responsibility and assist agriculture
in adapting to dramatically new conditions.

Agriculture—plant and animal domestication—arose in multiple locations around
the world in the early and middle Holocene
and thereafter spread globally (Larson et al.
2014). As a consequence of the transition
from hunting and gathering to agriculture,
human beings became increasingly dependent on the vagaries of weather, as sedentary farmers' livelihoods are more sensitive
to fluctuations in temperature and rainfall
than those of nomadic hunter-gatherers.

Over time, exposure to different environments, climates, and human cultures
helped generate and sustain enormous genetic diversity within the gene pools of the
different domesticated crops. According to
the statistics of the Food and Agriculture
Organization of the United Nations, wheat,
which originated in the Near East as a crop,
is now grown in 124 countries, and maize,
domesticated in the Americas, is cultivated
in 166 countries (FAO 2014). Thus, the ecological niche of agricultural subsistence is
wide, but climate change is about to alter
the climatic parameters of that niche in
an unprecedented way in many parts of

the world. There is evidence that agricultural production is already being substantially and negatively affected by changes in climate, and projections based on climate models indicate that later in this century many farming areas will have climates not experienced during the entire history of agriculture. We are seeing an increasing mismatch between our crops and the human-altered environment (Carroll et al. 2014).

One in eight persons on Earth today, or more than 795 million in total, are already considered food insecure (FAO, IFAD, and WFP 2016). Most undernourished people live in South and East Asia and Sub-Saharan Africa. Although the challenge of adapting agriculture to new environmental circumstances will be global, posing an obstacle to development goals at all levels, it will take on a particularly urgent humanitarian aspect where the poor and food insecure are concentrated, where the impacts of climate change on food production tragically will be most severe.

IMPACTS

Crop yield is the most studied aspect of the impacts of climate change on food security. A large number of studies document that effects are already evident in several regions and project increasingly negative effects on yield. Climate extremes including unusually hot night- and daytime temperatures as well as natural disasters attributed to anthropogenic activity have already negatively affected overall crop production (Porter et al. 2014). A study at the International Rice Research Institute (IRRI) found that a 1°C increase in minimum temperature during the dry growing season had the effect of reducing rice yields by 10 percent (Peng et al. 2004). Lobell et al. (2011) found that global maize production declined 3.8 percent and wheat declined 5.5 percent between 1980 and 2008 compared to a counterfactual without climate change.

There are two principal types of studies estimating effects of climate change on future crop yield: statistical models relying on historical yield and weather data, and biophysical or process-based models relying on data from experimental crop trials (Porter et al. 2014; Lobell and Burke 2010). A recent global meta-level study that synthesized projections from more than 1,700 published simulations to evaluate yield impacts of climate change concluded: "There is a majority consensus that yield changes will be negative from the 2030s onwards. More than 70 percent of the projections indicate yield decreases for the 2040s and 2050s, and more than 45 percent of all projections for the second half of the century indicate yield decreases greater than 10 percent" (Challinor et al. 2014, 289). The picture emerging from these models is dire. At a global aggregate level, yields of maize and wheat begin to decline with 1°C–2°C of warming in the tropics, while temperate maize and tropical rice yields are significantly affected with warming of 3°C–5°C (Porter et al. 2014).

These global estimates mask considerable regional differences, however. Studies projecting positive effects on crop production in temperate regions buffer some of the anticipated decreases in yields in tropical regions for which there is strong consensus (Challinor et al. 2014). But the latest report from the International Panel of Climate Change (IPCC) (Porter et al. 2014) says it is now less likely that moderate warming will raise crop yields at mid- to high latitudes than projected in their previous assessment. Instead, the IPPC projects that there will be more yield decreases than increases even under moderate warming.

Sub-Saharan Africa is emerging as a region that will be particularly hard hit by climate change. On the basis of national historical climate and crop production data, Schlenker and Lobell (2010) projected aggregate production losses of 17 percent for sorghum and 22 percent for maize in the region by midcentury (Schlenker and Lobell 2010). Lobell et al. (2011) analyzed experimental yield data from 20,000 maize trials across South and East Africa and projected that 1°C warming will lead to yield losses in 65 percent–100 percent of the regions' maize growing areas. They found significant varietal differences depending on the maturity period of the varieties and on whether the variety is hybrid or open pollinated (Lobell et al. 2011). At a local level in this region, Thornton et al. (2009) used a biophysical crop model to estimate effects in grids corresponding to ~18 km by 18 km in the landscape and found that between 9 percent and 33 percent of the assessed grids across East Africa are likely to see a potential yield loss of more than 20 percent for maize by 2050.

Impact studies generally agree that the effect of temperature increase will be more important than changes in precipitation and that the benefits of increased CO_2 in the atmosphere for photosynthesis will be outweighed by the negative effects of temperature increase (Lobell and Gourdji 2012; Thornton et al. 2009). Excessive heat can harm virtually all plant parts and affect plant growth at every stage in the life cycle. The life cycle and growth period are shortened. Leaves curl and wither, reducing light interception. Partial closure of stomata to reduce transpiration increases canopy temperature and reduces photosynthesis effectiveness. The quantity, quality, and timing of growth in reproductive tissue are reduced (Cairns et al. 2013; Cossani and Reynolds 2012). Thus, negative effects of heat both cause and compound already existing moisture deficiencies. Most yield projections omit consideration of pest, weed, and disease impacts on crop yields, but several studies conclude that changes in climate and CO_2 concentration will alter the distribution and probably increase the competitiveness of nematodes, insects, fungal diseases, and weeds in many areas (Ziska et al. 2011; Porter et al. 2014). Effects on other biotic factors important for crop production such as on soil communities are underresearched but also represent potential threats to agricultural production.

At 2°C of global warming, many crops will begin to enter a new, historically unprecedented climate territory. New climates will generate combinations of conditions and biological assemblages with no current analogues (Burke, Lobell, and Guarino 2009). Changes in climate extremes can have more adverse impacts on crop production than changes in means. Moreover, today's extremes might become tomorrow's normal—today's hot growing seasons, which often result in large declines in food production, could become the cool and "best" growing seasons of the future (Figure 27.1). The damage done by short-run events could become long-term trends without sufficient investments in adaptation (Battisti and Naylor 2009). A recent statistical study of global crop production and extreme weather events found that droughts and extreme heat have reduced national cereal production by 9 percent–10 percent and 8 percent–11 percent more damage in developed countries than in developing ones (Lesk, Rowhani, and Ramankutty 2016). This is dire evidence in light of the estimate that by 2050, the majority of African countries will experience climates over at least half their current crop area that lie outside the range currently experienced within the country (Burke, Lobell, and Guarino 2009).

Evidence presented in previous chapters of this book suggests substantial risks to wild plant biodiversity due to range shifts and contraction in habitats with suitable climates. In a seminal study, Thomas et al. (2004) predicted that 15 percent–37 percent of wild plant biodiversity is threatened with extinction due to climate change by 2050. Similar projections have been made for crop wild relatives. Jarvis et al. (2008) estimates that up to 61 percent of wild peanut (*Arachis*) species, 12 percent of potato (*Solanum*) species, and 8 percent of cowpea (*Vigna*) species could become extinct within 50 years due to climate change. Other ecological niche modeling studies have projected large contractions in distribution areas for wild relatives in the Cucurbitacea

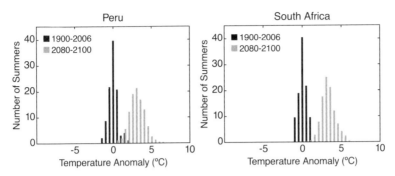

Figure 27.1. Historic and future average growing conditions in Peru and South Africa. (Data from Battisti and Naylor 2009.)

CLIMATE CHANGE AND FOOD SECURITY

Future food security will be affected not only by crop production levels but also by a host of socioeconomic factors that make it difficult to link climate change unambiguously to impacts on food security (Lobell and Gourdji 2012; HLPE 2012). Nevertheless, the IPCC highlights that climate change is likely to affect all aspects of food security, not only food availability, but also food access, utilization, and price stability (Porter et al. 2014). Nutritional quality of food and fodder (important for the utilization aspect of food security) is negatively affected by elevated CO_2 (Myers et al. 2014; Porter et al. 2014). Climate extremes have, together with other factors, played a role in periods of rapid food and cereal price increases (important for access and stability aspects of food security) in recent years. Yield losses in wheat and maize caused by already evident climate change between 1980 and 2008 resulted in an estimated average commodity price increase of 6.9 percent compared to a counterfactual with no climate change in the period (Lobell, Schlenker, and Costa-Roberts 2011). Projecting this trend into the future, the IPCC finds that changes in temperature and precipitation, without considering effects of CO_2, will contribute to increased global food prices in the range of 3 percent–84 percent by 2050 (Porter et al. 2014).

Periods of abnormally hot weather in the past provide an imperfect but important glimpse of how we might expect agricultural and social systems to react in the future unless food production keeps pace with the growing demand. Significant climate events due primarily, it is argued, to their impact on food production have been positively correlated with increased incidents of war and civil strife over the past 2,000 years (Büntgen et al. 2011; Hsiang, Burke, and Miguel 2013). As for food security, the effect of climate change on human security will heavily depend on political and economic pathways taken (Theisen, Gleditsch, and Buhaug 2013), but the link between climate-caused food scarcity and social instability is arguably an underlying trend in human history. Analyzing new conflict event data covering Asia and Africa in the period 1989–2014, von Uexkull et al. (2016) found that for groups depending on agriculture living in poor countries and experiencing political exclusion, droughts significantly increased the likelihood of violence.

family (to which pumpkins and squashes belong) in Mexico (Lira, Téllez, and Dávila 2009), on wild relatives of maize (*Zea*) in Mexico (Ureta et al. 2012), and wild Arabica coffee (*Coffea arabica*) in Ethiopia (Davis et al. 2012). Like wild plant biodiversity, crop wild relatives will face many compounding influences such as habitat degradation due to agriculture and human infrastructure, and shifts in pests and disease stress.

PREPARING FOR CLIMATE CHANGE THROUGH CROP ADAPTATION

As the twenty-first century progresses, a central question becomes whether, to what extent, and through which means agriculture will successfully adapt to climate change. This question is critical to food security, to global peace, and to the environment, because agricultural systems that function poorly are a threat to all.

Challinor et al. (2014) assess several common adaptation strategies in agriculture: cultivar adjustment, planting date adjustment, and irrigation and fertilizer optimization. Cultivar adjustment is estimated to be the most effective measure with a median potential to increase yield by 23 percent compared with situations where farmers do not change cultivars (Figure 27.2).

Cultivar adjustment refers to switching to a presumably better adapted variety, be it a local or an improved variety and can involve both formal and informal seed supply systems (Westengen and Brysting 2014). Recent typologies of adaptation distinguish between incremental, systemic, and transformative adaptation options (Vermeulen et al. 2013; Porter et al. 2014; Rippke et al. 2016) In this typology, cultivar adjustment involving local varieties and informal seed supply systems exemplifies incremental ad-

aptation as they are "based on long experience in dealing with a highly variable environment" (Vermeulen et al. 2013), while those involving improved varieties are classified either as incremental or systemic change, depending on whether the varieties are available off the shelves or are developed through targeted breeding programs. The cardinal question is whether farmers and breeders will be able to adapt based on the genetic variability in traditional and improved varieties. Early research into this question indicates that although there is great genetic potential in available genetic resources, there are challenges regarding access to the appropriate genetic resources, both for farmers and for breeders (Burke, Lobell, and Guarino 2009; Bellon, Hodson, and Hellin 2011; Bjørnstad, Tekle, and Göransson 2013).

Adaptation of crops to warmer new climates will not be reliably—and certainly not

Figure 27.2. The percentage benefit (yield difference between cases with and without the adaptation) for different crop management adaptations: cultivar adjustment (CA); planting date adjustment (PDA); adjusting planting date in combination with cultivar adjustment (PDA, CA); irrigation optimization (IO); fertilizer optimization (FO); other management adaptations (Other). The simulated median benefit is marked with a solid, dark gray, horizontal line and the whiskers indicate the 25th and 75th percentile. (Based on data from Challinor et al. 2014; Porter et al. 2014.)

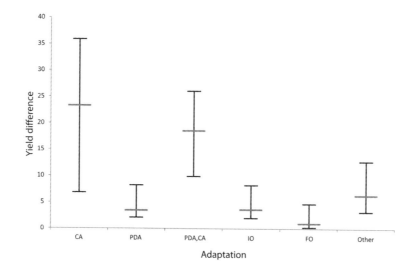

always—achieved by shifting production of current varieties to cooler climates. Such shifts involve changes in photoperiod, rainfall timing, pests and diseases, and involvement of other flora and soil types, making adaptation a nontrivial event. Incremental crop adaptation to environmental change by on-farm selection of local populations has dominated the 10,000-year history of agriculture; this is still the major pathway to crop adaptation for many crops in developing countries where informal seed systems dominate. Local varieties called landraces typically have a long history in their particular, often limited human and ecological communities and are adapted to those specific niches. But given the rapidity and scale of the projected climate change, landraces may experience a loss of the adaptive advantages they currently have. We cannot assume that landraces will routinely "evolve their way" into new climate regimes. Scientific plant breeders typically have access to a larger gene pool than farmers working with landraces, and therefore have a larger toolbox of genetic traits as well as technologies to draw on when developing new varieties. Integration of formal and informal seed system institutions and collaboration between farmers and breeders will be needed to harness the strengths of local and scientific knowledge (McGuire and Sperling 2013; Challinor et al. 2016; Westengen and Brysting 2014).

Because farmers and their crops are headed into uncharted territory, the employment not just of the primary but also of the secondary and tertiary gene pools of crops will likely be useful in deploying important adaptive traits. Crop wild relatives are increasingly recognized as a key resource for adapting crop species to climate change (McCouch et al. 2013). Successful utilization of genetic resources from other countries and from crop wild relatives in plant breeding relies on functioning policies, economic and scientific factors on the supply side as well as policies and socioeconomic factors on the demand side.

The alternative to adaptation on existing farmland will often be expansion of agriculture into nearby natural habitats. A recent study modeling climate change effects on global wine-grape production shows how agricultural adaptation may affect ecosystems negatively if farmers respond to declining climate suitability in traditional wine-growing areas by expanding into wildlife habitats (Hannah et al. 2013). Such indirect effects of climate change on biodiversity exacerbate direct impacts described in previous chapters of this book.

In addition to the five agricultural adaptation strategies described above, there are also a range of other adaptation strategies used by farmers around the world, some on-farm and some off-farm strategies. The options are widely different for a large corn farmer in Iowa, in the United States, compared to those available to a small-scale farmer in Morogoro, in Tanzania. Cultivar adjustment alone is unlikely to be a sufficient response, but without it other measures cannot succeed.

INSTITUTIONAL PREPAREDNESS: GAPS AND PROGRESS

Because new crop varieties typically take a decade or for some crops much longer to develop and deploy, the impact of climate change on agriculture and food becomes, to some extent, a question of how well prepared we are to begin to produce these new adapted varieties for the expected climate of 2025 and beyond. In other words, do we have access to the needed raw material?

Collecting of economic plants dates to 4500 BP. From the sixteenth century, botanical gardens were heavily involved. The geneticist N. I. Vavilov in the former Soviet Union organized the first systematic large-scale collections of crop diversity in the 1920s, which were conserved in St. Petersburg (then Leningrad) and used for plant breeding and research (Plucknett et al. 1987). The United States established its national genebank in 1954. Today, FAO records some 1,750 genetic resource collections housing more than 7 million samples. There is currently much duplication of conservation efforts, with the same samples held by several genebanks, and the FAO estimates that only 25 percent–30 percent of the total number of samples are unique (FAO 2010). This duplication of conservation efforts and proliferation of genebanks would not be a concern were it not for the fact that only a few of them actually offer secure long-term conservation and access (FAO 2010). A partial solution—an opportunity to solve this problem—is provided by the legal framework provided by the International Treaty on Plant Genetic Resources (ITPGRFA) and the institutional support provided by the Global Crop Diversity Trust, both crucial instruments for the development of an efficient and sustainable system of *ex situ* conservation called for by the FAO (2010).

Plant-breeding expertise and investment is focused on a small number of commercial crops, with maize at the top of the list. Perhaps half of all domesticated crops have never had a single Mendelian-trained plant breeder working with them (Fowler and Hodgkin 2004). Even crops of tremendous importance (e.g., banana, yam) have fewer than 10 plant breeders (Fowler and Hodgkin 2004). This would indicate a severe lack of capacity to develop adapted new varieties quickly. Furthermore, as many as a billion people live in farm families that are largely self-provisioning in terms of seeds; these are families mainly in developing countries with limited access to the genetic resources outside their own communities that might be needed for crop adaptation to significant climate change through on-farm selection and breeding. Genetic diversity and the diffusion of diversity such as took place with the massive government seed distribution in the United States in the 1800s (Fowler 1994) have heretofore enabled wide adaption, but no large-scale programs exist today to provide this diversity to farmers who

by default are the sole plant breeders for many crops.

Although plant genetic resource collections are numerically large globally, there are questions about the adequacy of collections for many individual crops (Fowler and Hodgkin 2004); moreover, no country is even remotely independent or self-reliant in terms of the resources they will need in the future for plant breeding. Every country is in some way dependent on others (Khoury et al. 2016). Figure 27.1 shows how dramatically the growing climates might change in individual countries, a reminder of countries' interdependency in terms of crop genetic resources needed to sustain breeding and production. Recognizing this interdependency, the ITPGRFA mandates "facilitated access" among its member states, but a recent survey shows that fewer than 40 percent are providing access; many genebanks do not respond to requests or deny them outright. Political obstacles cripple implementation of this key provision and even deter open acknowledgment of the problem by countries (Bjørnstad, Tekle, and Göransson 2013). This lack of access to genetic resources can be an impediment to development and deployment of traits needed for crop adaptation to climate change.

In recent years there has, however, been progress in securing the conservation of plant genetic resources. The Global Crop Diversity Trust established an endowment to provide ongoing funding to important collections. Norway constructed and—together with the Nordic Genetic Resource Centre and the Global Crop Diversity Trust—oversees the Svalbard Global Seed Vault, which functions as a safety net for the emerging international *ex situ* conservation system for seed crops (Westengen, Jeppson, and Guarino 2013; Fowler 2008). By the end of 2017, nearly 1,000,000 safety duplicates of accessions (varieties or populations) held in more than 60 genebanks were conserved at Svalbard.

Successful adaptation of crops to climate change will be required for the world to avoid significant increases in hunger, malnutrition, political instability, and environmental destruction. Additional investments will need to be made in crop diversity collection and conservation, in the screening of collections for useful traits, and particularly in plant-breeding programs, including for those crops for which there is little or no effort under way presently. Such efforts are a basic prerequisite for crop adaptation. If crops do not successfully adapt to climate change, neither will agriculture and neither will we.

REFERENCES

Battisti, David S., and Rosamond L. Naylor. 2009. "Historical warnings of future food insecurity with unprecedented seasonal heat." *Science* 323 (5911): 240–244.

Bellon, M. R., D. Hodson, and J. Hellin. 2011. "Assessing the vulnerability of traditional maize seed systems in Mexico to climate change." *Proceedings of the National Academy of Sciences* 108 (33): 13432–13437.

Bjørnstad, Åsmund, Selamawit Tekle, and Magnus Göransson. 2013. "'Facilitated access' to plant genetic resources: Does it work?" *Genetic Resources and Crop Evolution* 60 (7): 1959–1965.

Burke, M. B., D. B. Lobell, and L. Guarino. 2009. "Shifts in African crop climates by 2050, and the implications for crop improvement and genetic resources conservation." *Global Environmental Change: Human and Policy Dimensions* 19 (3): 317–325. https://doi.org/10.1016/j.gloenvcha.2009.04.003.

Büntgen, Ulf, Willy Tegel, Kurt Nicolussi, Michael McCormick, David Frank, Valerie Trouet, Jed O. Kaplan, Franz Herzig, Karl-Uwe Heussner, and Heinz Wanner. 2011. "2500 years of European climate variability and human susceptibility." *Science* 331 (6017): 578–582.

Cairns, Jill E., Jose Crossa, P. H. Zaidi, Pichet Grudloyma, Ciro Sanchez, Jose Luis Araus, Suriphat Thaitad, Dan Makumbi, Cosmos Magorokosho, and Marianne Bänziger. 2013. "Identification of drought, heat, and combined drought and heat tolerant donors in maize." *Crop Science* 53 (4): 1335–1346.

Carroll, Scott P., Peter Sogaard Jorgensen, Michael T. Kinnison, Carl T. Bergstrom, R. Ford Denison, Peter Gluckman, Thomas B. Smith, Sharon Y. Strauss, and Bruce E. Tabashnik. 2014. "Applying evolutionary biology to address global challenges." *Science* 346 (6207): art. 1245993.

Challinor, A. J., J. Watson, D. B. Lobell, S. M. Howden, D. R. Smith, and N. Chhetri. 2014. "A meta-analysis of crop yield under climate change and adaptation." *Nature Climate Change* 4 (4): 287–291.

Challinor, Andrew J., A.-K. Koehler, Julian Ramirez-Villegas, S. Whitfield, and B. Das. 2016. "Current warming will reduce yields unless maize breeding and seed systems adapt immediately." *Nature Climate Change* 6 (10): 954–958.

Cossani, C. Mariano, and Matthew P. Reynolds. 2012. "Physiological traits for improving heat tolerance in wheat." *Plant Physiology* 160 (4): 1710–1718.

Davis, Aaron P., Tadesse Woldemariam Gole, Susana Baena, and Justin Moat. 2012. "The impact of climate change on indigenous arabica coffee (*Coffea arabica*): Predicting future trends and identifying priorities." *PLOS One* 7 (11): e47981.

FAO. 2010. *The Second Report on the State of the World's Plant Genetic Resources.* FAO.

FAO. 2014. "FAOSTAT." http://faostat3.fao.org/home/index.html.

FAO, IFAD, and WFP. 2016. *The State of Food Insecurity in the World: Meeting the 2015 International Hunger Targets: Taking Stock of Uneven Progress.* Food and Agriculture Organization, International Fund for Agricultural Development, World Food Program.

Fowler, C., and T. Hodgkin. 2004. "Plant genetic resources for food and agriculture: Assessing global availability." *Annual Review of Environment and Resources* 29: 143–179. https://doi.org/10.1146/annurev.energy.29.062403.102203.

Fowler, Cary. 1994. *Unnatural Selection: Technology, Politics, and Plant Evolution.* Vol. 6. Gordon and Breach Yverdon.

Fowler, Cary. 2008. "The Svalbard seed vault and crop security." *BioScience* 58 (3): 190–191.

High Level Panel of Experts. 2012. *Food Security and Climate Change: A Report by the High Level Panel of Experts on Food Security and Nutrition of the Committee on World Food Security.* FAO.

Hsiang, Solomon M., Marshall Burke, and Edward Miguel. 2013. "Quantifying the influence of climate on human conflict." *Science* 341 (6151): 1235367.

Jarvis, Andy, Annie Lane, and Robert J. Hijmans. 2008. "The effect of climate change on crop wild relatives." *Agriculture Ecosystems and Environment* 126 (1–2): 13–23. https://doi.org/10.1016/j.agee,2008.01.013.

Khoury, Colin K., Harold A. Achicanoy, Anne D. Bjorkman, Carlos Navarro-Racines, Luigi Guarino, Ximena Flores-Palacios, Johannes M. M. Engels, John H. Wiersema, Hannes Dempewolf, and Steven Sotelo. 2016. "Origins of food crops connect countries worldwide." *Proceedings of the Royal Society B* 283: art. 20160792.

Larson, Greger, Dolores R. Piperno, Robin G. Allaby, Michael D. Purugganan, Leif Andersson, Manuel Arroyo-Kalin, Loukas Barton, Cynthia Climer Vigueira, Tim Denham, and Keith Dobney. 2014. "Current perspectives and the future of domestication studies." *Proceedings of the National Academy of Sciences* 111 (17): 6139–6146.

Lesk, Corey, Pedram Rowhani, and Navin Ramankutty. 2016. "Influence of extreme weather disasters on global crop production." *Nature* 529 (7584): 84–87.

Lira, Rafael, Oswaldo Téllez, and Patricia Dávila. 2009. "The effects of climate change on the geographic distribution of Mexican wild relatives of domesticated Cucurbitaceae." *Genetic Resources and Crop Evolution* 56 (5): 691–703.

Lobell, D. B., and M. B. Burke. 2010. "On the use of statistical models to predict crop yield responses to climate change." *Agricultural and Forest Meteorology* 150 (11): 1443–1452.

Lobell, D. B., M. Bänziger, C. Magorokosho, and B. Vivek. 2011. "Nonlinear heat effects on African maize as evidenced by historical yield trials." *Nature Climate Change* 1 (1): 42–45.

Lobell, David B., and Sharon M. Gourdji. 2012. "The influence of climate change on global crop productivity." *Plant Physiology* 160 (4): 1686–1697.

Lobell, David B., Wolfram Schlenker, and Justin Costa-Roberts. 2011. "Climate trends and global crop production since 1980." *Science* 333 (6042): 616–620.

McCouch, Susan, Gregory J. Baute, James Bradeen, Paula Bramel, Peter K. Bretting, Edward Buckler, John M. Burke, David Charest, Sylvie Cloutier, and Glenn Cole. 2013. "Agriculture: Feeding the future." *Nature* 499 (7456): 23–24.

McGuire, Shawn, and Louise Sperling. 2013. "Making seed systems more resilient to stress." *Global Environmental Change: Human and Policy Dimensions* 23 (3): 644–653.

Myers, Samuel S., Antonella Zanobetti, Itai Kloog, Peter Huybers, Andrew D. B. Leakey, Arnold J. Bloom, Eli Carlisle, Lee H. Dietterich, Glenn Fitzgerald, and Toshihiro Hasegawa. 2014. "Increasing CO_2 threatens human nutrition." *Nature* 510 (7503): 139–142.

Peng, Shaobing, Jianliang Huang, John E. Sheehy, Rebecca C. Laza, Romeo M. Visperas, Xuhua Zhong, Grace S. Centeno, Gurdev S. Khush, and Kenneth G. Cassman. 2004. "Rice yields decline with higher night temperature from global warming." *Proceedings of the National Academy of Sciences* 101 (27): 9971–9975.

Plucknett, Donald, Nigel J. H. Smith, J. T. Williams, and N. Murthi Anishetty. 1987. *Gene Banks and the World's Food.* Los Baños, Philippines: International Rice Resource Institute.

Porter, J. R., L. Xie, A. J. Challinor, K. Cochrane, S. M. Howden, M. M. Iqbal, D. B. Lobell, and M. I. Travasso. 2014. "Food security and food production systems." In *Climate Change 2014: Impacts, Adaptation, and Vulnerability. Part A: Global and Sectoral Aspects. Contribution of Working Group II to the Fifth Assessment Report of the Intergovernmental Panel on Climate Change*, ed. C. B. Field, V. R. Barros, D. J. Dokken, K. J. Mach, M. D. Mastrandrea, T. E. Bilir, M. Chatterjee, K. L. Ebi, Y. O. Estrada, R. C. Genova, B. Girma, E. S. Kissel, A. N. Levy, S. MacCracken, P. R. Mastrandrea, and L. L. White, 485–533. Cambridge University Press.

Rippke, U., J. Ramirez-Villegas, A. Jarvis, S. J. Vermeulen, L. Parker, F. Mer, B. Diekkrüger, A. J. Challinor, and M. Howden. 2016. "Timescales of transformational climate change adaptation in sub-Saharan African agriculture." *Nature Climate Change* 6: 605–609.

Schlenker, W., and D. B. Lobell. 2010. "Robust negative impacts of climate change on African agriculture." *Environmental Research Letters* 5 (1): 014010. https://doi.org/10.1088/1748-9326/5/1/014010.

Theisen, Ole Magnus, Nils Petter Gleditsch, and Halvard Buhaug. 2013. "Is climate change a driver of armed conflict?" *Climatic Change* 117 (3): 613–625.

Thomas, Chris D., Alison Cameron, Rhys E. Green, Michel Bakkenes, Linda J. Beaumont, Yvonne C. Collingham, Barend F. N. Erasmus, Marinez Ferreira De Siqueira, Alan Grainger, and Lee Hannah. 2004. "Extinction risk from climate change." *Nature* 427 (6970): 145–148.

Thornton, P. K., P. G. Jones, G. Alagarswamy, and J. Andresen. 2009. "Spatial variation of crop yield response to climate change in East Africa." *Global Environmental Change: Human and Policy Dimensions* 19 (1): 54–65. https://doi.org/10.1016/j.gloenvcha.2008.08.005.

Ureta, Carolina, Enrique Martínez-Meyer, Hugo R. Perales, and Elena R. Álvarez-Buylla. 2012. "Projecting the effects of climate change on the distribution of maize races and their wild relatives in Mexico." *Global Change Biology* 18 (3): 1073–1082.

Vermeulen, Sonja J., Andrew J. Challinor, Philip K. Thornton, Bruce M. Campbell, Nishadi Eriyagama, Joost M. Vervoort, James Kinyangi, Andy Jarvis, Peter Läderach, and Julian Ramirez-Villegas. 2013. "Addressing uncertainty in adaptation planning for agriculture." *Proceedings of the National Academy of Sciences* 110 (21): 8357–8362.

Von Uexkull, N., M. Croicu, H. Fjelde, and H. Buhaug. 2016. "Civil conflict sensitivity to growing-season drought." *Proceedings of the National Academy of Sciences* 113: 12391–12396.

Westengen, Ola T., and Anne K. Brysting. 2014. "Crop adaptation to climate change in the semi-arid zone in Tanzania: The role of genetic resources and seed systems." *Agriculture and Food Security* 3 (1): 3.

Westengen, Ola T., Simon Jeppson, and Luigi Guarino. 2013. "Global ex-situ crop diversity conservation and the Svalbard Global Seed Vault: Assessing the current status." *PLOS One* 8 (5): e64146.

Ziska, Lewis H., Dana M. Blumenthal, G. Brett Runion, E. Raymond Hunt Jr., and Hilda Diaz-Soltero. 2011. "Invasive species and climate change: an agronomic perspective." *Climatic Change* 105 (1–2): 13–42.

CHAPTER TWENTY-EIGHT

Saving Biodiversity in the Era of Human-Dominated Ecosystems

G. DAVID TILMAN, NIKO HARTLINE,
AND MICHAEL A. CLARK

INTRODUCTION

The most unique feature of Earth is the existence of life, and the most amazing feature of this life is its immense diversity. This biodiversity, which is the result of 3 billion years of evolution, is threatened by the explosive growth in the extent and intensity of human-caused habitat destruction, habitat fragmentation, pollution, hunting, overharvesting, and climate change. Each of these threats is, in its own right, a potentially major cause of extinctions. The combined impacts on biodiversity of these multiple threats are as yet poorly understood, but they are reasonably assumed to pose a large threat to biodiversity, perhaps similar in scale to the great mass extinction events that are evident in the fossil record (Barnosky et al. 2011). This chapter explores ways that the risks of extinction could be greatly decreased while still meeting current and future global food needs and while reducing the impacts of agriculture on greenhouse gas emissions.

Biodiversity loss matters for both ethical and pragmatic reasons. The ethics behind extinction prevention are clear. Pragmatically, biodiversity is a dominant factor determining the productivity, nutrient dynamics, invasibility, and stability of ecosystems (Cardinale et al. 2012; Tilman, Isbell, and Cowles 2014). More than a hundred biodiversity experiments and related theory have shown that biodiversity impacts ecosystem functioning precisely because species have different functional traits. In essence, each species performs best in certain environmental conditions or situations.

Higher biodiversity thus improves community and ecosystem functioning because a larger suite of species increases the productivity, stability, and carbon capture and storage abilities of ecosystems. It is for these reasons that the loss of biodiversity threatens the ability of ecosystems to provide humanity with valued services.

Most of humanity's threats to biodiversity, and most of our other negative environmental impacts, are the inadvertent affects of how we meet the food and energy demands of the global population (Matson et al. 1997). These environmental impacts have been dramatically escalating since about 1900 because of massive growth in both global population and per capita consumption. From 1900 to 2015 global population increased from 1.7 billion to 7.3 billion and global per capita purchasing power increased from about $700 to $8,500. This 50-fold increase in the global human economy (i.e., $[7.3 \times 8,500]/[1.7 \times 700] = 52$) has greatly increased global demand for food and energy.

The net effect of providing food and energy to the global population is that more than 4,200 vertebrate species are now highly threatened with extinction (Figure 28.1A). In particular, according to analyses done by the International Union for the Conservation of Nature (IUCN), 16 percent of all bird and mammal species are either endangered or critically endangered with extinction, as are 7 percent of amphibians and reptiles (Figure 28.1B). The major causes of these extinction risks are ag-

ricultural land clearing and land clearing for logging, as well as hunting and climate change (Figure 28.2).

Moreover, global demands for food and energy seem likely to continue increasing until the end of this century, though at a much slower rate than in the last century. The UN projects that global population will reach 9.7 billion people by 2050 and will then level off at about 11.2 billion by 2100. Although economic growth is slowing in richer nations, the vast majority of the global population lives in nations that currently are poor but have economies that are growing rapidly. When projections of global economic growth and population growth are combined, total global economic activity may be about four times higher in 2050, and eight times higher in 2100, than it was in 2015. Unless policies are implemented that decrease agricultural land clearing, logging, excessive hunting, and fossil-fuel combustion, the extinction risks faced by animal and plant species are likely to greatly increase throughout the rest of this century. Such policies, and their scientific bases, are the focus of this chapter.

Figure 28.1. (A) The number of vertebrate animal species that have been driven extinct by human actions, or that are classified as critically endangered with extinction or as endangered, by the IUCN Red List. (B) The percentage of all known species that are extinct or threatened in each the four groups, based on the IUCN Red List.

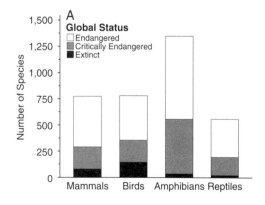

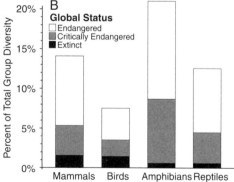

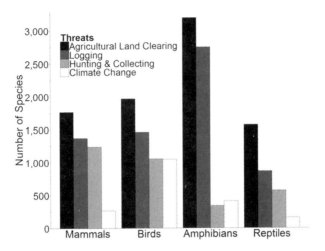

Figure 28.2. Four of the major factors that threaten vertebrate species and the number of species threatened by each factor, based on the IUCN Red List.

EXTINCTION FROM FOOD AND ENERGY PRODUCTION

People need food and energy; with 7 billion people now, heading toward 11 billion richer people by the end of the century, global demand for both food and energy will increase greatly. Global demand for agricultural crops is forecast to increase by 70 percent–110 percent from 2005 to 2050 (Tilman et al. 2011). About a third of this increase is from population growth; two-thirds is from dietary changes associated with higher incomes, especially increased per capita meat consumption, the animal feeds needed to produce it, and increased caloric consumption.

Agricultural land clearing is a major threat to biodiversity, with 8,530 species of mammals, birds, amphibians, and reptiles listed by the IUCN as threatened by this activity. Logging is also important, and it is often the initial step toward the conversion of forests into agricultural lands. Currently 4.9 billion hectares of land, which is 40 percent of the ice-free land area of Earth, is in agriculture, with 1.5 billion hectares in croplands and the remainder in pastures. An additional 0.5 billion hectares had been cleared and used for agriculture

but abandoned during the past half century. Moreover, because global crop demand is increasing more rapidly than yields, about 400 million–700 million additional hectares of cropland may be needed in the next 40 years to meet global food demand (Tilman et al. 2011; Tilman and Clark 2014). Because this land clearing is mainly occurring in high-biodiversity tropical nations (Gibbs et al. 2009), large numbers of additional species are likely to face risks of extinction.

Habitat destruction causes extinctions by two different mechanisms. First, as habitat is destroyed, there is the possibility that the act of destruction will lead to the direct death of the last living members of a species. Such direct extinction by habitat destruction is most likely to affect rare species and to cause the extinction of widespread and abundant species only as larger and larger proportions of a region are destroyed. The potential direct impact of habitat destruction on biodiversity is often modeled by using the species-area relationship or its derivatives (Rosenzweig 1995; He and Hubbell 2011).

Second, the fragmentation that commonly results from habitat destruction can cause the time-delayed extinction of those species that survived (e.g., Gibson et al. 2013). This has been termed the "extinction debt" (Tilman et al. 1994) and can lead to the eventual extinction of even those species that were, at the time of destruction, some of the most abundant species in the fragments (Kuussaari et al. 2009). In particular, the species most susceptible to time-delayed extinction from fragmentation are predicted to be species that are good local competitors but poor dispersers. The extinctions from destruction and from fragmentation are each predicted to be sharply increasing functions the amount of habitat that is destroyed (Tilman et al. 1994). Current theory suggests that the numbers of species lost by direct extinction and the numbers lost by fragmentation may be about equal, but the actual spatial patterning of destruction would have great influence on each type of extinction.

Climate change is a looming threat to biodiversity, and many species are already negatively impacted by climate change. The IUCN Red List has a total of 1,880 species of mammals, birds, amphibians, and reptiles threatened by climate change and severe weather events (Figure 28.2). As climate change accelerates, so will these threats. Because per capita energy demand is an increasing function of per capita incomes, global energy demand is forecast to increase at a greater rate than population (International Energy Agency 2015). Whether this demand will be met by combustion of fossil fuels or by renewable energy will depend on the adoption of global greenhouse gas policies. Moreover, the greenhouse gas releases from agriculture, which currently account for 30 percent of annual emissions via CO_2 from land clearing and soil tillage, methane from ruminants and rice, and nitrous oxide from nitrogen fertilizer, are also increasing.

Climate change can make ecosystems become unsuitable to the species living in them. During glacial cycles, most species successfully migrated so as to remain in regions with suitable climate, but the ability of species to do so depended on there being few barriers to such migration. Regions with barriers suffered extinctions. For instance, most tree species in Asia and North America successfully migrated toward the equator when the last glacial period started, and then to the north when glaciers retreated. However, because of barriers to migration, the glacial cycles contributed to tree extinctions in Europe, thus partially explaining its low tree diversity relative to Asia (Latham and Ricklefs 1993). Massive human-caused habitat fragmentation has literally changed the landscapes of all continents, imposing numerous barriers to the migration of plants and animals. The combination of rapid of human-driven climate change and barriers to migration may pose a severe threat to biodiversity. Finally, two other ways that people obtain food, hunting and fisheries, also pose extinction risks for some types of prey species and especially for other species, called bycatch in fisheries, that are not directly targeted but are inadvertently killed.

In total, increasing global demand for food and energy are greatly increasing the extinction risks facing life on Earth. Because agriculture and fossil energy use pose such major risks for biodiversity, the reduction or elimination of these threats will require changes in agricultural and energy systems worldwide. Fortunately, many such changes offer multiple benefits for biodiversity preservation, climate stabilization, and pollution reduction, and they can be achieved using current knowledge and technology. These changes provide a pathway toward a more sustainable Earth that would provide the food, energy, and livable environments that all of us, and all future generations, need.

MINIMIZING EXTINCTIONS ON A HUMAN-DOMINATED PLANET

What will determine how many species survive in remnant habitats around the world in 2050 and beyond? The greatest factors are how we obtain food and energy, and how much of each we demand. There are five pathways, all of them feasible because they use existing knowledge and technologies, each of which would reduce the risk of a massive wave of extinctions. These five actions are outlined below.

Prevent Extinctions by Increasing Yields of Developing Nations

Because global demand for crops is growing more rapidly than yields, from 1980 to 2000, about 30 million hectares of land were cleared annually for agriculture in developing nations, at the same time that about 15 million hectares per year globally were abandoned from agriculture (Gibbs et al. 2010). Future land clearing will depend on how global per capita demand for crops increases with income and on how rapidly the yields of developing nations increase. The past trends in both of these variables are not encouraging. Mainly because of rapid increases in meat consumption as incomes increase, crop demand may double. However, yield trends in developing nations show that many of these nations have yields that are growing slowly and currently are much below their potential, a phenomenon called the yield gap (Cassman 1999; Lobell, Cassman, and Field 2009).

In some of the least developed nations, crop yields are 20 percent–25 percent of the yields that could be obtained via intensification. In a large suite of additional developing nations, current yields are less than half of those that could be attained. Because demand for crops is rising rapidly in developing nations, most of which have such yield gaps, two to five times more land is being cleared every year that would be needed if these yield gaps were closed.

The net effect of current trends is that global land clearing could, for decades to come, remain at about the 20 million ha per year rate reported by Gibbs et al. (2009) for 1980–2000, which is similar to the 20 million ha per year of deforestation for 2000–2005 reported by Hansen, Stehman, and Potapov (2010). This land clearing would cause large numbers of extinctions

in some of the most diverse ecosystems on Earth.

Because some cropland is taken out of production each year because of conversion to urban, suburban, transit, or other uses and because of loss of fertility, the net increase in global cropland is likely to be less than the amount of land that is cleared. Tilman and Clark (2014) estimated that the total amount of global cropland might increase by from 400 million to 700 million hectares between 2010 and 2050 if yields in yield-gap nations were to continue to increase along their past trajectories.

A variety of analyses suggest that closing the yield gap could provide healthy diets for 9 billion people while requiring little more than the existing amount of global agricultural lands (Foley et al. 2011; Tilman et al. 2011; Mueller et al. 2012). As such, efforts to bring yields up to their potential in all nations could not only help provide all the people of the Earth with secure and nutritious diets but also greatly reduce, or perhaps eliminate, land clearing that is a major threat to global biodiversity. However, it is imperative that yields be increased via a much more sustainable approach to agricultural intensification, as described below.

Sustainable Intensification Raises Yields with Lower Environmental Impacts

Agriculture is of central importance to humanity, providing most of the 7 billion people of Earth with secure and nutritious supplies of food. However, as currently practiced, agriculture also has inadvertent but globally significant harmful impacts. It creates 30 percent of total GHG emissions, and pollutes aquifers, lakes, river, and nearshore marine ecosystems with agrochemicals such as nitrogen and phosphorus from fertilizers and various pesticides (Tilman et al. 2001; Foley et al. 2011). The GHG sources are carbon dioxide released as land is cleared and soils are tilled; the potent GHG, nitrous oxide, released from nitrogen fertilization; and methane released by cattle, sheep, and goat production and during rice cultivation. Moreover, both the GHG and the agrochemical impacts of agriculture are on trajectories to double within the next 40 years (Tilman et al. 2001, 2011). These impacts and trends mean that the methods of agricultural intensification used to close the yield gap must be much more sustainable than are current methods of agriculture.

A variety of practices have been identified that reduce the environmental impacts of intensive agriculture while still achieving high yields (Cassman 1999; Snyder, Bruulsema, and Fixen 2009; Robertson and Vitousek 2009; Vitousek et al. 2009). These practices are called sustainable intensification (Godfray et al. 2010; Tilman et al. 2011; Godfray and Garnett 2014). Chief among these are methods to obtain high crop yields that minimize the amount of nitrogen fertilizer applied, and thus its negative environmental impacts (e.g., Robertson and Vitousek 2009; Vitousek et al. 2009; Hoben et al. 2011; Meuller et al. 2014). A foundation of sustainable intensification is the timely application of the amount of fertilizers needed at various points during crop growth (Snyder, Bruulsema, and Fixen 2009). For instance, when all the nitrogen fertilizer needed for a year is applied at once, often before the crop is planted, 30 percent–40 percent of this nitrogen is lost from the system as ammonia that vaporizes into the air, as nitrous oxide, and as nitrate and nitrite that enter waters. In sustainable intensification, because fertilizers are applied several times during the growing season so as to meet current crop demands for nitrogen and other nutrients, most nutrients are taken up by the crop, and there is a much smaller pool of unused nutrient that can be lost from the system. Because of this, the same high yields have been attained using about 30 percent less nitrogen than in conventional intensive agriculture (Robertson and Vitousek 2009; Vitousek et al. 2009). Sustainable intensification thus offers large GHG and water-quality benefits (Godfray and Garnett 2014).

Prevent Extinctions with Healthier Diets

Adoption of healthy diets could also prevent future land clearing and resultant species extinctions (Tilman and Clark 2014; Tilman et al. 2017). The major reason why a 30 percent increase in global population is forecast to lead to a 70 percent–110 percent increase in crop demand it that diets change as incomes rise (Figure 28.3). The adoption of "Western diets," which is especially common in urban areas as developing nations industrialize, leads to increased per capita caloric, meat, processed food and sugar consumption, to great increases in production of crops for animal feeds, and to detrimental health impacts caused by this nutrition transition (Popkin 1994; Drewnowski and Popkin 1997; Popkin, Adair and Ng 2012).

If people were to adopt healthy diets, rather than typical Western diets, the lower demand for meat and thus animal feeds would greatly reduce the need for land clearing (Tilman and Clark 2014; Figure 28.4). For instance, the traditional Mediterranean diet, which is an omnivorous diet that has high consumption of vegetables and fruits, nuts, and whole grains, and low consumption of red meat, could eliminate the need for about 400 million more hectares of cropland by 2050. In general, diets that have low consumption of meat, especially beef and other ruminant meats, require less cropland and pasture per capita. In total, if per capita consumption of red meat were to decline in developed nations to healthier levels, and if the current income dependence of increased meat consumption were to stop in developing nations, much less land would have to be cleared while providing the world with healthy diets in 2050 (Figure 28.4).

Reduce Food Waste to Prevent Extinctions

Reduction in food waste could also reduce future land clearing and decrease the other environmental impacts of agriculture

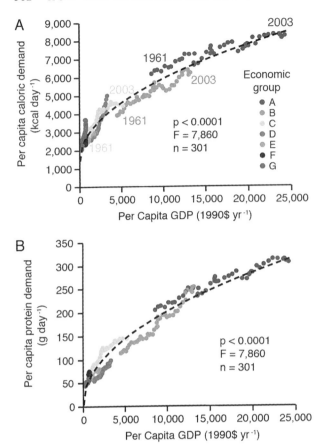

Figure 28.3. Per capita demand for protein (A) and for calories (B) increase as a function of per capita income (as estimated by per capita GDP). Each point is the average of 15 economically similar nations for a given year. Years shown are 1961 through 2005.

(Godfray and Garnett 2014). The amounts of crops and foods wasted and the reasons for the wastage vary among nations and regions (Gustavvsson, Cederberg, and Sonesson. 2011). On average, in developing nations, wastage occurs more from failure to fully harvest crops, or to preserve or adequately store them at the time of harvest, with less wasted once foods reach homes. In developed nations, wastage is often small during crop harvesting, storage, and processing, but high for perishables in grocery stores, and even higher in homes and restaurants. Although the reasons for wastage differ among nations, on average about 30 percent of crop production is

wasted rather than eaten. If the proportion of crop production that is wasted could be cut in half, about 300 million hectares less land would be needed to meet the currently projected 2050 global food demand. This estimate is based on a simple calculation; 15 percent less crop production would require 15 percent less cropland. With about 2 billion hectares of cropland needed in 2050, the savings would be 300 million hectares. A 15 percent reduction in waste would also proportionately reduce all of the other environmental impacts of agriculture.

In total, increasing yields in low-yielding nations via sustainable intensification, healthier diets, and reduced food waste would each decrease the cropland needed to feed the world of 2050. Because all three, in unison, would allow the world to be fed with less land than is currently used, partial

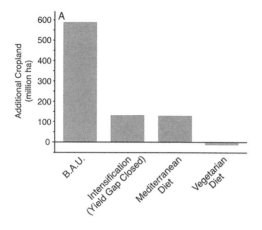
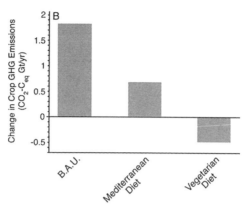

Figure 28.4. (A) Scenarios of additional global cropland that would be needed in 2050, relative to the amount of 2010. The BAU (business as usual) scenario uses forecasted food demand based on income-dependent global diets and current yield trends. The intensification scenario assumes the same food demand as BAU but assumes that yield gaps are closed via sustainable intensification. The Mediterranean diet scenario assumes global adoption of this diet current yield trends. The vegetarian diet scenario assumes global adoption of this diet current yield trends. (Based on Tilman et al. 2011 and Tilman and Clark 2014.) (B) Similar to part A but showing the resultant changes in GHG emissions from crop production (but not land clearing) associated with BAU, Mediterranean, and vegetarian scenarios.

adoption of each of these three approaches could still provide major biodiversity benefits. Each of these actions would also decrease nutrient pollution of aquifers, lakes, rivers, and oceans and lead to lower agricultural GHG emissions.

Prevent Extinctions by Reducing Climate Change

A sustainable world also requires implementation of policies and technologies that could first prevent any further increase in the annual rate of global GHG emissions, and then could progressively reduce GHG emissions below the current rate. Pacala and Socolow (2004) show that appropriate deployment of existing technologies could stop any further increase in the annual rate of global fossil-fuel-based GHG emissions. Such a stabilization of global GHG emis-

sions could prevent a doubling of preindustrial CO_2 levels, and thus avoid some of the worst climate scenarios and the extinctions they would cause. The greatest emitter of GHG is electric power generation, mainly from coal, that creates 40 percent of total global GHG emissions. Renewable power from wind and solar could displace much of current coal-based power and do more as power storage capabilities are increased. Energy-efficient appliances, buildings, and vehicles are another part of the Pacala and Socolow strategy. As discussed above, sustainable intensification of agriculture, healthier diets, and reduced food waste also provide significant GHG benefits and should be part of a global strategy to address climate change.

SUMMARY

With 7 billion people currently living on Earth, and with global population expected to reach about 11 billion by the end of this century, it is no longer possible to ignore the massive, though inadvertent, environmental impacts that humanity would cause should our ways of meeting our food and energy demands remain on their current trajectories. Here we have outlined five feasible steps that would change these trajectories and provide great long-term benefits for all the peoples and nations of the world. Efficient agricultural production, sustainable intensification of current croplands, healthy diets, efficient use of food and energy, and a focus on renewable energy sources are the keys to creating a global economy and culture that can nourish and support humanity and nature for millennia to come. There is no unavoidable need for any further land clearing, no unavoidable reasons for a massive wave of species extinctions, and no factors that must necessarily drive the Earth into unprecedented climatic changes.

REFERENCES

Barnosky, A. D., N. Matzkel, S. Tomiya, G. O. U. Wogan, B. Swartz, T. B. Quental, C. Marshall, et al. 2011. "Has the Earth's sixth mass extinction already arrived?" *Nature* 471: 51–57.

Cassman, K. C. 1999. "Ecological intensification of cereal production systems: Yield potential, soil quality, and precision agriculture." *Proceedings of the National Academy of Sciences* 96: 5952–5959.

Cardinale, B. J., J. Duffy, A. Gonzalez, D. Hooper, C. Perrings, P. Venail, A Narwani, et al. 2012. "Biodiversity loss and its impact on humanity." *Nature* 486: 59–67.

Drewnowski, A., and B. M. Popkin. 1997. "The nutrition transition: New trends in the global diet." *Nutrition Reviews* 55: 31–43.

Foley, J. A., N. Ramankutty, E. Bennett, K. Brauman, S. Carpenter, E. Cassidy, J. Gerber, et al. 2011. "Solutions for a cultivated planet." *Nature* 478: 337–342.

Gibbs, H. K., A. Ruesch, F. Achard, M. Clayton, P. Holmgren, N. Ramankutty, and J. Foley. 2009. "Tropical forests were the primary sources of new agricultural land in the 1980s and 1990s." *Proceedings of the National Academy of Sciences* 107: 16732–16737.

Gibson, L., A. J. Lynam, C. J. A. Bradshaw, F. He, D. P. Bickford, D. Woodruff, S. Bumrungsri, and W. F. Laurance. 2013. "Near-complete extinction of native small mammal fauna 25 years after forest fragmentation." *Science* 341: 1508–1510.

Godfray H. C. J., J. R. Beddington, I. R. Crute, L. Haddad, D. Lawrence, J. F. Muir, J. Pretty, S. Robinson, S. M. Thomas, and C. Toulmin. 2010. "Food security: The challenge of feeding 9 billion people." *Science* 327: 812–818. https://doi.org/10.1126/science.1185383.

Godfray, H. C. J., and T. Garnett. 2014. "Food security and sustainable intensification." *Philosophical Transactions of the Royal Society B* 369. https://doi.org/10.1098/rstb.2012.0273.

Gustavvsson, J., C. Cederberg, and U. Sonesson. 2011. *Global Food Losses and Food Waste*. FAO.

Hansen, M. C., S. V. Stehman, and P. V. Potapov. 2010. "Quantification of global gross forest cover loss." *Proceedings of the National Academy of Sciences* 107: 8650–8655.

He, F., and S. P. Hubbell. 2011. "Species-area relationships always overestimate extinction rates from habitat loss." *Nature* 463: 368–371.

Hoben, J. R. J. Gehl, N. Millar, P. R. Grace, and G. P. Robertson. 2011. "Nonlinear nitrous oxide (N_2O) response to nitrogen fertilizer in on-farm corn crops of the US Midwest." *Global Change Biology* 17: 1140–1152.

Hu, F. B. 2011. "Globalization of diabetes: The role of diet, lifestyle, and genes." *Diabetes Care* 34: 1249–1257.

International Energy Agency. 2015. *World Energy Outlook 2015*. International Energy Agency.

Kuussaari, M., R. Bommarco, R. K. Heikkinen, A. Helm, J. Krauss, R. Lindborg, E. Ockinger, et al. 2009. "Extinction debt: A challenge for biodiversity conservation." *Trends in Ecology and Evolution* 24: 564–570.

Latham, R. E., and R. E. Ricklefs. 1993. "Continental comparisons of temperate-zone tree species diversity." In *Species Diversity in Ecological Communities: Historical and Geographical Perspectives*, ed. R. E. Ricklefs and D. Schluter. University of Chicago Press.

Lobell, D. B., K. C. Cassman, and C. B. Field. 2009. "Crop yield gaps: Their importance, magnitude and causes." *Annual Review of Environment and Resources* 34: 1–26.

Matson, P. A., W. J. Parton, A. G. Power, and M. J. Swift. 1997. "Agricultural intensification and ecosystem properties." *Science* 277: 504–509.

Mueller, N. D., J. S. Gerber, M. Johnston, D. K. Ray, N. Ramankutty, and J. A. Foley. 2012. "Closing yield gaps through nutrient and water management." *Nature* 409: 254–257.

Mueller, N. D., P. C. West, J. S. Gerber, G. K. MacDonald, S. Polasky, and J. A. Foley. 2014. "A tradeoff frontier for global nitrogen use and cereal production." *Environmental Research Letters* 9. https://doi.org/10.1088/1748–9326/9/5/054002.

Pacala, S., and R. Socolow. 2004. "Stabilization wedges: Solving the climate problem for the next 50 years with current technologies." *Science* 305: 968–972.

Popkin, B. M. 1994. "The nutrition transition in low-income countries: An emerging crisis." *Nutrition Reviews* 52: 285–298.

Popkin, B. M., L. Adair, and S. W. Ng. 2012. "Global nutrition transition and the pandemic of obesity in developing countries." *Nutrition Reviews* 70: 3–21.

Robertson, P. G., and P. Vitousek. 2009. "Nitrogen in agriculture: Balancing the cost of an essential resource." *Annual Review of Environment and Resources* 34: 97–125.

Rosenzweig, M. L. 1995. *Species Diversity in Space and Time*. Cambridge University Press.

Snyder, C. S., T. W. Bruulsema, T. L. Jensen, and P. E. Fixen. 2009. "Review of greenhouse gas emissions from crop production systems and fertilizer management effects." *Agriculture, Ecosystems and Environment* 133: 247–266.

Tilman, D., C. Balzer, J. Hill, and B. Befort. 2011. "Global food demand and the sustainable intensification of agriculture." *Proceedings of the National Academy of Sciences* 108: 20260–20264.

Tilman, D., and M. Clark. 2014. "Global diets link environmental sustainability and human health." *Nature* 515: 518–522.

Tilman, D., M. Clark, D. R. Williams, K. Kimmel, S. Polasky, and C. Packer. 2017. "Future threats to biodiversity and pathways to their prevention." *Nature* 546: 73–81.

Tilman, D., J. Fargione, B. Wolff, C. D'Antonio, A Dobson, R. Howarth, D. Schindler, W. Schlesinger, D. Simberloff, and D. Swackhamer. 2001. "Forecasting agriculturally driven global environmental change." *Science* 292: 281–284.

Tilman, D., F. Isbell, and J. M. Cowles. 2014. "Biodiversity and ecosystem functioning." *Annual Review of Ecology, Evolution, and Systematics* 45: 471–493.

Tilman, D., R. M. May, C. L. Lehman, and M. A. Nowak. 1994. "Habitat destruction and the extinction debt." *Nature* 371: 65–66.

Vitousek, P., R. Naylor, T. Crews, M. B. David, L. E. Drinkwater, E. Holland, P. J. Johnes, et al. 2009. "Nutrient imbalances in agricultural development." *Science* 324: 1520–1521.

Contributors

David Ainley
Senior Ecologist
H. T. Harvey & Associates

Cecile Albert
Research Officer
Mediterranean Institute of Biodiversity and
Marine Continental Ecology

Craig D. Allen
Research Ecologist
Fort Collins Science Center
New Mexico Landscapes Field Station
United States Geological Survey

Katie Arkema
Lead Scientist
Natural Capital Project
Stanford Woods Institute for the
Environment
Stanford University

Richard B. Aronson
Professor and Head
Department of Biological Sciences
Florida Institute of Technology

Michael Avery
PhD Candidate
Department of Biology
Pennsylvania State University

Grant Ballard
Chief Science Officer
Point Blue Conservation Science

Céline Bellard
CNRS Researcher
Université Paris-Sud

Monika Bertzky
Independent Consultant
Biodiversity, Ecosystems, and Environment

John B. Bradford
Research Ecologist
Southwest Biological Science Center
United States Geological Survey

David D. Breshears
Regents' Professor
School of Natural Resources and the
Environment
Department of Ecology and Evolutionary
Biology
University of Arizona

Rebecca C. Brock
Senior Programme Officer
Climate Change and Biodiversity
United Nations Environment Programme
World Conservation Monitoring Centre
(UNEP-WCMC)

Olivier Broennimann
Staff Scientist
Department of Ecology and Evolution
Institute of Earth Surface Dynamics
University of Lausanne, Switzerland

Lauren B. Buckley
Associate Professor
Department of Biology
University of Washington

Aline Buri
Graduate Assistant
Institute of Earth Surface Dynamics
University of Lausanne, Switzerland

Kevin D. Burke
PhD Candidate
Nelson Institute for Environmental Studies
University of Wisconsin–Madison

Mark B. Bush
Professor
Department of Biological Sciences
Florida Institute of Technology

Lindsay P. Campbell, PhD
Biodiversity Institute
University of Kansas

William W. L. Cheung
Associate Professor
Institute for Oceans and Fisheries
University of British Columbia

Carmen Cianfrani
Postdoctoral Fellow
Department of Ecology and Evolution
University of Lausanne, Switzerland

Michael A. Clark
PhD Student
Natural Resources Science and
Management
University of Minnesota

Neil S. Cobb
Research Professor
Merriam-Powell Center for Environmental
Research
Northern Arizona University

Lee W. Cooper
Research Professor
Chesapeake Biological Laboratory
University of Maryland Center for Environ-
mental Science

Franck Courchamp
Director of Research
Centre National de la Recherche
Scientifique
Université Paris-Sud

Manuela D'Amen
Postdoctoral Fellow
Parco del Mincio
Piazza Porta Giulia

Valeria Di Cola
Postdoctoral Fellow
Department of Ecology and Evolution
University of Lausanne, Switzerland

Benjamin J. Dittbrenner
PhD Candidate
School of Environmental and Forest
Sciences
University of Washington

Maria Dumitru
Research Assistant
Gonzalez Laboratory
Department of Biology
McGill University

Rui Fernandes
PhD Student
Department of Ecology and Evolution
University of Lausanne, Switzerland

Jason P. Field
Research Associate
School of Natural Resources and the
Environment
University of Arizona

Cary Fowler
Chair, Board of Trustees
Rhodes College

Emily Fung
MSc Ecosystem and Hydrological Modeling
Unit
Forests, Biodiversity, and Climate Change
Program
Tropical Agricultural Research and Higher
Education Center (CATIE)

Andrew Gonzalez
Professor and Liber Ero Chair
Director Quebec Centre for Biodiversity
Science
Department of Biology
McGill University

Sarah M. Gray
Postdoctoral Fellow
Department of Biology–Ecology and
Evolution
University of Fribourg, Switzerland

Antoine A. Guisan
Professor
Department of Ecology and Evolution
Institute of Earth Surface Dynamics
University of Lausanne, Switzerland

Lee Hannah
Senior Scientist
Climate Change Biology
The Moore Center for Science Conservation
International

Niko Hartline
Bren School of Environmental Science and
Management
University of California–Santa Barbara

Jason R. Hartog
Research Scientist
Oceans and Atmosphere
Commonwealth Scientific and Industrial
Research Organization (CSIRO)

Jessica J. Hellmann
Director and Professor
Institute for the Environment
Department of Ecology, Evolution, and
Behavior
University of Minnesota

Janneke HilleRisLambers
Professor
Department of Biology
University of Washington

Elizabeth H. T. Hiroyasu
PhD Candidate
Bren School of Environmental Science and
Management
University of California, Santa Barbara

Alistair J. Hobday
Senior Principal Research Scientist
Oceans and Atmosphere
The Commonwealth Scientific and Indus-
trial Research Organization (CSIRO)

Ove Hoegh-Guldberg
Professor of Marine Studies
Director, Global Change Institute
University of Queensland

Pablo Imbach
Climate and Ecosystems Scientist
Climate Change, Agriculture, and Food
Security (CCAFS)
International Center for Tropical Agricul-
ture (CIAT)

David Inouye
Professor Emeritus
Department of Biology
University of Maryland
Rocky Mountain Biological Laboratory

Lauren Jarvis
MSc
University of Guelph

Miranda C. Jones
Post-Doctoral Researcher
Changing Ocean Research Unit
Institute for Oceans and Fisheries
University of British Columbia

Valerie Kapos
Head of Programme
Climate Change and Biodiversity
United Nations Environment Programme
World Conservation Monitoring Centre
(UNEP-WCMC)

Les Kaufman
Professor
Boston University Marine Program
Department of Biology
Boston University

Joan A. Kleypas
Scientist III
Climate and Global Dynamics
National Center for Atmospheric Research

Darin J. Law
Research Specialist
School of Natural Resources and the
Environment
University of Arizona

Joshua J. Lawler
Professor
School of Environmental and Forest
Sciences
University of Washington

Camille Leclerc
PhD Student
Laboratoire Ecologie, Systématique et
Evolution
Université Paris-Sud

Janeth Lessmann
PhD Student
Departmento de Ecología
Facultad de Ciencias Biológicas
Pontificia Universidad Católica de Chile
Instituto de Ecología y Biodiversidad (IEB)

Caitlin Littlefield
PhD Student
School of Environmental and Forest
Sciences
University of Washington

Thomas E. Lovejoy
Professor
Environmental Science and Policy
George Mason University

Michael C. MacCracken
Chief Scientist for Climate Change
Programs
Climate Institute

Edward Maibach
Director
Center for Climate Change Communication
George Mason University

Pablo A. Marquet
Professor
Departamento de Ecología
Facultad de Ciencias Biológicas
Pontificia Universidad Católica de Chile
Instituto de Ecología y Biodiversidad (IEB)
Laboratorio Internacional de Cambio
Global (LINCGlobal)
Centro de Cambio Global UC

Rubén G. Mateo
Postdoctoral Researcher
ETSI de Montes, Forestal y del Medio Natural
Technical University of Madrid

Kevin McCann
Professor
Department of Integrative Biology
University of Guelph

Guy Midgley
Professor
Global Change Biology
Stellenbosch University

Lera Miles
Senior Programme Officer
Climate Change and Biodiversity
United Nations Environment Programme
World Conservation Monitoring Centre
(UNEP-WCMC)

Erik Nelson
Associate Professor
Department of Economics
Bowdoin College

Daniel Nepstad
Executive Director and Chief Scientist
Earth Innovation Institution

Donald J. Noakes
Dean
Science and Technology
Vancouver Island University

Mary O'Connor
Associate Professor
Department of Zoology and Biodiversity
Research Centre
University of British Columbia

Jeffrey Park
Professor
Geology and Geophysics
Yale University

Camille Parmesan
Professor
School of Biological Sciences
Plymouth University

A. Townsend Peterson
University Distinguished Professor
Biodiversity Institute
University of Kansas

Eric Pinto
Department of Ecology and Evolution
University of Lausanne, Switzerland

Eric Post
Professor, Climate Change Biology
Department of Wildlife, Fish, and Conservation Biology
University of California, Davis

Jean-Nicolas Pradervand
Conservation Biologist
Swiss Ornithological Institute

Bronwyn Rayfield
Landscape Ecologist
Institut des sciences de la forêt tempérée

Brett R. Riddle
Professor
School of Life Sciences
University of Nevada, Las Vegas

Abdallah M. Samy
Fulbright Fellow
Lecturer and Research Scientist
Department of Entomology
Faculty of Science
Ain Shams University

Daniel Scherrer
Postdoctoral Fellow
Department of Ecology and Evolution
University of Lausanne, Switzerland

Oswald J. Schmitz
Director
Yale Institute for Biospheric Sciences
Yale University

Daniel B. Segan
Principal Natural Resource Analyst
Tahoe Regional Planning Agency

Pep Serra-Diaz
UMR Silva
AgroParisTech, Université de la Lorraine
BIOCHANGE Center for Biodiversity Dynamics in a Changing World
Aarhus University

M. Rebecca Shaw
Chief Scientist
World Wildlife Fund

Joshua Tewksbury
Global Hub Director
Future Earth

G. David Tilman
Regents Professor
University of Minnesota
Professor
University of California–Santa Barbara

Juan Camilo Villegas
Associate Professor
Escuela Ambiental, Facultad de Ingeniería
Universidad de Antioquia, Medellín, Colombia

Pascal Vittoz
Faculty of Geosciences and Environment
Institute of Earth Surface Dynamics
University of Lausanne, Switzerland

Isaline von Däniken
Scientific Collaborator
Institute of Earth Surface Dynamics
University of Lausanne, Switzerland

James E. M. Watson
Professor
School of Earth and Environmental
Sciences
University of Queensland
Director of Science and Research
Initiative
Wildlife Conservation Society

Ola Tveitereid Westengen
Associate Professor
Noragric / Department of International
Environmental and Development Studies
Norwegian University of Life Sciences

John (Jack) W. Williams
Professor
Department of Geography and Center for
Climatic Research
University of Wisconsin–Madison

John Withey
Faculty
Graduate Program on the Environment
Evergreen State College

Carlos Yañez-Arenas
Associate Professor
Laboratory of Conservation Biology
Parque Científico Tecnológico de Yucatán
Universidad Nacional Autónoma de México

Erika Yashiro
Staff Scientist
Department of Chemistry and Bioscience,
Section of Biotechnology
Aalborg University

Index

Page numbers followed by "f" or "t" indicate material in figures or tables. "Pl." refers to color plates.

8.2 ka event, 130t, 132–133

abiotic conditions, 42–44, 52, 80, 328; neotropical perspectives, 142, 146, 148–149
absence data, 159
abundance changes, 25–38, 67, 81, 294–296; inferential approach, 26; invasive species and distribution of existing populations, 259, 262; land use and, 26; long-term observations, 26, 27t, 28; marine, 170, 171–172; meta-analyses, 25, 28, 30t; microrefugia, 147, 149–150; observed changes in individual species and in communities, 31–32; phenological asynchrony, 32–33
acclimation, 201, 215
acidification, 20, 40, 100, 106, 114; benthic systems and, 190–191, 193; biogeochemical impacts, 189; CO_2 concentration as cause, 185; CO_2 equilibrium, 185–186, 186f; freshwater fishes and, 240t; geochemical effects, 186–187, 187f; Holocene, 130t; local and global, 61–63; marine biodiversity and, 168–169, 171, 175, 178, 185–195; mechanisms of biodiversity change, 191–192; multiple life stages and, 189; oxygen and capacity-limited thermal tolerance (OCLTT), 192; physiological and behavioral effects, 187–189; present-day observations, 190–191; primary production and, 187–188, 193; species competition and, 191–192
adaptation, 5–6, 128; adaptive genes, 101; assisted colonization of species, 34–35, 243; coral reefs and, 63; crop, 351–352, 351f; ecosystem-based, 6–7, 297–309; four pillars of, 9; green-gray, 7, 301; hard approaches, 298–299; human, 5–6, 297; to landscape, 67; marine species, 189; as nontrivial event, 352; soft approaches, 298–299; tropical forests, 201. See also ecosystem-based adaptation (EbA); range and abundance changes
adaptive evolution, 67–68, 73
Adélie penguins (Pygoscelis adeliae), 91–92
aerobic scope, 169, 173
aerosol injections, 12–13, 17
afforestation, 20, 314, 329
Africa, 160f; disease transmission, 270, 275–277; East Africa, 242–243, 270–280, 349; food insecurity, 347–349; North Africa, 130f, 133, 136f, 137–139; South Africa, 235, 350f; tropical forests, 196, 197
African Great Lakes, 242
African Humid Period, 137, 138
age–life stage model, 165

agriculture: conversion of forest to, 311, 358; crop adaptation, 351–352, 351f; crop wild relatives, 349–350; crop yields, 348–349, 358–361; cultivar adjustment, 351, 351f; ecosystem-based adaptation, 299–301, 300f; frost damage, 234–235; genebanks, 353; Holocene transition to, 347; nitrogen overuse, 4–5; policy making and, 347–355; production losses, 348; terrestrial mitigation practices, 313–314. See also food security
agroecosystems, 298, 329–330
albedo, 165, 301, 326
algae, 118, 191–192; Chlorella vulgaris, 247; macroalgae, 171, 188; Phaeocystis antarctica, 92; reef recovery and, 6–7
alpha and beta diversity, 80
alpine snowpack and glaciers, 239t
Amazon and Andean regions, 142; aridification, 143–144; dams, 318; fires, 142; hyperdominance, 142–143; refugia, 147; refugial hypothesis, 143. See also Amazon region; Andean region
Amazon region, 66, 84–85; carbon sequestration, 330; deforestation, 203; drought, 144, 199, 208, 209f; free-air concentration enrichment (FACE) experiment, 9; as planetary cooling system, 208; precipitation, 143, 201, 208; tipping point, postponing, 208–210; tree die-off and fire feedbacks, 81, 147. See also Amazon and Andean regions
Amazon River, 239t, 241, 318
Americas, 196–197
amphibians, 223f, 225–226, 242, 357
amplifying processes, 13
Andean region, 142, 183, pl. 8; amphibians, 226; aridification, 144; bofedales, 304; Bolivian Altiplano, 144, 149; megafauna, 147; temperatures, 143. See also Amazon and Andean regions
anoxia, 103, 104
Antarctica, 91–92, 105–106, 115; benthic communities, 115; deepwater sources, 106; future of marine life, 120–121; invasive species, 120–122; time stamp metaphor, 119–120; unglaciated, 131; western Antarctic Peninsula (WAP), 120–121
Antarctic bottom water (AABW), 106
Antarctic Circumpolar Current (ACC), 119
Anthropocene, 15, 73, 98–99, 105–108, 131, 330; onset, speed of, 131
anthropogenic climate change, 15, 25–26, 33–34; broad-scale changes, 91–92; carbon dioxide, 57; genetic signatures and, 69, 73; in tropical forests, to date, 199–201. See also carbon dioxide (CO_2)
ant species, 259, 261
Appalachian Mountains, 183–184, 323–325; amphibians, 226
Aptenodytes forsteri, 92, 175
aragonite, 60, 191

373

silicate rocks, chemical weathering of, 100, 103
snow: cover, 222; pollinator-plant associations and,
     234–235; snowmelt, 17, 27, 33, 43, 48, 217, 305;
     snowpack dynamics, 33
snow- and ice-albedo feedbacks, 13
Snowball Earth, 101–102
social marketing, 342
social norms, 344–345
soil organic carbon (SOC), 298, 299, 300f
soils, 13, 143, 223f; carbon sequestration, 312, 328, 332,
     334f, 335; microorganisms, 224–225, 298
solar radiation, 12, 13, 17, 80; management, 318
Solidago rugosa, 332–335
sorghum, 348
South Africa, 235, 350f
South America, 120, 304; amphibians, 226; freshwater
     fishes, 241–242; megafauna, 147; tropical forests, 197.
     See also Amazon and Andean regions; Amazon region;
     Andean region
Southeast Asia, 196, 197, 204, 329
Southern Hemisphere, 259
Southern Ocean, 91–92, 106, 190
Spain, 43
spatial and temporal scales: broad ecosystem changes,
     80; phylogeography, 68–70, 68f, 69t
spatial management, 255–257
species: climate change responses, 8; codistributed, 67,
     72; competition, 191–192; decoupling of coevolved,
     237, 240, 252, 263, 284; diagnostic patterns, 31; fit-
     ness, 27, 32, 42, 45, 78, 188, 190, 201, 212; foun-
     dational, dominant, and keystone, 80, 82–83, 87;
     individual shifts, 4; morphology, 170, 171; observed
     changes in individual species and in communities,
     31–32; polar, 169; population subdivision, 67; re-
     silience, 101, 103–104, 176, 289. *See also* endangered
     species; marine biodiversity; modeling species and
     vegetation distribution; range and abundance changes
species-area relationship, 358
species distribution models (SDMs): ecological niche
     models (ENMs) and disease transmission, 272–274; in
     context of global change, 160–164
species pumps, 71
species richness: marine biodiversity, 169, 170, 171, 173;
     mountain biodiversity, 224
species sensitivity, 163
Spheniscus magellanicus, 175
spider species, 260; *Phidippus clarus*, 332–335, 334f; *Phidip-
     pus rimator*, 252; *Pisaurina mira*, 252, 332–335, 334f
SRES A1B scenario, 176
star-of-Bethlehem (*Gagea lutea*), 49
steady-state balance, 100–101, 107, 129, 135
St. Lawrence Island, 39
St. Lawrence lowlands, 323–325
St. Matthew Island, 39
storms: convective, 16, 20–21; extinctions and, 175;
     freshwater fishes and, 237, 240; surges, 6, 302, 330;
     west-to-east movement, 16
stormwater-management techniques, 298, 302

stratigraphy, 98–99
stromatolites, 118
subtropical air masses, 16
sulfur dioxide, SO_2, 13
summers, Northern Hemisphere land areas, 15–16, 16f
Sundaland, 125
sunlight, absorption of, 13
surface, 12; changes in global surface air temperature,
     14–17, 16t, 21f; global average surface air tempera-
     ture, 21f, 99–101; reflectivity, 13; temperature in-
     crease of 1°C, 15; warming, 4. *See also* sea surface tem-
     peratures (SSTs)
suspension feeders, 116
suture zones, 71
Svalbard archipelago, 43
Svalbard Global Seed Vault, 353
Sweden, 299
Switch (Health and Health), 343
Switzerland, 133f, 135, 227
Symbiodinium, 60, 61
synchrony, shifts in, 33, 37, 77, 134, 173, 217, 217f, 218
syrphid flies, 48, 48f

teleconnections, 85, 87
teleosts, 116, 119, 120
temperate and boreal responses to climate change, 211–
     220, 330; biodiversity consequences, 218–219; eco-
     system responses, 215–216, 215f, 218; interacting and
     indirect stressors and feedbacks, 216–218; meta-anal-
     yses, 212–213, 216; organismal responses, 214–215;
     responses to climate change, 214–218; seasonality,
     212, 213–214, 213f; sensitivity to climate change and
     variability, 212–214; tree line, 215–216, 215f; warm-
     ing statistics, 211
temperature, pl. 1; 4°C rise predicted, 18, 20; absolute
     rate of rise, 31; beyond global average, 20–21; changes
     in global surface air, 14–17, 16t, 21f, 99–100; degrees
     of warming, 14–15, 18, 19f, 20, 21f, 173, 234, 271,
     284, 295, 326–327, 348–349, 350; elevation and, 143,
     145–146, 146f, 149; functions of, 247; global average,
     15, 20, 21f, 99–101; isocline modification, 225; land-
     surface, 199; latitudinal gradients, 201, 204, 211, 212;
     oceans, 57, 77, 168; reducing rate of global change to
     zero, 63; sea surface temperatures (SSTs), 57, 77, 173,
     175; species sensitivity to, 169–170, 247; summers,
     Northern Hemisphere land areas, 15–16, 16f; tropi-
     cal forests, 199; velocity of climate change (VoCC),
     31, 85, 86f, 107–108, 215, 284. *See also* climate change;
     food webs, asymmetrical thermal responses; glacial-
     interglacial cycling; global warming
temporal transplants, 47
tephritid flies, 235
terrestrial ecosystems, climate change mitigation, 310–
     322; agricultural land, 313–314; agriculture and bio-
     fuels, 315–317; connectivity and, 323–325; forest and
     nonforest ecosystems, 314–315; impact on ecosystems
     and biodiversity, 314–318; life-cycle impact assess-
     ments, 314, 318; nationally appropriate mitigation